The University of Chicago School Mathematics Project

Transition Mathematics

Second Edition

About the Cover The art on the cover was generated by a computer. The three interlocking rings signify the major themes of this book—algebra, geometry, and applied arithmetic.

Authors

Zalman Usiskin   Cathy Hynes Feldman

Suzanne Davis   Sharon Mallo   Gladys Sanders   David Witonsky

James Flanders   Lydia Polonsky   Susan Porter   Steven S. Viktora

ScottForesman

A Division of HarperCollins*Publishers*

Editorial Offices: Glenview, Illiois
Regional offices: Sunnyvale, California ● Tucker, Georgia
Glenview, Illinois ● Oakland, New Jersey ● Dallas, Texas

ACKNOWLEDGMENTS

Authors

Zalman Usiskin
Professor of Education, The University of Chicago

Cathy Hynes Feldman
Mathematics Teacher, The University of Chicago
Laboratory Schools

Suzanne Davis
Mathematics Supervisor, Pinellas County Schools,
Largo, FL (Second Edition only)

Sharon Mallo
Mathematics Teacher, Lake Park H.S., Roselle, IL
(Second Edition only)

Gladys Sanders
K–12 Mathematics Coordinator, Lawrence Public
Schools, Lawrence, KS (Second Edition only)

David Witonsky
UCSMP (Second Edition only)

James Flanders
UCSMP (First Edition only)

Lydia Polonsky
UCSMP (First Edition only)

Susan Porter
Evanston Township H.S., Evanston, IL
(First Edition only)

Steven S. Viktora
Chairman, Mathematics Department, Kenwood
Academy, Chicago Public Schools
(First Edition only)

Design Development

Curtis Design

UCSMP Production and Evaluation

Series Editors: Zalman Usiskin, Sharon L. Senk

Directors of First Edition Studies: Kathryn Sloane,
Sandra Mathison; Assistant to the Directors:
Penelope Flores

Directors of Second Edition Studies:
Geraldine Macsai, Gurcharn Kaeley

Technical Coordinator: Susan Chang

Second Edition Teacher's Edition Editor:
Suzanne Levin

Second Edition Consultants: Amy Hackenberg,
Mary Lappan

First Edition Managing Editor: Natalie Jakucyn

We wish to thank the many editors, production
personnel, and design personnel at ScottForesman
for their magnificent assistance.

We wish to acknowledge the generous support of
the Amoco Foundation and the Carnegie
Corporation of New York in helping to make it
possible for the First Edition of these materials
to be developed, tested, and distributed, and the
continuing support of the Amoco Foundation for
the Second Edition.

Multicultural Reviewers for ScottForesman

Marion Bolden
Newark Board of Education, Newark, NJ

Seree Weroha
Kansas State University, Manhattan, KS

Efraín Meléndez
Dacatoh St. School, Los Angeles, CA

Yvonne Wynde
Educator, Sisseton-Wahpeton Dakota Nation,
Waubay, SD

It is impossible for UCSMP to thank all the people who have helped create and test these books. We wish particularly to thank James Schultz and Glenda Lappan, who as members of the UCSMP Advisory Board, commented on early manuscripts; Carol Siegel, who coordinated the use of the test materials in schools; Liggy Chien and Kate Fahey of our editorial staff; Sara Benson, Anil Gurnarney, Jee Yoon Lee, Adil Moiduddin, Jeong Moon, Antoun Nabhan, Young Nam, and Sara Zimmerman of our technical staff; and Rochelle Gutiérrez, Nancy Miller, and Gerald Pillsbury of our evaluation staff.

A first draft of *Transition Mathematics* was written and piloted during the 1983–84 school year. After a major revision, a field trial edition was tested in 1984–85. These studies were done at the following schools:

Parkside Community Academy
Chicago Public Schools

Lively Junior High School
Elk Grove Village, Illinois

Kenwood Academy
Chicago Public Schools

McClure Junior High School
Western Springs, Illinois

Grove Junior High School
Elk Grove Village, Illinois

Wheaton-Warrenville Middle School
Wheaton, Illinois

Mead Junior High School
Elk Grove Village, Illinois

Glenbrook South High School
Glenview, Illinois

A second revision underwent a comprehensive nationwide test in 1985–86. The following schools participated in those studies:

Powell Middle School
Littleton, Colorado

16th St. Middle School
St. Petersburg, Florida

Walt Disney Magnet School
Gale Community Academy
Hubbard High School
Von Steuben Upper Grade Center
Chicago, Illinois

Hillcrest High School
Country Club Hills, Illinois

Friendship Junior High School
Des Plaines, Illinois

Bremen High School
Midlothian, Illinois

Holmes Junior High School
Mt. Prospect, Illinois

Sundling Junior High School
Winston Park Junior High School
Palatine, Illinois

Addams Junior High School
Schaumburg, Illinois

Edison Junior High School
Wheaton, Illinois

Golden Ring Middle School
Sparrows Point High School
Baltimore, Maryland

Walled Lake Central High School
Walled Lake, Michigan

Columbia High School
Columbia, Mississippi

Oak Grove High School
Hattiesburg, Mississippi

Roosevelt Middle School
Taylor Middle School
Albuquerque, New Mexico

Gamble Middle School
Schwab Middle School
Cincinnati, Ohio

Tuckahoe Middle School
Richmond, Virginia

Shumway Middle School
Vancouver, Washington

Since the ScottForesman publication of the First Edition of *Transition Mathematics* in 1990, thousands of teachers and schools have used the materials and have made additional suggestions for improvements. The materials were again revised, and the following teachers and schools participated in field studies in 1992–1993:

Charlotte Kulbacki
East Junior High School
Colorado Springs, Colorado

Patricia Gresko
Kerr Middle School
Blue Island, Illinois

Jane Sughrue
McCall Middle School
Winchester, Massachusetts

Audrey Reineck
Washington High School
Milwaukee, Wisconsin

Joseph Pierre-Lewis
Edison Middle School
Miami, Florida

Mary Fitzpatrick
Olson Middle School
Woodstock, Illinois

Cindy Urban
Forest Hills Central Middle School
Grand Rapids, Michigan

Barbara Pulliam
East Coweta Middle School
Senoia, Georgia

Kathy Hying
Northwood Middle School
Woodstock, Illinois

Garry J. Hopkins
Hillside Middle School
Kalamazoo, Michigan

Linda Ferreira
Osceola Middle School
Seminole, Florida

Karen Bloss
Northwest High School
Wichita, Kansas

Linda Kennley
C.A. Johnson High School
Columbia, South Carolina

THE UNIVERSITY OF CHICAGO SCHOOL MATHEMATICS PROJECT

The University of Chicago School Mathematics Project (UCSMP) is a long-term project designed to improve school mathematics in grades K-12. UCSMP began in 1983 with a 6-year grant from the Amoco Foundation. Additional funding has come from the National Science Foundation, the Ford Motor Company, the Carnegie Corporation of New York, the General Electric Foundation, GTE, Citicorp/Citibank, and the Exxon Education Foundation.

UCSMP is centered in the Departments of Education and Mathematics of the University of Chicago. The project has translated dozens of mathematics textbooks from other countries, held three international conferences, developed curricular materials for elementary and secondary schools, formulated models of teacher training and retraining, conducted a large number of large and small conferences, engaged in evaluations of many of its activities, and through its royalties has supported a wide variety of research projects in mathematics education at the University. UCSMP currently has the following components and directors:

Resources	Izaak Wirszup, Professor Emeritus of Mathematics
Elementary Materials	Max Bell, Professor of Education
Elementary Teacher Development	Sheila Sconiers, Research Associate in Education
Secondary	Sharon L. Senk, Associate Professor of Mathematics, Michigan State University Zalman Usiskin, Professor of Education
Evaluation	Larry Hedges, Professor of Education

From 1983 to 1987, the director of UCSMP was Paul Sally, Professor of Mathematics. Since 1987, the director has been Zalman Usiskin.

Transition Mathematics

The text *Transition Mathematics* has been developed by the Secondary Component of the project, and constitutes the core of the first year in a six-year mathematics curriculum devised by that component. The names of the six texts around which these years are built are:

Transition Mathematics
Algebra
Geometry
Advanced Algebra
Functions, Statistics, and Trigonometry
Precalculus and Discrete Mathematics

The content and questions of this book have been carefully sequenced to provide a smooth path from arithmetic to algebra, and from the visual world and arithmetic to geometry. It is for this reason that this book is entitled *Transition Mathematics*.

The first edition of *Transition Mathematics* introduced many features which have come now to be deemed essential in courses at this level. There is **wider scope,** including substantial amounts of geometry integrated with the arithmetic and algebra that is customary, to correct the present situation in which many students who finish algebra find themselves without enough prior knowledge to succeed in geometry. A **real-world orientation** has guided both the selection of content and the approaches allowed the student in working out exercises and problems, because being able to do mathematics is of little ultimate use to an individual unless he or she can apply that content. We require **reading mathematics,** because students must read to understand mathematics in later courses and must learn to read technical matter in the world at large. The use of **up-to-date technology** is integrated throughout, with *scientific calculators* assumed and *computer* exercises found in all content areas.

We continue to use two unique overall features designed to maximize performance. **Four dimensions of understanding** are emphasized: skill in carrying out various algorithms; developing and using mathematics properties and relationships; applying mathematics in realistic situations; and representing or picturing mathematical concepts. We call this the SPUR approach: **S**kills, **P**roperties, **U**ses, **R**epresentations. On occasion, a fifth dimension of understanding, the Culture dimension, is discussed.

The **book organization** is designed to maximize the acquisition of both skills and concepts. Ideas introduced in a lesson are reinforced through Review questions in the immediately succeeding lessons. This daily review feature allows students several nights to learn and practice important concepts and skills. At the end of each chapter, a carefully focused Progress Self-Test and a Chapter Review, each keyed to objectives in all the dimensions of understanding, are then used to solidify performance of skills and concepts from the chapter so that they may be applied later with confidence. Finally, to increase retention, important ideas are reviewed in later chapters.

Since the ScottForesman publication of the first edition of *Transition Mathematics* in 1990, the entire UCSMP secondary series has been completed and published. Thousands of teachers and schools have used the first edition and made additional suggestions for improvements. There have been advances in technology and in thinking about how students learn. We have attempted to utilize these ideas in the development of the second edition. We have found that many students are beginning this course with less experience with fractions than they should have, and so the sequence of some lessons has been changed, and there is more work with probability and fractions. At the same time, we assume a little more experience with measurement than we did in the first edition.

Those familiar with the first edition will note also a number of new features, including the following:

Activities have been incorporated into many of the lessons to help students develop concepts before or as they read. There are **projects** at the end of each chapter because in the real world much of the mathematics done requires a longer period of time than is customarily available to students in daily assignments. There are many more questions requiring student **writing,** because writing helps students clarify their own thinking, and writing is an important aspect of communicating mathematical ideas to others. Each of these features is designed to involve students more actively in their learning. In the area of technology, **spreadsheets** are now being used as a tool to help solve problems.

Comments about these materials are welcomed. Please address comments to UCSMP, The University of Chicago, 5835 S. Kimbark, Chicago, IL 60637.

CONTENTS

DECIMAL NOTATION

LARGE AND SMALL NUMBERS

CHAPTER 3   116

MEASUREMENT

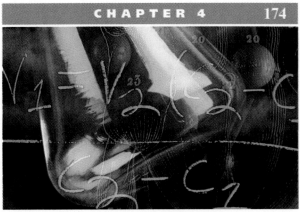

CHAPTER 4   174

USES OF VARIABLES

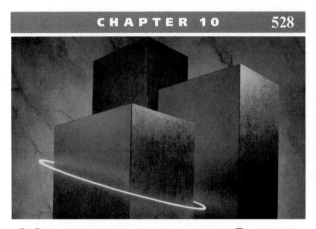

COORDINATE GRAPHS AND EQUATIONS

GETTING STARTED

Welcome to Transition Mathematics.
We hope you enjoy this book; it was written for you.

The Goals of *Transition Mathematics*

The goals of this book are to solidify the arithmetic you already know and to prepare you for algebra and geometry.

THERE IS ANOTHER important goal of this book: for you to be able to deal with the mathematics all around you—in newspapers, magazines, on television, in any job, and in school. To accomplish this goal you should try to learn from the reading in each lesson as well as from your teacher and your classmates. The authors, who are all experienced teachers, offer the following advice:

1 You can watch basketball hundreds of times on television. Still, to learn how to play basketball, you must have a ball in your hand and actually dribble, shoot, and pass it. Mathematics is no different. You cannot learn much mathematics just by watching other people do it. You must participate. Some teachers have a slogan: *Mathematics is not a spectator sport.*

Mathematics is not a spectator sport.

2 You are expected to read each lesson. Here are some ways to improve your reading comprehension:

Read slowly, paying attention to each word.

Look up the meaning of any word you do not understand.

Work examples yourself as you follow the steps in the text.

Reread sections that are unclear to you.

Discuss troublesome ideas with a fellow student or your teacher.

3 Writing can help you understand mathematics, too. So you will sometimes be asked to explain your solution to a problem or to justify an answer. Writing good explanations takes practice. Use the solutions to the examples in each lesson as a guide for your own writing.

4 If you cannot answer a question immediately, don't give up! Read the lesson, examples, or question again. If you can, go away from the problem and come back to it a little later. Do not be afraid to ask questions in class and to talk to others when you do not understand something.

What Tools Do You Need for This Book?

In addition to paper, pencils, and erasers, you will need the following equipment:

ruler with both centimeter and inch markings *(by Lesson 1-2)*

scientific calculator *(by Lesson 1-5)*

protractor to draw and measure angles *(by Lesson 3-6)*

The calculator should have the following keys:

`EE` or `EXP` to display very large or very small numbers in scientific notation

`±` or `+/-` (opposite)

`^`, `xʸ`, or `yˣ` (powering)

`√x` (square root)   `x!` (factorial)

`π` (pi)   `1/x` (reciprocal).

Your teacher may have more recommendations regarding what you need.

Getting Off to a Good Start

One way to get off to a good start is to spend some time getting acquainted with your textbook. The questions that follow are designed to help you become familiar with *Transition Mathematics*.

We hope you join the hundreds of thousands of students who have enjoyed this book. We wish you much success.

QUESTIONS

Covering the Reading

1. What are the goals of *Transition Mathematics*?

2. Explain what is meant by the statement "Mathematics is not a spectator sport."

3. Of the five things listed for improving reading comprehension, list the three you think are most helpful to you. Why do you think they are most helpful?

4. *Multiple choice.* Choose all correct answers. Which are good things to do if you cannot answer a homework question immediately?
   (a) Give up.
   (b) Look for a similar example in the reading.
   (c) Relax, go on to the next question, and come back to this question later.
   (d) Discuss the problem with a friend.
   (e) Read the relevant part of the lesson again.
   (f) Beg your teacher or parents to do it for you.
   (g) Go watch *Wheel of Fortune.*

Knowing Your Textbook

In 5 and 6, refer to the Table of Contents beginning on page *vi.*

5. Algebra uses variables. In which chapter does work in algebra begin?

6. What lesson would you read to learn how to measure angles?

In 7–12, refer to other parts of the book.

7. Look at several lessons. What are the four categories of questions in each lesson?

8. Suppose you have just finished the questions in Lesson 3-4. On what page can you find answers to check your work? What answers are given?

9. When you finish a Progress Self-Test, what is recommended for you to do?

10. What kinds of questions are in the Chapter Review at the end of each chapter?

11. Where is the glossary and what does it contain?

12. Locate the index.
    a. According to the index, where will you find information about David Robinson?
    b. What is this information?

David Robinson

CHAPTER 1

DECIMAL NOTATION

Virtually every culture, from ancient times until now, has had spoken names for the numbers from one to nine. Many cultures have represented these numbers with written symbols. Most of these symbols are quite different from the ones we use today in the United States. The chart below gives some examples.

Culture	1	2	3	4	5	6	7	8	9
Greek	α	β	γ	δ	ε	·ς	ζ	η	θ
Roman	I	II	III	IV	V	VI	VII	VIII	IX
Chinese (ancient)	—	=	≡	☰	⊠	八	+	)(	𝇈
Chinese (modern)	一	二	三	四	五	六	七	八	九

Among the earliest cultures to use the number 0 were the Mayas of Central America, and the peoples of India. The chart below shows how the ten **digits** we use, 0, 1, 2, 3, 4, 5, 6, 7, 8, and 9, have developed over the years.

The Development of Hindu-Arabic Numerals										
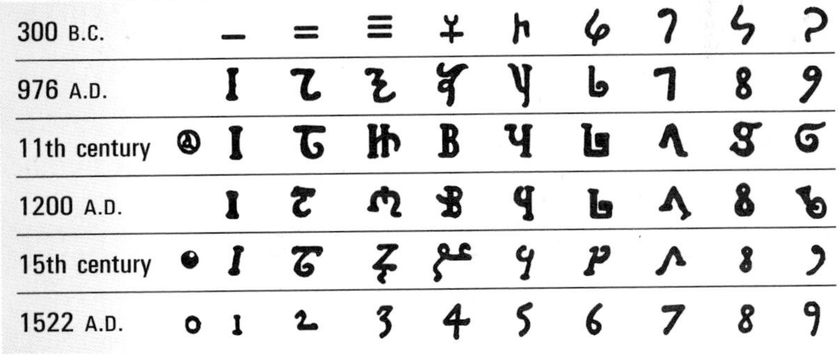 300 B.C.		—	=	≡	⊬	ɧ	𝟼	𝟽	𝟧	𝟸
976 A.D.		I	𝟤	𝟥	𝟦	𝖸	𝟨	𝟩	8	𝟫
11th century	④	I	𝟨	ℍ	B	𝟦	𝖻	∧	𝟪	𝟨
1200 A.D.		I	𝟤	𝔪	ℬ	𝟫	𝖻	∧	8	𝟨
15th century	●	I	𝟨	𝟥	𝜉	𝟫	𝑃	∧	𝟪	𝟤
1522 A.D.	○	I	𝟤	𝟥	4	𝟧	6	7	8	9

By the year 900 A.D., some Hindus were using a *decimal system* to denote numbers. The Arabs wrote about this system. Europeans did not learn about it until 1202 A.D., when Leonardo of Pisa, an Italian mathematician also known as Fibonacci, translated an Arabic manuscript into Latin. This system is now used by mathematicians throughout the world, and we call the numerals **Arabic** or **Hindu-Arabic numerals.**

LESSON
1-1

Decimals for Whole Numbers

A taste of big numbers. *The Taste of Chicago draws big crowds who sample foods from Chicago's numerous ethnic restaurants. The 1993 attendance over ten days was 2,500,000, about the same as the population of Brooklyn, New York.*

What Is the Decimal System?

The **decimal system** is a system of writing numbers based on the number ten. A number written in the decimal system is said to be in **decimal notation.** In decimal notation, the smallest ten **whole numbers** need only one digit. Today we write them as 0, 1, 2, 3, 4, 5, 6, 7, 8, and 9. But look again at the chart on page 5. Notice that some of the symbols we use today did not appear until the 1400s.

Two digits are needed to write the next ninety whole numbers as decimals.

  10   11   12   . . .   19 20 21   . . .   99

Notice that we call the whole numbers *decimals* even though there are no decimal points.

The next nine hundred whole numbers can be written with three digits.

  100   101   102   . . .   200 201   . . .   999

By using more digits, very large numbers can be written. The 1990 census estimate of the U.S. population was 248709873 people. To make this easier to read, groups of digits are separated by commas.

In Decimal Notation:	In English:
2 4 8 , 7 0 9 , 8 7 3	Two hundred forty-eight million, seven hundred nine thousand, eight hundred seventy-three

Usually decimal notation is shorter than English words.

Each digit in a decimal has a **place value.** Here are the place values for the nine digits of the number that represents the U.S. population in 1990.

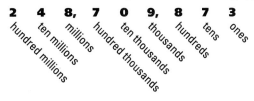

2 4 8, 7 0 9, 8 7 3
hundred millions / ten millions / millions / hundred thousands / ten thousands / thousands / hundreds / tens / ones

Counts and Counting Units

The U.S. population is an example of a **count.** The most basic use of numbers is as counts. For every count there is a **counting unit.** Counts are always whole numbers, never fractions between whole numbers. Here are some examples.

Phrase	Count	Counting Unit
0 eight-legged insects	0	eight-legged insects
28 letters in "antidisestablishmentarianism"	28	letters
U.S. population of 248,709,873	248,709,873	people

QUESTIONS

Covering the Reading

Use these questions to check your understanding of the reading. If you cannot answer a question, you should go back to the reading for help in obtaining an answer.

1. What is the modern Chinese symbol for the number eight?

2. What letter did the Romans use to stand for the number of fingers on one normal hand?

3. What people invented the decimal system? When was it invented?

4. Name two early cultures that used the number zero.

5. Name the smallest whole number.

6. **a.** Who was Fibonacci?
   b. What did he do that relates to the decimal system?
   c. When did he do it?

7. About when did all the symbols for the numbers zero through nine appear as we know them?

8. Written in the decimal system, how many whole numbers have three digits?

9. Name the digit in the indicated place of the number 568,249.
   a. thousands   **b.** ones   **c.** hundreds   **d.** tens

Why are dishes called China? *Porcelain dishes are often called china because porcelain was first made in China. The card shows the price of the dishes.*

10. A number is written as a decimal. Must there be a decimal point? Support your answer with an example.

11. Name one advantage of writing numbers as decimals rather than using English words.

12. A count is always what kind of number?

13. Name something for which the count is between one million and one billion.

Applying the Mathematics

These questions extend the content of the lesson. You should study the examples and explanations if you cannot get an answer. For some questions, you can check your answers with the ones in the back of the book.

14. Describe a situation with a count that is larger than the 1990 U.S. population.

15. Consider the number of stars on a U.S. flag.
    a. Name the count.   b. Name the counting unit.

16. One of the books of the Old Testament is called Numbers because it begins with a census of the adult males of the tribes of Israel. The ancients did not have our numerals. So they wrote out the population in words. ". . . of the tribe of Reuben, were forty and six thousand and five hundred." Write this number as a decimal.

17. Federal aid is often given on the basis of population: the greater the population, the greater the aid. Use these 1990 Census data.

    | Corpus Christi, Texas | 349,894 |
    | Pensacola, Florida | 344,406 |
    | Portsmouth-Dover-Rochester, New Hampshire | 350,078 |
    | Salinas-Seaside-Monterey, California | 355,660 |

    a. Which of these four metropolitan areas would get the most aid?

    b. Which of these four metropolitan areas would get the least aid?

18. Tell why the counts in Question 17 are estimates, even on the day they were made.

A glimpse of Portsmouth.
Portsmouth, New Hampshire, is a community which prides itself on its historical homes, its art and cultural centers, and an international harbor.

In 19–21, write as a decimal.

19. 6 hundred million   **20.** ten thousand   **21.** five hundred six

22. In decimal notation, what is the smallest seven-digit whole number?

23. In decimal notation, write the number that is one less than ten thousand.

Review

Every lesson contains review questions to give you practice on ideas you have studied earlier.

24. Danielle saved $17 in May, $16 in June, and $18 in July. Elena saved three times as much as Danielle.

 a. How much did Danielle save?
 b. How much did Elena save?
 c. How much did they save altogether?   *(Previous course)*

25. a. How many years are in a decade?
 b. How many years are in a century?   *(Previous course)*

Exploration

These questions ask you to explore mathematics topics related to the chapter. Sometimes you will need to use dictionaries or other sources of information.

26. In Europe, the decimal numeral for seven is sometimes written as shown in the cartoon. Why do you think this is done?

1-2

Decimals for Numbers Between Whole Numbers

What's next for Flo Jo? *Olympic track gold medalist Florence Griffith-Joyner holds the women's world record for the 200-meter sprint with a time of 21.34 seconds. In the summer of 1993, this popular athlete was named to co-chair the President's Council for Physical Fitness.*

Measuring is as common and important a use of numbers as counting. A **unit of measure** can always be split into smaller parts. This makes measures different from counts. For instance, you can split up measures of time.

Intervals and Tick Marks

The women's world record for the 200-meter run is between 21 and 22 seconds. This time **interval** is pictured on the **number line** below. The marks on the number line are called **tick marks.** The interval on this number line is one second, the distance between two neighboring tick marks.

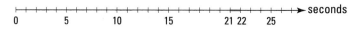

Notice the difference between intervals and tick marks. To divide a segment into three intervals requires two tick marks in the middle and two others at the ends.

	interval		interval		interval	
tick mark		tick mark		tick mark		tick mark

Measuring requires counting intervals, but some applications require counting tick marks.

To get more accuracy in the interval between 21 and 22 seconds, blow up the number line between 21 and 22. Then split that interval into ten equal parts. Nine new tick marks are needed. The interval on the new number line is **one tenth** of a second. The dot shown stands for the world record. Its location indicates that the time is between 21.3 and 21.4 seconds.

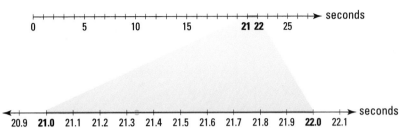

Florence Griffith-Joyner set this world record in 1988. Her time of 21.34 seconds is what is being graphed. To graph 21.34, blow up the line again. Split the interval between 21.3 and 21.4 into ten parts. The interval on the new number line is one tenth of one tenth, or one **one-hundredth** of a second. In 21.34, the digit 3 is in the **tenths place.** The digit 4 is in the **hundredths place.**

On the above number lines, we rewrote 21 seconds as 21.0 seconds, and 21.3 seconds as 21.30 seconds. A zero at the right of a decimal on the right side of the decimal point does not affect the value of the decimal. The same is true for money, for money is on the decimal system. $5 and $5.00 have the same value.

Who participates in the Special Olympics?
Everyone who is at least eight years old, has completed a medical application form showing cognitive delay or has at least 50% special education classes, and has trained with a coach for eight weeks is encouraged to participate.

Names for Decimal Places

Some situations require more places to the right of the decimal point. For instance, the famous number pi, written π, is the circumference of (distance around) a circle whose diameter is one unit. As a decimal, π never ends.

π = 3.14159265358979323846264338327950288419716939937510 . . .

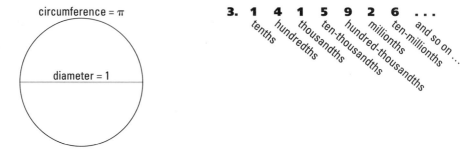

The names of the places to the right of the decimal point are similar to the names of the places to the left. Think of the ones place and the decimal point as the center. Then there is perfect balance of names to the right and to the left.

Today's uses often require many decimal places. Some grinding tools are accurate to within two millionths of an inch. (That's much less than the thickness of this page.) Supercomputers work at speeds often measured in billionths of a second.

In 1585, Simon Stevin, a Flemish mathematician, first extended the use of decimal places to the right of the ones place. Before then, fractions were used. Decimals are now more common than fractions for measurements. An advantage of decimals is that they are easier to put in order and compare.

Comparing Decimals

Most lessons in this book have examples. **What you might write in the solutions will look like this.** Whenever you can, try to answer the question in the example before reading the solution.

Example 1

Which is larger, 3.01 or 2.999?

Solution

Align the decimal points. "Align" means to put one above the other.

3.01
2.999

Start at the left of each number. 3 is larger than 2, so **3.01 is larger.**

As you know, with whole numbers, the decimal with more digits always stands for a larger number. This is not true, however, for numbers which are not whole numbers.

Example 2

Which is the largest?

0.0073   0.007294   0.00078

Solution 1

Again, align the decimal points.

$$0.0073$$
$$0.007294$$
$$0.00078$$

The bottom number is smallest because it has 0 thousandths while the others have 7 thousandths. To find out which of the first two is larger, compare the ten-thousandths place. The 3 is larger than the 2, so 0.0073 is largest.

Solution 2

Align the decimal points. Write all the numbers to show the same number of decimal places.

$$0.007300$$
$$0.007294$$
$$0.000780$$

Now it is easy to tell that 0.0073 is largest.

In Example 2, a zero appears to the left of the decimal point of each number. This zero can make it easier to order numbers. It also draws attention to the decimal point and corresponds to the display on most calculators. Also notice that two different solutions are given to answer the question of Example 2. When there is more than one way of getting the answer to a question, you should try to learn all the ways. Then you can use the way that is easiest for you.

QUESTIONS

Covering the Reading

In 1–3, consider the number 21.34.

1. Between what two consecutive whole numbers is this number?

2. What digit is in the tenths place?

3. What digit is in the hundredths place?

In 4–7, consider the number 654,987.123456789. What digit is in each place?

4. thousands

5. tenths

6. thousandths

7. hundred-thousandths

8. What digit is in the millionths place of π?

9. Name a kind of measurement that can require accuracy to billionths.

10. Who invented the idea of extending decimal places to the right of the decimal point, and when?

11. Name one advantage of decimals over fractions.

In 12–14, put the numbers in order from smallest to largest.

12. 0.033   0.015   0.024

13. 6.783   .6783   67.83

14. 4.398   4.4   4.4001

15. a. How many tick marks are needed to divide a number line into five intervals?
b. Draw a number line to represent any time from zero to five minutes.

In 16 and 17, use the number line drawn here. The tick marks are equally spaced.

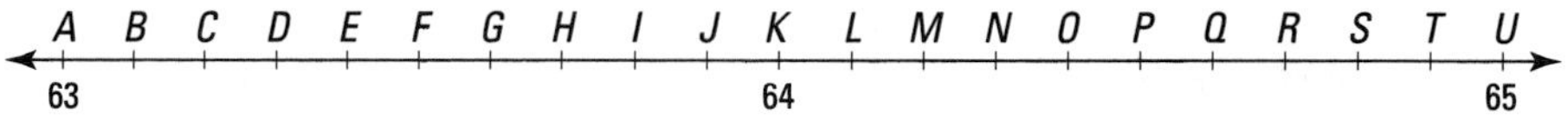

16. What is the length of each interval on this number line?

17. Which letter (if any) corresponds to the given number?
a. 63.4   **b.** 64.8   **c.** 64.80   **d.** 64.08

18. Explain the difference between a count and a measure.

19. a. What is the name of the number that is the circumference of a circle with diameter 1?
b. Give the first five decimal places of this number. ("Decimal places" refers to places to the right of the decimal point.)

Applying the Mathematics

Here are examples showing decimals translated into English.

3.5	three and five tenths
3.54	three and fifty-four hundredths
3.549	three and five hundred forty-nine thousandths

In 20–22, use the above examples to help translate the given number into English.

20. 5.9   **21.** 324.66   **22.** 0.024

23. To find a number between 8.2 and 8.3, write them as 8.20 and 8.30. Then any decimal beginning with 8.21, 8.22, and so on up to 8.29 is between them. Use this idea to find a number between 44.6 and 44.7.

24. A store sells 5 pairs of white tube socks for $16. Mel wants 1 pair and divides 16 by 5, using a calculator. The calculator shows 3.2. What should Mel pay?

25. In 1988, Florence Griffith-Joyner set a women's world record of 10.49 seconds in the 100-meter dash.
  a. Copy the number line below. Graph this number on your number line.

  b. If this record were lowered by a tenth of a second, what would the new record be?
  c. Graph your answer to part **b** on your number line.

In 26 and 27, order the numbers from smallest to largest.

26. three thousandths
four thousandths
three millionths

27. sixty-five thousandths
sixty-five thousand
sixty-five

28. Draw a number line six inches long. Indicate each inch with a tick mark.

29. A commuter train leaves Central Station every half hour from 6:30 A.M. to 9:00 A.M. How many morning trains could a commuter take? (Hint: Use a number line to represent this situation.)

Trains will keep you on track. *This photo shows commuters at Shinjuku station in Tokyo, Japan.*

Review

In 30 and 31, write as a decimal. *(Lesson 1-1)*

30. four hundred million   **31.** thirty-one thousand sixty-eight

32. Written as a decimal, the number one million is a 1 followed by how many zeros? *(Lesson 1-1)*

33. In "76 trombones," name the count and the counting unit. *(Lesson 1-1)*

34. The metric system uses prefixes which are related to place values in the decimal system. What is the meaning of the three most common prefixes: kilo-, centi-, milli-? *(Previous course)*

Exploration

35. Use an encyclopedia or a book on the history of mathematics to find out some additional information about the number π. Write a short report on what you find.

Estimating by Rounding Up or Rounding Down

How costly? *On August 24, 1992, Hurricane Andrew ripped through South Florida with winds of over 160 miles per hour. This photo of Florida City, Florida, shows some of the devastation caused by the hurricane. The damage was estimated to be $15 billion, and about 50,000 people were left homeless.*

Why Are Estimates Needed?

In many types of situations, an **estimate** may be preferred over an exact value.

1. *An exact value may not be worth the trouble* it would take to get it. Example: About 30,000 people attended the baseball game.
2. *An estimate is often easier to work with* than the exact value. Example: Instead of multiplying $169.95, let's use $170.
3. *It may be safer to use an estimate* than to try to use an exact value. Example: The home repairs will cost at least $18,000 as a result of the hurricane. So we will budget $20,000 to play it safe.
4. *An exact value may change* from time to time, forcing an estimate. Example: I estimate that the coin will land heads 5 times in 10 tosses.
5. Predictions of the future or notions about the past usually are estimates, since *exact values may be impossible to obtain.* Example: One estimate of the world population in the year 2025 is 12 billion.

What Kinds of Rounding Are There?

The most common method of estimating is **rounding.** There are three kinds of rounding: **rounding up, rounding down,** and rounding to the nearest. Some examples of rounding up and rounding down are on the next page. (Rounding to the nearest is discussed in Lesson 1-4.)

Rounding is almost always done with a particular decimal place in mind.

Example 1

A certain type of label is sold in packages of 100. If you need 1325 labels, how many labels must you buy? Give reasons for your answer.

Solution

You must buy more labels than you need. So you need to round up to the next 100. Since 1325 is between 1300 and 1400, you must buy 1400 labels.

The rounding of Example 1 can be pictured on a number line. Let the length of the intervals be 100. Rounding up 1325 to the next 100 means going to the higher endpoint of the interval. Rounding down would mean going to the lower endpoint, 1300.

Example 2

A store sells six bottles of fruit juice for $4.69. You want one bottle. So you divide 4.69 by 6 to get the cost. Your calculator shows 0.7816667. What will you probably pay for the bottle?

Solution

The store will probably round up to the next penny. Pennies are hundredths of dollars, so look at the hundredths place in 0.7816667. The hundredths place is 8. That means that 0.7816667 is between 0.78 and 0.79. You will probably have to pay $0.79, which is 79¢.

Some calculators round *down,* or **truncate,** all long positive decimals to the preceding millionth (the sixth decimal place).

Example 3

What will a calculator that truncates show for $\pi = 3.1415926535 \ldots$?

Solution

The sixth decimal place is 2. The calculator will show 3.141592.

Covering the Reading

1. Give an example of a situation where an estimate would be preferred over an exact value.

2. Name five reasons why estimates are often preferred over exact values.

3. The most common way of estimating is by ___?___.

4. Name three types of rounding.

5. Some drawing pencils are sold in packages of 10. A teacher needs one pencil for each student in a class of 32. How many pencils must be bought?

6. A store sells avocados at three for $1. You want one. So you divide $1.00 by 3 to get the cost. Your calculator shows 0.333333. How much will you probably have to pay for the avocado?

7. Refer to Example 2. If the store rounded down, what would you pay?

8. A store sells a dozen eggs for $1.09. You want a half dozen. To find out how much you will pay, you divide $1.09 by 2. You get 0.545. How much will you have to pay?

9. If a calculator rounds down to the preceding millionth, what will it show for 0.0123456?

What is guacamole? *Peel and mash an avocado. Add a chopped tomato, a clove of garlic, one teaspoon of lemon juice, and a half-teaspoon of salt. The result is guacamole, a tasty dip for chips.*

Applying the Mathematics

10. Explain what happens when a decimal is truncated.

11. Suppose a calculator rounds down to the preceding hundred-millionth (the eighth decimal place). What will it show for 0.97531246809?

12. **a.** Round $1795 up to the next ten dollars.
    b. Round $1795 down to the preceding ten dollars.
    c. Graph $1795 and your answers to parts **a** and **b** on the same number line.

13. **a.** Round 5280, the number of feet in one mile, up to the next 1000.
    b. Round 5280 down to the preceding 1000.
    c. Graph 5280 and your answers to parts **a** and **b** on the same number line.

14. There are exactly 30.48 centimeters in one foot.
    a. Round 30.48 up to the next tenth.
    b. Round 30.48 down to the preceding tenth.

15. There are exactly 1.609344 kilometers in one mile.
    a. Round 1.609344 up to the next thousandth.
    b. Round 1.609344 down to the preceding thousandth.

16. Suppose you have $97 in your bank account and you would like to withdraw as much as you can from a cash machine. Tell how much cash you can get from the machine if the only bill the machine dispenses is for the indicated amount.
    a. $5   **b.** $10   **c.** $20

In 17–20, tell whether a high or a low estimate would be preferred. Then explain why.

17. You are estimating how large a birthday cake to order for a party.

18. You are estimating how much money you should take on a trip.

19. You are estimating how much weight an elevator can carry without being overloaded.

20. You are estimating how many minutes it will take to do your math homework.

Whose money is in the machine? *Banks place money in cash machines. When you withdraw money from a machine, your bank account is charged for the amount you take out.*

Review

In 21 and 22, order the numbers from smallest to largest.   *(Lesson 1-2)*

21. 5.1   5.01   5.001   **22.** .29   0.3   .07

23. *Multiple choice.* Which number does not equal 0.86?   *(Lesson 1-2)*
    (a) 0.860   (b) .86   (c) .086

In 24 and 25, find a number that is between the two given numbers. *(Lesson 1-2)*

24. 5.8 and 5.9   **25.** 5.9 and 6

In 26 and 27, use this number line.   *(Lesson 1-2)*

```
 Z  Y  X  W  V  U  T  S  R  Q  P  O  N  M  L  K  J  I
 |--|--|--|--|--|--|--|--|--|--|--|--|--|--|--|--|--|
             2              3              4
```

26. What is the interval on the number line?

27. Which letter on the number line corresponds to the given number?
    a. 3.0   **b.** 2.8   **c.** 1.4

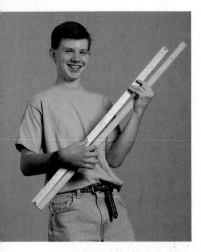

28. For this question, you need a ruler that measures in centimeters. The metric system measures length in meters, centimeters, and millimeters.

   a. Draw a segment 10 cm long. Indicate each centimeter with a tick mark.

   b. One millimeter equals one-tenth of a centimeter. Then how many millimeters equal one centimeter?

   c. On your segment, mark off a point that is 38 millimeters from one end.   *(Previous course, Lesson 1-2)*

29. Write this number as a decimal: four million, thirty thousand. *(Lesson 1-1)*

About the same? *The photo shows that the yardstick and meterstick are about the same length. A meter is a little more than three inches longer than a yard.*

Exploration

30. What is the dictionary definition of the word *truncate?* How does this relate to the way truncate is used in this lesson?

31. Look through a newspaper.
   a. Find a number which underestimates the value it is reporting.
   b. Find a number which overestimates the value it is reporting.

   Write down enough about the examples to be able to report your findings to the class.

32. On this and other computer questions in this book, it is possible that your computer will not act like other computers. If you get a strange message or no response, ask your teacher for help.

   a. Put your computer in programming mode, and type ?INT(4.57). The ? is short for PRINT. You could also type PRINT INT(4.57). Now press the RETURN key. What does the computer screen show?

   b. Try part **a** with the following:

```
?INT(115.68)
?INT(789)
?INT(3000.12345)
?INT(.995)
```

   Based on what the computer shows, what does INT() do to the number inside the parentheses?

1-4

Estimating by Rounding to the Nearest

What is a laser? Laser *is an acronym for Light Amplification by Stimulated Emission of Radiation. Because a laser beam can be focused more precisely than ordinary light, lasers are used in medical surgery, military range finders, bar code scanners, and compact disc players. Laser beams travel at the speed of light.*

If 38 is rounded *up* to the next ten, the result is 40. If 38 is rounded *down* to the preceding ten, the result is 30. Look at this number line.

```
                                                    38
 |            |            |            |         |++++++|            |
 0           10           20           30              40           50
```

The number 38 is nearer to 40 than to 30. So 40 is a better estimate of 38 than 30 is. When 38 is rounded to 40, we say that 38 has been **rounded to the nearest** 10.

Example 1

The speed of light is nearly 186,282.4 miles per second. Round this number to the nearest

a. whole number. **b.** ten. **c.** hundred.
d. thousand. **e.** ten thousand.

Solutions

a. 186,282.4 is between 186,282 and 186,283. Because .4 is less than .5, the number 186,282.4 is nearer to 186,282.

b. 186,282.4 is between 186,280 and 186,290. Because 82.4 is closer to 80 than to 90, round 186,282.4 down to 186,280.

c. 186,282.4 is between 186,200 and 186,300 and is closer to 186,300. So round up to 186,300.

d. The answer is 186,000.

e. The answer is 190,000.

The more accuracy that is needed, the closer one would want to be to the original value.

Example 2

To calculate interest at 8.237%, you may use the number 0.08237 as a multiplier. Round this number to the nearest

a. tenth.   **b.** hundredth.
c. thousandth.   **d.** ten-thousandth.

Solutions

a. 0.08237 is between 0.0 and 0.1. It is nearer to 0.1, so *0.1* is the answer.
b. 0.08237 is between 0.08 and 0.09 and is nearer to *0.08*.
c. 0.08237 is between 0.082 and 0.083 and is closer to *0.082*.
d. 0.08237 is between 0.0823 and 0.0824 and is nearer to *0.0824*.

Rounding is often used to estimate costs.

Example 3

One tape costs $10.49, and Kiko wants to buy 7. Explain how she can use rounding to estimate the cost.

Solution

She can round $10.49 to the nearest ten cents, $10.50, to estimate.
Kiko's estimate:

$$\begin{array}{r} \$10.50 \\ \times \quad 7 \\ \hline \$73.50 \end{array}$$

Tapes or CDs? *Although CDs are more expensive, they last longer and carry almost no background noise or reduction.*

In Example 3, the actual cost is $10.49 × 7, or $73.43. So, Kiko's estimate is only $0.07 more than the actual cost. Kiko could get a quicker but less accurate estimate by rounding $10.49 to the nearest dollar, $10. This would make it easy to estimate the cost in her head.

When Is There a Choice in Rounding?

There are situations in which there may be a choice in rounding. For instance, when rounding to the nearest dollar, you may round $10.50 up to $11 or down to $10, since $10.50 falls exactly halfway between the two. When there are many numbers that can be rounded either way, it makes sense to round up half the time and round down the other half of the time.

QUESTIONS

Covering the Reading

1. Round 43
   a. up to the next ten.
   b. down to the preceding ten.
   c. to the nearest ten.

2. Round 0.547
   a. up to the next hundredth.
   b. down to the preceding hundredth.
   c. to the nearest hundredth.

3. Round 88.8888 to the nearest
   a. hundredth.   b. tenth.   c. one.   d. ten.   e. hundred.

4. You wish to estimate the cost of 4 tapes at $8.69 each. What do you get for the rounded value, and what is your estimate of the cost in each situation?
   a. if you round $8.69 to the nearest ten cents
   b. if you round $8.69 to the nearest dollar

5. Pull-over shirts cost $19.95 each. Describe how to estimate the cost of 6 shirts. What is the estimated cost?

6. Round the speed of light to the nearest hundred thousand miles per second.

7. When is there a choice in rounding to the nearest? Explain why there is a choice, using an example.

8. The number 0.0325 is used in some calculations of the interest earned in a savings account. Round this number to the nearest thousandth.

9. When there are many numbers that can be rounded either up or down, what is the sensible thing to do?

Applying the Mathematics

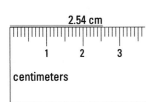

2.54 cm

1   2   3

centimeters

10. The U.S. Internal Revenue Service allows taxpayers to round all amounts to the nearest dollar. But half dollars must be rounded up. In figuring income tax, to what value can you round each amount?
    a. $89.46   b. $165.50   c. $100.91   d. $5324.28

11. Round 2.54, the number of centimeters in an inch, to the nearest tenth.

12. The total American Indian population in the United States in 1990, according to the census, was 1,878,285. Round this number to the nearest ten thousand.

13. Round 3.666666 to the nearest
    a. tenth. **b.** hundredth. **c.** thousandth. **d.** ten-thousandth.

14. Round 12.3500 to the nearest hundredth.

15. You buy items costing $4.99, $6.99, and $8.99 in a store.
    a. Add these quantities.
    b. How close would you be to the actual cost if you rounded each given quantity to the nearest dollar and then added?

16. Consider the addition problem 2.898765489 + 8.1898989898.
    a. Estimate the answer by first rounding both numbers to the nearest whole number and then adding the estimates.
    b. Explain how you could get a better estimate.

In 17–20, what is the answer rounded to the nearest whole number? (Hint: You should be able to do these mentally.)

17. 6 × $3.99

18. $11.95 divided by 2

19. 920.9994 − 0.0003992

20. 2.0123456789 + 3.0123456789

21. A number is rounded to the nearest hundred. The resulting estimate is 9,600.
    a. What is the smallest value the original number might have had?
    b. What is the largest value the original number might have had?

Where do American Indians live? *According to the 1990 census, about 739,000 of the American Indians in the U.S. live on reservations or in tribal designated areas, mainly in the states of Oklahoma, Arizona, and New Mexico. The accountants pictured above live in Texas.*

Review

22. Sarah and George have $10 to spend at the grocery store. They want to buy 3 grapefruits at $0.96 each and 5 oranges at $0.83 each.
    a. Use rounding to estimate the cost.
    b. Do they have enough money? How do you know?
    c. Calculate the exact cost.
    d. How much change will they receive? *(Lesson 1-3)*

23. If a calculator shows that you should pay $1.534 for something, what will a store probably charge you? *(Lesson 1-3)*

24. Round 1.008
    a. up to the next hundredth.
    b. down to the preceding hundredth. *(Lesson 1-3)*

25. Truncate 3.775 to one decimal place. *(Lesson 1-3)*

26. The Gardners are putting a picket fence along the sidewalk in front of their house. The fence has a post every 3 feet. If the front of the lot measures 45 feet, how many posts will they need? (Hint: Represent the fence with a number line.) *(Lesson 1-2)*

In 27–29, find a number that is between the two given numbers. *(Lesson 1-2)*

27. 3.2 and 3.4   **28.** 6.3 and 6.29   **29.** 14.23 and 14.230

30. Write one thousand, five hundred six and three tenths as a decimal. *(Lesson 1-2)*

31. John bought a dozen eggs. In "a dozen eggs," what is the count, and what is the counting unit? (Watch out!) *(Lesson 1-1)*

32. The number ten million consists of a one followed by how many zeros? *(Lesson 1-1)*

Exploration

33. a. Find the number that satisfies all of these conditions.

Condition 1: When rounded up to the next hundred, the number becomes 600.

Condition 2: When rounded down to the preceding ten, the number becomes 570.

Condition 3: When rounded to the nearest ten, the number is increased by 4.

Condition 4: The number is a whole number.

b. Are any of the conditions not needed? If so, which ones?

34. How many decimal places does your computer show when it does a computation? Do this activity to find out.

a. Type ?3/4 and press RETURN. What does the computer print, 0.75 or .75?

b. Type ?2/3 and press RETURN. How many decimal places does the computer print? Does the computer round to the nearest or does it truncate?

1-5

Knowing Your Calculator

Who uses calculators? *Just about anyone who uses numbers uses a calculator. From left to right, the calculators pictured are a four-function calculator, a scientific calculator, and a graphics calculator.*

Calculators make it easy to do arithmetic quickly and accurately. But they cannot help unless you know how and when to use them. Different calculators may give different answers even when the same buttons are pushed. With this book it is best if you have a **scientific calculator.**

When reading mathematics, you will learn and remember more by doing the problems yourself as you read. As you read this lesson, you will need a calculator, paper, and a pencil.

Clearing and Entering

When you turn on the calculator, 0 or 0. will appear in the **display.** We show this as ⎡ 0. ⎤. As you press keys, the display changes. If your calculator is already on, press the clear key C or CLEAR twice before doing calculations.

Activity 1

To find 29 + 54, follow the directions in the left column below. After pressing each key, compare your calculator display to the one in the right column.

Display shows

Press 2, then 9.	29.
Now press +.	29.
Next press 5, then 4.	54.
Now press =.	83.

Pressing a key on the calculator is called **entering** or **keying in.** The set of instructions in the left column of Activity 1 is called the **key sequence** for this problem. We will write the key sequence using boxes for everything pressed but the numbers.

$$29 \boxed{+} 54 \boxed{=}$$

Sometimes we show the display values underneath the last key pressed.

Key sequence:	29	$\boxed{+}$	54	$\boxed{=}$
Display:	29.	29.	54.	83.

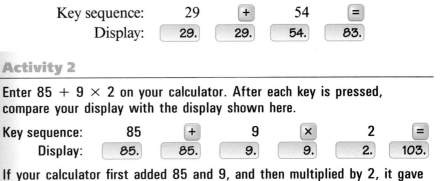

Activity 2

Enter $85 + 9 \times 2$ on your calculator. After each key is pressed, compare your display with the display shown here.

Key sequence:	85	$\boxed{+}$	9	$\boxed{\times}$	2	$\boxed{=}$
Display:	85.	85.	9.	9.	2.	103.

If your calculator first added 85 and 9, and then multiplied by 2, it gave you the incorrect answer 188. If you got 188, your calculator is not appropriate for this class.

Entering π

Most scientific calculators have a way of entering the number π. If you have a π key, simply press it. However, on some calculators, you must press two keys to display π. If there is a small π written above or below a key, two keys are probably needed. In this case, press $\boxed{\text{INV}}$, $\boxed{\text{2nd}}$, $\boxed{\text{SHIFT}}$, or $\boxed{\text{F}}$ before pressing the key with the π above or below it.

Activity 3

Enter π. What number is displayed? How many places after the decimal point does your calculator show?

How Does Your Calculator Round Decimals That Are Too Long to Be Displayed?

Calculators differ in the way they round decimals to the right of a decimal point. This usually does not make much of a difference, but you should know what *your* calculator does. To check your calculator, try this activity.

Activity 4

This book's calculator symbol for division is $\boxed{\div}$. Key in $2 \boxed{\div} 3 \boxed{=}$.

The actual answer to 2 divided by 3 is 0.666666666666666 . . . , where the digit 6 repeats forever. No calculator can list all the digits. So the calculator must be programmed to round. (Calculators *are* computers; each key triggers a program.) If the last digit your calculator displays is a 7, your calculator rounds to the nearest. If the last digit your calculator displays is a 6, your calculator truncates. Does your calculator round to the nearest or truncate?

Memory Keys

Scientific calculators have keys that will store a value in "memory" for later use. Such a key might be labeled as STO or M+. In order to use a value that has been stored, you need to press RCL or MR for memory "recall." You should explore the use of that key. The memory on most scientific calculators is cleared when the AC key is pressed.

Writing Key Sequences

For many homework questions in this book, your teacher may wish you to "show your work." When you use a calculator, this means writing the key sequence and the final display.

Example

Vince bought four books at $12.95 each at a store that gives a discount of $5 if you spend more than $50. What was the cost of the four books (before tax)?

Solution

You must multiply $12.95 by 4 and then subtract 5. If you use a calculator and want to show your work, you can write:

Key Sequence: 4 × 12.95 − 5 =
        Display: 46.8
The four books cost $46.80 before tax.

QUESTIONS

Covering the Reading

1. What is a *key sequence?*

2. *True or false.*   If you follow the same key sequence on two different calculators, you will always get the same answer.

3. What answer did you get for Activity 2?

4. Do the following key sequence on your calculator. Write down what is in the display after each key is pressed.

Key sequence: 8   +   7.2   ×   10   =
        Display:

5. **a.** Write a key sequence for the problem 15 ÷ 27.
   b. What is the final display?

6. **a.** What value does your calculator give for π?
   b. Compare the value for π your calculator gives to the one on page 12. Does your calculator truncate or round to the nearest?

7. Explain the purpose of the [STO] key.

Applying the Mathematics

8. **a.** Follow the directions of Question 4, using this key sequence.
   17.95 [STO] 6 [×] [RCL] [=] [+] 5 [×] [RCL] [=]
   b. What have you calculated in part **a**?

9. All calculators have a way of starting from scratch with a new calculation. How is this done on your calculator?

10. All calculators have a way of allowing you to correct a mistake in an entry. You press a key to replace one entry with another. On your calculator, what is this key called?

11. How many decimal places does your calculator display? Enter the following key sequence to find out.

    13717421 [÷] 333 [÷] 333667 [=]

In 12–15, do the arithmetic problem on your calculator.

12. 3.5625×512

13. $0.9 - 0.99 + 0.999$

14. $6 \times \pi$

15. $5 + 3 \times 17$

16. Perform the calculation and round the answer to the nearest tenth.
    $28.3 \div 5.1 - 3.71$

17. What is the largest number in decimal notation that your calculator can display?

18. Which is larger, $\pi \times \pi$ or 10?

19. Explore the memory keys on your calculator. Write at least one key sequence using those keys and explain what the calculator has done.

20. Use this information and your calculator. Jeffrey is saving money each week to buy presents for his family. He wants to save $150. How much must he save each week if he saves for the given amount of time?

    a. 3 weeks   **b.** 5 weeks   **c.** 8 weeks

    d. 12 weeks   **e.** 20 weeks

    (Hint: You may be able to save time by using a memory key.)

How old is the calculator?
The calculator shown is the first scientific hand-held calculator, the TI SR-10, invented at Texas Instruments in 1967. It is about 6" by 4" by 2" and weighs about 3 pounds. By 1973, Texas Instruments had introduced a commercial version that sold for about $150.

21. Round to the nearest thousandth.
   a. .44041   **b.** .44051   **c.** .40451   *(Lesson 1-4)*

22. a. Round 7.25 up to the next tenth.
   b. Round 7.25 down to the preceding tenth.   *(Lesson 1-3)*

23. You run a race in 53.7 seconds. Someone beats you by two tenths of a second. What was that person's time?   *(Lesson 1-2)*

24. Find a number between 2.36 and 2.37.   *(Lesson 1-2)*

25. Mark began his math homework in school and needed to complete questions 10–30 at home. How many questions does he need to complete at home?   *(Lesson 1-2)*

26. Consider the number of buttons on your calculator.
   a. Name the count.
   b. Name the counting unit.   *(Lesson 1-1)*

27. Translate into English words: 3,412,670.   *(Lesson 1-1)*

28. Three hundred thousand is written as a three followed by how many zeros?   *(Lesson 1-1)*

29. What number is three less than three hundred thousand?   *(Lesson 1-1)*

30. Sandy made purchases of $2.23, $4.07, and $2.49. She gave the clerk $10.04. If Sandy received one bill and one coin from the clerk, what bill and coin were they?   *(Previous course)*

31. Key in 5 ÷ 0 = on your calculator.
   a. What is displayed?
   b. What does the display mean?
   c. Why did this happen?

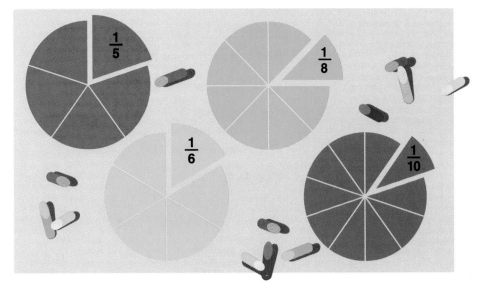

LESSON 1-6

Decimals for Simple Fractions

Fraction Vocabulary

The above picture shows *fractions* of four circles. A symbol of the form $\frac{a}{b}$ or a/b is a **fraction** with a **numerator** a and **denominator** b. The **fraction bar** — or **slash** / indicates division.

$$\frac{a}{b} = a/b = a \div b$$

In the language of division, the number a is the **dividend,** b is the **divisor,** and $\frac{a}{b}$ is the **quotient.** In $\frac{2}{3}$, 2 is the numerator or dividend, and 3 is the denominator or divisor. The fraction itself is the quotient, the result of dividing 2 by 3.

The fraction bar was first used by the Arabs and later by Fibonacci, but it was not widely used until the 1500s. A curved slash, $a|b$, was first used by the Mexican Manuel Antonio Valdes in 1784. In the 1800s this developed into the slash in a/b.

A **simple fraction** is a fraction with a whole number in its numerator and a nonzero whole number in its denominator. (Because division by zero is not allowed, zero cannot be in the denominator of a fraction.) Here are some simple fractions.

$$3/4 \qquad \frac{72}{8} \qquad \frac{3}{11} \qquad \frac{0}{135} \qquad \frac{4}{180} \qquad \frac{34}{19}$$

Fractions and Division

Fractions are very useful because they are related to division. But fractions are harder to order, round, add, and subtract than decimals. So it often helps to find a decimal that equals a given fraction. This is easy to do, particularly with a calculator.

Example 1

Find a decimal equal to $\frac{3}{5}$.

Solution

Key sequence: 3 $\div$ 5 $=$
        Display: 0.6
        Answer: 0.6.

Check 1

(a very rough check) $\frac{3}{5}$ is less than 1. So is 0.6. (This checks that the numbers were entered in the correct order.)

Check 2

The answer 0.6 is six tenths, or $\frac{6}{10}$. Examine the circles at the beginning of this lesson. Do three of the $\frac{1}{5}$ pieces equal six of the $\frac{1}{10}$ pieces? Yes, they do. So it checks.

Activity

Draw fractions of the circles at the beginning of this lesson to show Check 2 to Example 1.

Example 2

Find the decimal equal to $\frac{7}{4}$.

Solution 1

Key in: 7 $\div$ 4 $=$. The calculator displays the exact answer, 1.75.

Solution 2

Divide 4 into 7 using paper and pencil.

$$
\begin{array}{r}
1.75 \\
4\overline{)7.00} \\
\underline{4} \\
30 \\
\underline{28} \\
20 \\
\underline{20} \\
0
\end{array}
$$

Example 3

Find the decimal equal to $\frac{3}{11}$.

Solution

Key in: **3** ÷ **11** = . What the calculator shows depends on the way it rounds and the number of decimal places it displays. You might see 0.27272727 or 0.2727273 or 0.2727272, or something like this with fewer or more decimal places. This suggests that the 27 repeats again and again. That is the case.

$$\frac{3}{11} = 0.2727272727272727272727272727272727 \ldots,$$

where the 27 repeats forever. For practical purposes an abbreviation is needed. It has become the custom to write

$$\frac{3}{11} = 0.\overline{27}.$$

The bar over the 27 indicates that the 27 repeats forever. The digits under the bar are the **repetend** of this **infinite repeating decimal.**

You can use long division to verify that a decimal repeats. Here we again determine the decimal for $\frac{3}{11}$, this time using long division.

$$\begin{array}{r} .2727\ldots \\ 11)\overline{3.0000} \\ 2\,2 \\ \hline 80 \\ 77 \\ \hline 30 \\ 22 \\ \hline 80 \\ 77 \\ \hline 3 \end{array}$$

Notice that the remainders, after subtraction, alternate between 8 and 3. This shows that the digits in the quotient repeat forever.

Example 4

Find the decimal equal to $\frac{87}{70}$.

Solution

Key in: **87** ÷ **70** = . Display: 1.242857142 .

You cannot tell from this display whether or not the decimal repeats. Also, you cannot tell whether the last digit is rounded up or not. So write $\frac{87}{70}$ = 1.24285714 . . . . That is all you are expected to do at this time.

In Example 4, the decimal actually does repeat.

$$\frac{87}{70} = 1.2\overline{428571}$$

In fact, *all* simple fractions are equal to ending or repeating decimals.

What Decimals and Fractions Should You Know by Heart?

The common simple fractions in the chart below appear frequently in mathematics and in real-life situations. You need to know them.

Fourths and Eighths	**Thirds and Sixths**	**Fifths and Tenths**
$\frac{1}{8} = 0.125$	$\frac{1}{6} = 0.1\overline{6}$	$\frac{1}{10} = 0.1$
$\frac{1}{4} = \frac{2}{8} = 0.25$	$\frac{1}{3} = \frac{2}{6} = 0.\overline{3}$	$\frac{1}{5} = \frac{2}{10} = 0.2$
$\frac{3}{8} = 0.375$	$\frac{3}{6} = 0.5$	$\frac{3}{10} = 0.3$
$\frac{2}{4} = \frac{4}{8} = 0.5$	$\frac{2}{3} = \frac{4}{6} = 0.\overline{6}$	$\frac{2}{5} = \frac{4}{10} = 0.4$
$\frac{5}{8} = 0.625$	$\frac{5}{6} = 0.8\overline{3}$	$\frac{5}{10} = 0.5$
$\frac{3}{4} = \frac{6}{8} = 0.75$		$\frac{3}{5} = \frac{6}{10} = 0.6$
$\frac{7}{8} = 0.875$		$\frac{7}{10} = 0.7$
		$\frac{4}{5} = \frac{8}{10} = 0.8$
		$\frac{9}{10} = 0.9$

Some fractions are whole numbers in disguise.

Example 5

Find the decimal equal to $\frac{91}{13}$.

Solution

Key in: 91 ÷ 13 =.

Display:  7. . $\frac{91}{13} = 7$.

In Example 5, we say that 91 is **evenly divisible** by 13. Sometimes we merely say that 91 is **divisible** by 13.

QUESTIONS

Covering the Reading

1. Consider the fraction $\frac{15}{8}$.
   a. What is its numerator?
   b. What is its denominator?
   c. What is the symbol for division?
   d. Does it equal 15/8 or 8/15?
   e. Does it equal 8 ÷ 15?
   f. Does it equal 15 ÷ 8?
   g. What is the divisor?
   h. What is the dividend?
   i. Find the decimal equal to it.

2. Before the 1500s, who used the fraction bar?

3. Who first developed a slash symbol for fractions, and when?

4. Which of the following are simple fractions?
   a. $\frac{15}{8}$ b. 2   c. $\frac{3.5}{2.3}$ d. $6\frac{2}{3}$ e. $\frac{0}{9}$

5. Why is it helpful to be able to find a decimal for a fraction?

6. Draw fractions of the circles on page 31 to confirm Check 2 to Example 1.

In 7–10, find a decimal for each fraction.

7. $\frac{3}{20}$ 8. $\frac{23}{20}$ 9. $\frac{4}{7}$ 10. $\frac{1}{27}$

11. In $86.\overline{27}$, what is the repetend?

12. In $0.39\overline{8}$, what is the repetend?

In 13 and 14 write the first ten decimal places.

13. $9.8\overline{7}$ 14. $0.\overline{142}$

15. If you do not know the decimals for these fractions, you should learn them now. Try to find each decimal without looking at page 34.

   a. $\frac{1}{10}$ b. $\frac{2}{10}$ c. $\frac{3}{10}$ d. $\frac{4}{10}$ e. $\frac{5}{10}$

   f. $\frac{6}{10}$ g. $\frac{7}{10}$ h. $\frac{8}{10}$ i. $\frac{9}{10}$ j. $\frac{1}{5}$

   k. $\frac{2}{5}$ l. $\frac{3}{5}$ m. $\frac{4}{5}$ n. $\frac{1}{2}$ o. $\frac{1}{4}$

   p. $\frac{3}{4}$ q. $\frac{1}{8}$ r. $\frac{3}{8}$ s. $\frac{5}{8}$ t. $\frac{7}{8}$

   u. $\frac{1}{3}$ v. $\frac{2}{3}$ w. $\frac{4}{6}$ x. $\frac{1}{6}$ y. $\frac{5}{6}$

In 16–21, give a simple fraction for each decimal. You should be able to find each fraction without looking at page 34.

16. 0.4   17. .25   18. $.\overline{3}$

19. 0.60   20. 0.7   21. $.\overline{6}$

Who uses fractions? *Both salespersons and customers need to know decimal equivalents for common fractions. Someone requesting 1/4 pound of cheese should know that 0.25 pound is the same weight.*

22. $\frac{92}{23} = 4$. So we say that 92 is __?__ by 23.

Applying the Mathematics

---

23. Carpenters often measure in sixteenths of an inch.
    a. Change $\frac{3}{16}''$ to a decimal.
    b. Is $\frac{3}{16}''$ shorter or longer than $\frac{1}{5}''$?

24. Rewrite $\frac{1}{14}$ as a decimal rounded to the nearest thousandth.

25.
   0 ├─────────────────────────────────────┤ 1

    a. Trace this line segment onto your paper.
    b. Graph the fractions $\frac{1}{2}, \frac{1}{3}, \frac{2}{3}, \frac{1}{4}, \frac{2}{4}$, and $\frac{3}{4}$ on it.

26. Order 3/10, 1/3, and 0.33 from smallest to largest.

27. Order $\frac{2}{9}, \frac{2}{11}, \frac{2}{7}$ from smallest to largest.

28. Write a letter to a younger student explaining how to change a fraction into a decimal. Include in your explanation how to do this with and without a calculator.

Review

29. a. Do this arithmetic on your calculator: 8.868×6.668. What does your calculator display?
   b. Round your answer to part **a** to the nearest thousandth.
   (Lessons 1-4, 1-5)

30. Round 3,522,037, the population of Puerto Rico in 1990 according to the U.S. census, to the nearest hundred thousand. *(Lesson 1-4)*

31. Round 9.8978675645 to the nearest ten-thousandth. *(Lesson 1-4)*

32. Find a number between 0.036 and 0.0359. *(Lesson 1-2)*

33. Which is larger, 34.000791 or 34.0079? *(Lesson 1-2)*

34. What temperature is shown by this thermometer? *(Lesson 1-2)*

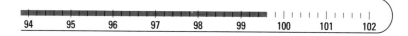

35. The Sanchez Dairy Farm sold 479 gallons of milk one day and 493 gallons the next. On the third day they sold twice as many gallons as on the previous two days. What was the total number of gallons sold on those three days? *(Previous course)*

Exploration

Nothing can ever come between us.

36. a. Find the decimals for $\frac{1}{9}, \frac{2}{9}, \frac{3}{9}, \frac{4}{9}, \frac{5}{9}, \frac{6}{9}, \frac{7}{9},$ and $\frac{8}{9}$.
   b. Based on the pattern you find, what fraction should equal $.\overline{9}$?
   c. Is the cartoon true?

37. a. Write down the decimals for $\frac{1}{2}, \frac{1}{3}, \frac{1}{4}, \frac{1}{5},$ and $\frac{1}{6}$.
   b. Find the decimals for $\frac{1}{7}, \frac{1}{8}, \frac{1}{9}, \frac{1}{10}, \frac{1}{11},$ and $\frac{1}{12}$.
   c. If you keep going, to what number are these decimals getting closer and closer?

38. a. Explore the decimals for all simple fractions between 0 and 1 whose denominator is 7.
   b. Use your results from part **a** to give the first twelve decimal places for each of these fractions.

What can you get for $2.75? *A hamburger, a tube of toothaste, a magazine, or 2 gallons of gas are examples of items that could each be purchased with $2.75 or less.*

The number $2\frac{3}{4}$ consists of a whole number and a fraction. It is called a **mixed number.** This mixed number is the sum of the whole number 2 and the fraction $\frac{3}{4}$. Only the plus sign is missing.

Graphing Mixed Numbers on a Number Line

Mixed numbers are common in measurement. The line segment below is three inches long and is divided into intervals each one inch long.

If each of these intervals is divided into two equal parts, the new intervals will be $\frac{1}{2}$ inch long.

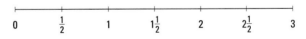

Dividing each half-inch interval into two equal parts creates intervals $\frac{1}{4}$ inch long.

A line segment $2\frac{3}{4}$ inches long consists of 2 whole inches plus $\frac{3}{4}$ of an inch.

You can see that the segment is longer than 2 inches and shorter than 3 inches. This is because $2\frac{3}{4}$ is between 2 and 3.

Changing Mixed Numbers to Decimals

On the previous page, we thought of $2\frac{3}{4}$ as $2 + \frac{3}{4}$. This suggests how the decimal for a mixed number can be found. First calculate the decimal for the simple fraction. Then add that decimal to the whole number.

Example

Express $2\frac{3}{4}$ in decimal notation.

Solution 1

Remember or calculate: $\frac{3}{4} = 0.75$.

Now add: $2\frac{3}{4} = 2 + \frac{3}{4} = 2 + 0.75 = 2.75$.

Solution 2

Use a scientific calculator. Key in 2 $\boxed{+}$ 3 $\boxed{\div}$ 4 $\boxed{=}$. On simpler calculators, the division must be done first, then the 2 must be added.

Display: $\boxed{2.75}$

Solution 3

Think money. You might write the following:

Fourths are quarters. Two and three-fourths is like two dollars and three quarters. Two dollars and three quarters is $2.75, which includes the correct decimal, 2.75.

QUESTIONS

Covering the Reading

1. Consider the mixed number $5\frac{3}{4}$.
   a. Between what two whole numbers is this number?
   b. Identify the whole number part of this mixed number.
   c. Rewrite this number in decimal notation.
   d. Draw a line segment $5\frac{3}{4}$ inches long.

2. What is it about the mixed number of Question 1 that enables a person to think about it in terms of money?

3. Consider the mixed number $4\frac{1}{3}$.
   a. What is the largest whole number less than $4\frac{1}{3}$?
   b. Identify the fraction part of this number.
   c. Write this number in decimal notation.
   d. Draw a number line 6 units long with intervals of 1 unit. Locate $4\frac{1}{3}$ on your number line.

In 4–7, change the mixed number to a decimal. (The fraction parts are ones you should know, so try to do these without a calculator.)

4. $2\frac{1}{2}$ **5.** $7\frac{2}{5}$ **6.** $1\frac{3}{10}$ **7.** $17\frac{5}{6}$

In 8–11, change the mixed number to a decimal.

8. $12\frac{5}{16}$ **9.** $4\frac{1}{11}$ **10.** $20\frac{8}{15}$ **11.** $5\frac{7}{8}$

Applying the Mathematics

12. Most stock prices are measured in eighths. What is each value in dollars and cents?

     a. A stock's price goes up $4\frac{1}{4}$ dollars a share.

     b. A stock's price goes up $1\frac{3}{8}$ dollars a share.

13. Order from the smallest to the largest: $2\frac{3}{5}$ $5\frac{2}{3}$ $3\frac{2}{5}$.

14. Mouse A is $2\frac{3}{10}$ inches long. Mouse B is $2\frac{1}{4}$ inches long. Which mouse is longer?

15. Round $12\frac{8}{15}$ to the nearest thousandth.

16. The Preakness, a famous horse race, is $1\frac{3}{16}$ miles long. Convert this length to a length in decimals.

17. A shelf is measured to be $35\frac{11}{32}$ inches long. Is this shorter or longer than $35\frac{1}{3}$ inches?

18. Trace this segment onto your paper. Graph each of the indicated numbers.

     a. $\frac{1}{2}$ **b.** $\frac{1}{4}$ **c.** $\frac{1}{8}$ **d.** $\frac{2}{8}$

     e. $1\frac{1}{2}$ **f.** $2\frac{1}{4}$ **g.** $2\frac{7}{8}$ **h.** 2.5

19. Draw a line segment 3.375 inches long. (Hint: What fraction equals 0.375?)

Review

20. In parts **a–h,** give the decimal for each number.   *(Lesson 1-6)*

     a. $\frac{1}{8}$ **b.** $\frac{2}{8}$ **c.** $\frac{3}{8}$ **d.** $\frac{4}{8}$

     e. $\frac{5}{8}$ **f.** $\frac{6}{8}$ **g.** $\frac{7}{8}$ **h.** $\frac{8}{8}$

     i. Explain how the answer to part **a** enables a person to find the answers to all the other parts.

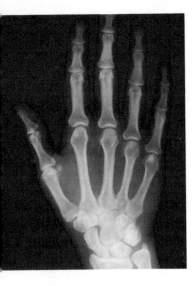

What's inside?
Discovered by Wilhelm Roentgen in 1895, X-rays play a vital role in determining the structure of matter and in treating diseases such as cancer.

21. **a.** Give an example of a simple fraction.
    b. Write a fraction that is not a simple fraction.   *(Lesson 1-6)*

22. What digit is in the eleventh decimal place of $7.\overline{8142}$?   *(Lesson 1-6)*

23. Estimate $16.432893542050 + 83.5633344441$ to the nearest whole number.   *(Lesson 1-4)*

24. Find a number between 2 and 2.1.   *(Lesson 1-2)*

25. *True or false.* $5 = 5.0$ *(Lesson 1-2)*

26. In decimal notation, write the whole number that is one less than one million.   *(Lesson 1-1)*

27. Consider the following sentences. Each human hand has 27 small bones. Together the hands have over $\frac{1}{4}$ of the 206 bones in the whole body.
    a. Name the counts.   **b.** Name the counting units.   *(Lesson 1-1)*

28. Mark has four $20 bills to buy school clothes. He bought a shirt for $16.97, a pair of jeans for $21.63, and a pair of sneakers for $33.78. How much change should Mark receive when he pays for the clothes?   *(Previous course)*

Exploration

29. **a.** Examine a stock market page from a daily newspaper. Estimate how many mixed numbers there are on the page.
    b. Explain how you found your estimate in part **a.**

30. Refer to the Example in this lesson. Write a brief explanation telling why you think three solutions are given instead of just one. Which solution do you prefer for solving the problem? Why is this your preference?

*Negative
Numbers*

Is the price right? *In this store, a shopper is considering unusual clothing and decorative items imported from all over the world. The amount the customer pays for these items must be greater than the store owner's cost in order for the store to avoid losses and remain in business.*

Situations with Negative Numbers

On every item a store sells, the store can make money, lose money, or break even. Here are some of the possibilities.

Using words and numbers	Using numbers only
make $3	3
make $2	2
make $1	1
make $0.50	.5
break even	0
lose $1	-1
lose $1.50	-1.5
lose $2	-2
lose $3	-3

The numbers along the number line describe the situation without using words. Higher numbers on the line are larger and mean more profits. The - (negative) sign stands for *opposite of.* The opposite of making $1.50 is losing $1.50. Similarly, the opposite of 1.50 is -1.50, and vice versa. The numbers with the - sign are called **negative numbers.**

Most people know negative numbers from temperatures. But they are found in many other situations. On TV bowling programs, -12 means "behind by twelve pins." The symbol +12, called **positive** 12, means "ahead by twelve pins." The numbers 12 and +12 are identical. Since it is shorter and simpler to leave off the + (positive) sign, positive 12 is usually written as 12.

Negative numbers can be used when a situation has two opposite directions. Either direction may be picked as positive. The other is then negative. Zero stands for the starting point. The table below gives some situations that often use negative numbers.

Situation	Negative	Zero	Positive
savings account	withdrawal	no change	deposit
time	before	now	after
games	behind	even	ahead
business	loss	break even	profit
elevation	below sea level	sea level	above sea level

For instance, the shore of the Dead Sea in Israel, the lowest land on Earth, is 396 meters below sea level. This can be represented by -396 meters.

There are three common ways in which the - sign for negatives is said out loud.

write	say	
-3	negative 3	correct
-3	opposite of 3	correct
-3	minus 3	very commonly used, but can be confusing since there is no subtraction here

Does anything live in the Dead Sea? *Brine shrimp and a few plant species are able to live in the Dead Sea, which is nine times as salty as the ocean. The salty nature of the Dead Sea increases the density of the water. Because of this density, people do not sink.*

Example 1

A space shuttle is to be launched. Represent each of these times (in seconds) by positive or negative numbers.
a. 4.3 seconds before the launch
b. 1 minute after the launch
c. the time of the launch

Solution

a. Since the time is *before* the launch, it is negative.
The time is -4.3 seconds.
b. The time is *after* the launch, so it is positive. Since 1 minute = 60 seconds, The time is 60 seconds.
c. The time of the launch is represented by 0 seconds.

Example 2

The price of a share of stock went down $1\frac{7}{8}$ dollars.

a. How will this be written on the stock market pages of a newspaper?
b. Write the answer to part **a** as a decimal.
c. Round the answer to part **b** to the nearest penny.

Solution

a. Since the price went down, the change will be written as $-1\frac{7}{8}$.
(Newspapers omit the dollar sign.)

b. $\frac{7}{8} = 0.875$, so $-1\frac{7}{8} = -1.875$.

c. -1.875 is halfway between -1.88 and -1.87. The answer is: either $-\$1.88$ or $-\$1.87$.

Graphing Negative Numbers on a Number Line

On a horizontal number line, negative numbers are almost always placed at the left. The numbers identified on the number line drawn below are the **integers**. The **positive integers** are the numbers 1, 2, 3, ... The **negative integers** are $-1, -2, -3,$ . . . . Zero is an integer but is neither positive nor negative. All numbers to the right of 0 are **positive numbers.** All numbers to the left of 0 are **negative numbers.**

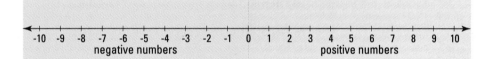

Example 3

Graph 1.3 and -1.3.

Solution

Use a number line from -2 to 2. To graph 1.3, split the interval from 1 to 2 into tenths. To graph -1.3, split the interval from -2 to -1 into tenths. Then graph each number by placing a dot on the appropriate tick mark.

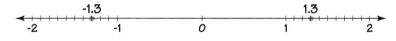

Entering Negative Numbers into a Calculator

All scientific calculators have a way to enter negative numbers. This is done by the opposite key +/− or ± or (−). For example, 7 ± keys in -7 on some calculators.

Covering the Reading

1. Translate -4 into English words in two different ways.

2. Next to a bowler's name on TV is the number -8. Is the bowler ahead or behind?

3. On a horizontal number line, negative numbers are usually to the __?__ of positive numbers.

4. On a vertical number line, negative numbers are usually __?__ positive numbers.

5. Graph -7, $-3\frac{1}{2}$, $3\frac{1}{2}$, 0, and 7 on the same horizontal number line.

6. Graph a profit of $4, a loss of $2.10, breaking even, and a profit of $10 on a vertical number line.

7. Write a key sequence for entering -3.5 on a calculator.

In 8–10, three words or phrases relating to a situation are given. Which would usually be considered positive? which negative? which zero?

8. Football: losing yardage, gaining yardage, no gain

9. Time: tomorrow, today, yesterday

10. Stock market: no change, gain, loss

11. The price of a stock went down $2\frac{1}{8}$ dollars a share.
    a. How will this be written on the stock market pages of a newspaper?
    b. Write the answer to part **a** as a decimal.
    c. Round the answer to part **b** to the nearest penny.

12. *Multiple choice.* Which of the following numbers is not an integer?
    (a) 5   (b) 0   (c) -5   (d) .5

13. Give an example of an integer that is neither positive nor negative.

14. Give an example of a negative number that is not an integer.

Pins to spare. *For a spare, you get 10 pins plus the number of pins knocked down by the first ball of your next turn. For a strike, you get 10 pins plus the pins from your next 2 balls. In each case, the total number of pins is added in the frame where the spare or strike occurred.*

Applying the Mathematics

15. Suppose time is measured in days and 0 stands for today.
    a. What number stands for yesterday?
    b. What number stands for tomorrow?
    c. What number stands for the day before yesterday?
    d. What number stands for the day after tomorrow?

In 16–19, use the number line drawn here. Which letter, if any, corresponds to the given number?

A B C D E F G H I J K L M N O P Q R S T U

-10   -9   -8

16. -9.1   **17.** -8.4   **18.** -9.0   **19.** -10.1

20. Pick the two numbers that are equal: -43.3   -43.03   -43.30   43.3

21. Which of -1, 0, $\frac{1}{2}$, $\frac{2}{2}$, and $\frac{3}{2}$ are not positive integers?

In 22–24, round to the nearest integer.

22. -1.75   **23.** $-\frac{3}{5}$ **24.** -0.4273

In 25–27, round to the nearest tenth.

25. -60.52   **26.** -43.06   **27.** $-9\frac{1}{3}$

Review

28. Write each number as a decimal. *(Lessons 1-6, 1-7)*
   a. $8\frac{3}{4}$ **b.** $-\frac{2}{3}$ **c.** $5\frac{3}{5}$ **d.** $\frac{5}{16}$

29. Write the key sequence and the display after each calculator entry for $6 + 3 \times 4 \div 2$. *(Lesson 1-5)*

30. Round $28.47
   a. up to the next dollar.   **b.** down to the preceding dollar.
   c. to the nearest dollar. *(Lessons 1-3, 1-4)*

31. Suppose the points are equally spaced on the number line drawn here. If *E* is 1 and *L* is 2, what number corresponds to *C?* *(Lesson 1-2)*

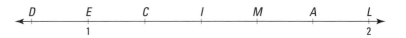

D   E   C   I   M   A   L

1   2

32. Order from smallest to largest: 5.67   5.067   5.607   5.60 *(Lesson 1-2)*

33. Write as a decimal: four hundred sixty-two thousand and one hundredth. *(Lesson 1-2)*

Exploration

34. Use an almanac to find the place in the United States with the lowest elevation. What number represents this elevation?

35. Examine a page of a newspaper that has prices of stocks. Copy or cut out some rows with negative numbers.

36. Find an example of negative numbers that is not given in this lesson.

Comparing Counts

Counts are frequently compared. For instance, in 1990 there were about 4,179,000 births in the United States. In 1991 there were about 4,111,000 births. To indicate that there were fewer births in 1991 we write

$$4{,}111{,}000 < 4{,}179{,}000.$$

The symbol < means **is less than.** The symbol > means **is greater than,** so we could also write

$$4{,}179{,}000 > 4{,}111{,}000.$$

Comparison of populations is useful in knowing whether more schools or hospitals should be built, or how many people could buy a particular item or watch a television program.

The symbols < and > are **inequality symbols.** Thomas Harriot, an English mathematician, first used < and > in the early 1600s. You can remember which symbol is which because each symbol points to the smaller number: $3 < 5$ and $5 > 3$.

Comparing Measures

Measures can also be compared. You probably have compared your height and weight to those of other people.

Example 1

Mario is 5'6" tall. Setsuko is 4'10" tall.
a. Compare these numbers using the > sign.
b. Compare these numbers using the < sign.

Solution

a. Mario is taller than Setsuko.

$$5'6" > 4'10"$$

b. Setsuko is shorter than Mario.

$$4'10" < 5'6"$$

Comparing Positive and Negative Numbers

Numbers can be compared whether they are positive, negative, or zero. You can always think of temperature. For instance, to compare 0 and 4, think: a temperature of 0°C is colder than one of 4°C. In symbols,

$$0 < 4.$$

A temperature of -7°C is colder than either of these temperatures.

$$-7 < 0$$
$$-7 < 4$$

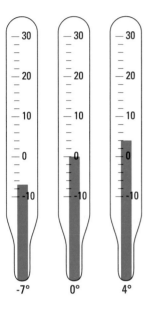

-7°   0°   4°

Numbers graphed on a number line are also easy to compare. Smaller numbers are usually to the *left* or *below* larger numbers. The numbers -7, 0, and 4 are shown on the vertical thermometers and the horizontal number line below.

When numbers are in order, inequalities can be combined. For the numbers pictured here, you could write either **-7 < 0 < 4** or **4 > 0 > -7**.

Caution: Even though 10 is greater than 5, -10 is less than -5. In symbols, -10 < -5. For example, -10 could mean a lower temperature or bigger loss than -5.

Example 2

Rewrite the following sentence using inequality symbols: 12 is between 10 and 15.

Solution

Since 12 is between 10 and 15, it is larger than 10. So 12 > 10.
Since 12 is between 10 and 15, it is smaller than 15. So 15 > 12.
These inequalities can be combined into one sentence: 15 > 12 > 10.
This is read ''fifteen is greater than twelve which is greater than ten.''
Another sentence uses the < sign: 10 < 12 < 15.

Caution: Do not use > and < in the same sentence. For instance, do not write $5\frac{3}{4} > 3 < 4\frac{1}{3}$. A sentence comparing $5\frac{3}{4}$, 3, and $4\frac{1}{3}$ could be written as $5\frac{3}{4} > 4\frac{1}{3} > 3$ or $3 < 4\frac{1}{3} < 5\frac{3}{4}$.

Comparing Fractions

When numbers are written as fractions, it is not always easy to compare them. But if you convert them to decimals, they can be compared more easily.

Example 3

Klaus ate 2 of the 3 pieces of a small pizza. Sonja ate 5 of the 8 pieces of a pizza of the same size. Compare $\frac{2}{3}$ and $\frac{5}{8}$ to find out who ate more.

Solution

Use the decimal equivalents. $\frac{2}{3} = .\overline{6}$ and $\frac{5}{8} = .625$.

Since $.625 < .\overline{6}$, $\frac{5}{8} < \frac{2}{3}$. So, Klaus ate more.

QUESTIONS

Covering the Reading

1. Give an example in which it would be useful to compare counts.

2. What is the meaning of the symbol $<$?

3. What is the meaning of the symbol $>$?

4. *Multiple choice.* Which is true?
   (a) $2 < -5$ (b) $2 = -5$ (c) $2 > -5$

In 5–8, rewrite the sentence using one of the symbols $<$, $>$, or $=$.

5. -5 is less than -3.   6. 6 is greater than -12.   7. $4'11''$ is shorter than $5'$.

8. A height of 4.5 feet is the same as a height of $4\frac{1}{2}$ feet.

In 9–11, write a sentence with the same meaning, using the other inequality symbol.

9. $2 < 2\frac{1}{10}$ 10. $-18 < 0$ 11. $0.43 < 0.432 < 0.44$

12. Write this inequality in two different ways: 0 is between -2 and 2.

13. Marissa ate 2 of the 6 pieces of a small pizza. Mal ate 3 of the 8 pieces of another pizza of the same size. Who ate more?

In 14–16, translate into English words.

14. $-3 < 3$ 15. $17 > -1.5$ 16. $-4 < -3\frac{1}{2} < -3$

In 17 and 18, put the numbers in a mathematical sentence, using $<$, $>$, or $=$.

17. A temperature of $-6°C$ is colder than a temperature of $15°C$.

18. Shawn Bradley, whose height is $7'6''$, is taller than David Robinson, whose height is $7'1''$.

19. On a horizontal number line, larger numbers are usually graphed to the __?__ of smaller numbers.

20. On a vertical number line, larger numbers are usually graphed __?__ smaller numbers.

Meet Mr. Robinson.
As a child, David was interested in music, electronics, and sports. He played basketball in eighth grade—but did not like it! With his father's encouragement, David attended the U.S. Naval Academy in Annapolis, MD, and earned a degree in mathematics. He began his professional basketball career in 1989.

Applying the Mathematics

In 21 and 22, two numbers are given.
a. Graph the numbers on a number line.
b. Put the numbers in a mathematical sentence.

21. A profit of $1 is more than a loss of $2000.

22. An elevation 300 ft below sea level is higher than an elevation 400 ft below sea level.

In 23–30, choose the correct symbol: $<$, $>$, or $=$.

23. $0.305 \underline{\ ?\ } 0.3046$ **24.** $\frac{8}{8} \underline{\ ?\ } \frac{3}{3}$ **25.** $3.\overline{1515} \underline{\ ?\ } 3.\overline{15}$

26. $10.\overline{8} \underline{\ ?\ } 10.8$ **27.** $\frac{10}{11} \underline{\ ?\ } \frac{67}{73}$ **28.** $-14 \underline{\ ?\ } -14\frac{1}{2}$

29. $-0.75 \underline{\ ?\ } 0.75$ **30.** $0.5 \underline{\ ?\ } -0.6$

In 31 and 32, put the three numbers into one sentence with two inequality symbols.

31. $62.1 \quad 6.21 \quad 0.621$ **32.** $2\frac{1}{2} \quad 2\frac{2}{3} \quad 2\frac{2}{5}$

Day 1   Day 2   Day 3

33. The thermometers pictured at the left show Joanne's body temperature on three consecutive days of a cold. Put the three numbers into one sentence connected by inequality symbols.

Review

34. Name all integers between -4 and 3.   *(Lesson 1-8)*

35. Name all the positive integers less than 5.   *(Lesson 1-8)*

36. Suppose time is measured in years and 0 stands for this year. Tell what number each year stands for.   *(Lesson 1-8)*
a. next year   **b.** last year   **c.** 2010   **d.** 1925

37. Round $2 \times \pi$ to the nearest thousandth.   *(Lessons 1-4, 1-5)*

In 38 and 39, estimate each sum to the nearest whole number. Do not use a calculator.   *(Lesson 1-4)*

38. $70.0392 + 6.98234$ **39.** $\$14.95 + \$2.99 + \$7.89$

Exploration

40. Truncating decimal places in a positive number is the same as rounding down. When decimal places are truncated in a negative number, is the result a larger number, a smaller number, or equal to the original number?

1-10

Equal Fractions

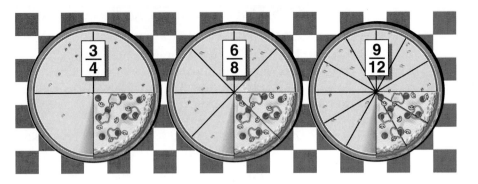

Eat 3 of 4 pieces, or 6 of 8 pieces, or 9 of 12 pieces of a pizza. You will have eaten the same amount.

The numbers $\frac{3}{4}$, $\frac{6}{8}$, and $\frac{9}{12}$ are graphed in the same place on a number line.

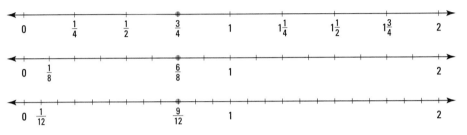

The above situations picture $\frac{3}{4} = \frac{6}{8} = \frac{9}{12}$.

You can verify this equality by changing each fraction to a decimal. You will get 0.75. The fractions $\frac{3}{4}$, $\frac{6}{8}$, and $\frac{9}{12}$ are **equal fractions.**

How Can You Create Equal Fractions?

Notice that $\frac{6}{8} = \frac{3 \times 2}{4 \times 2}$ and $\frac{9}{12} = \frac{3 \times 3}{4 \times 3}$. This suggests a way to create many fractions equal to a given fraction. Just multiply both numerator and denominator by the same nonzero number.

Example 1

Find three other fractions equal to $\frac{2}{3}$.

Solution

You can multiply the numerator and denominator by any nonzero number. We pick 6, 15, and 1000.

$$\frac{2 \times 6}{3 \times 6} = \frac{12}{18} \qquad \frac{2 \times 15}{3 \times 15} = \frac{30}{45} \qquad \frac{2 \times 1000}{3 \times 1000} = \frac{2000}{3000}$$

Check

You know that $\frac{2}{3} = 2 \div 3 = .\overline{6}$. Do the other fractions equal $.\overline{6}$? Divide to find out.

$12 \div 18 = 0.666 \ldots \qquad 30 \div 45 = 0.666 \ldots \qquad 2000 \div 3000 = 0.666 \ldots$

Yes they do.

What happens if you divide the numerator and denominator of $\frac{2000}{3000}$ by the same number?

$$\frac{2000 \div 100}{3000 \div 100} = \frac{20}{30} \qquad \frac{2000 \div 500}{3000 \div 500} = \frac{4}{6} \qquad \frac{2000 \div 1000}{3000 \div 1000} = \frac{2}{3}$$

The results are fractions equal to $\frac{2}{3}$. They are equal because you are essentially multiplying or dividing by 1.

> **Equal Fractions Property**
> If the numerator and denominator of a fraction are both multiplied (or divided) by the same nonzero number, then the resulting fraction is equal to the original one.

In the previous lesson, the fractions $\frac{2}{3}$ and $\frac{5}{8}$ were compared by changing each fraction to a decimal. You can also compare $\frac{2}{3}$ and $\frac{5}{8}$ by using equal fractions. First find a number that is divisible by both 3 and 8. We use 24. Then change the fractions to twenty-fourths.

$$\frac{2}{3} = \frac{2 \times 8}{3 \times 8} = \frac{16}{24} \qquad \frac{5}{8} = \frac{5 \times 3}{8 \times 3} = \frac{15}{24}$$

Since $\frac{16}{24}$ is greater than $\frac{15}{24}$, $\frac{2}{3} > \frac{5}{8}$.

Using Factors to Write Fractions in Lowest Terms

A number that divides evenly into another number is called a **factor** of that number. For example, 3 and 8 are factors of 24. To find all the factors of 24, write all the whole number multiplication problems that equal 24. Be careful not to miss any.

$$1 \times 24 = 24 \qquad 2 \times 12 = 24 \qquad 3 \times 8 = 24 \qquad 4 \times 6 = 24$$

So the factors of 24 are 1, 2, 3, 4, 6, 8, 12, and 24.

Of the many fractions equal to $\frac{2000}{3000}$, the simple fraction with the smallest whole numbers is $\frac{2}{3}$. We say that $\frac{2000}{3000}$, written in **lowest terms,** equals $\frac{2}{3}$. To write a fraction in lowest terms, look for a factor of both the numerator and denominator.

Example 2

Write $\frac{20}{35}$ in lowest terms.

Solution

5 is a factor of both 20 and 35. It is the largest whole number that divides both 20 and 35. Divide both numerator and denominator by 5.

$$\frac{20}{35} = \frac{20/5}{35/5} = \frac{4}{7}$$

Check

You should verify that $\frac{4}{7}$ and $\frac{20}{35}$ equal the same decimal.

There are many ways to simplify a fraction. All correct ways lead to the same answer.

Example 3

Simplify $\frac{60}{24}$. (To simplify means to write in lowest terms.)

Solution 1

Vince saw that 4 is a factor of both 60 and 24. Here is his work.

$$\frac{60}{24} = \frac{60/4}{24/4} = \frac{15}{6}$$

Since 3 is a factor of both 15 and 6, $\frac{15}{6}$ is not in lowest terms. Vince needed a second step.

$$\frac{15}{6} = \frac{15/3}{6/3} = \frac{5}{2}$$

$\frac{5}{2}$ is in lowest terms. So $\frac{60}{24} = \frac{5}{2}$.

Solution 2

Heather knew that 3 is a factor of both 60 and 24. Here is her work.

$$\frac{60}{24} = \frac{60/3}{24/3} = \frac{20}{8}$$

Then she saw that 2 is a factor of 20 and 8.

$$\frac{20}{8} = \frac{20/2}{8/2} = \frac{10}{4}$$

Dividing numerator and denominator again by 2, she got the same answer Vince got.

$$\frac{10}{4} = \frac{10/2}{4/2} = \frac{5}{2}$$

Solution 3

Karen got the same answer in just one step.

$$\frac{60}{24} = \frac{60/12}{24/12} = \frac{5}{2}$$

In Solution 3, Karen makes simplifying $\frac{60}{24}$ a one-step problem by dividing numerator and denominator by 12. We call 12 the **greatest common factor,** or GCF, of 60 and 24. To find the GCF of 60 and 24, write all the factors of 24:

$$1, 2, 3, 4, 6, 8, 12, 24.$$

Next write all the factors of 60:

$$1, 2, 3, 4, 5, 6, 10, 12, 15, 20, 30, 60.$$

The factors they have in common are 1, 2, 3, 4, 6, and 12. So the greatest common factor is 12.

Covering the Reading

1. *Multiple choice.* Which of the following is not equal to $\frac{3}{4}$?
(a) $\frac{6}{8}$ (b) $\frac{8}{12}$ (c) 0.75   (d) $\frac{1.5}{2}$

2. *Multiple choice.* Which fraction, if any, is not equal to the others?
(a) $\frac{24}{36}$ (b) $\frac{48}{72}$ (c) $\frac{4.8}{7.2}$
(d) $\frac{24 \text{ million}}{36 \text{ million}}$ (e) All are equal.

3. a. Picture $\frac{1}{3} = \frac{2}{6} = \frac{3}{9}$ with circles.
    b. Picture this equality with number lines.

4. Find a fraction equal to $\frac{21}{12}$ that has a bigger numerator.

5. *True or false.* Sixteen and twenty are both factors of 80.

6. How can you tell when a fraction is in lowest terms?

7. a. Write the factors of 48.   **b.** Write the factors of 60.
    c. Name all common factors of 48 and 60.
    d. Name the greatest common factor of 48 and 60.
    e. Write $\frac{48}{60}$ in lowest terms.

8. a. Find the GCF of 32 and 48.   **b.** Write $\frac{32}{48}$ in lowest terms.
    c. Write $\frac{48}{32}$ in lowest terms.

In 9–14, a fraction is given.
a. Name a common factor of the numerator and denominator.
b. Rewrite the fraction in lowest terms.

9. $\frac{21}{12}$ **10.** $\frac{15}{20}$ **11.** $\frac{12}{32}$

12. $\frac{52}{64}$ **13.** $\frac{180}{16}$ **14.** $\frac{240}{72}$

Applying the Mathematics

15. a. Convert $\frac{3}{11}$ and $\frac{2}{7}$ to fractions with the same denominator to decide which is larger.
    b. Check by converting the fractions to decimals.

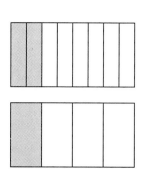

16. A carpenter finds that the height of a door is $75\frac{12}{16}''$. Reduce this mixed number to lowest terms.

17. What equality of fractions is pictured at the left?

18. As you know, $\frac{13}{1} = 13$. Find three other fractions equal to 13.

19. Find three fractions equal to 1.

20. Find a fraction equal to 8 that has 3 as its denominator.

In 21 and 22, write the number as a fraction in lowest terms.

21. fourteen eighths   **22.** seventy-five hundredths

23. Sixty people are expected at a meeting and you are to arrange the chairs.
   a. If you want the chairs in rows with equal numbers of chairs in each row, what choices do you have?
   b. What idea in this lesson is applied in part **a**?

Review

24. A 10th century Chinese mathematician, Liu Hui, used two approximations to π, $\frac{157}{50}$ and $\frac{3927}{1250}$. Fill in the blanks with $<$, $=$, or $>$.   *(Lessons 1-2, 1-7, 1-9)*
   a. $\pi \underline{\quad ? \quad} \frac{157}{50}$ **b.** $\pi \underline{\quad ? \quad} \frac{3927}{1250}$

25. Find three different numbers between 0 and -1.   *(Lesson 1-8)*

In 26–28, estimate to the nearest tenth.   *(Lessons 1-4, 1-5)*

26. four thousand sixty-two times three thousandths

27. $\pi \times 567.34$ **28.** $18 + 1.8 - 0.18$

In 29 and 30, round *up* to the next hundredth.   *(Lessons 1-3, 1-7, 1-8)*

29. $4\frac{2}{17}$ **30.** -0.00785

31. Try to find the decimal equal to each fraction without using a calculator or looking it up.   *(Lesson 1-6)*
   a. $\frac{4}{10}$ **b.** $\frac{5}{8}$ **c.** $\frac{3}{4}$ **d.** $\frac{1}{3}$ **e.** $\frac{5}{6}$

Exploration

32. There are easy ways to tell whether 2, 3, 5, and 9 are factors of numbers.
   a. What about the digits of an integer tells you that 2 is a factor?
   b. What about the digits of an integer tells you that 5 is a factor?
   c. 3 is a factor of an integer when the sum of the digits of the integer is divisible by 3. Which of these numbers is not divisible by 3?
        321   2856   198   4444
   d. 9 is a factor of an integer when the sum of the digits of the integer is divisible by 9. Which of these numbers is *not* divisible by 9?
        198   44442   267   87561
   e. Find a 5-digit number that is divisible by 5 and 9, but not by 2.

33. Some calculators display fractions and can rewrite fractions in lowest terms. If you have such a calculator, use it to write $\frac{323}{493}$ in lowest terms.

A project presents an opportunity for you to extend your knowledge of a topic related to the material of this chapter. You should allow more time for a project than you do for typical homework questions.

PROJECTS
1
CHAPTER ONE

1 Numbers in the Newspaper

Pick one of the news pages of a daily or weekly newspaper. Make certain your page has at least five newspaper

articles on it. Find every number on this page, including those in advertisements. (Remember that some numbers may be written with their word names.) How many numbers are on this page? (If there are over a few hundred numbers, you may have to estimate.) Make a list of all of the counting units and all of the measure units used on the page. Write a paragraph describing what you have found and what you think is most interesting on this page.

2 Comparing Estimates

Take a survey of at least 25 people, asking them to estimate some quantity of your choosing. For instance, you might ask for an estimate of the population of a particular foreign country (such as India, where the population in 1990 was about 844,000,000). Or you might ask for an estimate of the year that the first Ph.D. in mathematics was awarded to a black woman (1949, to Marjorie Lee Browne at the University of Michigan and to Evelyn Boyd Granville at Yale University). Or you might ask for an estimate of the year we will run out of oil if we continue to use it at current rates and do not discover new sources (a 1989 estimate by British Petroleum is that we would run out in 2033). It is preferable if you pick some other thing to estimate. Then, once you have the estimates, write a paragraph about them. Are they usually too high or too low? Did anyone pick the exact amount? And so on.

Evelyn Boyd Granville

Marjorie L. Browne

3 Which Fractions Equal Repeating Decimals?

From an Exploration in Lesson 1-6, you may have learned that the repetends of $\frac{1}{7}$, $\frac{2}{7}$, $\frac{3}{7}$, $\frac{4}{7}$, $\frac{5}{7}$, and $\frac{6}{7}$ have the same sequence of digits 142857 but they all start at different places in the sequence. This happens with other denominators and enables a person to determine the repetends of fractions even when the repetends have a large number of digits. Use your calculator to make a chart of the decimals for the fractions $\frac{1}{2}$, $\frac{1}{3}$, and so on, up to $\frac{1}{25}$. Which of these decimals terminate? Which of them repeat? For those that are repeating, determine the entire repetend by finding the decimal equivalents for other fractions with that denominator. (Sometimes the repetend has many places, but never more places than the value of the denominator.) Try to find a general pattern that will tell a person which fractions equal repeating decimals and which equal terminating decimals.

4 The Advantages of Rounding

Some people have suggested that pennies be taken out of circulation. Suppose all pennies were taken out of circulation and a food store changed its prices as follows: prices that end in 0 or 5 stay the same; prices that end in 1 or 6 are rounded down a penny to the nearest nickel; prices that end in 2, 3, 4, 7, 8, and 9 are rounded up to the next nickel. Does this make much of a difference? Go to a large store and try to estimate how many people shop there. Estimate how many items might be bought in which the store would round up, and then estimate how much the store might gain from rounding up. Report about what you have found.

5 Other Numeration Systems

Many cultures have written numbers differently from the way we write numbers now. Among these are the ancient Greeks of Europe, Aztecs of Mexico,

the Yoruba and Egyptians of Africa, the Hebrews, the Babylonians, and all the cultures of Southern and Eastern Asia. Read about the numeration systems of one of these cultures or some other culture with a different numeration system. Write a short report explaining how the numbers 1 through 10, 25, 100, and larger numbers were written by these peoples.

SUMMARY

Today, by far the most common way of writing numbers is in the decimal system. In this chapter, decimals are used for whole numbers, for numbers between whole numbers, for negative numbers, for fractions, and for mixed numbers.

Decimals are easy to order. This makes it easy to estimate them. We estimated decimals by rounding up, rounding down, and rounding to the nearest decimal place. By dividing a number line into smaller intervals using tick marks, decimals can be represented on the number line.

Calculators represent numbers as decimals. So, if you can write numbers as decimals, then you can make a calculator work for you. By changing fractions to decimals, you can order them and tell whether two fractions are equal.

VOCABULARY

You should be able to give a general description and a specific example of each of the following ideas.

Lesson 1-1
digit
Arabic numerals, Hindu-Arabic numerals
decimal system, decimal notation
whole number
place value, ones place, tens place, hundreds place, thousands place, and so on
count, counting unit

Lesson 1-2
measure, unit of measure, interval
number line
tick marks
tenths place, hundredths place, thousandths place, and so on

Lesson 1-3
estimate
rounding up, rounding down
truncate

Lesson 1-4
rounding to the nearest

Lesson 1-5
scientific calculator
display
enter, key in, key sequence
RCL or STO
INV , 2nd , O

Lesson 1-6
fraction, fraction bar, simple fraction
numerator, denominator
dividend, divisor, quotient
infinite repeating decimal, repetend
evenly divisible, divisible

Lesson 1-7
mixed number

Lesson 1-8
negative number, positive number
integer, positive integer, negative integer
+/- or ±

Lesson 1-9
inequality symbols
< (is less than), > (is greater than)

Lesson 1-10
equal fractions
Equal Fractions Property
factor
lowest terms
greatest common factor, GCF

PROGRESS SELF-TEST

Take this test as you would take a test in class. You will need a calculator. Then check your work with the solutions in the Selected Answers section in the back of the book.

In 1–4, write as a decimal.

1. seven hundred thousand

2. forty-five and six tenths

3. $\frac{1}{4}$ 4. $15\frac{13}{16}$

5. What digit is in the hundredths place of 1234.5678?

6. Write 0.003 in English.

7. Consider the four numbers .6, .66, $.\overline{6}$, and .606. Which is the largest?

8. Consider the four numbers $\frac{1}{2}$, $\frac{2}{5}$, $\frac{1}{3}$, and $\frac{3}{10}$. Which is the smallest?

9. Round 98.76 down to the preceding tenth.

10. Round 98.76 to the nearest integer.

11. Translate into mathematics: An elevation 80 ft below sea level is higher than an elevation 100 ft below sea level.

In 12 and 13, use the number line pictured.

12. Which letter corresponds to the position of $\frac{1}{2}$?

13. Which letter corresponds to the position of -1.25?

14. Give a number between 16.5 and 16.6.

15. Give a number between -2.39 and -2.391.

In 16–18, which symbol, $<$, $=$, or $>$, goes between the numbers?

16. 0.45 __?__ 0.4500000001

17. -9.24 __?__ -9.240

18. $4\frac{4}{15}$ __?__ 4.93

19. -4 __?__ -5

20. What fraction is equal to 0.6?

21. Give an example of a number that is not an integer.

22. A store sells grapes on sale at 69¢ a pound. You need a quarter pound. So you divide by 4 on your calculator. The display shows $\boxed{0.1725}$. What will you have to pay?

23. Estimate 3.012012012 + 9.08888888888888 to the nearest integer.

24. Indicate the key sequence for evaluating 3.456 × 2.345 on a calculator.

25. Use your calculator to estimate 6 × π to the nearest integer.

26. What is the repetend on the repeating decimal 24.247474747 . . . ?

27. Graph the numbers 7, 7.7, and 8 on the same number line.

28. Graph these temperatures on the same vertical number line: 5°, -4°, 0°.

29. Give a situation where an estimate must be used because an exact value cannot be obtained.

30. Write a sentence containing a count and a counting unit. Underline the count once and the counting unit twice.

31. Which is largest: one tenth, one millionth, one billionth, or one thousandth?

32. Find all the factors of 18.

33. Find a fraction equal to 6 with 5 as its denominator.

34. Rewrite $\frac{12}{21}$ in lowest terms.

35. *Multiple choice.* When was the decimal system developed?
    (a) between 2000 B.C. and 1000 B.C.
    (b) between 1000 B.C. and 1 B.C.
    (c) between 1 A.D. and 1000 A.D.
    (d) between 1000 A.D. and today

After taking and correcting the Self-Test, you may want to make a list of the problems you got wrong. Then write down what you need to study most. If you can, try to explain your most frequent or common mistakes. Use what you write to help you study and review the chapter.

CHAPTER REVIEW

Questions on SPUR Objectives

SPUR stands for **S**kills, **P**roperties, **U**ses, and **R**epresentations. The Chapter Review questions are grouped according to the SPUR Objectives for this chapter.

SKILLS DEAL WITH THE PROCEDURES USED TO GET ANSWERS.

Objective A: *Translate back and forth from English into the decimal system.* *(Lessons 1-1, 1-2)*

In 1–6, write the number as a decimal.

1. four thousand three
2. seventy-five hundredths
3. one hundred twenty million
4. three and six thousandths
5. seventy-five and six tenths
6. nine hundred two thousand and nine hundred two thousandths

In 7–10, translate into English words.

7. 500,400
8. 0.001
9. 3.041
10. 71,026,985

Objective B: *Order decimals and fractions.* *(Lessons 1-2, 1-6, 1-7, 1-9)*

In 11 and 12, which of the four given numbers is largest, which smallest?

11. 400,000   400,001   -.40000000001   0.4
12. .34   .3$\overline{4}$.$\overline{34}$.$\overline{343}$

In 13–18, order from largest to smallest.

13. 0   -0.2   0.2   0.19
14. -586.36   -586.363   -586.34
15. $\frac{1}{7}$ $\frac{1}{11}$ $\frac{1}{9}$
16. $\frac{2}{3}$ $\frac{6}{10}$ .66
17. $3\frac{1}{3}$ $2\frac{2}{3}$ $4\frac{1}{6}$
18. 5.3   5.$\overline{3}$ 4.33

Objective C: *Give a number that is between two decimals.* *(Lessons 1-2, 1-8)*

In 19–23, give a number between the two numbers.

19. 73 and 73.1
20. -1 and -2
21. 6.99 and 7
22. 3.40 and 3.$\overline{40}$
23. -4 and 2

Objective D: *Round any decimal place up or down or to the nearest value of a decimal place.* *(Lessons 1-3, 1-4, 1-8)*

24. Round 345.76 down to the preceding tenth.
25. Round 5.8346 up to the next hundredth.
26. Round 39 down to the preceding ten.
27. After six decimal places, Joan's calculator truncates. What will the calculator display for 0.59595959595959 . . . ?
28. Round 34,498 to the nearest thousand.
29. Round 6.81 to the nearest tenth.
30. Round 5.55 to the nearest integer.
31. Round -2.47 to the nearest tenth.
32. Round -0.129 up to the next hundredth.
33. Round -14 down to the preceding ten.

In 34 and 35, estimate to the nearest integer without a calculator.

34. 58.9995320003 + 2.86574309
35. 6 × 7.99

Objective E: *Use a calculator to perform arithmetic operations.* *(Lessons 1-5, 1-6, 1-8)*

36. Find 35.68×123.4.

37. Find $555 + 5.55 + .555 + 0.50$.

38. Find $73 - \pi$ to the nearest ten-thousandth.

39. What is the key sequence for entering -5 on a calculator you use?

40. Give the key sequence for converting $\frac{77}{8.2}$ to a decimal.

Objective F: *Convert simple fractions and mixed numbers to decimals.* *(Lessons 1-6, 1-7, 1-8)*

In 41–44, write the number as a decimal.

41. $\frac{11}{5}$

42. $\frac{-16}{3}$

43. $6\frac{4}{7}$

44. $5\frac{1}{4}$

Objective G: *Know by memory the common decimal and fraction equivalences between 0 and 1.* *(Lesson 1-6)*

In 45–48, give the decimal for the fraction.

45. $\frac{3}{4}$

46. $\frac{2}{3}$

47. $\frac{1}{5}$

48. $\frac{1}{6}$

In 49–52, give a simple fraction for the decimal.

49. $.8$

50. $.\overline{3}$

51. 0.25

52. 0.625

PROPERTIES DEAL WITH THE PRINCIPLES BEHIND THE MATHEMATICS.

Objective H: *Use the $<$ and $>$ symbols correctly between numbers.* *(Lesson 1-9)*

In 53 and 54, choose the correct symbol $<$, $=$, or $>$.

53. $2.0 \underline{\ ?\ } 0.2$

54. $0.1 \underline{\ ?\ } 0.\overline{1}$

55. Arrange the numbers $\frac{2}{3}$, $.6$, and $.667$ in one sentence with two $>$ symbols between them.

56. Arrange the numbers -1, -2, and $-\frac{3}{2}$ in one sentence with two $<$ symbols between them.

Objective I: *Correctly use the raised bar symbol for repeating decimals.* *(Lesson 1-6)*

57. Give the 13th decimal place in $.\overline{1428}$.

58. Give the 10th decimal place in $71.5\overline{36}$.

In 59 and 60, write the repeating decimal using the repetend symbol.

59. $6.8999\ldots$ (9 repeats)

60. $-0.002020202\ldots$ (02 repeats)

Objective J: *Use the Equal Fractions Property to rewrite fractions.* *(Lesson 1-10)*

61. Find two other fractions equal to $\frac{2}{7}$.

62. Find two other fractions equal to $\frac{280}{72}$.

63. Rewrite $\frac{80}{60}$ in lowest terms.

64. Rewrite $-1\frac{12}{16}$ in lowest terms.

USES DEAL WITH APPLICATIONS OF MATHEMATICS IN REAL SITUATIONS.

Objective K: *Deal with estimates in real situations.* *(Lessons 1-3, 1-4)*

65. According to U.S. census data, the population of Albany, Oregon was 29,462 in 1990. Round the population of Albany to the nearest thousand.

66. To quickly estimate the cost of 5 records at $8.95 each, what rounding can you do?

67. A store sells 6 granola bars for $2.99. You want 1 bar. Dividing on your calculator gives 0.4983333. What will the bar cost?

68. Give a situation where an estimate would be used for a safety reason.

69. Name a reason other than safety for needing an estimate.

Objective L: *Correctly interpret situations with two directions as positive, negative, or corresponding to zero.* *(Lesson 1-8)*

70. 350 meters below sea level corresponds to what number?

71. Translate into mathematics: A loss of $75,000 is worse than a gain of $10,000.

72. An auto mechanic estimates the cost to fix your car. What number could stand for

  a. an estimate $25 below the actual cost of repair?

  b. an estimate $40 higher than the actual cost of repair?

  c. an estimate equal to the actual cost of repair?

REPRESENTATIONS DEAL WITH PICTURES, GRAPHS, OR OBJECTS THAT ILLUSTRATE CONCEPTS.

Objective M: *Graph and read numbers on a number line.* *(Lessons 1-2, 1-7, 1-8)*

In 73–76, graph the numbers on the same number line.

73. $\frac{1}{3}$, $\frac{2}{3}$, and $1\frac{1}{3}$

74. 0, $\frac{1}{2}$, $\frac{1}{4}$, -1, and 1

75. 6, 6.4, and 7

76. $-3°$, $1°$, and $-5°$

77. Use this number line.

```
←——+——+——+——+——•——+——+——+——+——+——+——+——+——+——→
   4              5                          6
```

  a. What is the distance between tick marks?

  b. The dot is the graph of what number?

78. Use this number line.

```
←——+——+——+——+——•——+——+——+——+——+——+——+——+——+——→
 -10        -9        -8         -7          -6
```

  a. What is the distance between tick marks?

  b. The dot corresponds to what number?

CULTURE DEALS WITH THE PEOPLES AND THE HISTORY RELATED TO THE DEVELOPMENT OF MATHEMATICAL IDEAS.

Objective N: *Give people and rough dates for key ideas in the development of arithmetic.* *(Lessons 1-1, 1-2, 1-6)*

79. *Multiple choice.*   Our symbols for 0, 1, 2, 3, 4, 5, 6, 7, 8, and 9 did not all appear until about what date?
   (a) 2000 B.C.   (b) 1000 A.D.
   (c) 1400 A.D.   (d) 1900 A.D.

80. Name a culture that used symbols quite different from the ones we use today in the United States.

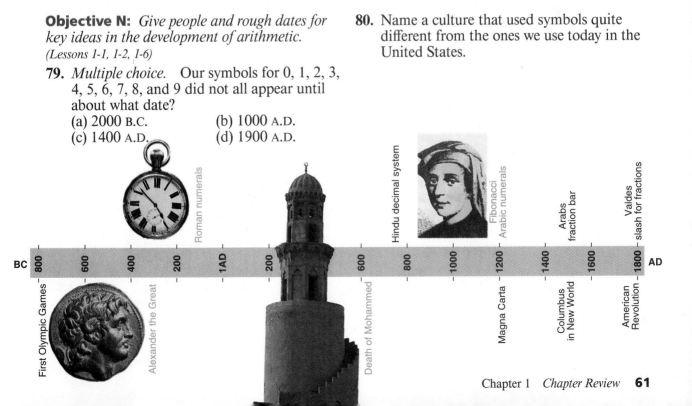

CHAPTER

2

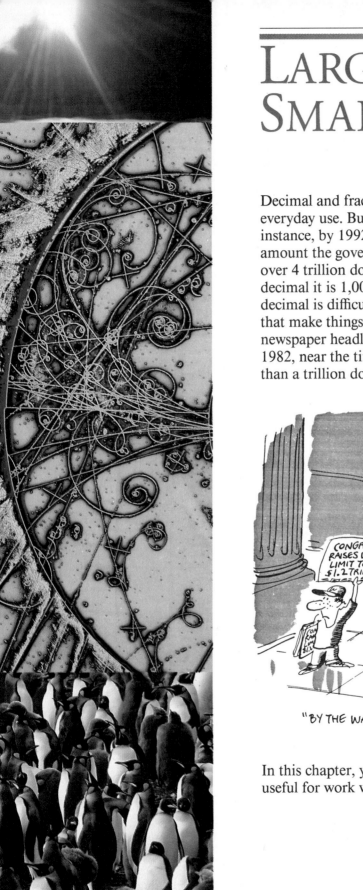

LARGE AND SMALL NUMBERS

Decimal and fraction notations work well for most numbers in everyday use. But some uses require very large numbers. For instance, by 1992 the national debt of the United States, the amount the government has borrowed to finance its work, was over 4 trillion dollars. A trillion is a very large number. As a decimal it is 1,000,000,000,000. With so many digits, this decimal is difficult to work with. But, there are other notations that make things easier. One of these other notations is in the newspaper headline in the cartoon. This cartoon was drawn in 1982, near the time that the national debt first became more than a trillion dollars.

"BY THE WAY, WHAT COMES AFTER A TRILLION?"

In this chapter, you will study notations that are particularly useful for work with large and small numbers.

Multiplying by 10, 100, . . .

This was not a rocky start. *A National League regular season record of 80,227 fans showed their support of the new expansion team, the Colorado Rockies, at Mile High Stadium during its 1993 home opener.*

Multiplying by 10

On April 9, 1993, the baseball game between the Montreal Expos and the Colorado Rockies was attended by 80,227 (eighty thousand, two hundred twenty-seven) people. In this number, the 7 is in the ones place. There is a 2 in the tens place (so it stands for 20). There is also a 2 in the hundreds place (so it stands for 200). The 0 is in the thousands place. And the 8 is in the ten-thousands place (so it stands for 80,000). In this way, each place value is ten times the value of the place to its right. This makes it easy to multiply the number by 10. This is what you would do if the average price of a ticket were $10, and you wanted to know the total gate receipts for this game.

$$\text{Attendance} \qquad 8\ 0,2\ 2\ 7$$
$$80,227 \times 10 \ = 8\ 0\ 2,2\ 7\ 0$$

The total gate receipts would be $802,270.

There is another way to think about this. Write 80,227 with a decimal point and a zero following it. (Remember that you can always insert a decimal point followed by zeros to the right of a whole number without changing its value. For example, $5 = $5.00.)

$$\text{Attendance} \qquad 8\ 0,2\ 2\ 7.0$$
$$80,227.0 \times 10 \ = 8\ 0\ 2,2\ 7\ 0.$$

> To multiply a number in decimal notation by 10, move the decimal point one place to the right.

For example, $62.58 \times 10 = 625.8$, and $.0034 \times 10 = .034$.

Multiplying by 100, 1000, 10,000, and so on

This simple idea is very powerful. Suppose you want to multiply a number by 100. Since $100 = 10 \times 10$, multiplying by 100 is like multiplying by 10 and then multiplying by 10 again. So move the decimal point *two* places to the right. For example, $59.072 \times 100 = 5907.2$.

The same idea can be extended to multiply by 1000, 10,000, and so on.

To multiply by 10, 100, 1000, and so on, move the decimal point as many places to the right as there are zeros.

$$10 \times 47.3 = 473.$$
$$100 \times 47.3 = 4730.$$
$$1000 \times 47.3 = 47,300.$$
$$10,000 \times 47.3 = 473,000.$$

Word Names for Some Large Numbers

In English, the numbers 10, 100, 1000, have the word names ten, hundred, thousand. Here are other numbers and their word names.

Decimal	Word name
1,000,000	million
1,000,000,000	billion
1,000,000,000,000	trillion
1,000,000,000,000,000	quadrillion
1,000,000,000,000,000,000	quintillion

Now look again at the cartoon on page 63. The newsboy is holding up a newspaper mentioning a debt limit of 1.2 trillion dollars. The phrase 1.2 trillion means 1.2 *times* a trillion. Since a trillion has 12 zeros, move the decimal point 12 places to the right.

$$1.2 \text{ trillion} = 1.2 \times 1,000,000,000,000$$
$$= 1,200,000,000,000$$

Notice how much shorter and clearer 1.2 trillion is than 1,200,000,000,000. For these reasons, it is common to use word names for large numbers in sentences and charts.

Example

A Bureau of Labor Statistics report in 1993 listed 118.451 million people in the United States as employed in February. Write this number in decimal notation (without words).

Solution

$$118.451 \text{ million} = 118.451 \times 1,000,000$$
$$= 118,451,000$$

Check

This is easy to check. Because 118.451 is between 118 and 119, you would expect 118.451 million to be between 118 million and 119 million.

QUESTIONS

Covering the Reading

1. In the number 81,345, the place value of the digit 1 is __?__ times the place value of the digit 3.

In 2–5, multiply the number by 10.

2. 634   3. 2.4   4. 0.08   5. 47.21

6. Give a general rule for multiplying a decimal by 10.

7. Give a general rule for multiplying a decimal by 100.

In 8–11, multiply the number by 100.

8. 113   9. .05   10. 7755.2   11. 6.301

12. Give a general rule for multiplying a decimal by 1000.

In 13–16, calculate.

13. 32 × 1000

14. 1000 × 0.02

15. 1.43 × 10,000

16. 100,000 × 21.146

17. In your own words, write one rule for multiplying by 10, 100, 1000, etc.

In 18–23, give the word name for the decimal 1 followed by:

18. 3 zeros   19. 6 zeros   20. 9 zeros

21. 12 zeros   22. 15 zeros   23. 18 zeros

Leaner, meaner burgers. *Some fast-food corporations have begun offering leaner hamburgers as an alternative to the kind pictured here. The leaner burgers are lower in fat and cholesterol due to the use of fat substitutes, such as oat bran and a seaweed derivative, carrageenan.*

24. According to "What America Eats" from *Parade Magazine*, November 24, 1991, "Sales in 1990 reached $15 billion, making pizza second only to hamburgers as America's most popular fast food." Write the underlined number in decimal notation.

In 25 and 26, the information is from the *Information Please Almanac*, 1993. Write the underlined numbers in decimal notation.

25. In the United States, there are 92.1 million homes with a TV. There are 90.8 million homes with a color TV.

26. In 1988–1989, total funding for public elementary and secondary education in the United States was 191.21 billion dollars.

Applying the Mathematics

27. How can rounding help you check your answer to Question 15?

28. 98.765 times what number equals 98,765?

29. According to the U.S. Department of the Treasury, 9,324,386,076 pennies were minted in 1991. Round this number of pennies to:
   a. the nearest hundred million.   b. the nearest ten million.

In 30–33, use the graph to estimate world population for the given year to the nearest tenth of a billion. Write this number in decimal notation.

30. 1950 **31.** 1965 **32.** 1980 **33.** 1990

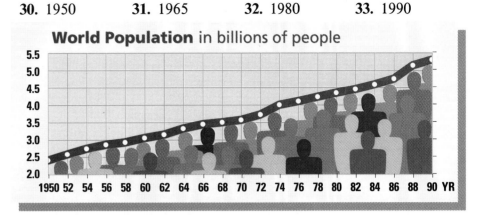

World Population in billions of people

In 34–36, write the number as it might appear in a newspaper.

34. An estimated 15,000,000 people live in the Republic of Ghana.

35. Chile's gross domestic product was reported to be $27,800,000,000.

36. The star *Proxima Centauri* is 24,800,000,000,000 miles from Earth.

37. Answer the question in the cartoon on page 63.

Review

38. Translate into mathematics: Negative ten is less than nine. *(Lesson 1-8)*

39. Change each number to a decimal. *(Lesson 1-8)*
  a. $\frac{1}{8}$ **b.** $\frac{1}{10}$ **c.** $\frac{1}{50}$ **d.** $\frac{1}{100}$

40. The letters are equally spaced on the number line below.

A B C D E F G H I J K
0   1

  a. What number corresponds to *F*?
  b. What number corresponds to *B*?
  c. What letter corresponds to $\frac{3}{5}$?
  d. Between what neighboring letters is 0.0085? *(Lessons 1-2, 1-6)*

41. Round 2.6494 *down* to thousandths. *(Lesson 1-3)*

Exploration

42. Locate, in a newspaper or magazine, at least two numbers written with a decimal followed by a word name (like those in the Example). Copy the complete sentences that contain the numbers.

43. In England, the word *billion* does not always mean the number 1 followed by 9 zeros. What number does the word *billion* often represent in England?

Where is it used? *Unlike our base 10 system which has ten digits (0, 1, 2, . . ., 9), the base 2 system has only two digits (0 and 1). Computers, laser discs, and CDs rely on the simple process of recognizing a 0 or 1 to generate sounds or pictures like the one above.*

What Are Powers?

In Lesson 2-1, you multiplied by 100. This was explained as multiplying by 10, followed by multiplying by 10. Repeated multiplication is so common that there is a shorthand for it. We write

$$10^2$$

(say "10 to the 2nd **power**" or "the second power of 10") to mean 10×10, or 100. In 10^2, the number 10 is called the **base,** and 2 is called the **exponent.** Similarly, 10^3 (say "10 to the 3rd power") means $10 \times 10 \times 10$, or 1000, and 10^4 (say "10 to the 4th power") means $10 \times 10 \times 10 \times 10$, or 10,000. In this book, only integers are used as exponents. With integer exponents, any number can be used as a base.

Words	Exponential form		Repeated multiplication		Decimal
3 to the 2nd power =	3^2	=	3×3	=	9
8 to the 3rd power =	8^3	=	$8 \times 8 \times 8$	=	512
1.3 to the 5th power =	1.3^5	=	$1.3 \times 1.3 \times 1.3 \times 1.3 \times 1.3$	=	3.71293
1 to the 7th power =	1^7	=	$1 \times 1 \times 1 \times 1 \times 1 \times 1 \times 1$	=	1

Some Important Powers to Know

We say that a number "to the first power" equals the number itself. So "3 to the first power" means 3^1, which equals 3. Also, $10^1 = 10$ and $\left(\frac{3}{4}\right)^1 = \frac{3}{4}$.

It is useful to know some powers of small numbers without having to calculate them every time. Here are the smallest positive integer powers of 2:

$$2^1 = 2 \qquad 2^2 = 4 \qquad 2^3 = 8 \qquad 2^4 = 16 \qquad 2^5 = 32$$
$$2^6 = 64 \qquad 2^7 = 128 \qquad 2^8 = 256 \qquad 2^9 = 512 \qquad 2^{10} = 1024$$

If you do not know the powers of 2, you have to calculate them.

Positive Integer Powers of Ten

The powers of 10 are very special in the decimal system. You can calculate them in your head. We have already noted that $10^1 = 10$, $10^2 = 100$, $10^3 = 1000$, and $10^4 = 10,000$. The next power, 10^5, is found by multiplying 10,000 by 10. So $10^5 = 100,000$. Notice the pattern: when written as decimals, 10^1 is a 1 followed by 1 zero, 10^2 is a 1 followed by 2 zeros, 10^3 is a 1 followed by 3 zeros, and so on. So 10^{12} is a 1 followed by 12 zeros; it is another way of writing one trillion.

You have now seen three different ways of representing the place values in the decimal system.

Power of 10	Word name	Written as decimal
10^1	ten	10
10^2	hundred	100
10^3	thousand	1,000
10^6	million	1,000,000
10^9	billion	1,000,000,000
10^{12}	trillion	1,000,000,000,000
10^{15}	quadrillion	1,000,000,000,000,000
10^{18}	quintillion	1,000,000,000,000,000,000

Multiplying by Positive Integer Powers of Ten

You already know a quick way to multiply by 10, 100, 1000, and so on. It is just as quick to multiply by these numbers when they are written as powers.

$$53 \times 10^5 = 53 \times 100,000 = 5,300,000$$
$$2.38 \times 10^4 = 2.38 \times 10,000 = 23,800.$$

The decimal point moves to the right one place for each power of 10.

To multiply by a positive integer power of 10
Move the decimal point to the right the same number of places as the value of the exponent.

Finding Other Powers

Besides powers of 10, only a few powers of small numbers are easy to calculate by hand. Powers of other small numbers can be quite large. For example, $9^8 = 43,046,721$. Usually it is quicker and more accurate to use a calculator. A scientific calculator has a special key labeled x^y or y^x or $\wedge$, the powering key. (If this label is in small print above or below a key, you will need to press INV or 2nd or F before pressing the powering key.) For example, to evaluate 5^7, enter the following:

Key sequence: 5 y^x 7   =
Display: 5.   5.   7. 78125.

So $5^7 = 78,125$.

Note: Whether the key is labeled x^y or y^x or $\wedge$, the base is entered before the exponent.

Activity

Use your calculator to verify the value for 9^8 given above.

QUESTIONS

Covering the Reading

1. Consider 4^6.
   a. Name the base.
   b. Name the exponent.
   c. This number is _?_ to the _?_th _?_.

2. Calculate 3^1, 3^2, 3^3, 3^4, 3^5, and 3^6.

3. Give the values of 2^1, 2^2, 2^3, 2^4, 2^5, and 2^6.

4. Calculate 7^6.

5. Calculate 2^{20} and 20^2.

6. Calculate 1^{994}.

7. Calculate 1.08^3. (This kind of calculation is found in money matters.)

8. In decimal notation, 10^7 is a 1 followed by _?_ zeros.

9. Write 10^6 in the indicated form.
   a. decimal   b. word name

10. Write one thousand in the indicated form.
   a. decimal   b. as a power of 10

11. Write each number as a power of 10.
   a. million   b. billion   c. trillion

12. According to the table in this lesson, 10 to the first power is equal to what number?

In 13–16, write the number as a decimal.

13. 5×10^2

14. 8×10^3

15. 3.7×10^4

16. 0.246×10^6

17. Write the general rule for multiplying by a positive integer power of 10.

18. To multiply by 10 to the first power, you should move the decimal point how many places to the right?

Applying the Mathematics

19. What number is 1 less than 10^3?

20. Which is larger, 2^3 or 3^2?

21. The table in this lesson skips from 10^3 to 10^6. Fill in the two rows that are missing.

22. Ten million is the __?__ power of 10.

23. a. Give the next number in this pattern of powers:
256, 64, 16, __?__.
b. The smallest positive integer powers of what integer are in part **a?**

24. *Multiple choice.* $3^{10} - 2^{10}$ is between
(a) 1 and 100.
(b) 100 and 10,000.
(c) 10,000 and 1,000,000.

25. Two ways 72 can be written as the sum or difference of powers are $72 = 9^2 - 3^2$ and $72 = 6^2 + 6^2$. Write 100 as a sum and a difference of powers.

26. a. Enter the key sequence 3 ⊠ 2 y^x 5 ⊟ on your calculator. What number results?

b. Some calculators do the multiplication first, then take the power. Other calculators take the power first, then multiply. What did your calculator do first?

27. *True or false.* $2^{10} > 10^3$.

28. The object of a puzzle known as Rubik's Cube™ is to mix up the cube, then return it to its original position. There are 43,252,003,274,489,856,000 possible positions. Write this number in English. (This shows how much easier decimal notation is than English words.)

Are you a puzzlemeister?
The object of many puzzles is to mix them up and then return them to their original position. Rubik's Cube™, invented by Hungarian professor Erno Rubik, is shown at the bottom left.

29. According to a 1993 estimate, the world population had grown by 2.5 billion people since 1960. Write 2.5 billion as a decimal. *(Lesson 2-1)*

In 30–32, multiply in your head. *(Lesson 2-1)*

30. $100 \times 10,000$ **31.** $180 \times 10,000$ **32.** 20×400

In 33–35, write the positive or negative decimal suggested by the situation. *(Lesson 1-8)*

33. Barry Sanders ran for thirteen yards on the first play of the game.

34. The deepest lake in the world, Lake Baikal in Siberia, has a point that is eleven hundred eighty-one meters below sea level.

35. Absolute zero is four hundred fifty-nine and sixty-seven hundredths degrees Fahrenheit below zero.

36. Tell whether the number is or is not an integer. *(Lesson 1-8)*

   a. -3 **b.** 4.7 **c.** 4.0 **d.** $\frac{15}{3}$

   e. $-17\frac{1}{3}$ **f.** $\frac{89}{5}$ **g.** 23

37. What digit is in the thousandths place when $\frac{15}{7}$ is rewritten as a decimal? *(Lessons 1-2, 1-6)*

38. What number results from this key sequence? *(Lesson 1-5)*

$$8 \div 9 \div 4 \times 200 =$$

39. Suppose you buy 3 shirts at \$8.95 each. What multiplication can you do to estimate the cost to the nearest dollar? *(Lesson 1-4)*

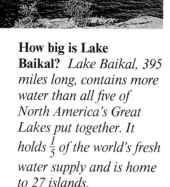

How big is Lake Baikal? *Lake Baikal, 395 miles long, contains more water than all five of North America's Great Lakes put together. It holds $\frac{1}{5}$ of the world's fresh water supply and is home to 27 islands.*

40. A *googol* is one of the largest numbers that has a name. Look in a dictionary or other reference book to find out something about this number.

41. Calculate 3^5 using a computer. (You will need to type a symbol, usually either $\wedge$ or **, between the 3 and the 5.)

   a. What did you type to get the computer to calculate 3^5?

   b. Use this computer to compute 6^7. Compare this result with the result given on your calculator.

   c. Repeat part **b** for the number 15^{10}.

Scientific Notation for Large Numbers

Striking information. *Every year about 100 people in the U.S. are injured by lightning which carries 10 to 30 million volts of electricity. So during a lightning storm, go indoors, get in a car, or lie down in a ditch.*

You should have a scientific calculator and some paper to use as you read this lesson.

Light travels at the speed of about 186,282.4 miles per second. Since there are 60 seconds in a minute, light travels $60 \times 186,282.4$ miles per minute.

Activity 1

Multiply 60 by 186,282.4 on your calculator. Record the answer.

To find out how far light travels in an hour, multiply by 60 again.

$$60 \times 186,282.4 \times 60 = 670,616,640 \text{ miles per hour}$$

Calculator Displays of Large Numbers

When the multiplication $60 \times 186,282.4 \times 60$ is done on a calculator, the calculator will do one of three things.

1. It may display all 9 digits. | 670616640. |

2. It may display an error message, like one of those below. This means that the number is too big for the calculator. The calculator will not do anything until you clear the number.

    | E 6.7061664 | | ERROR 0 | | ERROR |

3. It may display the number in *scientific notation.* The display usually looks like one of those shown here.

    | 6.7061664 08 | | 6.7062 08 | | 6.7062 × 10 8 |

Scientific notation is the way that scientific calculators display very large and very small numbers. Each of the above displays stands for the number 6.7061664×10^8 or a rounded value of that number. The user is expected to know that the 8 (or 08) stands for 10^8. So to convert the number into decimal notation, move the decimal point 8 places to the right. (The display at the right above contains $\times 10$ and is clearest.)

Example 1

Calculate how far light travels in a day.

Solution

The key sequence below includes all the calculations done so far in this lesson and a multiplication by 24, the number of hours in a day.

Key sequence: 60 [×] 186,282.4 [×] 60 [×] 24 [=]

The exact product is 16,094,799,360.

Many calculators must round the answer to 1.60948×10^{10} because they do not have enough room to display all digits.

Display: [1.60948   10]

To write 1.60948×10^{10} as a decimal, perform this multiplication without a calculator. Move the decimal point 10 places to the right.

Answer: Light travels about 16,094,800,000 miles in a day.

Activity 2

Use your calculator to determine how far light travels in a 365-day year. This is the *distance* known as a **light-year**. Show your work as in Example 1.

What Is Scientific Notation?

The number 1.60948×10^{10} in Example 1 is in scientific notation.

> In scientific notation, a number greater than or equal to 1 and less than 10 is multiplied by an integer power of 10.
> $$decimal \times 10^{exponent}$$

Here are some more numbers written in scientific notation.

Decimal or word name	Scientific notation
670,620,000	6.7062×10^{8}
340.67	3.4067×10^{2}
2,380,000,000	2.38×10^{9}
60 trillion	6×10^{13}

Notice that there are three parts to scientific notation:
a number from 1 to 10, but less than 10
a multiplication sign
a power of 10.

Calculators often omit the multiplication sign and show the exponent instead of the power of 10.

Converting Decimals into Scientific Notation

Examples 2 and 3 show how to convert decimals into scientific notation.

Example 2

The distance from Earth to the Sun is about 150,000,000 km. Write this number in scientific notation.

Solution

First, move the decimal point to get a number between 1 and 10. In this case, the number is 1.5 and this tells you the answer will look like this:

$$1.5 \times 10^{exponent}.$$

Second, find the power of 10. The exponent of 10 is the number of places you must move the decimal in 1.5 to the *right* in order to get 150,000,000. You must move it 8 places, so the number in scientific notation is:

$$1.5 \times 10^{8}.$$

Example 3

Write 45,678 in scientific notation.

Solution

Ask yourself: 45,678 equals 4.5678 times what power of 10? The answer to the question is 4. So

$$45,678 = 4.5678 \times 10^{4}.$$

Numbers with word names are easy to convert to scientific notation. Use the power of 10 equivalent to the word name.

Example 4

The population of India is expected to exceed 1.4 billion by the year 2025. This number is 1,400,000,000 and has too many digits for most calculators. Write it in scientific notation so that it can be entered into a calculator and used.

Solution

Since 1 billion $= 10^{9}$, 1.4 billion $= 1.4 \times 10^{9}$. This is already in scientific notation.

You can enter a number in scientific notation directly into your calculator.

India has two Delhis. *The busy street pictured here is Chandi Chowk Street of Old Delhi. Just three miles south of Old Delhi is New Delhi, the capital of India and a modern, carefully planned city.*

Activity 3

Enter 1.4 billion on your calculator, using one of the key sequences:

$$1.4 \boxed{\text{EE}} \ 9 \ \text{or} \ 1.4 \boxed{\text{Exp}} \ 9.$$

Either key sequence enters 1.4×10^{9}.
a. Identify the key sequence you need to use on your calculator.
b. Check by entering 1.4 $\boxed{\times}$ 10 $\boxed{y^x}$ 9 $\boxed{=}$. Do you get the same answer?

Using the ⟨EE⟩ key is more efficient than pressing the *three* keys ⟨×⟩ 10 ⟨y^x⟩. The ⟨EE⟩ key is programmed to mean "times 10 to the power of."

On some calculators and computers, the number 1.4 billion may be written as 1.4 E 9 or even 1.4 E + 9. In this case the E means "exponent of 10" and does not indicate an error. The + sign indicates a positive number. In Lesson 2-9, you will learn about scientific notation with negative exponents.

QUESTIONS

Covering the Reading

1. **a.** How many miles does light travel in a 365-day year?
   b. What is this distance called?

2. How does your calculator display the number 9.01×10^{25}?

3. In scientific notation, a number greater than or equal to __?__ and less than __?__ is multiplied by an integer __?__ of 10.

In 4–11, rewrite the number in scientific notation.

4. 800   5. 804   6. 63.21   7. 765.4

8. 3,500,000 square miles, the approximate land area of the U.S.

9. 6,588,000,000,000,000,000,000 tons, the approximate mass of Earth

10. 59.85 million people, the number of passengers handled by Chicago's O'Hare Field in 1991

11. 4 trillion dollars, the approximate U.S. national debt in 1993

12. Indicate how to enter 1.4×10^9 into your calculator
    a. using a key for scientific notation.
    b. using a powering key.

In 13 and 14, what does your calculator display for the number?

13. 49×10^{14} 14. 60 trillion

Applying the Mathematics

15. The approximate weight of the largest blue whale ever measured is 1.9×10^5 kg. Multiply this number by 2.2 to estimate this weight in pounds.

16. How many seconds are there in a 365-day year?

17. **a.** Using scientific notation, what is the largest number you can display on your calculator?
    b. What is the smallest number you can display? (Hint: The smallest number is negative.)

18. a. Which is larger, 1×10^{10} or 9×10^9?
   b. How can you tell?

19. Show that the $\boxed{\text{EE}}$ key and the $\boxed{y^x}$ key are different by writing a question that can be answered

   a. using either key.
   b. by using the $\boxed{\text{EE}}$ key but not the $\boxed{y^x}$ key.
   c. by using the $\boxed{y^x}$ key but not the $\boxed{\text{EE}}$ key.

Review

20. Calculate the first, second, and third powers of 8. *(Lesson 2-2)*

In 21 and 22, calculate mentally. *(Lesson 2-1)*

21. $0.0006 \times 10,000$ **22.** 523×100

23. On the number line below, what letter corresponds to each number?
   (Lessons 1-2, 1-6, 1-8)

   a. $\frac{3}{5}$ **b.** $\frac{-3}{5}$ **c.** $.8$ **d.** $-.8$

24. Arrange from smallest to largest: $14\frac{3}{5}$ 14.66 $14.\overline{61}$ $14.\overline{6}$.
   (Lessons 1-6, 1-7)

25. Name a fraction whose decimal is 0.9. *(Lesson 1-6)*

26. Batting "averages" in baseball are calculated by dividing the number of hits by the number of at-bats and usually rounding to the nearest thousandth. In 1970, Alex Johnson and Carl Yastrzemski had the top two batting averages in the American League. *(Lessons 1-4, 1-6)*

   a. Johnson had 202 hits in 614 at-bats. What was his average?
   b. Yastrzemski had 186 hits in 566 at-bats. What was his average?
   c. The player with the higher average was the batting champion. Which player was this?

Exploration

27. Kathleen entered 531×10^{20} on her calculator using the following key sequence:

531	EE	20
$\boxed{\text{531.}}$	$\boxed{\text{531. 00}}$	$\boxed{\text{531. 20}}$

   Then she pressed $\boxed{=}$.
   a. What was the final display?   **b.** What did pressing $\boxed{=}$ do?

2-4

Multiplying by $\frac{1}{10}$, $\frac{1}{100}$, ...

When was 43% enough? *Bill Clinton won the 1992 presidential election with 43.01% of the popular vote. He is shown here at inaugural festivities in Arlington, VA, getting a hug from daughter Chelsea as wife Hillary looks on.*

The numbers 10, 100, 1000, and so on, get larger and larger. By contrast, the numbers $\frac{1}{10}$, $\frac{1}{100}$, $\frac{1}{1000}$, and so on, get smaller and smaller. You can see that they get smaller by looking at the decimals for them.

$$\frac{1}{10} = .1 \qquad = \text{one tenth}$$
$$\frac{1}{100} = .01 \qquad = \text{one hundredth}$$
$$\frac{1}{1000} = .001 \qquad = \text{one thousandth}$$
$$\frac{1}{10,000} = .0001 = \text{one ten-thousandth}$$
$$\text{and so on}$$

Multiplying by $\frac{1}{10}$

To multiply a decimal by 10, remember that you move the decimal point one place to the *right*. Multiplication by $\frac{1}{10}$ undoes multiplication by 10. So, to multiply by $\frac{1}{10}$ or 0.1, you move the decimal point one place to the *left*.

$$\frac{1}{10} \times 15.283 = 1.5283$$

You can estimate to check. A tenth of $15 is $1.50.

Multiplying by $\frac{1}{100}$ and $\frac{1}{1000}$

Multiplying by $\frac{1}{100}$ is equivalent to multiplying by $\frac{1}{10}$ and then multiplying by $\frac{1}{10}$ again. So, to multiply by $\frac{1}{100}$ or .01, move the decimal point *two* places to the *left*.

$$\frac{1}{100} \times 15.283 = 0.15283$$

The pattern continues. Multiplication by $\frac{1}{1000}$ is equivalent to multiplying by $\frac{1}{10}$ three times. So move the decimal point three places to the left.

$$\frac{1}{1000} \times 15.283 = 0.015283$$

Activity

Check the above multiplications on your calculator by using decimals instead of fractions.
a. Multiply: $.1 \times 15.283$
b. Multiply: $.01 \times 15.283$
c. Multiply: $.001 \times 15.283$

What Does % Mean?

If your calculator has a **percent key** %, enter 15.283 and press %. You should find the same display as when you multiplied by .01. This is because the **percent sign %** means to multiply by one hundredth. In general, to change a percent to a decimal, just multiply the number in front of the percent sign by $\frac{1}{100}$ or .01. Consider 50%.

As a fraction, $50\% = 50 \times \frac{1}{100} = \frac{50}{100} = \frac{1}{2}$.
As a decimal, $50\% = 50 \times .01 = .50 = .5$.

This confirms the idea that 50% of something is half of that thing.

The percent symbol is very common. Here are four examples of its use.

Example 1

Suppose a pair of designer jeans is on sale at 20% off. Rewrite 20% as a decimal.

Solution

We can use $\frac{1}{100}$ for the percent sign.

$$20\% = 20 \times \frac{1}{100} = .20 = .2$$

Recall that $.2 = \frac{1}{5}$. So $20\% = \frac{1}{5}$. Thus 20% off is the same as $\frac{1}{5}$ off.

Example 2

In the 1992 U.S. presidential election, an estimated 55% of the registered voters voted in the election. Rewrite 55% as a decimal.

Solution

Here we use .01 for the percent sign.

$$55\% = 55 \times .01 = .55$$

Here we multiplied by .01. Multiplying by $\frac{1}{100}$ would give the same answer. Use whichever is easier for you.

Example 3

A savings account earns a $3\frac{1}{4}$% interest rate. Rewrite $3\frac{1}{4}$% as a decimal.

Solution

First change the fraction to a decimal.

$$3\frac{1}{4}\% = 3.25\%$$

Now multiply by .01 instead of using the % sign.

$$= 3.25 \times .01$$
$$= 0.0325$$

The number 0.0325 is the number you can multiply by to determine the amount of interest paid in a savings account paying $3\frac{1}{4}$% interest.

Example 4

A baby weighs about 200% more at the age of one year than it does at birth. Rewrite 200% as a decimal.

Solution

$$200\% = 200 \times \frac{1}{100} = 2$$

A baby who weighs 200% *more* on its first birthday than it did at birth has increased its weight by a factor of two. If you want to determine the baby's weight, you need to add this increase to its original weight. So its final weight is 3 times its birth weight.

Graphing Percents on a Number Line

In Example 1, 20% was seen to equal the simple fraction $\frac{1}{5}$. Other percents equal commonly found simple fractions. Some of these are graphed on the number line below. You should learn these.

In Example 4, 200% was found to equal the integer 2. 100% = 1. Below is a different number line with some other percents indicated on it.

The idea of percent is old. The word *percent* comes from the Latin words *per centum,* meaning "through 100." (Sometimes it is useful to think of percent as "out of 100.") The symbol for percent is much newer. In 1650 the symbol $\frac{0}{0}$ was used. People have used the symbol % only in the last 100 years. Many writers still write "per cent" as two words. Either one word or two words are correct.

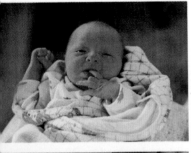

3 days old

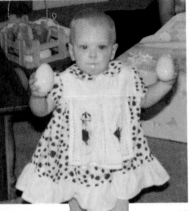

1 year old

What would happen if . . .? *If babies continued to gain weight at the rate of 200% per year, as they do during their first year, a baby with a birth weight of eight pounds would weigh 4,251,528 pounds by age 12.*

You should try to do all questions in this lesson without a calculator and without paper and pencil calculation.

Covering the Reading

In 1 and 2, write as a decimal.

1. one tenth

2. $\frac{1}{10,000}$

In 3 and 4, give a general rule for multiplying a decimal by the given number.

3. .01

4. $\frac{1}{10,000}$

In 5–8, write the answer as a decimal.

5. $46 \times \frac{1}{10}$

6. $.01 \times 6$

7. $\frac{1}{1000} \times 52$

8. $250,000 \times .0001$

9. Multiplying by $\frac{1}{100}$ is like multiplying by $\frac{1}{10}$ and then multiplying by __?__.

10. To multiply by $\frac{1}{1000}$ using a calculator, what decimal can you use?

11. The symbol % is read __?__ and means __?__.

12. About how old is the symbol %?

13. To change a % to a decimal, __?__ the number in front of the % symbol by __?__.

In 14–22, rewrite the percent as a decimal.

14. 50%

15. 25%

16. 5%

17. 2%

18. 1.5%

19. 5.75%

20. 300%

21. 150%

22. 82.3%

In 23–31, convert the percent to the indicated form.
a. decimal
b. simple fraction in lowest terms

23. 25%

24. 75%

25. 10%

26. 20%

27. 30%

28. $33\frac{1}{3}$%

29. $66\frac{2}{3}$%

30. $87\frac{1}{2}$%

31. 90%

In 32–37, change the fraction to a percent.

32. $\frac{1}{2}$

33. $\frac{1}{4}$

34. $\frac{1}{8}$

35. $\frac{3}{8}$

36. $\frac{3}{5}$

37. $\frac{1}{3}$

38. Trace this number line. Graph 10%, 20%, 30%, 40%, 50%, 60%, 70%, 80%, 90%, and 100% on it.

0 ————————————————————————————————→ 1

Applying the Mathematics

39. By what number can you multiply 46.381 to get 0.46381?

40. Betty weighs 87.5 pounds. She weighed about a tenth of that at birth. To the nearest pound, what did she weigh at birth?

41. Multiply $\frac{1}{10^3}$ by 43.87.

In 42 and 43, rewrite each underlined number as a fraction and as a decimal.

42. The teachers wanted a 7% raise, and the school board offered 4%.

43. In 1983, the president of Brazil's central bank said, "We cannot live with 150% inflation." Yet in 1990, inflation was 1,795%.

44. *Multiple choice.* Which is equal to 0.3?

   (a) 300%   (b) 30%   (c) 3%   (d) 0.3%

45. *Multiple choice.* Which is closest to $0.\overline{3}$?

   (a) 333%   (b) 33%   (c) 3%   (d) .3%

46. Between what two integers is 5.625%? Explain your thinking.

47. Change 0.1% to a decimal and to a fraction.

In 48–51, use the graph below:

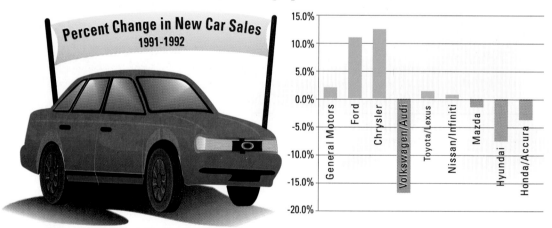

Percent Change in New Car Sales 1991-1992

Source: *Automotive News,* May 26, 1993

48. Which automaker had the largest percent increase in sales?

49. Which automaker had the largest percent decrease in sales?

50. Which automaker had an increase, but the smallest percent increase in sales?

51. What percent would indicate no change in sales?

52. Multiply 2.3 by each number. *(Lessons 2-1, 2-4)*
   a. 1000   **b.** 10   **c.** 10^4 **d.** $\frac{1}{100}$ **e.** 0.0001

53. a. Write the three underlined numbers as decimals. *(Lesson 2-1)*

   The number of students enrolled in grades 9–12 dropped from <u>fourteen million, five hundred seventy thousand</u> in 1980 to <u>twelve million, four hundred seventy-two thousand</u> in 1990. An estimate for the year 2000 is <u>fifteen million, two hundred thousand</u>.

   b. Write the underlined numbers of part **a** in scientific notation. *(Lesson 2-3)*

54. Which number is smallest, 9×10^4, 8.2×10^5, or 3.01×10^9? *(Lesson 2-3)*

55. Which is greater, 8×3 or 8^3? *(Lesson 2-2)*

56. Consider the numbers -1.4, -14, and 0.14.
   a. Order the numbers from smallest to largest.
   b. Use the numbers in a sentence with two inequality symbols.
   c. Graph the numbers on the same number line. *(Lessons 1-2, 1-9)*

57. In the previous lessons, the largest power of 10 named was quintillion. But there are larger powers of 10 with names. Look up the given words in a dictionary. (You may need a large dictionary.) Write each number as a decimal and as a power of 10.

   a. sextillion   **b.** octillion   **c.** nonillion   **d.** decillion

58. Money rates are often given as percents. Find the following rates by looking in a daily newspaper or weekly magazine.

   a. the prime interest rate charged by banks to companies with good credit ratings
   b. a local mortgage rate on a new home purchase
   c. the interest rate on an account at a local savings institution

2-5

Percent of a Quantity

What percent graduate? *According to 1990 data, 71.2% of high school freshmen enrolled in U.S. public schools stayed in school and graduated. Vermont, Minnesota, and North Dakota had the highest rates.*

Why Are There So Many Ways to Write Numbers?

You have learned to convert percents to fractions and to decimals. You have learned to put some decimals in scientific notation. The purpose of all this rewriting is to give you *flexibility*. Sometimes it's easier to use fractions. Sometimes decimals are easier. Sometimes percent or scientific notation is needed.

But why does all of this work? Why can you use .01 in place of $\frac{1}{100}$, or 3×10^6 instead of 3 million, or 0.4 instead of $\frac{2}{5}$ or 40%? The reason is due to the general idea called the *Substitution Principle*.

> **Substitution Principle**
> If two numbers are equal, then one can be substituted for the other in any computation without changing the results of the computation.

Using the Substitution Principle with Percents

The Substitution Principle is used in many places. Here it is used to find percents of a quantity. The phrase "percent of" is a signal to multiply. Use a calculator as you work through the following examples.

Example 1

Suppose 30% of the 2000 students in a high school are freshmen. How many students is this?

Solution

Change "of" to "times."
Now use the Substitution Principle,
rewriting 30% as .3.

$$30\% \text{ of } 2000 \text{ students}$$
$$= 30\% \times 2000 \text{ students}$$
$$= .3 \times 2000 \text{ students}$$
$$= 600 \text{ students}$$

Example 2

Suppose 100% of the 2000 students live within the school district. How many students is this?

Solution

Change "of" to "times." Now use the Substitution Principle, rewriting 100% as 1.

$$100\% \text{ of } 2000 \text{ students}$$
$$= 100\% \times 2000 \text{ students}$$
$$= 1 \times 2000 \text{ students}$$
$$= 2000 \text{ students}$$

So, *all* of the students live within the school district.

Using Percents

From Example 2, you can see that "100% of" means "all of." Since 0% = 0, 0% of the students is 0 × 2000 students. So "0% of" means "none of." Since 50% is equal to $\frac{1}{2}$, "50% of" means "half of."

Percents are often used when items are put on sale.

Example 3

A sofa sells for $569.95. It is put on sale at 20% off.
a. How much will you save if you buy the sofa during the sale?
b. What will the sale price be?

Solution

a. You save 20% of $569.95.

$$20\% \text{ of } \$569.95$$
$$= .2 \times \$569.95$$
$$= \$113.99, \text{ the amount you save}$$

b. Subtraction tells you what the sale price will be.

$$\$569.95 \text{ original price}$$
$$-113.99 \text{ amount saved}$$
$$\$455.96 \text{ sale price}$$

Check

If you save 20%, you are paying 80% of the original cost.

$$80\% \text{ of } \$569.95$$
$$= .8 \times \$569.95$$
$$= \$455.96, \text{ the sale price}$$

If you do not have a calculator, you might choose to estimate the answer to Example 3a. One estimate is found by rounding $569.95 up to $570. Because $570 is so close to the actual price, the estimate is very accurate.

$$20\% \times \$570$$
$$= .2 \times \$570$$
$$= \$114, \text{ only one penny off the actual amount saved!}$$

How Percents Describe Inflation

Inflation is the general amount by which goods or services increase in price. Inflation is usually reported as a percent.

Example 4

In 1991, the U.S. inflation rate was 3.1%. Assume a house cost $60,000 at the beginning of 1991. If it increased in value at the rate of inflation, what was it worth at the end of 1991?

Solution

$$3.1\% \text{ of } \$60,000$$
$$= 3.1\% \times 60,000$$

Change 3.1% to a decimal. $= .031 \times 60,000$

$$= \$1860, \text{ the amount of the increase}$$

The value of the house at the end of 1991 was
$60,000 + $1860 = $61,860.

QUESTIONS

Covering the Reading

1. Why is it useful to have many ways of writing numbers?

2. State the Substitution Principle.

3. *True or false.*   (Hint: Recall that $20\% = \frac{1}{5}$ and $30\% = \frac{3}{10}$.)
   a. You can substitute $\frac{1}{5}$ for 20% in any computation and the answer will not be affected.
   b. $20\% + 30\% = \frac{1}{5} + \frac{3}{10}$
   c. $20\% \times \$6000 = \frac{1}{5} \times \$6000$

4. a. In calculating 30% of 2000, when is the Substitution Principle used?
   b. Calculate 30% of 2000.

5. Match each percent at the left with the correct phrase at the right.

100% of	none of
50% of	all of
0% of	half of

In 6–9, determine the answer in your head.

6. 50% of 6000

7. 100% of 12

8. 0% of 50

9. 150% of 30

10. A store normally sells futons at 2 for $899.
    a. To estimate how much the beds would cost at 25% off, what value can be used in place of $899?
    b. Estimate the price for the two beds at a "25% off" sale.

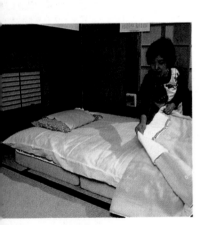

What's a futon?
A futon is a bed. It is a padded quilt generally placed on a mat or wooden frame. Originally used in Japan, futons are particularly functional in small homes as the futon can be rolled up and put away.

11. Use the information in Example 4. Suppose the cost of a vacation increased at the rate of inflation. If a vacation cost $1200 at the beginning of 1991, what would you expect its cost to be at the end of 1991?

12. Suppose the price of a CD increased 10% in the last year. What would you expect to pay for a CD that cost $10.95 a year ago?

Applying the Mathematics

In 13 and 14, what does the remark mean?

13. "We are with you 100%!"   **14.** "Let's split it 50-50."

In 15 and 16, assume the U.S. population to be about 250,000,000.

15. In your head, figure out what 10% of the U.S. population is. Use this to figure out **a.** 20%, **b.** 30%, **c.** 40%, and **d.** 50% of the population.

16. The U.S. population is now increasing at the rate of about 1.06% a year. How many people is this?

17. In Bakersfield, California, it rains on about 10% of the days in a year. About how many days is this?

18. In store *A* you see a $600 stereo at 25% off. In store *B* the same stereo normally costs $575 and is on sale at 20% off. Which store has the lower sale price?

19. During the 1992–1993 season, the Detroit Pistons basketball team won 48.8% of its games. Did the Pistons win or lose more often? Justify your answer.

20. An interest penalty is charged on credit card purchases if you do not pay on time. Suppose you have $1000 in overdue bills. If the penalty is 1.5% per month, how much interest will you have to pay the first month?

21. The population of Los Angeles, California, was about 100,000 in 1900 and increased 1800% from 1900 to 1950. How many people is that increase?

22. The caption below says that 46% of all revenues from sales of computers in 1991 were for personal computers. Did the writer round up or down to obtain this percent?

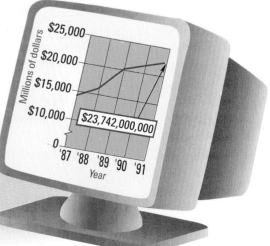

In 1991, total factory revenues from sales of computers were 50,885 million dollars. The percentage of the total for personal computers was 46%, up from just one-third of total revenues in 1987.

Source: Dataquest Inc., San Jose, CA, Consolidated Data Base   Lesson 2-5 *Percent of a Quantity*   **87**

23. Multiply 990 by: **a.** 1 **b.** .1 **c.** .01 **d.** .001. *(Lesson 2-4)*

24. Change to a percent: **a.** $\frac{1}{4}$ **b.** $\frac{1}{3}$ **c.** $\frac{5}{4}$. *(Lesson 2-4)*

25. Write as a power of 10: **a.** quadrillion **b.** ten thousand. *(Lesson 2-2)*

In 26–30, write the number in decimal notation. *(Lessons 1-1, 1-7, 2-1, 2-2)*

26. 5^4 **27.** 8.3 million **28.** 2.56×10^8 **29.** $4\frac{4}{5}$

30. three billion, four hundred thousand

31. Write $\frac{48}{100}$ in lowest terms. *(Lesson 1-10)*

32. The highest mountain in the world is Mt. Everest, in the Himalayas on the Tibet-Nepal border. Its peak is about 29,028 feet above sea level.

  a. Should you call its height 29,028 feet, or ⁻29,028 feet?

  b. Round the height to the nearest 100 feet.

  c. Round the height to the nearest 1000 feet.

  d. Round the height to the nearest 10,000 feet. *(Lessons 1-4, 1-8)*

Would you climb it? *Mt. Everest is known in the Tibetan language as* Chomolungma, *"Goddess Mother of the World." It was first climbed in 1953 by Australian Edmund Hillary and Tenzing Norgay, a Sherpa of Nepal.*

33. Give the value of $\frac{\pi}{4}$ truncated to ten-thousandths. *(Lesson 1-3)*

Exploration

34. a. Find a use of percent in a newspaper or magazine.
  b. Make up a question about the information you have found.
  c. Answer the question you made up.

LESSON

2-6

From
Decimals to
Fractions
and
Percents

Words without consonants. *In the words "ai," "oe," "aa," and "ae," 0% of the letters are consonants. All of these words are legal in some board games.*

In the word *power*, $\frac{3}{5}$ of the letters are consonants. Since $\frac{3}{5} = 60\%$, we could say 60% of the letters are consonants. Since 60% = .6, we could say .6 of the letters are consonants. Of these three choices, the decimal .6 is least used. It is usually converted to the fraction or the percent.

How to Convert a Terminating Decimal to a Fraction

If a terminating, or ending, decimal has only a few decimal places, just read the decimal in English and write the fraction.

Example 1

Convert 4.53 to a mixed number.

Solution

Read the decimal 4.53 as "four and fifty-three hundredths."
That tells you $4.53 = 4\frac{53}{100}$.

Another way to convert a decimal to a fraction is to write the decimal as a fraction over 1. Multiply the numerator and denominator by a power of 10 large enough to eliminate the decimal point in the numerator. This is an application of the Equal Fractions Property from Lesson 1-10.

Example 2

Write 0.036 as a fraction in lowest terms.

Solution

Write 0.036 as $\frac{0.036}{1}$.

Multiply both numerator and denominator by 1000. This moves the decimal point three places to the right in both the numerator and denominator to get $\frac{36}{1000}$. To write $\frac{36}{1000}$ in lowest terms, notice that 4 is a factor of both the numerator and the denominator. Divide each by 4.

$$0.036 = \frac{0.036}{1} = \frac{0.036 \times 1000}{1 \times 1000} = \frac{36}{1000} = \frac{36 \div 4}{1000 \div 4} = \frac{9}{250}$$

Check

Convert $\frac{9}{250}$ to a decimal using a calculator. You should get 0.036.

How to Convert a Fraction with Decimals in It

Using the idea from Example 2, you can convert a fraction with decimals in it to a simple fraction.

Example 3

Find a simple fraction equal to $\frac{2.5}{35}$.

Solution

Remember that a simple fraction is a fraction with integers in its numerator and denominator. Multiply the numerator and the denominator by 10.

$$\frac{2.5 \times 10}{35 \times 10} = \frac{25}{350}$$

So $\frac{25}{350}$ is one answer. To put this fraction in lowest terms, notice that 25 is a factor of 25 and 350. Divide both numerator and denominator by 25.

$$\frac{2.5}{35} = \frac{25}{350} = \frac{25 \div 25}{350 \div 25} = \frac{1}{14}$$

Check

Use a calculator to verify that $\frac{2.5}{35} = .071428\ldots$ and $\frac{1}{14} = .071428.\ldots$

How to Convert a Decimal to a Percent

Now think about changing decimals to percents. Remember that to convert a percent to a decimal, move the decimal point two places to the *left*. For example, 53% = 0.53, 1800% = 18, and 6.25% = 0.0625.

To convert decimals to percents, reverse the procedure. Move the decimal point two places to the *right*. Here are a few examples.

$$
\begin{aligned}
0.036 &= 3.6\% \\
0.46 &= 46\% \\
3 &= 300\% \\
0.0007 &= 0.07\%
\end{aligned}
$$

How to Convert a Fraction to a Percent

One way to convert fractions to percents is to convert them to decimals first. Then convert the decimals to percents.

Example 4

A worker receives time and a half for overtime. What percent of a person's pay is this?

Solution

time and a half = $1\frac{1}{2}$ = 1.5 = 150%

Example 5

A photographer reduces the dimensions of a photo to $\frac{5}{8}$ their original size. The final width is what percent of the original width?

Solution

First convert $\frac{5}{8}$ to a decimal. Then change the decimal to a percent.

$$\frac{5}{8} = 0.625 = 62.5\%$$

QUESTIONS

Covering the Reading

In 1–3, a decimal is given.
a. Write the decimal in English words.
b. Convert the decimal to a fraction or mixed number.

1. 8.27   **2.** 630.5   **3.** 0.001

4. The probability of a single birth being a boy is about .51. Convert this number to a percent.

In 5 and 6, convert to a percent.

5. 0.724   **6.** 8

7. A decimal approximation to π is 3.14. Convert 3.14 to a mixed number.

8. Explain how to change a fraction to a percent.

9. About $\frac{1}{4}$ of all families with two children are likely to have two girls. What percent of families is this?

10. Write $\frac{4.2}{1.04}$ as a fraction in lowest terms.

Applying the Mathematics

11. a. The cost of living in 1993 was about $3\frac{1}{2}$ times what it was in 1970. What percent is this?
   b. Suppose a slice of pizza cost 45¢ in 1970. Using the cost of living information in part **a,** estimate the cost of a slice in 1993.

In 12 and 13, use this information. Just as fractions can be represented by parts of circles, so can percents. Notice $\frac{1}{10}$ of a circle is 10%.

$\frac{1}{10}$ or 10%

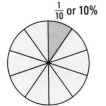

12. a. Draw a circle with a part representing 25%.
   b. What percent does the unshaded part of the circle represent?

13. Draw a circle. Shade a part that represents $\frac{2}{5}$. What percent does this part represent?

How big are Bugs and Daffy? *Cartoons are enlarged and shrunk to fit on clothing, on comic pages, and so on. Thus, the same cartoon in one newspaper may appear larger or smaller in another newspaper.*

In 14–19, complete the table. Put fractions in lowest terms.

	Fraction	Decimal	Percent
	$\frac{1}{2}$	0.5	50%
14.	?	?	9.8%
15.	?	3.2	?
16.	$\frac{5}{16}$	?	?
17.	?	0.27	?
18.	$\frac{3}{20}$	?	?
19.	?	?	44%

20. Ariel put 6.5 gallons of gasoline in a 12-gallon gas tank. So the tank is $\frac{6.5}{12}$ full. What simple fraction of a full tank is this?

21. A money market account at a local bank pays 3.75% interest. What fraction is this?

Review

Who lives in the Sahara?
Over 2 million people live in the Sahara Desert. Most are nomads who herd sheep, goats, camels, and cattle near an oasis like the one pictured here.

22. **a.** Leslie used 30% of her $5.00 allowance to buy a magazine. How much did the magazine cost?
b. Explain how you could do part **a** in your head. *(Lesson 2-5)*

23. Would you prefer to buy something at 30% off or $\frac{1}{3}$ off? *(Lesson 2-4)*

In 24–26, calculate. *(Lessons 2-1, 2-2, 2-5)*
24. 1918.37 × 10,000   25. 14% of 231   26. 3^5

In 27 and 28, write the number as a decimal. *(Lessons 2-2, 2-4)*
27. 2.4×10^5 28. 3200%

In 29–31, rewrite the number in scientific notation. *(Lesson 2-3)*

29. 8,800,000 square kilometers, the approximate area of the Sahara Desert

30. 5280^3, the number of cubic feet in a cubic mile

31. 525,600 minutes, the number of minutes in a 365-day year

In 32 and 33, give the word name for the decimal. *(Lesson 2-1)*
32. One followed by nine zeros   33. One followed by fifteen zeros

34. Order from smallest to largest: .011 1/10 1/100.   *(Lessons 1-2, 1-6)*

Exploration

35. If your calculator has a % key, try the following sequence:
500 [+] 7 [%] [=]. Explain what the calculator has done.

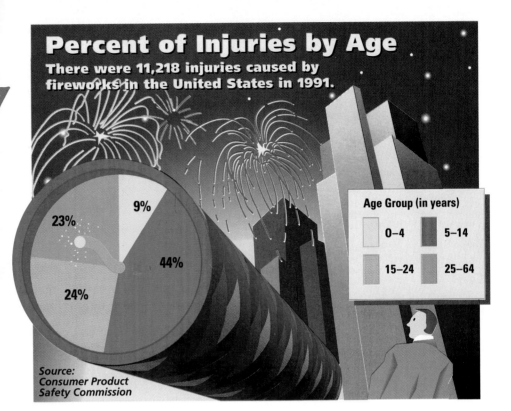

Percent of Injuries by Age
There were 11,218 injuries caused by fireworks in the United States in 1991.

9%

23%

44%

24%

Age Group (in years)	
0–4	5–14
15–24	25–64

Source:
Consumer Product
Safety Commission

Reading a Circle Graph

Percents of a whole are often pictured in a **circle graph** or **pie chart.** In a circle graph, you can quickly see the relationship of the parts. For example, the largest "pie piece" is the largest part of the whole.

Example 1

Use the information displayed in the circle graph above.
a. The entire circle represents what percent of the whole?
b. For the smallest piece of the circle, what does the 0–4 mean?
c. Which age group had the most injuries?
d. What percent of injuries are not in the 15–64 age groups?

Solution

a. Since the circle represents the whole quantity, it represents 100% of the injuries by fireworks.
   Notice that 9% + 23% + 24% + 44% = 100%.
b. The 0–4 refers to children under the age of 5.
c. The 5–14 age group had the most injuries.
d. The percent in the 15–64 age groups is 24% + 23% = 47%. Therefore,
   100% – 47% = 53% are not in the 15–64 age group.

From information given in a circle graph, or next to the graph, you may be able to determine other information.

Example 2

Find the number of injuries by fireworks for the 5–14 age group.

Solution

To determine the number of injuries in any age group, multiply its percent of injuries by 11,218, the total number of injuries.

$$44\% \times 11,218$$
$$= .44 \times 11,218$$
$$= 4935.92$$
$$\approx 4936$$

There were 4936 injuries in the 5–14 age group.

Since it does not make sense to have .92 injuries, we rounded 4935.92 up to 4936. The symbol $\approx$ means **approximately equal to.**

Check

44% is less than 50%. The answer should be less than $\frac{1}{2}$ of 11,218, which is 5,609. It is.

Vocabulary for Circles

To discuss circle graphs, it is useful to know the names of parts of circles. A **circle** is the set of points which are all the same distance (its **radius**) from a certain point (its **center**). The plural of radius is **radii.** The distance across the circle is twice the length of its radius. That distance is the circle's **diameter.**

The words radius and diameter are used to name both distances and segments. The circle below at the left has center C and is named circle C. One radius of the circle is the line segment CD, written $\overline{CD}$. The radius of the circle is 5. One diameter of the circle is $\overline{DE}$. The diameter is 10.

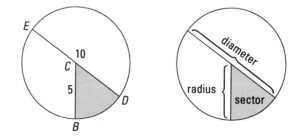

A part of a circle graph that looks like a slice of pie is called a **sector.** A sector is bounded by two radii and an *arc* of the circle. An **arc** is a part of a circle connecting two points (its endpoints) on the circle. The sector in the circle above bounded by $\overline{CB}$ and $\overline{CD}$ is named sector BCD or sector DCB. Sector BCD is also bounded by arc BD (written $\overset{\frown}{BD}$ or $\overset{\frown}{DB}$).

Making a Circle Graph with Fractions or Percents

Fractions and percents are helpful in approximating the size of a sector on a circle graph.

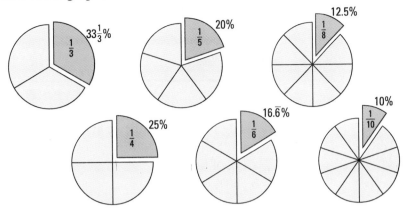

Example 3

Sixty students were asked to choose one dessert at a carnival. Draw a circle graph that displays the percent of students picking each dessert.

Dessert	Number of Students Choosing that Dessert
Ice Cream	30
Popcorn	15
Fruit	9
None	6

Solution

$\frac{30}{60}$ of the students chose ice cream. This is $\frac{1}{2}$, or 50%, of the students. So the sector for ice cream is half the circle.

Popcorn was chosen by $\frac{15}{60}$, or $\frac{1}{4}$, or 25%, of the students. So the sector for popcorn is $\frac{1}{4}$ of the circle.

$\frac{6}{60}$ or $\frac{1}{10}$, or 10%, of the students chose none. The sector for none is $\frac{1}{10}$ of the circle. The circle piece for $\frac{1}{10}$ above can be used to approximate the sector for none.

The remaining piece or unshaded part represents fruit. It is $\frac{9}{60}$, or $\frac{3}{20}$, or 15% of the circle.

A circle graph usually has the actual number or percent in each sector rather than the fraction.

Circle graphs were first constructed by William Playfair in the late 1700s. Today circle graphs are found in many newspapers and magazines. There is software for most computers that will automatically construct circle graphs.

QUESTIONS

Covering the Reading

1. Using the circle at the right, state the number of sectors that represent
   a. $\frac{12}{40}$ of the circle.
   b. 40% of the circle.

In 2–4, use the circle graph on fireworks, page 93.

2. Which age group had the fewest injuries?

3. What percent of injuries are not in the 5–14 age group?

4. Find the number of injuries for the 15–24 age group.

5. What is the sum of the percents on a circle graph?

6. Who first made a circle graph and when?

7. What is another name for circle graph?

In 8–10, use the circle graph on dessert choices, page 95.

8. The total number of people choosing to have desserts is __?__.

9. What fraction of people wanted fruit?

10. Explain why the circle divided in 10 pieces helped in drawing the circle graph.

11. What is a circle?

In 12–14, use the circle at the right.

12. Name a radius.

13. a. Name a sector.
    b. Identify the radii and arc that bound the sector named in part **a.**

14. The length of the diameter is __?__.

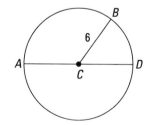

In 15–18, use the graph below.

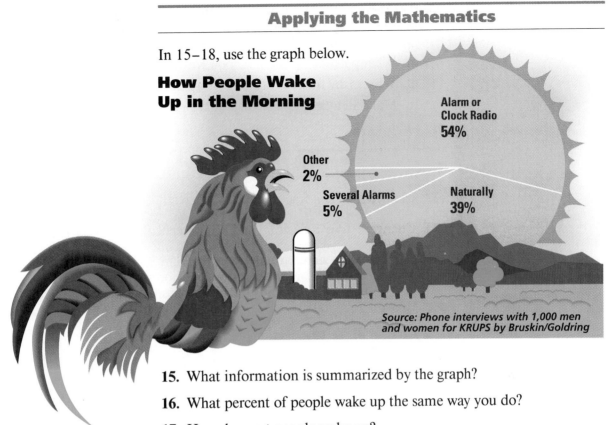

How People Wake Up in the Morning

Alarm or Clock Radio **54%**

Other **2%**

Several Alarms **5%**

Naturally **39%**

Source: Phone interviews with 1,000 men and women for KRUPS by Bruskin/Goldring

15. What information is summarized by the graph?

16. What percent of people wake up the same way you do?

17. How do most people wake up?

18. How do you think the 2% of people included in "other" wake up in the morning?

19. Refer to the table for dessert choices in this lesson. Suppose the number of students choosing ice cream was 40 and popcorn was 5.

   a. The sector for ice cream should now be what fraction of the circle?

   b. What percent of the circle is for ice cream?

   c. Draw a circle on your paper. Shade the part that represents ice cream.

20. Use the circle graph below. It shows how many days the Williams family spent on their recent vacation in four states.

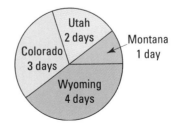

Give the percent of time they spent in each state.
   a. Utah
   b. Montana
   c. Wyoming
   d. Colorado

21. For their elective course, the students at Martin Luther King High School signed up as follows: 20 took Music, 10 took Woodshop, 5 took Drawing, and 15 took Keyboarding.

   a. What percent of students took Keyboarding?

   b. What is the fraction of students who took either Music or Drawing?

   c. Make a circle graph to display the given information.

Review

22. Change .7 to **a.** a fraction and **b.** a percent.   *(Lesson 2-6)*

23. Change $\frac{7}{20}$ to **a.** a decimal and **b.** a percent.   *(Lessons 2-6, 1-6)*

24. In one store a stereo usually sells for $179.95 but is on sale for 30% off. In a second store the same stereo is on sale for 25% off its normal price of $169.95. Which store has the better sale price? Explain how you arrived at your answer.   *(Lesson 2-5)*

25. Change 12.5% to a decimal and to a fraction.   *(Lesson 2-4)*

26. Give a general rule for multiplying a decimal by 10.   *(Lesson 2-1)*

27. State the Equal Fractions Property in your own words.   *(Lesson 1-10)*

28. What numbers are the integers?   *(Lesson 1-8)*

29. Write the English words for 5.34.   *(Lesson 1-1)*

Exploration

30. **a.** Find an example of a pie or circle graph involving percent in your local newspaper.

   b. Make up a question about the information you have found and answer your question.

31. Find a software package that enables a computer to display a circle graph on a monitor. If you have access to a computer, learn how to use it and the software to draw a circle graph. Draw the circle graph shown on the first page of this lesson.

King's Dream. *Martin Luther King, Jr., shared his dream of people "not being judged by the color of their skin but by the content of their character," during his 1963 "I Have a Dream" speech. The speech, given after the March on Washington, was intended to urge Congress to pass the Civil Rights Bill.*

More Powers of Ten

The power of an abacus. *The picture shows an abacus used in a porcelain shop in Kweilin, China. Each column in an abacus represents a power of ten. Numbers are represented by properly positioning the beads in each column.*

What Is Exponential Form?

A number that is written with an exponent, such as 4^3 or 10^9, is said to be in **exponential form.** Exponential form is a short way of writing some large numbers. For example, $2^{20} = 1,048,576$. The decimal 1,048,576 is longer than the exponential form 2^{20} for the same number. Powers of 10 such as 100, 1000, and 10,000 are easily written in exponential form as 10^2, 10^3, and 10^4. Thus exponential form makes it possible to rewrite *large* numbers in the shorter scientific notation. Now we consider how to use exponential form to write *small* numbers.

What Is the Zero Power of a Number?

Examine this pattern closely. The numbers in each row going across are equal. Before turning the page, guess what should be the next entry in each column.

exponential form		decimal form		word name
10^6	=	1,000,000	=	million
10^5	=	100,000	=	hundred thousand
10^4	=	10,000	=	ten thousand
10^3	=	1,000	=	thousand
10^2	=	100	=	hundred
10^1	=	10	=	ten

In the left column, the exponents decrease by 1, so the next entry should be 10^0. In the middle column, each number is $\frac{1}{10}$ the number above it. So the next entry should be $\frac{1}{10}$ of 10, or 1. In the right column are word names. Next to the tens place is the ones place. Here is the next row.

exponential form		decimal form		word name
10^0	=	1	=	one

Negative Integer Powers of Ten

If the above pattern is continued, negative exponents will appear in the left column. This is exactly what is needed in order to represent small numbers.

10^{-1}	=	0.1	=	one tenth	=	$\frac{1}{10}$
10^{-2}	=	0.01	=	one hundredth	=	$\frac{1}{100}$
10^{-3}	=	0.001	=	one thousandth	=	$\frac{1}{1000}$
10^{-4}	=	0.0001	=	one ten-thousandth	=	$\frac{1}{10,000}$
10^{-5}	=	0.00001	=	one hundred-thousandth	=	$\frac{1}{100,000}$

and so on

Multiplying by Negative Integer Powers of Ten

To multiply a decimal by 10^{-1}, use the Substitution Principle. Since $10^{-1} = 0.1$, think of multiplying by 0.1, and move the decimal point one unit to the left.

$$829.43 \times 10^{-1} = 82.943$$

To multiply a decimal by 0.01 or $\frac{1}{100}$, you know you only have to move the decimal point two places to the left. Since $10^{-2} = 0.01$, the same goes for multiplying by 10^{-2}.

$$829.43 \times 10^{-2} = 8.2943$$

Do you see the simple pattern?

To multiply by a negative integer power of 10
Move the decimal point to the left as many places as indicated by the exponent.

Example 1

Write 72×10^{-5} as a decimal.

Solution

To multiply by 10^{-5}, move the decimal point five places to the left.

$72 \times 10^{-5} = 72.0 \times 10^{-5} = .00072$

The names of the negative integer powers of 10 correspond to the names of the positive integer powers.

10^1 = ten	10^{-1} = one tenth
10^2 = one hundred	10^{-2} = one hundredth
10^3 = one thousand	10^{-3} = one thousandth
10^6 = one million	10^{-6} = one millionth
10^9 = one billion	10^{-9} = one billionth
10^{12} = one trillion	10^{-12} = one trillionth
10^{15} = one quadrillion	10^{-15} = one quadrillionth
10^{18} = one quintillion	10^{-18} = one quintillionth

Example 2

Write 6 trillionths as a decimal.

Solution

Think: one trillion $= 10^{12}$, so one trillionth is 10^{-12}.

6 trillionths $= 6 \times 10^{-12} = 0.00000\ 00000\ 06$

(When a decimal has many digits, we put a space after each five digits, starting at the decimal point, to make it easier to count the digits.)

QUESTIONS

Covering the Reading

1. *Multiple choice.* Which number is written in exponential form?
   (a) $\frac{2}{3}$ (b) million   (c) 2×10^6 (d) 7^3 (e) 1.45

2. Write $10 \times 10 \times 10 \times 10$ in exponential form.

3. Give the place value name for each power.
   a. 10^5 **b.** 10^4 **c.** 10^3 **d.** 10^2 **e.** 10^1

4. Continue the pattern of Question 3 to give the place value name for each power:
   a. 10^0 **b.** 10^{-1} **c.** 10^{-2} **d.** 10^{-3} **e.** 10^{-4}

5. What keys can you press on your calculator to verify that $10^0 = 1$?

6. Write each number as a decimal.
   a. 10^{-2} **b.** 10^{-3}

7. Tell whether each number is positive, negative, or equal to zero.
   a. 10^0 **b.** 10^{-1}

8. $10^7 =$ ten million. What is a word name for 10^{-7}?

In 9–11, write the number as a power of 10.
   9. one thousandth   **10.** one millionth   **11.** one trillionth

In 12–17, write the number as a decimal.

12. 3×10^{-2} **13.** 3.45×10^{-4}

14. 41.3×10^0 **15.** four thousandths

16. sixty billionths   **17.** five millionths

18. What is the general rule for multiplying by a negative integer power of 10?

Applying the Mathematics

19. *Multiple choice.* Which of (a) to (e) does not equal the others?
   (a) 1%   (b) .01   (c) $\frac{1}{100}$ (d) 10^{-2}
   (e) one hundredth   (f) All of (a) through (e) are equal.

20. Explain why 0^{10} does not equal 10^0.

21. Describe the results you obtained for Activity 2 in this lesson.

22. Write 10×10^{-7} as a decimal.

23. **a.** Calculate $10^4 \times 10^{-4}$.
   b. What is a hundred times one one-hundredth?
   c. Make a general conclusion from parts **a** and **b**.

24. Arrange from smallest to largest: 1 10^{-5} 0 10^2.

25. In the metric system, all the prefixes have meaning. For example, kilo- means 1000, centi- means $\frac{1}{100}$ or .01, and milli- means $\frac{1}{1000}$ or .001. Using this information, explain the meaning of:
   a. kilometer   **b.** centimeter   **c.** millimeter.

26. An electron microscope can magnify an object 10^5 times. The length of a poliomyelitis virus is 1.2×10^{-8} meter. Multiply this length by 10^5 to find how many meters long the virus would appear to be when viewed through this microscope.

In 27–29, write the number in exponential form. (For example, $64 = 4^3$.)

27. 81   **28.** 144   **29.** 32

In 30–33, refer to the four circle graphs below. These graphs show how various Hispanic populations are distributed among the four regions of the United States: W = West, S = South, M = Midwest, and NE = Northeast. For instance, the left circle is about the 13.50 million people of Mexican heritage in the U.S. It shows that 58.0% of these people live in the West, 32.2% live in the South, 8.5% in the Midwest, and 1.3% in the Northeast. (Source: *Statistical Abstract of the United States: 1992*)

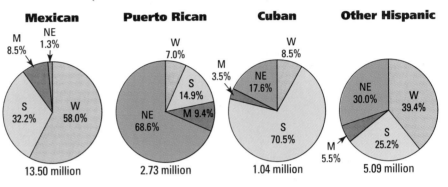

Mexican — 13.50 million
Puerto Rican — 2.73 million
Cuban — 1.04 million
Other Hispanic — 5.09 million

30. **a.** In which region do most people of Cuban heritage live?
    b. Why do you think more people of Cuban heritage live in this region than any other?   *(Lesson 2-7)*

31. How many people of Mexican heritage live in the West?
    (Lessons 2-5, 2-7)

32. **a.** Are there more people of Puerto Rican heritage or of Mexican heritage in the Midwest?
    b. Explain your answer to part **a.**   *(Lessons 2-5, 2-7)*

33. Fill in the blanks. Over $\frac{2}{3}$ of people of Other Hispanic heritage live in the __?__ or __?__ regions.   *(Lessons 2-5, 2-6, 2-7)*

34. Change 1.03 to: **a.** fraction.   **b.** a percent.   *(Lesson 2-6)*

35. **a.** Change 0.1875 to a fraction in lowest terms.
    b. Change 0.1875 to a percent.   *(Lessons 1-10, 2-6)*

36. Store *C* has a $50 videotape marked 30% off. Store *D* has this tape for $35. Which store offers the better buy?   *(Lesson 2-5)*

37. Write a rule for finding 10% of a number.   *(Lessons 2-4, 2-5)*

38. 93,000,000 miles is the approximate distance from Earth to the Sun. Write this number in scientific notation.   *(Lesson 2-3)*

39. 512 is what power of 2?   *(Lesson 2-2)*

It's a celebration. *This Bolivian folklore group at a New York street festival is celebrating* Día de la Raza, *a Columbus Day observance of "the Day of the Race." People from Bolivia are among the people classified by the Census Bureau as "Other Hispanics."*

40. Large computers are able to do computations in nanoseconds. Look in a dictionary for the meaning of *nanosecond*.

IN·CLASS
ACTIVITY

*Scientific
Notation
for Small
Numbers*

Work with a partner. One of you (person A) needs to have a
scientific calculator. The other (person B) needs paper.

1 **Person A:** divide 1 by 10.
Person B: record the display.

2 Divide the answer by 10. Record this second display.
Your paper should look something like this

Calculation	Display	Decimal
$1 \div 10$	0.1	.1
Ans $\div$ 10		
Ans $\div$ 10		
⋮		

3 Continue dividing the answers and recording the displays
until you have 12 rows.

4 At some time, your display should shift into scientific
notation. Together, determine what these answers are in
scientific notation and what they are as decimals.

LESSON
2-9

Scientific Notation for Small Numbers

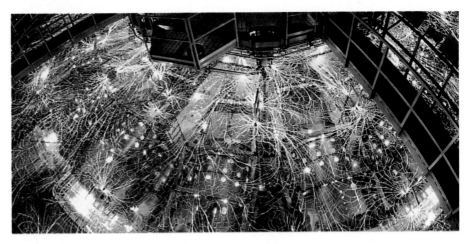

Crash! *At Sandia National Laboratories in Albuquerque, NM, subatomic particles with masses of 1.673×10^{-24} gram (at rest) are accelerated to nearly the speed of light. The particles collide to produce up to 900 billion electron volts. Operation of the particle accelerator creates the electrical discharges shown here.*

Converting Small Positive Numbers into Scientific Notation

The mass of one atom of hydrogen, the lightest element, is

.00000 00000 00000 00000 00016 75 gram.

(By comparison, a piece of notebook paper weighs more than a gram!) A number this small in decimal notation is difficult to read and use in computation. To help read this number, we have put a space after every fifth digit of the decimal. But there is a shorter way to write the number, namely, scientific notation.

> A small positive number written in scientific notation is a number greater than or equal to 1 and less than 10, multiplied by a negative integer power of 10.
>
> $$decimal \times 10^{negative\ integer\ exponent}$$

For instance, 1.91×10^{-5} is a small positive number written in scientific notation. The same ideas are involved in putting small numbers into scientific notation as are used with large numbers.

Example 1

Write the mass of one atom of hydrogen in scientific notation.

Solution

First place the decimal point between the 1 and the 6. This gives you 1.675, a number between 1 and 10. Now find the power of 10 by counting the number of places you must move the decimal to the *left* to change 1.675 into .00000 00000 00000 00000 00016 75. (The movement to the left is in the negative direction and signals the negative exponent.) The move is 24 places to the left. So the mass, in scientific notation, is:

$$1.675 \times 10^{-24} \text{ gram.}$$

Entering Small Positive Numbers into a Calculator

To enter 1.675×10^{-24} into a calculator, you may have two choices. One choice is to use the powering key.

Key sequence: 1.675 $\boxed{\times}$ 10 $\boxed{y^x}$ 24 $\boxed{\pm}$ $\boxed{=}$

A second choice is to use the scientific notation key.

Key sequence: 1.675 $\boxed{\text{EE}}$ 24 $\boxed{\pm}$

The scientific notation key is usually easier to use. Using either sequence, you should see displayed the following or some equivalent.

$$\boxed{1.675 \quad -24}$$

Activity 1

Enter 3.97×10^{-8} into a calculator.

Example 2

Change 3.97×10^{-8} to a decimal.

Solution

Recall that to multiply by 10^{-8}, move the decimal point 8 places to the left.

$$3.97 \times 10^{-8} = 0.00000\ 00397$$

To enter a small positive number written as a decimal into your calculator, first convert it into scientific notation.

Activity 2

Enter 0.00000 00000 6993 into a calculator.

From this activity, you should see $\boxed{6.993 \quad -11}$.

QUESTIONS

Covering the Reading

1. A small positive number in scientific notation is a number greater than or equal to __?__ and less than __?__ multiplied by a __?__ power of __?__.

2. Why are small numbers often written in scientific notation?

3. Give an example of a small quantity that is usually written in scientific notation.

4. Write a key sequence that will display 3.97×10^{-8} on your calculator
   a. using the power key.
   b. using the key for scientific notation.

5. What number in scientific notation is given by this key sequence?

$$6.008 \boxed{\text{EE}}\ 5 \boxed{\pm}$$

In 6–8, rewrite the number in scientific notation.

6. 0.00008052 second, the time needed for TV signals to travel 15 miles

7. 0.28 second, the time needed for sound to travel the length of a football field

8. 0.00000 00000 00000 00000 0396 gram, the mass of one atom of uranium

9. Suppose a decimal is multiplied by a negative power of 10. Should its decimal point be moved to the right or to the left?

10. What key sequence did you use for Activity 2?

Applying the Mathematics

11. Rewrite the number as a decimal part of a centimeter.
   a. 1×10^{-8} centimeter, the angstrom (a unit of length)
   b. 0.529 angstrom, the radius of a hydrogen atom

12. Write the number in *decimal* notation given by the key sequence

$$4.675 \boxed{\text{EE}}\ 7 \boxed{\pm}.$$

13. Calculate $\frac{3 \times 10^{-8}}{6 \times 10^{-7}}$. Give the answer in the indicated form.
   a. scientific notation   **b.** decimal

In 14 and 15, choose one of the symbols $<$, $=$, or $>$.

14. 5.37×10^{-5} __?__ 5.37×10^{-4}

15. 49×10^{-9} __?__ 4.9×10^{-8}

16. When an object is magnified at "100×" under a microscope, it appears 100 times as large as its actual size. Measurements of objects too small to be seen with the human eye are often written in scientific notation. Suppose the longest known virus, *citrus tristeza*, with length 2×10^{-5} meters, is viewed with a microscope at a magnification of 20,000×. How long does the virus appear to be?

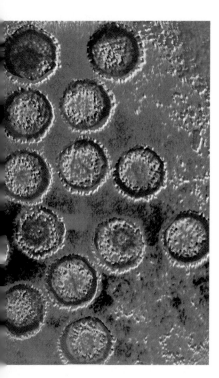

The photo shows Epstein-Barr virus particles, magnified 96,800 times.

Review

17. Write as a power of 10.
   a. one million   **b.** one millionth   *(Lessons 2-2, 2-8)*

In 18 and 19, order from smallest to largest.

18. $-.6$ $-.66$ $-.666$ $-.656$ $-2/3$ *(Lessons 1-6, 1-8)*

19. kilometer, millimeter, centimeter   *(Lesson 2-8)*

20. **a.** What percent is equal to $\frac{3}{4}$?

b. Show this percent as part of a circle. *(Lessons 2-6, 2-7)*

21. The Skunks baseball team lost 60% of its games. Did the team win or lose more often? *(Lesson 2-5)*

In 22–24, write the number as a decimal. *(Lessons 1-7, 2-5)*

22. 3%

23. $-19\frac{7}{10}$

24. 150%

25. Newspaper columnist Georgie Anne Geyer once wrote about receiving a tax bill for $0.01. The payment was due on June 30th.

a. If Ms. Geyer did not pay her bill by June 30th, she would have to pay a penalty of 10% of her bill. How much is this?

b. Also, if she paid late, she would have to pay an additional interest penalty of 1% of her bill. How much is this?

c. What would be the exact total she owned if she paid the bill late?

d. Round your answer to part **c** to the nearest penny. *(Lessons 1-4, 2-5)*

26. Between what two integers is 3.4%? (Watch out! Many students miss this one!) *(Lesson 2-4)*

27. Write 4,500,500,500 in scientific notation. *(Lesson 2-3)*

28. Calculate 5 to the 7th power. *(Lesson 2-2)*

29. There is a street in Great Britain that is only $19\frac{5}{16}$ inches wide.

a. Accurately place $19\frac{5}{16}$ on this number line.

b. If you were 19.3 inches wide, could you walk down this street? *(Lessons 1-7, 1-9)*

What is "Squeeze-Belly Alley?" *That is the name given to the passageway pictured above and described in Question 29. It is Great Britain's narrowest street and is located in Port Isaac, Cornwall.*

30. Change 76.23 to
a. a percent.   **b.** a fraction. *(Lesson 2-6)*

In 31 and 32, tell whether a high or a low estimate would be preferred and why. *(Lesson 1-3)*

31. An airline estimates how much baggage an airplane can carry without being overloaded.

32. A caterer estimates how much food to prepare for a graduation party.

Exploration

33. **a.** On your calculator, what is the smallest positive number that can be displayed?
b. What is the largest negative number that can be displayed?

A project presents an opportunity for you to extend your knowledge of a topic related to the material of this chapter. You should allow more time for a project than you do for typical homework questions.

PROJECTS
CHAPTER TWO

1 One Million Pennies

How tall might a stack of one million pennies be?

a. Guess at the height, and record your guess.

b. Now try to determine the height without guessing. Of course, you will not be able to actually use a million pennies, so you must use an indirect method. Measure the thickness of ten pennies. Based on this measurement, how tall is a stack of one million pennies?

c. Write a paragraph in which you tell how you made your determination, and compare the calculated height with your guess.

d. Think of some related questions. If the pennies were put down next to each other, how far would the million pennies stretch? Imagine a million of some other object. Indicate how high you think they would be if placed in a stack.

2 Using the Symbols 1, 2, 3, and a Decimal Point

How many different decimal numbers can you make with the digits 1, 2, and 3 and a decimal point, if you can use any or all of the digits? (One such number is 1.23, and obviously there are many others.) Make a list of your numbers from smallest to largest, and explain how you know that you have found all the possibilities. Then explore what would happen if you also included the digit 4.

3 Percents in Print

Look through a newspaper or magazine for twenty-five uses of the word *percent* or the symbol %. Make sure all the percents are different. Order the percents (now all different) from smallest to largest, and describe how each one was used. (For instance, you might write: "3.75%—rate on a savings account.") In a paragraph, summarize what you have found.

4 Powers of 10

A book entitled *Powers of Ten* is published by Scientific American Books, Inc. Find a copy of this book in your school or public library, and write a book report on it. (The book is based on a short film also entitled *Powers of Ten*. You may be able to find that, also.)

Michael Jordan scored a total of 2.1541×10^4 points in 6.67×10^2 games—and led the NBA in scoring for 7 consecutive seasons.

5 Scientific Notation

Explore what would happen if we wrote all numbers in scientific notation. Write a story, or copy a published story that has many numbers (at least 10) in it. It is best if some of the numbers are big, some small. Change every number in the story to scientific notation. For instance, if you were writing about Shiing-Shen Chern, one of the greatest mathematicians of the 20th century, you would write that he was born in the year 1.911×10^3.

Groucho, Harpo, and Chico appeared as the Marx Brothers in 1.3×10^2 movies.

Lucille Ball appeared in 1.92×10^2 episodes for a total of 6.15×10^3 minutes of I Love Lucy.

Mother Teresa received the Pope John XXIII Peace Prize in 1.971×10^3 for her work with the poor.

Al Unser, Jr., set the world record for a 5.00×10^2–mile car race in 1990 with an average speed of 1.897×10^2 mph.

SUMMARY

In Chapter 1, you learned three advantages of the decimal system. (1) All of the most common numbers can be written as decimals. (2) Decimals are easy to order. (3) Decimals are used by calculators to represent numbers. In this chapter, decimals are combined with other notations to write large and small numbers.

Positive integer powers of numbers arise from repeated multiplication. For instance, $7^3 = 7 \times 7 \times 7$. The numbers 10, 100, 1000, . . . (or 10^1, 10^2, 10^3, . . .) are positive integer powers of 10. The numbers $\frac{1}{10}, \frac{1}{100}, \frac{1}{1000}$ . . . or their decimal equivalents 0.1, 0.01, 0.001, . . . (or 10^{-1}, 10^{-2}, 10^{-3}, . . .) are negative integer powers of 10. The number one is the zero power of 10.

The decimal system is based on powers of the number 10. So it is easy to multiply decimals by powers of 10. To multiply a decimal by a positive (or negative) power of 10, move the decimal point to the right (or left) the same number of places as the value of the exponent.

Percent means multiply by $\frac{1}{100}$, so decimals can easily be converted to percents. To find a percent of a number, multiply the percent by that number. Percents, decimals, and fractions are all used often and sometimes interchangeably. So it is useful to be able to convert from one form to another.

Information involving percents and fractions is often represented visually using circle graphs. Sectors in the circle graph represent parts of a whole quantity.

Large and small numbers are also often written in exponential form. Scientific notation combines exponential form with decimal notation. Scientific notation is a standard way used all over the world to express very large or very small numbers.

VOCABULARY

You should be able to give a general description and a specific example of each of the following ideas.

Lesson 2-1
million, billion, trillion, and so on

Lesson 2-2
base, exponent, power
x^y, y^x

Lesson 2-3
scientific notation for large numbers, light-year
EE, exp

Lesson 2-4
percent, %

Lesson 2-5
Substitution Principle

Lesson 2-6
terminating decimal

Lesson 2-7
circle graph, pie chart
$\approx$
circle, center, radius, diameter
sector, arc

Lesson 2-8
exponential form
millionth, billionth, trillionth, and so on

Lesson 2-9
scientific notation for small numbers

PROGRESS SELF-TEST

Take this test as you would take a test in class. Then check your work with the solutions in the Selected Answers section in the back of the book.

In 1–10, write as a single decimal.

1. $100{,}000{,}000 \times 23.51864$

2. 34% of 600

3. 32 billionths

4. 824.59×0.00001

5. $\frac{1}{1000} \times 77$

6. 3456.8910×10^5

7. 2.816×10^{-3}

8. 10^{-7}

9. 8%

10. 6^3

In 11–13, consider 125^6.

11. 125 is called the __?__ and 6 the __?__.

12. a. What key sequence can you use to calculate this on your calculator?

   b. What is the resulting display?

13. Give a decimal estimate for 125^6.

14. What is the distance from a point on a circle to the center of the circle called?

15. What percent does a whole circle graph represent?

In 16 and 17, write in scientific notation.

16. 21,070,000,000

17. 0.00000 008

In 18 and 19, write a key sequence you can use to enter the number on your calculator.

18. 4.5×10^{13}

19. 0.00000 01234 56

20. Order from smallest to largest:
   4^4 5^3 3^5.

21. Between what two integers is 40%?

22. Rewrite 4.73 as:

   a. a simple fraction   **b.** a percent.

23. As a fraction, $33\frac{1}{3}\% =$ __?__.

24. Consider the circle graph below.

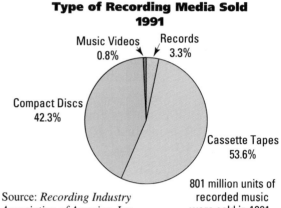

Type of Recording Media Sold 1991

Music Videos 0.8%
Records 3.3%
Compact Discs 42.3%
Cassette Tapes 53.6%

Source: *Recording Industry Association of America, Inc.*

801 million units of recorded music were sold in 1991.

   a. Which type of recording medium had the most sales?

   b. What percent of recorded media sold was not on compact disc?

   c. About how many records were sold in 1991?

25. A recent survey of 150 chefs reported that 30% of the chefs think broccoli is the top vegetable. How many chefs is this?

26. A computer system is on sale at 25% off. If the regular price is $1699, what is the sale price to the nearest dollar?

27. Julio correctly answered 80% of the items on a 20-item test. How many did he miss?

28. Why is 22.4×10^3 not in scientific notation?

29. What power of 10 equals one million?

30. According to the Substitution Principle, $\frac{3}{5} - \frac{1}{10} =$ __?__ % − __?__ %.

31. It is estimated that a swarm of 250 billion locusts descended on the Red Sea in 1889. Write this number in scientific notation.

After taking and correcting the Self-Test, you may want to make a list of the problems you got wrong. Then write down what you need to study most. If you can, try to explain your most frequent or common mistakes. Use what you write to help you study and review the chapter.

CHAPTER REVIEW

Questions on SPUR Objectives

SPUR stands for **S**kills, **P**roperties, **U**ses, and **R**epresentations. The Chapter Review questions are grouped according to the SPUR Objectives for this chapter.

SKILLS DEAL WITH THE PROCEDURES USED TO GET ANSWERS.

Objective A. *Multiply by 10, 100, 1000, and so on.* *(Lessons 2-1, 2-2)*

1. $32 \times 10,000 = \underline{\ ?\ }$
2. $100 \times 7.5 = \underline{\ ?\ }$
3. $1,000,000 \times 0.025 = \underline{\ ?\ }$
4. What number is 3.5 multiplied by to get 3500?
5. $3 \times 10^7 = \underline{\ ?\ }$ 6. $0.42 \times 10^5 = \underline{\ ?\ }$

Objective B. *Convert word names for numbers to decimals.* *(Lessons 2-1, 2-8)*

7. In 1991, Boeing had sales of 29.314 billion dollars. Write this number as a decimal.
8. Write 2 trillion as a decimal.
9. Write 4.6 millionths as a decimal.
10. A supercomputer can process a bit of information in a billionth of a second. Write this number as a decimal.

Objective C. *Convert powers to decimals.* *(Lesson 2-2)*

11. Convert 4^3 to a decimal.
12. Convert 72.5^2 to a decimal.

In 13 and 14, estimate by a decimal.
13. 12^8
14. 3.86^4

Objective D. *Give decimals and English word names for positive and negative integer powers of 10.* *(Lessons 2-2, 2-8)*

In 15–18, write the number as a decimal.
15. 10^5
16. 7.34×10^0
17. 10^{-4}
18. 10^1

19. In English, 10^9 is $\underline{\ ?\ }$.
20. In English, 10^{-2} is $\underline{\ ?\ }$.
21. One trillion is what power of ten?
22. 0.0001 is what power of ten?

Objective E. *Multiply by 0.1, 0.01, 0.001, and so on, and by $\frac{1}{10}, \frac{1}{100}, \frac{1}{1000}$, and so on.* *(Lesson 2-4)*

23. $2.73 \times 0.00000001 = \underline{\ ?\ }$
24. $495 \times 0.1 = \underline{\ ?\ }$
25. $75 \times \frac{1}{1000} = \underline{\ ?\ }$
26. $2.1 \times \frac{1}{100} = \underline{\ ?\ }$
27. $80 \times 10^{-1} = \underline{\ ?\ }$
28. $68.3 \times 10^{-4} = \underline{\ ?\ }$

Objective F. *Convert large and small positive numbers into scientific notation.* *(Lessons 2-3, 2-9)*

In 29–32, write the number in scientific notation.
29. 480,000
30. 9,000,000,000,000,000
31. 0.00013
32. 0.7
33. The number of nonhuman living things on Earth is estimated at 3×10^{33}. Write this number as a decimal.
34. A piece of paper is about 0.005 inches thick. What is that in scientific notation?

Objective G. *Convert and operate with percents as decimals.* *(Lessons 2-4, 2-5)*

In 35–38, write the percent as a decimal.

35. 15%
36. 5.25%
37. 9%
38. 200%
39. What is 50% of 150?
40. What is 3% of 3?
41. What is 100% of 6.2?
42. What is 7.8% of 3500?

Objective H. *Know common fraction and percent equivalents.* *(Lesson 2-4)*

43. Change $\frac{1}{2}$ to a percent.
44. Change $\frac{4}{5}$ to a percent.
45. What fraction equals 30%?
46. What fraction equals $66\frac{2}{3}\%$?

Objective I. *Convert terminating decimals to fractions, and either of these to percents.* *(Lesson 2-6)*

47. Find a simple fraction equal to 5.7.
48. Find the simple fraction in lowest terms equal to 0.892.
49. Convert 0.86 to percent.
50. Convert 3.2 to percent.
51. Convert $\frac{3}{7}$ to percent.
52. Convert $\frac{11}{8}$ to percent.

Objective J. *Identify the center, radius, and diameter of a circle.* *(Lesson 2-7)*

In 53–55, consider the circle below.

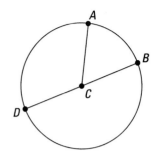

53. Name a radius.
54. Name the diameter.
55. Name the center.
56. Name a sector.
57. Name an arc of the circle.
58. The diameter of a circle is ___?___ as long as the radius of that circle.

Shown are irrigation circles in San Luis Valley, Colorado. They are formed by water sprayed from a pivotal irrigation system.

PROPERTIES DEAL WITH THE PRINCIPLES BEHIND THE MATHEMATICS.

Objective K. *Know and apply the Substitution Principle.* *(Lesson 2-5)*

59. State the Substitution Principle.
60. How is the Substitution Principle used in evaluating 75% of 40?
61. Name two numbers that could be substituted for 50%.
62. According to the Substitution Principle, $\frac{1}{2} + \frac{1}{4} =$ ___?___ % + ___?___ %.

Objective L. *Identify numbers as being written in scientific notation.* *(Lessons 2-3, 2-9)*

63. Why is 23×10^4 not in scientific notation?
64. In scientific notation, a number greater than or equal to ___?___ and less than ___?___ is multiplied by an ___?___ power of 10.

USES DEAL WITH APPLICATIONS OF MATHEMATICS IN REAL SITUATIONS.

Objective M. *Find percents of quantities in real situations.* *(Lesson 2-5)*

65. At a "40%-off" sale, what will you pay for a $26.50 sweater?

66. Bill Clinton received about 43.01% of the votes cast in the 1992 presidential election. About 104,400,000 votes were cast. About how many votes did Clinton get?

67. The value of a one-carat colorless flawless diamond reached $64,000 in 1980. By October of 1990, the price had lost 61% of its 1980 value. What was the value in 1990?

68. In one town, sales tax is 7.75%. What sales tax will you pay on a $20 purchase?

REPRESENTATIONS DEAL WITH PICTURES, GRAPHS, OR OBJECTS THAT ILLUSTRATE CONCEPTS.

Objective N. *Indicate key sequences and displays for large and small positive numbers on a calculator.* *(Lessons 2-3, 2-9)*

69. What key sequence will enter 32 billion on your calculator?

70. What key sequence will enter one trillionth on your calculator?

71. a. What key sequence will enter 2^{45} on your calculator?

b. Write an approximation to 2^{45} in scientific notation.

72. If a calculator displays ‖4.73   08‖, what decimal is being shown?

In 73–76, use a calculator to find the answer.

73. Estimate $1,357,975 \times 24,681,086$.

74. Estimate $.0025 \times .00004567$.

75. What key sequence will enter 1.93×10^{-15}?

76. What key sequence will enter 3×10^{21}?

Objective O. *Interpret circle graphs.* *(Lesson 2-7)*

In 77–80, consider the circle graph below.

77. What percent of adults sleep 9.5 hours or more?

78. What percent sleep more than 6.5 hours?

79. What does the largest sector represent?

80. What does the whole circle graph represent?

Number of Hours Adults Sleep at Night

Z-Z-Z-Z

22% 6.5 hours or less

26% 6.5 to 7.5 hours

6% 9.5 hours or more

9% 8.5 to 9.5 hours

37% 7.5 to 8.5 hours

MEASUREMENT

The first units of length were based on the human body. Some of these units are shown in the picture below. For instance, a "hand" was the width of a person's palm. So the size of a hand differed from person to person. Initially, these rough units were sufficient for most purposes. But as time went on, more accurate units were needed. So units began to be *standardized*.

According to tradition, the *yard* originally was the distance from the tip of the nose of King Henry I of England (who reigned from 1100 to 1135) to the tips of his fingers. The *foot* is supposedly based on the foot of Charlemagne (who ruled France and neighboring areas from 768 to 814).

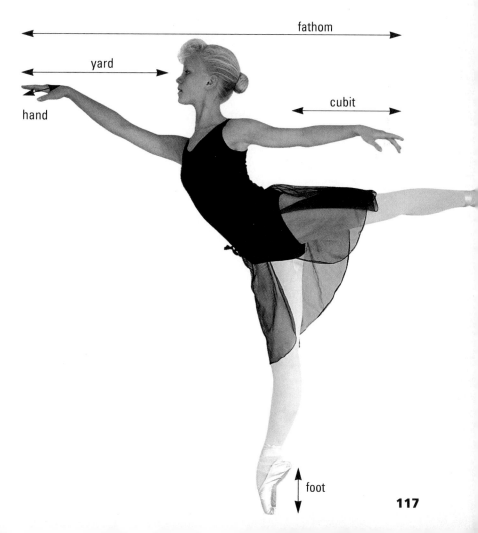

hand
yard
fathom
cubit
foot

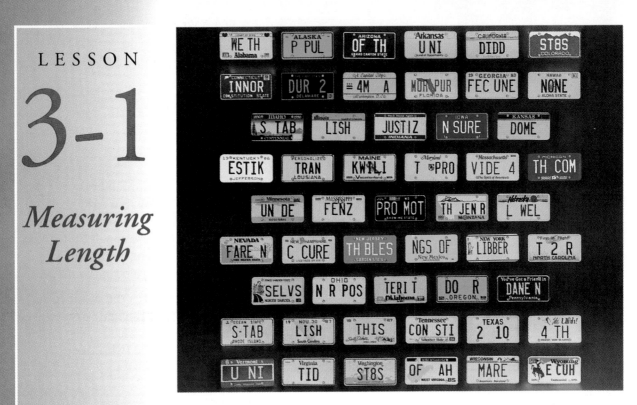

Poetic license. *As far back as the 1780s, when the preamble to the U.S. Constitution was being written, the nation's leaders were concerned with standardizing units of measure. Today, license plates have standardized dimensions. The automobile plates for all states must be 12″ by 6″.*

How Did Standardized Measuring Units Come to Be?

Around the year 1600, scientific experimentation began and still more accurate measurement was necessary. Scientists from different countries needed to be able to communicate with each other about their work. With the manufacturing of lenses and clocks, accurate measurement was needed outside of science. Around 1760, the Industrial Revolution began. Hand tools were replaced by power-driven machines. Accurate, consistent measurement was needed everywhere.

The writers of the U.S. Constitution in 1787 recognized the need for standardized units. One paragraph in the Constitution reads:

> *The Congress shall have power . . . to fix the standard of weights and measures.*

In 1790, Thomas Jefferson proposed to Congress a measuring system based on the number 10. This would closely relate the measuring system to the decimal system. Five years later the *metric system,* based on the number 10, was established in France. We could have been the first country with a measuring system based on the decimal system. But we were emotionally tied to England, which was at that time an enemy of France. So we adopted the system of measurement used in England instead. Not until 1866 did the metric system become legal in the United States.

At first the metric system was used mainly in science. But as years have gone by, it has been used in more and more fields and in more and more countries. The United States is the only large country in the world that has not officially converted to the metric system. The old "English system" or "British Imperial system" has evolved into the **U.S. system** or **customary system of measurement.** Today in the U.S., we measure in both the metric and U.S. systems.

All systems of measurement have units of length. For the metric system, the base unit of length is the **meter.** A **centimeter** is $\frac{1}{100}$, or .01, of a meter. For the U.S. system, the base unit of length is the **inch.** Here are the actual lengths of these units.

one inch   one centimeter

Measuring Lengths in Centimeters and Millimeters

The ruler pictured below is scaled in centimeters on the top and inches on the bottom.

Since a centimeter is .01 of a meter, there are 100 centimeters in one meter. On the ruler, each centimeter is divided into 10 parts, each a **millimeter,** or .001 meter. Since a millimeter is .001 meter, 1000 millimeters equal 1 meter. Since 1000 millimeters and 100 centimeters both make up a meter, 10 millimeters equal 1 centimeter. Each small interval on the top scale of the above ruler is 1 millimeter.

The three segments drawn below have lengths of about 6, 6.3, and 7 centimeters.

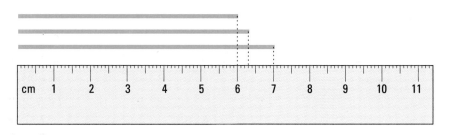

Example 1

Find the length of this small paper clip
a. to the nearest centimeter, and
b. to the nearest millimeter.

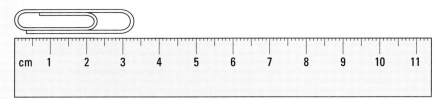

Solution

a. The numbered tick marks are 1 centimeter apart. So the length of the clip is **3 centimeters**, to the nearest centimeter.

b. The small tick marks are 0.1 centimeter apart. So the length of the clip is 3.3 centimeters, to the nearest tenth of a centimeter. To the nearest millimeter, the length is **33 millimeters**.

Activity 1

a. Measure the segment below to the nearest tenth of a centimeter.

——————

b. How many millimeters is this? (You should get a length somewhere between 20 and 30 millimeters.)

Measuring Lengths in Inches

On rulers, inches are usually divided into halves, fourths, eighths, and sixteenths. On most rulers, the tick marks have different lengths. The shortest interval is sixteenths.

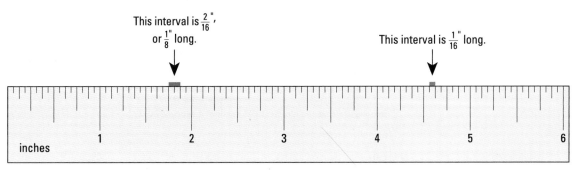

These equal fractions, $\frac{4}{16} = \frac{2}{8} = \frac{1}{4}$, $\frac{8}{16} = \frac{4}{8} = \frac{2}{4} = \frac{1}{2}$, $\frac{12}{16} = \frac{6}{8} = \frac{3}{4}$, and so on, are used in measuring inches. The segments drawn at the top of page 121 have lengths of about 1, $1\frac{1}{2}$, $1\frac{5}{8}$, and $1\frac{3}{4}$ inches.

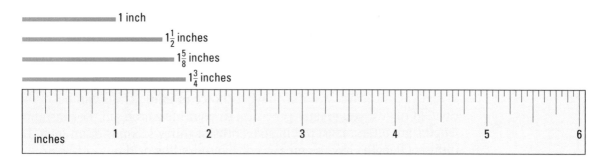

1 inch

$1\frac{1}{2}$ inches

$1\frac{5}{8}$ inches

$1\frac{3}{4}$ inches

inches   1   2   3   4   5   6

The above ruler is marked off in inches and fractions of inches.

Look at the segment that is $1\frac{3}{4}$ inches long. How do we know its length? The ruler's numbers show that its length is between 1 and 2 inches. The $\frac{3}{4}$ is not so easy to see. Notice that there are 16 intervals between 1 inch and 2 inches. So each interval is $\frac{1}{16}$ of an inch long. The segment reaches 12 intervals past 1 inch. So its length is $1\frac{12}{16}$ inches. Written in lowest terms, this is $1\frac{3}{4}$ inches.

Example 2

Find the height of this photograph:

a. to the nearest $\frac{1}{2}$ inch.

b. to the nearest $\frac{1}{4}$ inch.

c. to the nearest $\frac{1}{8}$ inch.

Modern Robin? *The photo is of some of the characters from 20th Century Fox's* Robin Hood: Men In Tights. *In this comedic version of the legend, Robin Hood is in the center, portrayed by Cary Elwes.*

Solution

The tick marks are $\frac{1}{16}$ in. apart, so the height is a little more than $2\frac{2}{16}$ in., or a little more than $2\frac{1}{8}$ in.

a. Rounded to the nearest $\frac{1}{2}$ in., $2\frac{1}{8}$ in. is **2 in.**

b. Rounded to the nearest $\frac{1}{4}$ in., a little more than $2\frac{1}{8}$ in. is $2\frac{1}{4}$ **in.**

c. Rounded to the nearest $\frac{1}{8}$ in., a little more than $2\frac{1}{8}$ in. is $2\frac{1}{8}$ **in.**

Activity 2

Measure the height of this page to the nearest eighth inch.

When you measure, you may have a choice of units. Units can be divided to give you greater precision in your measurement. For instance, lengths are often measured in sixteenths or thirty-seconds of an inch. In industry, lengths may be measured to hundredths or thousandths or even smaller parts of an inch. Whatever unit you work with, you are rounding to the nearest. The rounding unit (tenth of a centimeter, for example) indicates how precise your measurement is.

Example 3

Give a U.S. unit and a metric unit that would be appropriate for measuring the length of a pencil.

Solution

A pencil would probably be measured in inches or centimeters. A millimeter would be too tiny a unit; feet, yards, miles, meters, and kilometers would be too large.

A man for any era.
Thomas Jefferson (1743–1826) was an American revolutionary leader, a political philosopher, author of the Declaration of Independence, and third president of the U.S. He also pushed for public schools and libraries.

QUESTIONS

Covering the Reading

1. The first units of length were based on __?__.

2. The heights of horses are sometimes measured in hands. How was the length called a hand originally determined?

3. The yard originally was the distance from the __?__ to the __?__ of what king of England?

4. Whose foot is said to have been the foot from which today's foot originated?

5. Why did units become standardized?

6. About when did accurate lengths become needed everywhere?

In 7–10, *true or false.*

7. Thomas Jefferson wanted the United States to adopt the English system of measurement.

8. The metric system was established in France in 1795.

9. The metric system became legal in the United States over 100 years ago.

10. Congress has the power to set standards of measurement in the United States.

11. The base unit of length in the metric system is the __?__.

12. In Activity 1, give the length of the segment.
　　a. in centimeters　　　　　**b.** in millimeters

In 13–15, measure the length of the segment to the nearest millimeter.

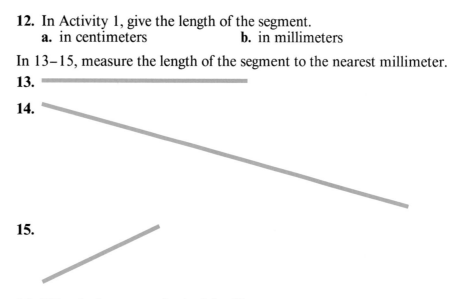

13.

14.

15.

16. What is the answer in Activity 2?

In 17–19, use this ruler to find the length of the segment to the nearest eighth of an inch. Write the length in lowest terms.

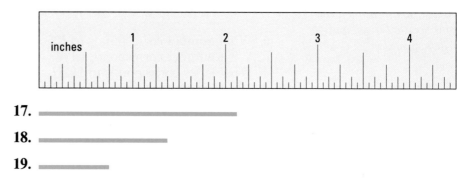

17.

18.

19.

Applying the Mathematics

In 20–24, *multiple choice.*　Match the quantity with an appropriate metric unit of measure.

20. length of a swimming pool　　　　　　　(a) millimeter

21. distance from London to Paris　　　　　　(b) centimeter

22. thickness of the cover of your math book　(c) meter

23. height of your desk　　　　　　　　　　(d) kilometer

24. length of your thumb

25. Draw a segment that is 6 cm long. Use your ruler to estimate its measure in inches.

26. Draw a vertical segment with length 3.5 inches.

27. Draw a horizontal segment with length 12.4 centimeters.

From top to bottom, the athletes shown are Andre Agassi; Kristi Yamaguchi; and Scottie Pippen and Michael Jordan.

28. Some people believe that if a sports team or individual appears on the cover of *Sports Illustrated,* then the team or individual becomes jinxed and will suffer a decline in performance. Researchers at the University of Southern California examined 271 covers. They found that 57.6% of the cover subjects improved in performance.

    a. Did more cover subjects improve or decline in performance?

    b. How many of the 271 cover subjects improved in performance? *(Lessons 2-4, 2-5)*

29. Since $\frac{1}{20} = 5\%$, what percent is equal to $\frac{9}{20}$? *(Lesson 2-4)*

30. Which is larger, 5×10^6 or 6×10^5? *(Lesson 2-3)*

31. Which is larger, 5 to the 6th power, or 6 to the 5th power? *(Lesson 2-2)*

32. Use the number line drawn here. The letters refer to points that are equally spaced. *(Lesson 1-2)*

```
      12.2            12.25
 ◄──────┬─────────────┬──────────────────►
     A B C D E F G H I J K L M N O P
```

    a. What number corresponds to *E*?
    b. What number corresponds to *N*?
    c. What letter corresponds to 12.3?
    d. 12.213 is graphed between which two points?

Exploration

33. There are many units of length that have specialized uses. Find out something about each of these specialized units of length.
    a. pica   **b.** ell
    c. link   **d.** chain

34. Look at the picture that opens this chapter. Find out what these lengths are today.
    a. fathom   **b.** cubit

3-2

Converting Lengths

A whale of a catch. *The coho salmon shown at the left is about 25 in., or 2 ft 1 in., or about $\frac{2}{3}$ yd long. It weighs about 22 lb. The largest coho salmon ever caught weighed 33 lb 4 oz.*

Converting Lengths Within the U.S. System

Carlos is 5 feet 3 inches tall. Theresa is 65 inches tall. Who is taller? To compare these measures, the units of measure need to be consistent.

First you need to know how these units of length are related. Three relationships are used so often you should memorize them if you do not know them already. We call them **conversion equations** because they have an equal sign and they are used for converting from one unit to another.

> **Conversion Equations for Customary Units of Length**
> 1 foot (ft) = 12 inches (in.)
> 1 yard (yd) = 3 feet
> 1 mile (mi) = 5280 feet

Except for the inch, it is now recommended that periods not be used in abbreviations for units of measure. Periods may be confused with decimals.

The conversion equations make it possible to convert from one unit to another by multiplying. The property that follows is extraordinarily useful.

> **Multiplication Property of Equality**
> When equal quantities are multiplied by the same number, the resulting quantities are equal.

Example 1

How many feet are in 1.7 miles?

Solution

Start with the conversion equation relating feet and miles.

$$1 \text{ mi} = 5280 \text{ ft}$$

Multiply both sides by 1.7.

$$1.7 \times 1 \text{ mile} = 1.7 \times 5280 \text{ ft}$$

$$1.7 \text{ miles} = 8976 \text{ feet.}$$

There is also an Addition Property of Equality. When the same quantity is added to equal quantities, the resulting sums are equal.

Example 2

Of the two people mentioned on page 125, who is taller, Carlos or Theresa?

Solution

Carlos's height is 5 feet 3 inches, so we convert this into inches.

Start with the conversion equation relating feet and inches.

$$1 \text{ foot} = 12 \text{ inches}$$

Multiply both sides by 5.

$$5 \times 1 \text{ foot} = 5 \times 12 \text{ inches}$$

$$5 \text{ feet} = 60 \text{ inches}$$

Add 3 inches to both sides.

$$5 \text{ feet } 3 \text{ inches} = 63 \text{ inches}$$

Since Theresa's height is 65 inches, she is taller than Carlos.

Sometimes more than one conversion is needed in a problem.

Example 3

How many inches are in a yard?

Solution

Start with the conversion equation relating yards to feet.

$$1 \text{ yard} = 3 \text{ feet}$$

Now convert 3 feet into inches.
Use the conversion equation relating feet and inches.

$$1 \text{ foot} = 12 \text{ inches}$$

Multiply both sides by 3.

$$3 \times 1 \text{ foot} = 3 \times 12 \text{ inches}$$

$$3 \text{ feet} = 36 \text{ inches}$$

This shows 1 yard = 3 feet = 36 inches. So there are 36 inches in a yard.

Converting from Inches to Centimeters

Because the metric system is now the worldwide system, the inch is officially defined (even in the U.S.) in terms of centimeters.

> **1 inch = 2.54 centimeters**

Because 1 inch is about 2.5 centimeters, 2 inches are about 5 centimeters. In general, it is easy to convert any number of inches to centimeters.

Activity

Determine exactly how many centimeters are in 4 inches. (Your answer should be close to 10.)

QUESTIONS

Covering the Reading

1. Name four units of length in the customary system.

In 2–6, complete the relationship.

2. 1 ft = __?__ in.

3. 1 yd = __?__ ft

4. 1 yd = __?__ in.

5. 1 mi = __?__ ft

6. 1 in. = __?__ cm

7. What value did you get for the Activity?

In 8–10, convert.

8. .62 mile to feet

9. 4 yards to feet

10. 10 inches to centimeters

11. Who is taller if Natalie says she is 4 feet 6 inches tall, and Nathan gives his height as 50 inches?

12. Why is it recommended to abbreviate feet *ft* rather than *ft.*?

13. Why is the inch defined in terms of the centimeter?

Applying the Mathematics

14. **a.** Measure the height of your eyes from the ground, in inches, by standing next to a wall and marking the height.
    b. Measure this height in centimeters.
    c. Convert the inches to centimeters to check whether your measurements are correct.

15. a. In the U.S. system, a *rod* is defined as $5\frac{1}{2}$ yards. How many feet are in one rod?

   b. A furlong is equal in length to 40 rods. How many feet is this?

   c. The Kentucky Derby, held every May in Louisville, Kentucky, was originally thought of as a 10-furlong horse race. How many feet long is the Kentucky Derby?

16. Some city blocks are $\frac{1}{8}$ mile long. Convert this to feet.

17. If your running stride is 2.5 feet long, how many strides will you take in running 100 yards?

18. Convert $7\frac{1}{2}$ yards to feet.

19. Convert 7 yards, 2 feet, 6 inches to inches.

20. a. How many inches are in a mile?
   b. How many centimeters are in a mile?

Review

21. Measure this segment
   a. to the nearest $\frac{1}{8}$ of an inch, and
   b. to the nearest centimeter.   *(Lesson 3-1)*

---

22. Measure the length of a dollar bill as indicated.
   a. to the nearest inch
   b. to the nearest fourth of an inch
   c. to the nearest eighth of an inch   *(Lesson 3-1)*

23. The National Autonomous University of Mexico has about 250,000 students. Since tuition had not been raised since 1948 and was equal to about six U.S. cents, in 1992 administrators announced a tuition increase of about 1 million percent!

   a. Write one million percent as a decimal.   *(Lesson 2-4)*

   b. Estimate the new tuition.   *(Lesson 2-5)*

   c. After the tuition raise was finalized, about how much total tuition did the university collect? Give your answer in both scientific notation and as a decimal.   *(Lesson 2-3)*

24. Which is larger, $\frac{3}{10}$ or $\frac{29}{97}$?   *(Lesson 1-8)*

25. a. Order -1, -2, and -1.5 from smallest to largest.
   b. Write the numbers in part **a** on one line with inequality signs between them.   *(Lesson 1-6)*

26. What number does the key sequence 89 +/- yield?   *(Lesson 1-5)*

27. Estimate 896.5555555555 + 7.96113 to the nearest hundred.
   (Lesson 1-4)

The National Autonomous University in Mexico City, originally opened in 1551, is Mexico's foremost center of higher learning. The Central Library, shown here, was built in 1951 and is completely covered with brilliant mosaics that include symbolic devices from early Mexican civilizations.

28. Having a frame of reference for a given measure can help you picture the length of a quantity. For example, an adult finger is about half an inch wide. Find something with a length that is about the same as the given unit.

a. inch   **b.** foot

c. yard   **d.** mile

e. millimeter   **f.** centimeter

g. meter

29. The computer program below instructs a computer to convert a length in miles to one in feet. The line numbers 10, 20, and so on, at left must be typed. The computer executes the program in the order of the line numbers, which can be any positive integers.

a. Type in the following.

```
NEW
10 PRINT "WHAT IS LENGTH IN MILES?"
20 INPUT NMILES
30 NFEET = 5280 * NMILES
40 PRINT "THE NUMBER OF FEET IS "NFEET
50 END
```

To see your program, type LIST and press RETURN. You can change any line by typing it over. You need not type the entire program again.

b. To run your program, type RUN and press RETURN. The computer will execute line 10 and ask you to input a number. Input 5 and press RETURN. The computer will then execute the rest of the program. What does the computer screen show?

c. Run the program a few times, with values of your own choosing. Write down the values you input and the answers the computer gives.

LESSON

3-3

Weight and Capacity in the Customary System of Measurement

It's worth its weight in gold. *Gold, like silver, is sold by the ounce. The price varies. In 1980, the price of gold was over $600 an ounce. In late 1993, the price was about $350 an ounce.*

In the customary or U.S. system of measurement, there are many units. So people refer to tables to check relationships between unfamiliar units. Still, you should know some relationships. Units of length were in the last lesson. Here are units of weight and capacity.

Conversion Equations for Customary Units of Weight and Capacity

For weight:
$$1 \text{ pound (lb)} = 16 \text{ ounces (oz)}$$
$$1 \text{ short ton} = 2000 \text{ pounds}$$

For capacity (liquid or dry volume):
$$1 \text{ cup (c)} = 8 \text{ fluid ounces (fl oz)}$$
$$1 \text{ pint (pt)} = 2 \text{ cups}$$
$$1 \text{ quart (qt)} = 2 \text{ pints}$$
$$1 \text{ gallon (gal)} = 4 \text{ quarts}$$

Notice that the word *ounce* appears in measures of both weight and capacity. To avoid confusion, the capacity measure is usually given in *fluid ounces*. The weight of one fluid ounce of water is one ounce, but this is not true for every liquid. You can tell whether weight or capacity is being measured by the context of the problem.

Example 1

A punch recipe calls for $3\frac{1}{2}$ pints of different juices. How many 1-cup servings will it make?

Solution

Start with the conversion equation relating pints and cups.   1 pint = 2 cups

Multiply both sides by 3.5.   3.5 × 1 pint = 3.5 × 2 cups

3.5 pints = 7 cups

3.5 pints = 7 cups, so there are 7 servings.

Sometimes many conversion equations are needed in the same problem.

Example 2

How many cups are in a gallon?

Solution

Start with the conversion equation for gallons and quarts.	1 gallon = 4 quarts

Now convert 4 quarts to pints.
Multiply both sides by 4.

1 quart = 2 pints
4 x 1 quart = 4 x 2 pints
4 quarts = 8 pints

Lastly, convert 8 pints to cups.

1 pint = 2 cups
8 x 1 pint = 8 x 2 cups
8 pints = 16 cups

This shows that 1 gal = 4 qt = 8 pt = 16 cups. There are 16 cups in a gallon.

When there are many conversions in the same question, you may find it useful to organize your work as shown in Example 3.

Example 3

Mineral water comes in half-gallon jugs or a six-pack of 12-ounce bottles. Which contains more water? How much more?

Solution

To compare, we need both quantities in the same units. To avoid fractions, choose ounces. The six-pack of 12-ounce bottles contains a total of 6 × 12, or 72 ounces. For the half-gallon jugs, we need to convert gallons to quarts to pints to cups to (fluid) ounces.

1 gallon = 4 quarts, so
$\frac{1}{2}$ gallon = 2 quarts.

Since 1 quart = 2 pints,
2 quarts = 4 pints.

Now 1 pint = 2 cups,
so 4 pints = 8 cups.

Finally, 1 cup = 8 fluid ounces,
so 8 cups = 64 fluid ounces.

This shows that $\frac{1}{2}$ gal = 2 qt = 4 pt = 8c = 64 oz.

The six-pack contains 72 fluid ounces; the jug contains 64 fluid ounces. Thus the six-pack contains 8 fluid ounces more water.

What's in mineral water? *Mineral water often contains such minerals as calcium, chlorine, iron, magnesium, potassium, sodium, and sulphur. Mineral water is consumed as a beverage and is used to help cure ailments.*

QUESTIONS

Covering the Reading

In 1 and 2, consider the customary system of measurement.

1. Name five units in which an amount of milk could be measured.

2. Name three units of weight.

In 3–6, copy and complete the conversion equation.

3. 1 gallon = _?_ quarts

4. 1 quart = _?_ pints

5. 1 lb = _?_ oz

6. 1 short ton = _?_ pounds

7. How many cups are in 3 quarts?

In 8–11, do the conversions.

8. 7 tons to pounds

9. 8.3 gallons to quarts

10. $1\frac{1}{2}$ cups to fluid ounces

11. $1\frac{1}{2}$ pounds to ounces

12. What sequence of conversions would change gallons to ounces?

Applying the Mathematics

13. In Great Britain, one *gross ton* = one *long ton* = 2240 pounds. The heaviest ship ever built is the oil tanker *Jahre Viking* which weighs 622,420 long tons when fully loaded. How many pounds does the *Jahre Viking* weigh?

14. A quart contains how many fluid ounces?

In 15–17, name the appropriate customary unit for each quantity.

15. the amount of flour in a recipe for cookies

16. the weight of an elephant

17. the amount of gas in a car gas tank

18. You have a one-pint measuring vessel. How many times should you use it to fill a 10-gallon tank?

19. A warning sign on a bridge is at the right. Can a truck weighing 8000 pounds safely cross the bridge?

20. Which weighs more, a quarter-pound or a 6-ounce hamburger?

WARNING
MAXIMUM LOAD
$3\frac{1}{2}$ TONS

21. Refer to the cartoon below. What should the chief cook tell Zero?

22. Convert 6 yards $2\frac{1}{2}$ feet to inches. *(Lesson 3-2)*

23. How many feet are in a half mile? *(Lesson 3-2)*

24. a. Find a door. Measure its height to the nearest inch.
   b. Will 6-foot 6-inch Michael Jordan be able to walk through without ducking? *(Lessons 3-1, 3-2)*

25. To the nearest $\frac{1}{16}$ of an inch, how long is this pencil? *(Lesson 3-1)*

In 26 and 27, write the number in scientific notation. *(Lessons 2-3, 2-9)*

26. 117,490,000—the estimated population of Pakistan in 1991

27. 0.013837 in.—the length of 1 point in typesetting

28. 0% of the 500 students in a school are traveling to a game by bus. How many students is this? *(Lesson 2-5)*

In 29–32, calculate in your head. *(Lessons 2-1, 2-4)*

29. 0.052×100 **30.** $3.446 \times .0001$

31. $15.36 \times .1$ **32.** $640 \times 10,000$

Exploration

The time is right. *The National Institute of Standards of Technology has an atomic clock so accurate that it will neither gain nor lose one second in a million years. The clock is shown here with one of the developers, John P. Lowe. He is wearing an eyepiece that makes the clock's laser light visible.*

33. All systems of measurement in common use today have the same units for time: 1 hour = 60 minutes, 1 minute = 60 seconds.
   a. How many seconds are in an hour?
   b. How many seconds are in a day?
   c. How many seconds are in a 365-day year?
   d. How many minutes are in a year?
   e. If your heart beats 70 times a minute, how many times does it beat in a year?
   f. If a heart beats 70 times a minute, how many times will it beat in 79 years, the average lifetime of a woman in the U.S.?

34. An old song goes, "I love you, a bushel and a peck, a bushel and a peck and a hug around your neck, a hug around your neck and a barrel and a heap, a barrel and a heap and I'm talking in my sleep about you. . . ." Three units of capacity of fruits and grains are in the words to the song. What are they and how are they related?

3-4

The Metric System of Measurement

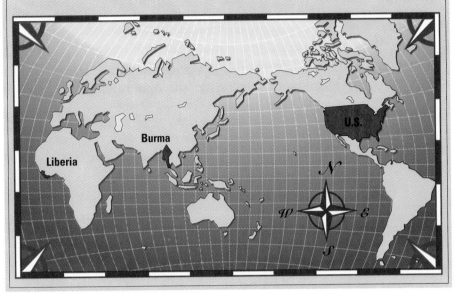

Countries Still Using Customary Units

Not much company. *The only countries still using the customary system are Burma, Liberia, and the U.S. But even in the U.S., all new NASA projects and $1 billion in new construction for the General Services Administration use metrics.*

Why Was the Metric System Developed?

The U.S. system of measurement has three major weaknesses. First, the many units have names that do not help you know how the units are related. Second, the units are multiples of each other in no consistent manner. To see this, look again at the conversion equations in Lessons 3-2 and 3-3. You see the numbers 12, 3, 5280, 16, 2000, 8, 2, and 4. Third, in the decimal system these numbers are not as easy to work with as powers of 10 such as 100, 1000, 10,000, or .1, .01, and .001.

Other older measurement systems had the same weaknesses. So, in the late 1700s, a movement arose to design a better measurement system. The system devised is called the **international** or **metric system of measurement.** It is based on the decimal system and is by far the most widely used system of measurement in the world.

Some Important Units in the Metric System

Metric prefixes have fixed meanings related to place values in the decimal system. The table on page 137 identifies many of the prefixes. You have seen the three most common prefixes in units for measuring length: kilo- (1000), centi- $\left(\frac{1}{100} \text{ or } .01\right)$, and milli- $\left(\frac{1}{1000} \text{ or } .001\right)$.

The base unit of length is the **meter,** abbreviated **m.**

$$1 \text{ kilometer} = 1 \text{ km} = 1000 \text{ m}$$
$$1 \text{ centimeter} = 1 \text{ cm} = \frac{1}{100}, \text{ or } .01 \text{ m}$$
$$1 \text{ millimeter} = 1 \text{ mm} = \frac{1}{1000}, \text{ or } .001 \text{ m}$$

Units of mass are multiples of the **gram,** abbreviated **g.** In everyday usage, the gram is also used to measure weight.

$$1 \text{ kilogram} = 1 \text{ kg} = 1000 \text{ g}$$
$$1 \text{ milligram} = 1 \text{ mg} = \frac{1}{1000}, \text{ or } .001, \text{ g}$$

The **liter,** abbreviated **L,** and milliliter (**mL**) are used to measure capacity or volume. Soft drinks today are often sold in 2-liter bottles. Smaller amounts are measured in milliliters.

$$1 \text{ milliliter} = 1 \text{mL} = \frac{1}{1000}, \text{ or } .001, \text{ L}$$

Converting Within the Metric System

All conversions within the metric system can be done without a calculator because the multiples are powers of 10.

Example 1

How many meters are in 3.46 kilometers?

Solution

Start with the conversion equation for 1 km. It is easy to remember because kilo- means 1000.	1 km = 1000 m
Multiply both sides by 3.46.	3.46 × 1 km = 3.46 × 1000 m
Calculate.	3.46 km = 3460 m

Example 2

Change 89 milligrams to grams.

Solution

Start with the conversion equation for 1 mg.	
Remember that milli- means $\frac{1}{1000}$ or .001.	1 mg = .001 g
Multiply both sides by 89.	89 × 1 mg = 89 × .001 g
	89 mg = .089 g

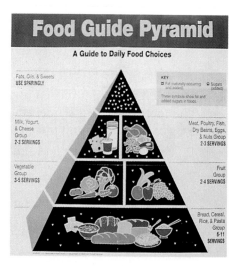

Every gram counts. *The Food Guide Pyramid suggests that everyone eat a variety of foods each day. Particularly good news for cereal lovers is that cereals are highly recommended in the grains category, as long as they contain no more than 2 grams of fat or 6 grams of sugar.*

Many of the questions refer to this table. You should study it before reading the questions. The most commonly used units are in **bold** type.

The International or Metric System of Measurement

place value	thousands	hundreds	tens	ones	tenths	hundredths	thousandths
power of 10	10^3	10^2	10^1	10^0	10^{-1}	10^{-2}	10^{-3}
unit of length	**kilometer**	hectometer	dekameter	**meter**	decimeter	**centimeter**	**millimeter**
unit of mass	**kilogram**	hectogram	dekagram	**gram**	decigram	centigram	**milligram**
unit of capacity	kiloliter	hectoliter	dekaliter	**liter**	deciliter	centiliter	**milliliter**

Some Other Prefixes:

place value	trillions	billions	millions		millionths	billionths	trillionths
power of 10	10^{12}	10^9	10^6		10^{-6}	10^{-9}	10^{-12}
prefix	tera-	giga-	mega-		micro-	nano-	pico-

Some Common Units:

Length: **kilometer (km)**—used for distances between towns and cities. 1 km ≈ 0.62 mi. The length of 10 football fields is about 1 kilometer.
meter (m)—used for rooms, heights, and fabrics. A doorknob is about 1 m high.
centimeter (cm)—used for small items. The diameter of an aspirin tablet is about 1 cm.
millimeter (mm)—used for very small items. The thickness of a dime is about 1 mm.

Mass: **kilogram (kg)**—used for heavier weights. A quart of milk weighs about 1 kg.
gram (g)—used for light items. An aspirin tablet weighs about 1 g.
milligram (mg)—used for very light items, such as vitamin content in food. A speck of sawdust weighs about 1 mg.

Capacity: **liter (L)**—used for soft drinks and other liquids. 1 liter ≈ 1.06 qt.
milliliter (mL)—used for small amounts such as perfume. 1 teaspoon ≈ 5 mL.

QUESTIONS

Covering the Reading

1. Name two weaknesses of the U.S. system of measurement.

2. Another name for the metric system is __?__.

3. Name three common units of length in the metric system.

4. Name three common units of mass in the metric system.

5. Name two common units of capacity in the metric system.

In 6–8, give the meaning of the prefix.

6. kilo-   **7.** milli-   **8.** centi-

In 9–11, write the abbreviation for the unit.

9. centimeter   **10.** kilogram   **11.** milliliter

In 12–17, convert.

12. 90 cm to meters   **13.** 345 mL to liters   **14.** 5 kg to grams

15. 10 km to meters   **16.** 48 mm to meters   **17.** 60 mg to grams

18. Name something weighing approximately the given weight.
     a. 1 kg   **b.** 1 g   **c.** 1 mg

19. Name something about as long as the given length.
     a. 1 m   **b.** 1 km   **c.** 1 mm

20. Name something with about as much liquid in it as 1 liter.

In 21–23, give the power of 10 associated with the prefix.

21. centi-   **22.** milli-   **23.** kilo-

Applying the Mathematics

In 24–26, choose the one best answer.

24. A high school freshman might weigh:
     (a) 50 g   (b) 50 mg   (c) 50 kg   (d) 500 g

25. A high school freshman might be how tall?
     (a) .7 m   (b) 1.7 m   (c) 2.2 m   (d) 5.6 m

26. A common dimension of camera film is:
     (a) 35 km   (b) 35 cm   (c) 35 m   (d) 35 mm

27. Should most students be able to walk one kilometer in an hour? Explain your answer.

What country? *Signs like this one are posted in Australia, the native home of koalas.*

28. The atomic bomb that the U.S. exploded on Hiroshima, Japan, in 1945 had a force equivalent to about 12.5 kilotons of TNT. How many tons is this?

29. A millisecond is how many seconds?

30. The United States was the first country in the world (1792) to have a money system based on decimals. In this system:
$$1 \text{ dollar} = 100 \text{ cents}$$
or equivalently, $1 \text{ cent} = \frac{1}{100}$, or .01, dollar.

     a. Convert 56¢ to dollars.

     b. Convert $13.49 to cents.

     c. On September 17, 1983, UPI reported that a truck loaded with 7.6 million new pennies overturned on Interstate 80 in the mountains north of Sacramento, California. How many dollars is this?

31. You have a 1-cup container. How many times would you have to use it to fill a half-gallon jug? *(Lesson 3-3)*

In 32–35, complete each with the correct symbol <, =, or >.
(Lessons 3-2, 3-3)

32. $2\frac{1}{2}$ pints __?__ 4 quarts

33. $3\frac{1}{2}$ feet __?__ 2 yards

34. 2850 feet __?__ 1 mile

35. 1 lb __?__ 16 oz

36. a. Find the length of the segment below to the nearest 0.1 cm.

b. Write the length in millimeters. *(Lesson 3-1)*

37. Which measurement is more precise, one made to the nearest $\frac{1}{16}$ of an inch or one made to the nearest $\frac{1}{10}$ of an inch? Explain why.
(Lessons 1-4, 3-1)

38. Order from smallest to largest: 10^{-4} 0 $\frac{1}{100}$. *(Lessons 1-2, 2-8)*

39. The results of a survey of teenagers are shown at the right. Suppose there are 30 teenagers in your class, and they are representative of the students in the survey. How many of them would you expect to own each item?

Survey Results	
	Percent of teenagers who own:
bicycle	85%
camera	80%
designer clothes	72%
TV set	52%

    a. bicycle

    b. camera

    c. designer clothes *(Lesson 2-5)*

40. Order from smallest to largest: 5^2 2^5 10^1. *(Lesson 2-2)*

41. Write $\frac{2}{1000}$ cm, the approximate diameter of a cloud droplet, as a decimal. *(Lesson 1-8)*

42. Almost every country in the world today has a decimal money system. Given are relationships between monetary units. Name a country in which these units are used.

    a. 1 franc = 100 centimes   **b.** 1 centavo = 100 pesos

    c. 1 ruble = 100 kopecks   **d.** 1 yuan = 100 fen

    e. 1 dinar = 1000 fils   **f.** 1 rupee = 100 paise

    g. 1 cedi = 100 pesewas

3-5

Converting Between Systems

A sporting chance. *Wrestlers in the Grand Sumo tournament at Kokugikan Hall, Tokyo, are known for their strong, immense bodies. Seven sumo wrestlers (rikishi) together, at 135 kg (about 300 lb) or more each, weigh more than a ton.*

Because using the metric system is so easy, almost every country in the world has adopted it. In the United States, science, medicine, and photography are almost all metric. Carpentry and other building trades usually use the U.S. system. The trend is to use metric units more often as time goes by. Some automobiles are manufactured with parts that conform to metric units. Others use customary units. Auto mechanics need tools for each system.

With two systems in use today, it is occasionally necessary to change from units in one system to units in the other. This change is called *converting between systems.* Converting between systems is like converting within one system. However, the numbers are usually not whole numbers.

Converting from Metric Units to Customary Units

You already know 1 inch = 2.54 centimeters exactly. Other conversions between the metric and U.S. systems are approximate. Remember that ≈ means "is approximately equal to."

Conversion Relations for Metric and U.S. Units		
1 km	≈	.62 mi
1 kilogram	≈	2.2 pounds
1 liter	≈	1.06 quarts

Conversions may be done with these approximate relationships just as they are done with the exact relationships you saw in earlier lessons.

These Swedish girls might write to their pen pals and describe their enjoyment of St. Lucia Day, the Festival of Light.

Example 1

Your Swedish pen pal writes you that she weighs 50 kg. How much is this in pounds?

Solution

Start with the conversion relation relating pounds and kilograms.	1 kg ≈ 2.2 pounds
Multiply both sides by 50.	50 × 1 kg ≈ 50 × 2.2 pounds
	50 kg ≈ 110 pounds

Example 2

A 10K race is ten kilometers long. How long is this in miles?

Solution

Start with the conversion relation between kilometers and miles.	1 km ≈ 0.62 mi
Multiply both sides by 10.	10 × 1 km ≈ 10 × 0.62 mi
Compute.	10 km ≈ 6.2 mi

A 10K race is about 6.2 miles long.

You can use approximate and exact conversions in the same problem.

Example 3

Soft drinks often come in 2-liter bottles. About how many fluid ounces are in a 2-liter bottle?

Solution

The only conversion relation for capacity that you have seen is between liters and quarts.

1 L ≈ 1.06 qt

Multiply both sides by 2.

2 L ≈ 2.12 qt

Now change 2.12 quarts to fluid ounces. This is like what was done in Lesson 3-3.

Since   1 qt = 2 pt,
    2.12 qt = 4.24 pt.

       Since 1 pt = 2 c,
        4.24 pt = 8.48 c.

            Since 1 c =   8 fl oz,
            8.48 c = 67.84 fl oz.

So 2 liters ≈ 2.12 quarts = 4.24 pints = 8.48 cups = 67.48 ounces. There are about 67.48 fluid ounces in a 2-liter bottle.

Converting from One Currency to Another

Most countries have their own *monetary systems* with their own units for money. For instance, Mexico uses pesos. Canada uses the Canadian dollar, whose value is different from the U.S. dollar. If you travel to these or other countries, you will see prices in their units. Most American travelers want to know what these prices are in U.S. dollars, so they know what they are spending.

Example 4

In March of 1993, one Mexican peso was equal in value to about 32¢. If a serape cost 80 pesos at that time, what was its cost in dollars?

Solution

From the given information, the conversion relation is

$$1 \ peso \approx \$.32.$$

Multiply both sides by 80.

$$80 \times 1 \ peso \approx 80 \times \$.32$$

$$80 \ pesos \approx \$25.60$$

The cost of the serape was about $25.60.

Celebrate! *This Latino youth is participating in the Mexican* Cinco de Mayo *celebration. Serapes, like the brightly-colored one the youth is wearing, were once part of the* charro *or horseman's costume.*

QUESTIONS

Covering the Reading

1. Name two professions in which a person in the U.S. would use the metric system more than the U.S. system.

2. Name two professions in which a person in the U.S. currently would use U.S. units more than metric units.

3. What does the $\approx$ sign mean?

4. What metric unit is a little larger than two pounds?

5. What U.S. unit is most like a liter?

6. One relationship between the U.S. and metric system is exact. What relationship is it?

In 7–9, give a relationship between:

7. kilograms and pounds.   8. liters and quarts.

9. kilometers and miles.

10. 30K is the length of the longest cross-country skiing race for women in the Olympics. Convert this distance to miles.

11. The lightest class for weightlifting in the Olympics is for people who weigh under 52 kg. How many pounds is this?

12. Is 900 milliliters more or less than a quart?

13. In March of 1993, one Canadian dollar was equal in value to about .80 U.S. dollars. If an item cost $50 in a Canadian store at that time, estimate its cost in U.S. dollars.

Applying the Mathematics

In 14–19, which is larger?

14. a pound or a kilogram

15. a quart or a liter

16. a meter or a yard

17. a centimeter or an inch

18. a kilometer or a mile

19. a gram or an ounce

20. In March 1993, how many pesos equaled 1 dollar? Answer to the nearest whole number.

21. a. Write $\frac{5}{16}$ as a decimal.
  b. A person needs a drill bit with a diameter of approximately $\frac{5}{16}$ in. If a bit with a metric measure must be used, what diameter is needed?

22. An adult human brain weighs about 1.5 kg. Convert this to ounces.

23. Convert 3 liters to cups.

24. The monetary unit of Kenya is the shilling. In April of 1993, one shilling was worth about .020 U.S. dollar. At that rate, what was the cost in U.S. dollars of a hotel room for one night if the charge was 1500 shillings a night?

Review

25. How many grams are in 4 kilograms?   *(Lesson 3-4)*

26. One centimeter is what percent of a meter?   *(Lessons 2-6, 3-4)*

27. A 5-lb bag of cat food has how many ounces of food in it?
(Lesson 3-3)

28. A bucket holds 8 gallons. How many quarts will it hold?   *(Lesson 3-3)*

29. How many inches are in a yard?   *(Lesson 3-2)*

30. Measure this segment to the nearest millimeter.   *(Lesson 3-1)*

31. From the circle graph below:
- **a.** Which age bracket has the fewest people?
- **b.** Which age bracket has the most people?
- **c.** How many people are between the ages of 5 and 19?
- **d.** To the nearest 10%, what percent of people are 40 years of age or older? *(Lesson 2-7)*

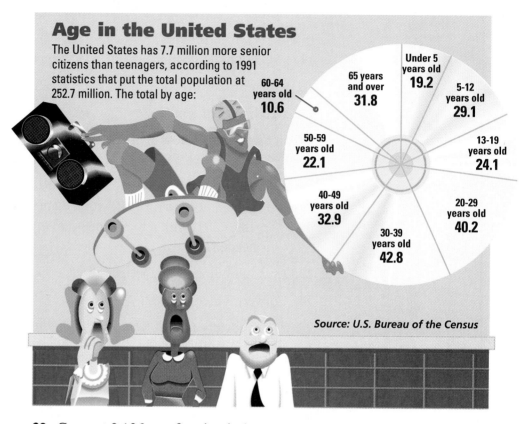

Age in the United States

The United States has 7.7 million more senior citizens than teenagers, according to 1991 statistics that put the total population at 252.7 million. The total by age:

60-64 years old **10.6**

65 years and over **31.8**

Under 5 years old **19.2**

5-12 years old **29.1**

50-59 years old **22.1**

13-19 years old **24.1**

40-49 years old **32.9**

20-29 years old **40.2**

30-39 years old **42.8**

Source: U.S. Bureau of the Census

32. Convert 0.136 to a fraction in lowest terms. *(Lesson 2-6)*

33. What is 50% of 50? *(Lesson 2-5)*

Exploration

---

34. On every cereal box, the amount of protein per serving is listed.
- **a.** Find a cereal box. How much protein per serving is listed?
- **b.** Is this amount given in U.S. units, metric units, or both?
- **c.** Is the weight of the box given in U.S. units, metric units, or both?

35. Find information about the current values of various units of currency from other countries. Has the value of a peso increased or decreased since March of 1993?

36. Examine the computer program in Question 29 of Lesson 3-2. Modify that program so that it converts a length in inches to a length in centimeters. Run your program a few times with values of your own choosing.

3-6

Measuring Angles

What Are Angles?

Think of rays of light coming from the sun. Each **ray** has the same starting point, called its **endpoint.** Each ray goes on forever in a particular direction. Only a part of any ray can be drawn.

Identified below are rays *SB* and *SA,* written $\overrightarrow{SB}$ and $\overrightarrow{SA}$. Notice that the first letter written when naming a ray is the endpoint of the ray. The second letter names another point on the ray. Two other rays are not identified.

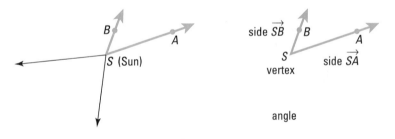

The union of two rays with the same endpoint is an **angle.** The rays are the **sides** of the angle. The endpoint is the **vertex** of the angle. The sides of an angle go on forever; you can draw only part of them.

The symbol for angle is ∠. This symbol was first used by William Oughtred in 1657.

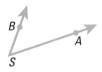

The angle above may be written as ∠*S,* ∠*ASB,* or ∠*BSA.* When three letters are used, the middle letter is the vertex. If an angle shares its vertex with any other angle, you must use three letters to name it. For instance, in the first drawing above, you should not name any angle ∠*S.* Which angle you meant would not be clear.

How Are Angles Measured?

Over 4000 years ago, the Babylonians wrote numbers in a system based on the number 60. So they measured with units based on 60. Even today, we use Babylonian ideas to measure time. That is why there are 60 minutes in an hour and 60 seconds in a minute.

We also use Babylonian ideas in measuring angles. The Babylonians divided a circle into 360 equally spaced units which we call **degrees.** (The number 360 equals 6 × 60 and is close to the number of days in a year.)

The measures of angles range from 0° to 180°. Here are angles with measures from 10° to 180°, in multiples of 10°.

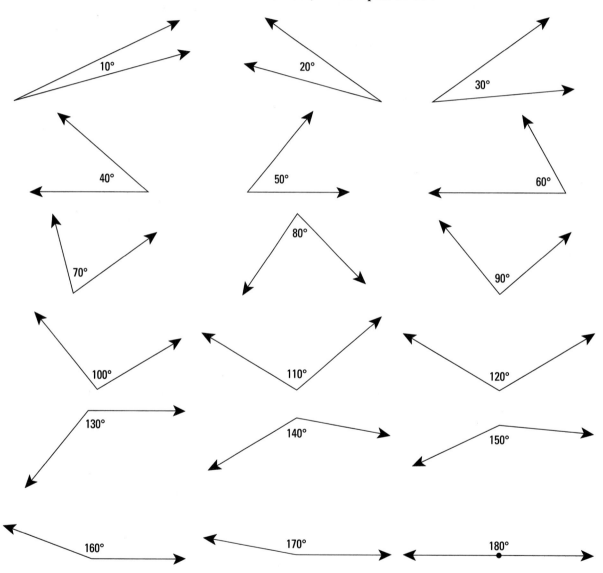

Caution: The measure of an angle has nothing to do with the lengths of its sides. The measure is entirely determined by the difference in the directions of its rays.

Measuring Angles with a Protractor

A **protractor** is the most common instrument used for measuring angles. Many protractors look like the one pictured below, covering only half a circle. When protractors cover half of a circle, the degree measures on the outside go from 0° to 180°.

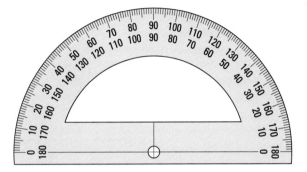

Every protractor has a segment connecting the 0° mark on one side to the 180° mark on the other. This segment is on the **base line** of the protractor. The middle point of this segment is called the **center** of the protractor. On the protractors in this lesson, the center is named *V*. On a protractor, *V* is usually marked by a hole, an arrow, or a + sign. There are almost always two curved scales on the outside of the protractor. One goes from 0° to 180°. The other goes from 180° to 0°.

The picture below shows how the protractor is placed on an angle.

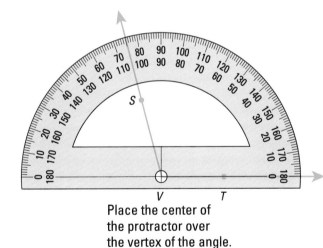

Align the baseline of the protractor with one side of the angle.

Place the center of the protractor over the vertex of the angle.

Example 1

What is the measure of ∠*SVT* drawn above?

Solution

$\overrightarrow{VS}$ crosses the protractor at 105 and 75. It seems that the measure of ∠*SVT* could be either 105° or 75°. Which measure is correct? The side $\overrightarrow{VT}$ crosses the *inner* scale at 0°. Therefore, the inner scale should be used to determine the measure of the angle. So, The measure of ∠SVT is 105°.

The measure of ∠A is written **m∠A.** The measure of ∠DEF is written **m∠DEF.** So in Example 1, m∠SVT = 105°.

Example 2

What is m∠EVF?

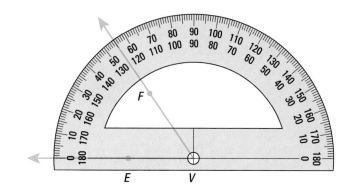

Solution

Examine where $\overrightarrow{VF}$ crosses the protractor. There are two numbers, 125° and 55°. Since side $\overrightarrow{VE}$ of the angle crosses the outer scale at 0°, pick the number in the outer scale. This is 55. Thus **m∠EVF = 55°.**

Activity

Measure the angle drawn here to the nearest degree.

If you get a measure between 60° and 70°, you probably know what you are doing.

Example 3

Draw a 30° angle.

Solution

One solution is shown below.

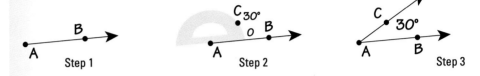

Step 1   Step 2   Step 3

1. Draw a ray. We call our ray $\overrightarrow{AB}$. Then place the center of a protractor at A, with the base line of the protractor on $\overrightarrow{AB}$.
2. Using the scale that crosses $\overrightarrow{AB}$ at 0°, put a point at 30°. We call this point C.
3. Draw $\overrightarrow{AC}$. Then m$\angle CAB$ = 30°.

QUESTIONS

Covering the Reading

In 1–4, use the angle at the left.

1. Name the sides.   **2.** Name the vertex.

3. Which of the following are correct names for the angle?
   (a) $\angle ACE$ (b) $\angle C$ (c) $\angle ECA$ (d) $\angle CBD$
   (e) $\angle ECB$ (f) $\angle DBC$ (g) $\angle ACD$ (h) $\angle ACB$

4. Which has the larger measure, $\angle BCE$ or $\angle ACD$?

5. Why did the Babylonians measure with units based on 60?

In 6–9, use the drawing.

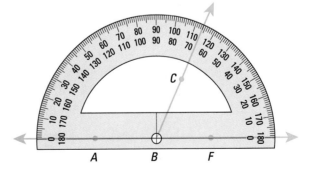

6. Name the base line of this protractor.

7. What point is at the center of this protractor?

8. What is the measure of $\angle ABC$?

9. What is the measure of $\angle CBF$?

10. m∠*AVB* stands for the __?__ of __?__ *AVB*.

11. What measure did you get for the angle in the Activity of this lesson?

In 12–14, use a protractor. Measure the angle to the nearest degree. (You may have to trace the angles and extend their sides.)

12.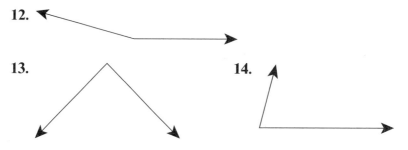

13.

14.

15. Who first used the symbol ∠ for an angle and when was this done?

In 16 and 17, draw an angle with the given measure.
Use a protractor and ruler.

16. 55°   **17.** 162°

Applying the Mathematics

18. Point *X* is on $\overrightarrow{UV}$ drawn here. *True or false:* $\overrightarrow{UX}$ is the same ray as $\overrightarrow{UV}$.

19. Which angle below has the largest angle measure, ∠*JGI*, ∠*IGH*, or ∠*JGH*?

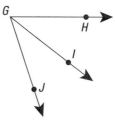

20. How many angles with vertex *E* are drawn below? (Be careful. Many students' answers are too low.)

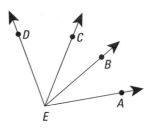

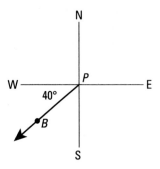

In 21–23, the ray $\overrightarrow{PB}$ represents the direction of a tornado seen 40°
South of West. Copy the drawing at the left but make your drawing
larger. Using a protractor, add the ray to your drawing.

21. $\overrightarrow{PX}$ to represent a UFO seen 5° East of North

22. $\overrightarrow{PY}$ to represent a whale sighted 15° South of East

23. $\overrightarrow{PZ}$ to represent a tanker observed 20° North of West

---

Review

---

24. While driving through Canada, Kirsten saw the sign below. About
how many miles away was Toronto? *(Lesson 3-5)*

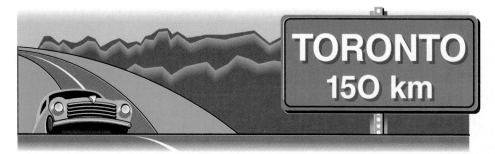

25. Convert 82 mm to meters. *(Lesson 3-4)*

26. a. In the metric system, the amount of water in a bathtub could be
measured in __?__. *(Lesson 3-4)*
   b. In the U.S. system, the amount of water in a bathtub could be
measured in __?__. *(Lesson 3-3)*

27. Draw a line segment with a length of 7 cm. *(Lesson 3-1)*

In 28–30, use the circle at the right.
Name the following. *(Lesson 2-7)*

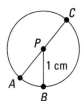

28. a radius

29. a diameter

30. The center

31. Write 41.6 million in scientific notation. *(Lesson 2-3)*

32. Roy found that 51 out of 68 people he polled liked the posters he
made. Rewrite $\frac{51}{68}$ as a fraction in lowest terms. *(Lesson 1-10)*

Exploration

---

33. Angles are not always measured in degrees. Two other units for
measuring angles are the grad and the radian. Find out something
about at least one of these units.

Kinds of Angles

IN-CLASS
ACTIVITY

A **central angle** of a circle is an angle with its vertex at the center of the circle. At the right, ∠*AOB* is a central angle. It is the central angle of the shaded sector *AOB*. By drawing central angles, you can make circle graphs.

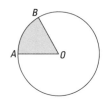

1 Take a survey of the favorite subjects of the students in your class. Then display the information in a circle graph.

2 For instance, suppose you survey 20 people. You might find the following preferences: 8, Math; 4, English; 5 Science; 3, Social Studies. So $\frac{8}{20}$, or 40%, prefer Math, and so on.

Put the numbers and percents into a table. Here is our example.

	Math	English	Science	Social Studies
Number	8	4	5	3
Percent	40%	20%	25%	15%

3 Display the information in a circle graph. Here is our display.

Since there are 360° in a circle, to find the number of degrees in 40% of a circle, find 40% of 360°. This is 144°. So we draw a 144° central angle for the Math sector. For the other central angles, the calculations are similar.

English:
20% of 360° = 0.20 × 360° = 72°

Science:
25% of 360° = 0.25 × 360° = 90°

Social Studies:
15% of 360° = 0.15 × 360° = 54°

4 Check. The sum of the measures of the central angles should be 360°.

Math 40%
144°
English 20%
72°
Social Studies 15%
54°
90°
Science 25%

French connections. *The street called the Champs Elysée, as seen from the top of the Arc de Triomphe in Paris, France, intersects the circular road around the Arc at a right angle. Notice that the other streets in this photo also meet this road at right angles.*

Right Angles

Angles can be classified by their measures. If the measure of an angle is 90°, the angle is called a **right angle.** Some right angles are drawn below. The sides of this page form right angles at the corners. Many streets intersect at right angles.

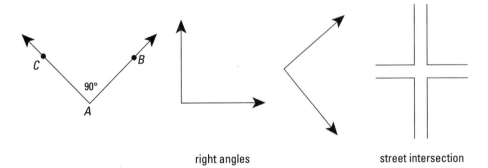

right angles   street intersection

Rays, segments, or lines that form right angles are called **perpendicular.** Above, $\overrightarrow{AC}$ is perpendicular to $\overrightarrow{AB}$. The streets drawn above are also perpendicular. Each long side of this page is perpendicular to each short side.

Acute and Obtuse Angles

If the measure of an angle is between 0° and 90°, the angle is called an **acute angle.** An **obtuse angle** is an angle whose measure is between 90° and 180°. Most of the time, you can tell whether an angle is acute or obtuse just by looking. If you are unsure, you can measure.

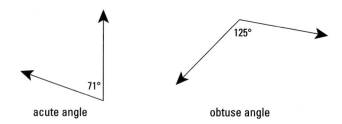

acute angle   obtuse angle

Triangles

A **triangle** gets its name because it is formed by parts of three angles. The triangle *TEN* drawn here has angles *T, E,* and *N.* Angle *T* is obtuse while angles *E* and *N* are acute. △*AOK* below (△ is the symbol for triangle) is called a **right triangle** because one of its angles is a right angle.

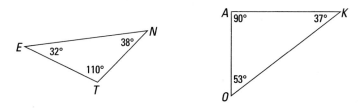

QUESTIONS

Covering the Reading

In 1–3, give a definition for the phrase.

1. acute angle   **2.** obtuse angle   **3.** right triangle

4. Explain what is wrong with this sentence: Two angles are perpendicular if they form right angles.

In 5–8, state whether an angle with the given measure is acute, right, or obtuse.

5. 40°   **6.** 9°   **7.** 140°   **8.** 90°

In 9–12, without measuring, tell whether the angle looks acute, right, or obtuse.

9.   **10.**   **11.**   **12.**

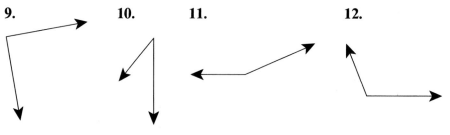

13. *Multiple choice.* Which triangle looks like a right triangle?

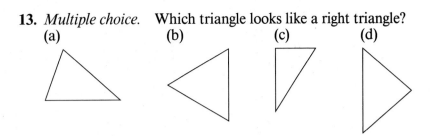

  (a)   (b)   (c)   (d)

In 14 and 15, use the figure.
a. Tell whether the angle is acute or obtuse.
b. Give the measure of the angle.

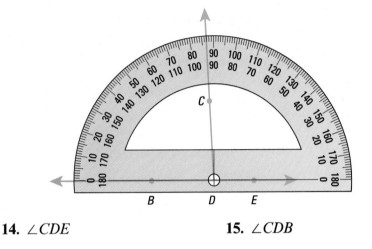

14. ∠CDE   **15.** ∠CDB

Applying the Mathematics

In 16–18, refer to the In-class Activity on page 152 and the circle graph drawn there.

16. Define: central angle.

17. What is the measure of the smallest central angle?

18. Which central angles are acute, which right, which obtuse?

19. Sixty students were asked to name their favorite pet. Their choices are given in the table below. Represent this information on a circle graph.

The purrfect pet.
Like many Americans, this boy prefers a cat for his pet. There are about 58 million pet cats in U.S. homes today. A bunch of cats is sometimes called a clowder of cats.

Pet	Number of students choosing that pet
Cat	25
Dog	24
Hamster	5
Other	6

20. Find two examples of right angles different from those mentioned in this lesson.

21. Copy the line. Then using a ruler and protractor, draw a line perpendicular to the given line.
a.   **b.**

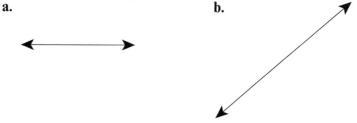

In 22–24, the picture is a closeup of the markings on a giraffe.
a. Tell whether the angle is acute, right, or obtuse.
b. Measure the angle.

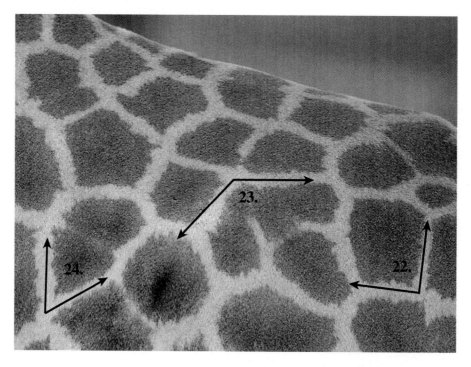

25. Name the type of angle and give the measure of the angle formed by the minute and hour hands of a watch at:
  a. 1:00.   **b.** 4:00.
  c. 9:00.   **d.** 6:30. (Be careful!)

26. Find the measures of the three angles of $\triangle XYZ$. (You may want to copy the triangle and extend the lines of the sides before measuring.)

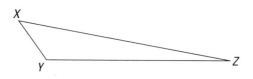

27. Accurately draw a 106° angle.   *(Lesson 3-6)*

In 28–31, use <, =, or > to complete the relationship.
(Lessons 1-6, 3-4, 3-5)

28. 2 meters __?__ 1 yard

29. 1 kg __?__ 10,000 g

30. 1 kg __?__ $4\frac{2}{3}$ lb

31. 2 liters __?__ 1 gal

In 32–34, complete the statement by using a reasonable metric unit.
(Lesson 3-4)

32. In one day we rode 40 __?__ on our bikes.

33. A cup can hold about 0.24 __?__ of water.

34. The meat she ate weighed 350 __?__.

35. Measure the longest side of triangle *XYZ* in Question 26 to the nearest $\frac{1}{4}$ inch.   *(Lesson 3-1)*

36. Tungsten wire four ten-thousandths of an inch in diameter is used to make filaments for light bulbs.
a. Write this number as a decimal.
b. Write this number in scientific notation.   *(Lessons 1-2, 2-9)*

37. According to one survey, teenage boys spend an average of 32% of their allowance on food. Teenage girls spend an average of 26% on food. If a boy and girl each receives $20, on average how much more does the boy spend on food?   *(Lesson 2-5)*

38. A person has acute appendicitis.
a. What does this mean?
b. Does this use of the word *acute* have any relation to the idea of acute angle?

39. Look up the meaning of the word *obtuse* in the dictionary.
a. What nonmathematical meaning does this word have?
b. Is the nonmathematical meaning related to the idea of obtuse angle?

40. a. Name a street intersection near your home or school in which the streets do not intersect at right angles.
b. Approximately what are the measures of the angles formed by the streets?
c. These kinds of intersections are usually not as safe as right angle intersections. Is anything done at the intersection you named to increase its safety?

Shopping today? *Rental fees for retail stores, such as those shown here in the Mall of America in Bloomington, MN, are based upon location and area of the floorspace. In upscale malls, yearly rents may vary from $25 to $40 a square foot.*

The visual impact of a circle graph is based on *area*. The larger the percent being graphed, the larger the area of the sector.

What Is Area?

Area is a measure of the space inside a two-dimensional (flat) figure. You can think of area as a measure of how much is shaded within the figures drawn here.

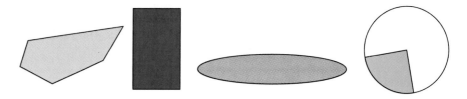

How Is Area Measured?

Regardless of how a figure is shaped, it is customary to measure its area in square units. Recall that a **square** is a four-sided figure with four right angles and four sides of equal length. The common units for measuring area are squares with sides of unit length.

1 cm
1 cm

1 square centimeter
(actual size)

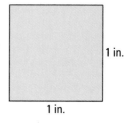

1 in.
1 in.

1 square inch
(actual size)

Squares may be of any size. Large amounts of land in the U.S. are measured in square miles. Elsewhere in the world, area is measured in metric units. A **square kilometer** is the space taken up by a square whose sides are 1 km long. A **hectare** is the space taken up by a square whose sides are 100 meters long.

Figures with different shapes can have the same area. Each of these figures has shaded area equal to 2 square centimeters. But the lengths of their sides are quite different.

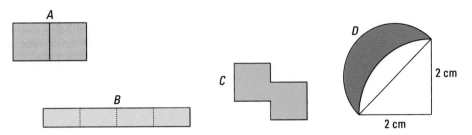

How Can Area Be Calculated?

There are three ways to find areas of figures. One way is to count square units. This will work in Figure *A* above. Another way is to cut and rearrange parts of figures. This will work in figures *B* and *C*. But if a figure is complicated, like the shaded part of *D*, formulas are needed. The simplest formula is that for the area of a square.

Each side of the square at the right has length 5 units. Counting shows that there are 25 square units.

Each side of this square has length 5.5 units. Counting shows 25 whole square units, 10 half squares (which equal 5 whole squares), and an extra quarter square. This totals 30.25 square units.

Notice that $5^2 = 25$ and $5.5^2 = 30.25$.

> **The area of a square equals the second power of the length of one of its sides.**

For this reason, 5 to the second power, 5^2, is often read "5 squared." Also, for this reason, you can write square inches as in^2, square centimeters as cm^2, and square kilometers as km^2. You can write the equation:

$$\text{Area of a square} = (\text{length of side})^2$$

The letter A is often used as an abbreviation for area. An abbreviation for the length of a side is s. Using these abbreviations, the area of a square can be described in a very short way.

$$A = s^2$$

Example

Find the area of a city block 220 yards on a side.

Solution 1

$$
\begin{aligned}
\text{Area} &= (220 \text{ yd})^2 \\
&= 220 \text{ yd} \times 220 \text{ yd} \\
&= 48{,}400 \text{ square yards} \\
&= 48{,}400 \text{ yd}^2
\end{aligned}
$$

Solution 2

Most scientific calculators have a $\boxed{x^2}$ key.
(On some calculators you must press $\boxed{\text{INV}}$ or $\boxed{\text{F}}$ before pressing this key.)

Key sequence: 220 $\boxed{x^2}$ $\boxed{=}$
Display: 48,400
Answer: The area is 48,400 yd^2.

QUESTIONS

Covering the Reading

1. What does area measure in a figure?

2. Suppose length is measured in centimeters. Area will most likely be measured in what units?

3. What is a *square?*

4. Which of the following seem to be pictures of squares?
   (a)   (b)   (c)   (d)

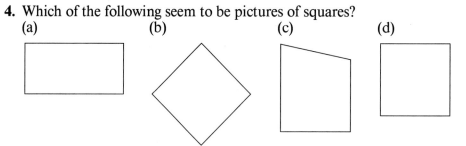

5. A square has a side of length 6.7 meters. Write a key sequence to find the area.

6. Give an example of a square you might find outside a mathematics class.

7. A square is a ___?___ -dimensional figure.

8. Give three ways to find the area of a figure.

In 9–11, the length of a side of a square is given. Find the area of the square. Be sure to include the correct unit.

9.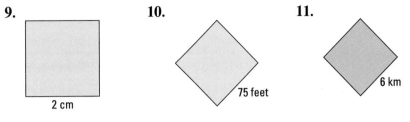

10.

11.

2 cm

75 feet

6 km

12. 40^2 may be read "40 to the second power" or "40 __?__."

13. Consider the sentence $A = s^2$.
  a. What does A represent?
  b. What does s represent?
  c. Write this sentence in words.

14. Find the area of a square that is 1.5 inches on a side.

Applying the Mathematics

---

15. The area of the figure drawn below at the left is how many square inches?

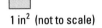

1 in^2 (not to scale)

16. a. Make an accurate drawing of 1 in^2.
  b. Shade 0.5 in^2.
  c. On another drawing, shade $\frac{1}{4}$ in^2.
  d. On still another drawing, shade 0.6 in^2.

17. Remember, there are 3 feet in a yard.
  a. Picture a square yard and split it up into square feet.
  b. How many square feet are in a square yard?

18. A baseball diamond is really a special diamond that is square in shape. The distance from home to 1st base is 90 ft. What is the area of the square?

19. Name the unit of measure in which the floor area of a room would most likely be given.
  a. in the metric system
  b. in the U.S. system

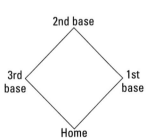

2nd base

3rd base

1st base

Home

An acre is a unit of area equal to 43,560 square feet. Use this fact in Questions 20 and 21.

20. How many square feet are in 10 acres?

21. How many square feet are in a half-acre lot?

22. a. How many hectares are in 1 square kilometer?
  b. In the Republic of the Congo, commonly known as the Congo, is the 3 million hectare Ndoki (en-doe-key) rain forest. Is the Ndoki larger or smaller than Maryland? The state of Maryland has an area of about 27,100 km².

Review

23. An angle has measure 19°. Is it acute, right, or obtuse?   *(Lesson 3-7)*

24. Measure this angle to the nearest degree.   *(Lesson 3-6)*

25. a. Sixty kilograms is about how many pounds?   *(Lesson 3-5)*
  b. Sixty kilograms is how many grams?   *(Lesson 3-4)*

26. Every spring, Indiana University holds a team bicycle race called the "Little 500." The women's course is 40.3 km long. About how many miles is this?   *(Lesson 3-5)*

27. Measure the length of this printed line (from the *M* in "Measure" to the *o* in "to") to the nearest half centimeter.   *(Lesson 3-1)*

28. Which is larger, 0 or 10^0?   *(Lesson 2-8)*

29. A school has 600 students.   *(Lessons 2-4, 2-5, 2-6)*
  a. Ten percent of the students is how many students?
  b. Use your answer in part **a** to find 20%, 40%, and 70% of the student body without doing another percent calculation.

30. Round 2^{30} to the nearest million.   *(Lessons 1-4, 2-2)*

31. Complete the statement with <, =, or >. 13.26 __?__ $13\frac{4}{13}$.
  (Lessons 1-6, 1-9)

Exploration

32. How many squares are in the pattern drawn below?

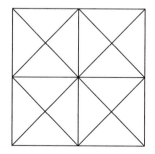

Where few have gone before. *The Ndoki rain forest in northern Congo is one of the last unexplored rain forests on Earth. Surrounded by swamps and the unnavigable Ndoki river, the rain forest is almost impossible to reach. So few humans have been to the Ndoki that the animals here, unlike in other parts of Africa, are not afraid of humans.*

3-9

Measuring Volume

Space is no problem here. *The Vehicle Assembly Building (VAB) at the Kennedy Space Center, Cape Canaveral, FL, is where space shuttles are prepared for launch. The VAB is 525 ft tall and has 4 bays with doors 460 ft high. Its capacity of 129.5 million cubic feet is greater than that of any other scientific building.*

What Is Volume?

Volume measures the space inside a three-dimensional or solid figure. Think of volume as measuring the amount a box, jar, or other container can hold. This is also called the container's **capacity.** Or think of volume as telling you how much material is in something that is solid. Whatever the shape of a figure, its volume is usually measured in **cubic units.** A **cube** is a figure with six faces, each face being a square. Sugar cubes, number cubes, and dice are examples of cubes. The most common units for measuring volume are cubes with edges of unit length.

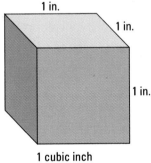

1 in.
1 in.
1 in.

1 cubic inch

In a **cubic centimeter,** each edge has length 1 centimeter. Each face is a square with area 1 square centimeter.

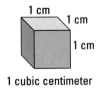

1 cm
1 cm
1 cm

1 cubic centimeter

Notice that volume is quite different from area. Area is two-dimensional. The amount of paper it would take to cover the outside surface of the cube is an area, the *surface area* of the cube. Volume is three-dimensional. The *volume* of the cube indicates how much sand can be poured into the cube.

How Can Volume Be Calculated?

Volume can be calculated by counting cubes, by putting cubic units together, or with a formula. Below is a cube with edges of length 2 units. The picture is partially transparent and shows that there are two layers. The top layer has 2 × 2, or 4, cubes. So does the bottom layer. In all, there are 2 × 2 × 2 or 2^3, or 8 cubes. So, the volume is 8 cubic units.

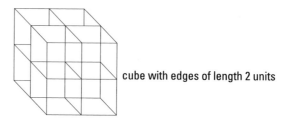

cube with edges of length 2 units

> **The volume of a cube equals the third power of the length of one of its edges.**

For this reason, 2^3, 2 to the third power, is often read "2 cubed." Also, for this reason, you can write cubic inches as in^3, cubic centimeters as cm^3, cubic meters as m^3, and so on. You can write the equation:

$$\text{Volume of a cube} = (\text{length of edge})^3$$

The letter V is an abbreviation for volume. An abbreviation for the length of an edge is e. Using these abbreviations, the above equation becomes

$$V = e^3.$$

The Liter as a Unit of Volume

You have already learned that the liter is a metric unit of capacity. Since capacity is another word for volume, the liter is a unit of volume. It is defined as the volume of a cube that is 10 centimeters on each side.

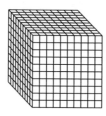

$$\text{Volume of cube} = (10 \text{ cm})^3$$
$$= 1000 \text{ cm}^3$$
$$= 1 \text{ liter}$$

Soft drinks are often sold in 2-liter bottles. Since 1 liter = 1000 cm^3, 2 liters = 2000 cm^3. Thus a 2-liter bottle has a volume of 2000 cm^3. Also, since 1 liter = 1000 cm^3, 1 milliliter = $\frac{1}{1000} \cdot 1000 \text{ cm}^3$. A milliliter is another name for cubic centimeter.

QUESTIONS

Covering the Reading

1. What does volume measure in a 3-dimensional figure?

2. Suppose length is measured in meters. Volume will most likely be measured in what unit?

3. What is a *cube?*

4. Draw a cube.

5. Give an example of a cube you might find outside a math class.

6. A cube is a __?__-dimensional figure.

7. Name three ways to find the volume of a figure.

8. What does the volume of a cube equal?

In 9–11, find the volume of a cube with an edge of the given length. Use the correct unit in your answer.

9. 4 inches   10. 40 cm   11. 7 yards

12. A liter is defined as the volume of a cube with an edge of length __?__.

13. How many cubic centimeters equal one liter?

14. A 2-liter bottle has a volume of __?__ cubic centimeters.

Bon voyage! *The U.S.S. Abraham Lincoln is a nuclear-powered, U.S. Navy aircraft carrier. Besides having 4.5 acres of flight deck, the carrier is one of several warships with the largest full-load displacements in the world, namely 100,846 tons.*

Applying the Mathematics

15. **a.** Find the volume of a cube with edge 3.25 feet.
    b. The correct answer to part **a** is accurate to a millionth of a cubic foot. This is too precise for many uses. Round your answer to the nearest hundredth, the same precision as the original measure.

16. Calculate six cubed plus five squared.

17. Arrange from smallest to largest: 1 liter, 89 cubic centimeters, and the volume of a cube with edge 9 cm.

In 18 and 19, use the following information. You have learned that the liter (a unit of capacity) and the centimeter (a unit of length) are related. In the metric system, these two units are also related to the mass of water. Specifically, 1 cm^3 of water equals 1 mL of water and has a mass of 1 gram at 4°C. Its weight at sea level is 1 gram.

18. How much does a liter of water weigh?

19. Suppose an aquarium is a cube 50 cm on a side.
    a. How much water will it hold?
    b. How much will the water weigh in kilograms?
    c. How much will the water weigh in pounds?

20. You may want to draw a picture to help with these questions.
  a. How many square inches are there in a square foot?
  b. How many cubic inches are there in a cubic foot?

Review

10 cm

21. Give the area of the square drawn at the left. *(Lesson 3-8)*

22. A band hired to play at a dance thinks that an area of at least 2000 m^2 is needed for the band and for dancing. Explain whether or not the school recreation room, a square 46.5 m on each side, will be large enough. *(Lesson 3-8)*

23. 25 people were asked, "What part of the day do you like best?"

  4 answered midnight to 6 A.M.
  3 answered 6 A.M. to noon
  7 answered noon to 6 P.M.
  11 answered 6 P.M. to midnight

  a. Calculate the percent that gave each response.
  b. Put this information into a circle graph. *(Lessons 2-5, 3-7)*

24. Change 45 km to feet. *(Lesson 3-5)*

25. *Multiple choice.* A 10-year-old boy who weighs 75 kg is likely to be:
  (a) underweight   (b) about the right weight
  (c) overweight. *(Lesson 3-4)*

26. 10^3 meters is how many kilometers? *(Lessons 2-2, 3-4)*

27. Change 6 yards to inches. *(Lesson 3-2)*

28. Rewrite 3.4×10^{-4}
  a. as a decimal.   **b.** as a fraction in lowest terms.
  (Lessons 2-6, 2-8)

29. In the last 6 months of 1991, an average of 15,353,982 copies of *TV Guide* were sold weekly, the most of any weekly magazine in the U.S.
  a. Round this number to the nearest hundred thousand.
  b. Estimate how many copies were sold over the entire year 1991.
  c. Write your estimate in scientific notation. *(Lessons 1-4, 2-3)*

30. *Multiple choice.* Which is not an integer? *(Lessons 1-5, 1-8, 2-2)*
  (a) -4   (b) $\frac{8}{4}$ (c) 52   (d) 0   (e) $\frac{1}{2}$

Exploration

31. a. What happens to the volume of a cube if the length of an edge is doubled? Try some examples to see what happens. Caution: The volume is *not* doubled.

  b. Can you predict what will happen to the volume of a cube if the length of an edge is tripled? multiplied by 10?

A project presents an opportunity for you to extend your knowledge of a topic related to the material of this chapter. You should allow more time for a project than you do for typical homework questions.

1 Metric Units

There are metric units other than the ones we have used in this chapter. From an almanac, a science book, or some other source, find out what the seven base units for the metric system are. Identify all of the prefixes that are used (they range from 10^{18} to 10^{-18}). Name some other metric units that are defined from these units, and tell where they are used.

2 Other Traditional Units

The U.S., British, and metric systems are not the only systems of measurement that have ever been devised. For instance, until recently, in East Africa, the Swahili used a system with the following measures: shibiri, mkono, pima, kibaba, kisaga, pishi, wakia, ratli, frasili. Look in the book *Africa Counts,* by Claudia Zaslavsky, or in some other source to find out what these units represent. Write a report about this or some other measurement system different from the systems mentioned in this chapter.

3 Weighing a Collection

Often people collect things for charity drives, for recycling or for discard. Pick an item that is sometimes collected (for instance, clothes, newspapers, coins, bottle caps). If everyone in your school were to collect a certain number of these (you pick the number) what would the collection weigh? How much would it be worth? Repeat your analysis if every family in your community were to collect these. Discuss how you have determined your answers, and the problems (if any) that might occur if you tried to transport the collection.

AFRICA COUNTS
Number and Pattern in African Culture
CLAUDIA ZASLAVSKY

4 Large Edifices

An edifice is a building or temple or monument or some other impressive structure. Over the ages, people have built many large edifices, including the pyramids of the Mayas, the Babylonians, or the Egyptians; the temples Angkor Wat in Cambodia or Machu Picchu in Peru; the Pentagon building in Washington; and the Kremlin in Russia. Pick one of these or some other famous large edifice. Look in encyclopedias for information regarding the edifice you picked. Draw an accurate picture of the edifice, labeling important points. Describe its area and volume, or areas and volumes related to it. Write a paragraph describing how and when the structure was built and what its use was or is.

5 Circle Graph Summary

Choose a topic of interest to you or to your class, such as lunch food, football, recycling, or homework. Write some questions about this topic that have a number of reasonable choices. (Think of the kind of question you would hear on the television program *Family Feud.*) Survey 25 or more people using these questions, and illustrate the distribution of answers to at least one of the questions in a circle graph. Then explain in words what you found.

6 Computer Drawing Programs

Locate drawing software for a computer that can draw angles and measure them, and that can draw angles of a specific measure. Explore this software with angles and triangles, and print out examples of your angles and measurements. After you have done some exploration, write instructions to a classmate on how to draw and measure angles with this software. If you can, test your instructions on a classmate who has never used the software to make sure that your instructions are clear.

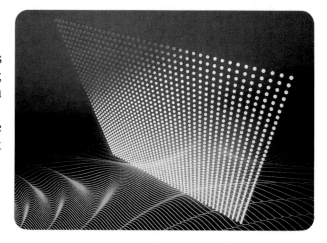

7 Units on Food

Search the containers of all of the foods in your house for as many different units as you can find. Consider both U.S. customary and metric units. List all the quantities you find with different units and where they were found. Then write a summary that includes how you did your search and whether anything you found surprised you.

SUMMARY

The most common uses of numbers are as counts or measures. In this chapter, measures of length, area, volume (or capacity), mass (or weight), and angles are discussed.

There are two systems of measurement in use today in the United States. One is the metric system. The basic units in the metric system are the meter for length, the kilogram for mass, and the liter for capacity. The other system is called the U.S. or customary system. It uses inches, feet, miles, and so on, for length; ounces, pounds, and so on, for weight; ounces, cups, pints, quarts, gallons, and so on, for capacity. The metric system is generally easier to work with because its units are related to each other by powers of 10. So it is closely related to the decimal system. Conversion within one system or between systems can be done by beginning with known conversion equations and using the Multiplication Property of Equality.

Angles are measured the same way in both the metric and the U.S. systems. The degree, the common unit for angle measure, is based on splitting a circle into 360 equal parts. Thus, by using angle measures, circle graphs can be drawn. Angles can be classified by their measure. Units for area are usually squares based on units of length. Units for volume are usually cubes based on units of length.

VOCABULARY

You should be able to give a general description and a specific example of each of the following ideas.

Lesson 3-1
U.S. or customary system of measurement
inch
meter, centimeter, millimeter

Lesson 3-2
foot (ft), yard (yd), mile (mi),
inch (in.)
Multiplication Property of Equality

Lesson 3-3
ounce (oz), pound (lb), short ton,
cup (c), fluid ounce (fl oz)
pint (pt), quart (qt), gallon (gal)

Lesson 3-4
international or metric system of measurement
milli-, centi-, kilo-
meter (m)
gram (g)
liter (L)

Lesson 3-6
ray, $\overrightarrow{AB}$, endpoint of ray
angle, $\angle ABC$, sides of an angle,
    vertex of an angle
degree (°)
protractor, base line of protractor,
    center of protractor

Lesson 3-7
central angle
right angle, acute angle, obtuse angle
perpendicular
triangle, right triangle

Lesson 3-8
area
square, square units
hectare
in^2, cm^2, km^2, and so on

Lesson 3-9
volume, capacity
cube, cubic units
in^3, cm^3, m^3, and so on

PROGRESS SELF-TEST

After taking and correcting the Progress Self-Test, use your errors to help you decide what to study and review in this chapter.

1. Measure this segment to the nearest eighth of an inch.

_____

2. a. Draw a segment 6.4 centimeters long.

   b. How long is your segment in millimeters?

3. Give the exact relationship between inches and centimeters.

4. Name the appropriate metric unit for measuring the weight or mass of a person. Explain why the unit you named is appropriate.

5. How many feet are there in $\frac{3}{4}$ of a mile?

6. Give an example of the use of the Multiplication Property of Equality.

7. A kiloton is how many tons?

8. 1103 mg = _?_ g

9. 3 quarts = _?_ cups

10. 3.2 pounds = _?_ ounces

11. In the Olympics, there is a race called the 50K walk. To the nearest mile, how many miles is this?

12. To convert kilograms to pounds, what estimate can be used?

13. Which is larger, 10 quarts or 9 liters?

14. The U.S. measuring system is derived from a system from what country?

15. Measure angle C to the nearest degree.

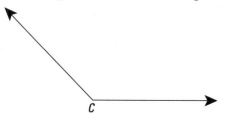

16. An acute angle has measure between _?_ and _?_ degrees.

17. Measure $\angle MNL$ to the nearest degree.

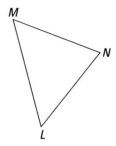

18. An angle has measure 90°. Is this angle right, acute, or obtuse?

19. Draw an angle with a measure of 80°.

20. Which angles of the triangle below seem to be acute, which right, which obtuse?

21. The circle graph shows the number of players in a string orchestra. If the graph is drawn correctly, what is the measure of the central angle for the violins sector?

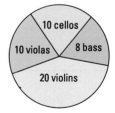

22. Find the area of a square with side 4.5 in.

23. If the side of a square is measured in cm, the area is usually measured in _?_.

24. How are liters and cubic centimeters related?

25. Give the volume of the cube drawn at the right.

5 cm
5 cm

26. How many blocks, 2 cm on each edge, can be stored in a box 10 cm on each edge?

CHAPTER REVIEW

Questions on SPUR Objectives

SPUR stands for **S**kills, **P**roperties, **U**ses, and **R**epresentations. The Chapter Review questions are grouped according to the SPUR Objectives for this chapter.

SKILLS DEAL WITH THE PROCEDURES USED TO GET ANSWERS.

Objective A. *Measure lengths to the nearest inch, half inch, quarter inch, or eighth of an inch, or to the nearest centimeter, or tenth of a centimeter. (Lesson 3-1)*

1. Measure the length of this segment to the nearest quarter inch.

   ▬▬▬▬▬▬▬▬

2. Measure the length of the above segment to the nearest tenth of a centimeter.

3. Measure the segment below to the nearest centimeter.

   ╱

4. Measure the segment above to the nearest eighth of an inch.

Objective B. *Measure angles to the nearest degree using a protractor. (Lesson 3-6)*

In 5–7, give the measure of the indicated angle in the drawing below.

5. ∠Q   6. ∠R   7. ∠PSR

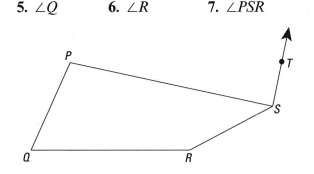

8. Jan van Eyck's *Portrait of a Man in a Red Turban* is contained very much within certain angles, as the diagram below shows. Give the measure of ∠ADC.

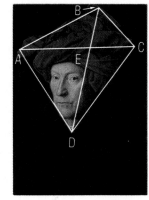

Jan van Eyck, *Portrait of a Man in a Red Turban*

Objective C. *Distinguish between acute, right, and obtuse angles by sight. (Lesson 3-7)*

9. Name an obtuse angle in the diagram over van Eyck's painting above.

In 10–12, use the figure of Questions 5–7.

10. Which angles seem to be acute?
11. Does ∠RST seem to be obtuse, acute, or right?
12. Name all angles that seem to be right angles.

Objective D. *Find areas of squares and volumes of cubes given the length of one side or edge.* *(Lessons 3-8, 3-9)*

13. Find the area of the square below.

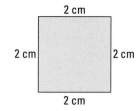

2 cm

2 cm | 2 cm

2 cm

14. Find the area of a square with side 6.5 in.

15. If length is measured in meters, area is most easily measured in what unit?

16. Find the volume of a cube with one edge of length 4 mm.

17. To the nearest integer, find the volume of a cube with edge of length 3.75 in.

PROPERTIES DEAL WITH THE PRINCIPLES BEHIND THE MATHEMATICS.

Objective E. *State and apply the Multiplication Property of Equality.* *(Lesson 3-2)*

18. State the Multiplication Property of Equality.

19. If 1 kg ≈ 2.2 lb, what does $\frac{1}{2}$ kg approximately equal?

20. If 1 franc is worth about 19¢, how much are 30.5 francs worth?

21. *Multiple choice.* Below are three steps in converting 25 centimeters to meters. Which step uses the Multiplication Property of Equality?
(a)   1 cm = .01 m
(b) 25 · 1 cm = 25 · .01 m
(c)   25 cm = .25 m

USES DEAL WITH APPLICATIONS OF MATHEMATICS IN REAL SITUATIONS.

Objective F. *Give appropriate units for measuring mass, length, and capacity in the U.S. or metric system of measurement.*
(Lessons 3-2, 3-3, 3-4)

In 22–25, give an appropriate unit for measuring each quantity: **a.** in the metric system, **b.** in the U.S. system.

22. the distance from New York to London, England

23. the length of your foot

24. the weight of a pencil

25. the capacity of a fish tank

Objective G. *Convert within the U.S. system of measurement.* *(Lessons 3-2, 3-3)*

26. Give a relationship between pints and quarts.

27. How many ounces are in 1 pound?

28. How are feet and miles related?

29. How many inches are in 2.5 yards?

30. Convert 7.3 gallons into quarts.

31. A road sign says: "Bridge ahead. Weight limit five tons." How many pounds is this?

Objective H. *Convert within the metric system.*
(Lesson 3-4)

32. What is the meaning of the prefix milli-?

33. How are centimeters and liters related?

34. 1 kilogram = __?__ grams

35. Convert 200 cm to meters.

36. Convert 5.8 km to meters.

37. Convert 265 mL to liters.

38. Convert 600 mg to grams.

Objective I. *Convert between different systems.*
(Lessons 3-2, 3-5)

39. Give an approximate relationship between pounds and kilograms.

40. How are centimeters and inches related?

41. Which is longer, a mile or a kilometer?

42. Which is longer, a meter or a yard?

43. How many centimeters are in 2 feet?

44. In a guide book the distance between Paris and London is given as 343 km. How many miles is this?

45. How many quarts are in 6.8 liters?

46. In March of 1993, one Israeli shekel was worth about 37¢. So a blouse that cost 80 shekels was worth about how many dollars?

47. In March of 1993, one Irish punt was worth about $1.49. If a shillelagh sold for 2.98 punts, how much was that in U.S. dollars?

Objective J. *Find areas of squares or volumes of cubes in real contexts.* *(Lessons 3-8, 3-9)*

48. A square table has a side of length 2.5 feet. Will a square tablecloth with an area of 6 square feet cover the table?

49. How many cubes, 1 cm on an edge, will fit in a cubical container 12 cm on an edge?

REPRESENTATIONS DEAL WITH PICTURES, GRAPHS, OR OBJECTS THAT ILLUSTRATE CONCEPTS.

Objective K. *Draw a line segment of a given length.* *(Lesson 3-1)*

50. Draw a line segment with length 3.5 cm.

51. Draw a line segment with length $2\frac{1}{4}$ inches.

52. Draw a vertical line segment with length 4.375 inches.

53. Draw a horizontal line segment with length 7.8 cm.

Objective L. *Draw an angle with a given measure.* *(Lesson 3-6)*

54. Draw an angle with a measure of 90°, in which neither side lies on a horizontal line.

55. Draw an angle with a measure of 37°.

56. Draw an angle with a measure of 145°.

57. Draw an angle with a measure of 100°.

Objective M. *Read, make, and interpret circle graphs.* *(Lesson 3-7)*

In 58 and 59, the graph pictures the spending of $40.

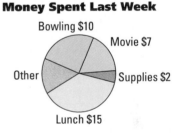

Money Spent Last Week

58. How much was spent on "other" things?

59. How can you tell that more than 25% of the money spent was for lunch?

60. Make a circle graph from the following information. The areas (in millions of square miles) of the continents of the world are: North America, 9.4; South America, 6.9; Europe, 3.8; Asia, 17.4; Africa 11.7; Australia, 3.3; Antarctica, 5.4.

CULTURE DEALS WITH THE PEOPLES AND THE HISTORY RELATED TO THE DEVELOPMENT OF MATHEMATICAL IDEAS.

Objective N. *Give countries and approximate dates of origin of current measuring ideas.*
(Lessons 3-1, 3-4, 3-6)

61. When and where did the metric system originate?

62. How was the length of a yard first determined?

63. Our system for measuring angles is based on measuring done by what people?

CHAPTER

4

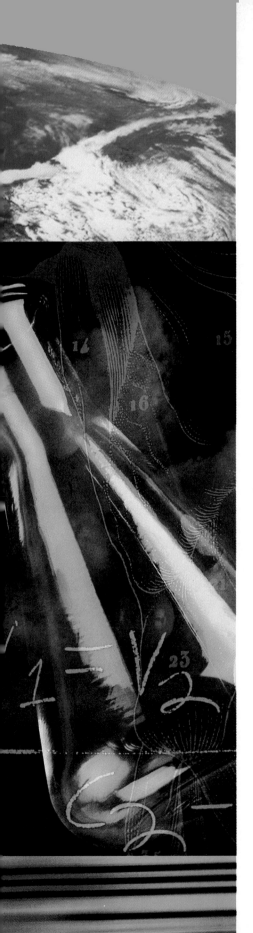

USES OF VARIABLES

Mathematics is, in many ways, a language. It is used to communicate information. It uses letters and symbols just as English, Spanish, Chinese, Hindi, Swahili, and other languages do. In the last 500 years, the symbols of mathematics have become an international language used in virtually all countries of the world. Just as our language changes over time, so does the language of mathematics. The table below shows some of the most important symbols of arithmetic, and who invented them. You can see that the language of mathematics has been developed by people from many different countries. The symbols invented since 1950 are used in computers and calculators worldwide.

Operation	Symbol	Name for Symbol	Inventor of Symbol (Year)	Name for Result
addition	$+$	plus	Johann Widman (German, 1498)	sum
subtraction	$-$	minus	Johann Widman (German, 1498)	difference
multiplication	$\times$	times	William Oughtred (English, 1631)	product
	$\cdot$	dot	Gottfried Leibniz (German, 1698)	product
	$*$	asterisk	(since 1950)	product
division	$-$ as in $\frac{2}{3}$	bar or vinculum	al-Hassar (Arab, late 1100s)	fraction
	$\overline{)}$	into	Michael Stifel (German, 1544)	quotient
	$\div$	divided by	Johann Rahn (Swiss, 1659)	quotient
	$:$ as in 2:3	colon	Gottfried Leibniz (German, 1684)	ratio
	$/$ as in 2/3	slash	Manuel A. Valdes (Mexican, 1784)	fraction
powering	3 as in 2^3	exponent	René Descartes (French, 1637)	power
	$\uparrow$ as in $2 \uparrow 3$	up arrow	(since 1950)	power
	$**$ as in $2**3$	double*	(since 1950)	power
	$\wedge$	carat	(since 1950)	power

In this chapter, you will learn about parentheses and variables. Parentheses ( ) are symbols for grouping. Variables are letters that stand for numbers or other objects, and have many uses.

Order in the kitchen. *Cooks know the importance of mixing ingredients in the proper order. When baking a cake with egg whites, a cook must whip them and fold them into the batter just before baking. Changing the order may ruin the cake.*

Numerical Expressions

On the previous page is a table of symbols for the operations of arithmetic and algebra. There are many symbols, each with a precise meaning.

A **numerical expression** is a combination of symbols for numbers and operations that stands for a number. The **value** of a numerical expression is found by performing the operations. Performing the operations is called **evaluating the expression.** For example, the seven expressions here all have the same value, 2.

$$2 \qquad 5 - 3 \qquad 20/10 \qquad 5 \times 2 - 8 \qquad 2^1 \qquad 347.8 - 345.8 \qquad 10^0 + 10^0$$

The Need for Rules for Order of Operations

The meaning of a numerical expression should be the same for everybody. But recall Lesson 1-5, where you were asked to evaluate $85 + 9 \times 2$ on your calculator. You keyed in

$$85 \;\boxed{+}\; 9 \;\boxed{\times}\; 2 \;\boxed{=}\;.$$

If you have a scientific calculator, the calculator multiplied first, making the problem $85 + 18 = 103$. A nonscientific calculator will probably add first; it will find $94 \times 2 = 188$.

Here is another confusing situation. Suppose you have $25, spend $10, and then spend $4. You will wind up with

$$25 - 10 - 4 \text{ dollars.}$$

The situation tells you that the value of this expression is $11. But someone else evaluating $25 - 10 - 4$ might first subtract the 4 from the 10. This would leave $25 - 6$, or $19.

Calculating powers can also be confusing. Consider the expression 2×3^4. Some people might do the multiplication first, getting 6^4, which equals 1296. Others might first calculate the 4th power of 3, getting 2×81, which equals 162. These values are not even close to each other. To avoid confusion, rules are needed.

Rules for Order of Operations

The rules for order of operations tell the order in which operations should be done. These rules were not stated until this century, but they are now used throughout the world.

Rules for Order of Operations
1. Calculate all powers in order, from left to right.
2. Next do multiplications or divisions in order, from left to right.
3. Then do additions or subtractions in order, from left to right.

To avoid mistakes, it is helpful to write each step clearly on a separate line. The examples below are done this way. After each example, you should check that your calculator evaluates expressions following these rules.

Example 1

Which value, 1296 or 162, is correct for the expression 2×3^4 discussed above?

Solution

Write the original expression.	2×3^4
Calculate the power before doing the multiplication.	$= 2 \times 81$
	$= 162$

162 is the correct value.

Activity 1

Evaluate 2×3^4 on your calculator.

Scientific notation follows the rules for order of operations. An expression like 3.5×10^6 is evaluated by first taking the power, then multiplying.

$$3.5 \times 10^6$$
$$= 3.5 \times 1,000,000$$
$$= 3,500,000$$

The next example applies the second of the rules for order of operations.

Example 2

Evaluate $100 \div 20 \times 2$.

Solution

Write the original expression.
Multiplications and divisions have equal
priority, so work from left to right, doing
the division first.

$$100 \div 20 \times 2$$
$$= 5 \times 2$$
$$= 10$$

Activity 2

Do Example 2 on a calculator.

All multiplications or divisions are done before all additions or subtractions.

Example 3

Evaluate $85 + 9 \times 2$.

Solution

Write the original expression.
Multiply before adding.

$$85 + 9 \times 2$$
$$= 85 + 18$$
$$= 103$$

Activity 3

Verify Example 3 on your calculator.

Example 4

Evaluate $8 \times 12 - 3 \times 12$.

Solution

Write the original expression.
Do *both* multiplications before the subtraction.

$$8 \times 12 - 3 \times 12$$
$$= 96 - 36$$
$$= 60$$

Activity 4

Evaluate $8 \times 12 - 3 \times 12$ using a calculator.

The next example combines three operations and provides a good test of your calculator.

Example 5

Evaluate $10 + 3 \times 4^2$.

Solution

Write the original expression.	$10 + 3 \times 4^2$
First evaluate the power.	$= 10 + 3 \times 16$
Now multiply.	$= 10 + 48$
Now add.	$= 58$

Activity 5

Evaluate $10 + 3 \times 4^2$ on your calculator using the key sequence indicated, and record the results. Do both key sequences produce the same results?

a. 10 [+] 3 [×] 4 [y^x] 2 [=]

b. 10 [+] 3 [×] 4 [x^2] [=]

Noted mathematician and philosopher. *René Descartes (1596-1650), a French mathematician and philosopher, was the first to use a raised number (exponent) to show powers, as in Example 5. His philosophy is known for the statement, "I think, therefore I am."*

QUESTIONS

Covering the Reading

1. **a.** $85 + 9 \times 2$ is an example of a __?__ expression.
   b. What is the value of this expression?

2. Finding the value of an expression is called __?__ the expression.

3. Why is there a need to have rules for order of operations?

4. Paulo had $8.50 this morning. He spent $3.75 for lunch and then spent $2.09 for school supplies.
   a. Use these numbers in a numerical expression telling how much money he had at the end of the day.
   b. Evaluate the expression.

In 5–8, an expression contains only the two given operations. Which one should you perform first?

5. division and addition

6. a power and subtraction

7. multiplication and division

8. addition and subtraction

In 9–18, evaluate the expression. Show your work.

9. $55 - 4 \times 7$

10. $16 - 9 + 7$

11. $200 \div 10 \div 2$

12. $6 + .03 \times 10$

13. $1000 - 3 \times 17^2$

14. $1 \div 9 + 1 \div 7$

15. $3 \times 9 - 2$

16. $4^2 + 8^3$

17. $2 \div 6 \times 9$

18. 6×2^3

19. In which of Activities 1–4 in the lesson did your calculator give a value different from the example above it?

20. What results did you find for Activity 5?

In 21–27, refer to the table on page 175.

21. Name three symbols that are used for multiplication.

22. Translate "three divided by nine" into symbols in 5 different ways.

23. How many years ago were the symbols we use for addition and multiplication invented and by whom?

24. When you add, what is the result called?

25. When you subtract, what is the result called?

26. When you multiply, what is the result called?

27. When you divide, what is the result called?

Applying the Mathematics

In 28 and 29, write the numerical expression, then evaluate it.

28. the sum of 11 and 4.2

29. the product of 6 and 0.3

30. Which would you prefer: an allowance equal to the sum of the digits in the page number for this page or one equal to the product of those digits?

31. Explain why the following sentence is confusing.
Calculate the sum of 2 and 4 divided by 8.

In 32–35, the expression is written in a computer language. Evaluate.

32. $2 * 3 + 8$ **33.** $120 - 3 * 4/4$

34. $200/2 * 10 - 4$ **35.** $17 + 16 * 3 + 2$

36. Why do you think mathematicians invent symbols?

37. *Multiple choice.* Which is largest?
(a) the sum of .1 and .2
(b) the product of .1 and .2
(c) .1 divided by .2
(d) the second power of .1

38. Find two numbers whose product is less than their sum.

39. Find the volume of a cube with an edge of length 5 inches. *(Lesson 3-9)*

40. Use this angle.

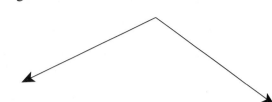

a. Does the angle appear to be acute, right, or obtuse? *(Lesson 3-7)*
b. Measure the angle to the nearest degree. *(Lesson 3-6)*

41. Draw a line segment with length 14.3 cm. *(Lesson 3-1)*

42. Write 0.00000 00000 06543 in scientific notation. *(Lesson 2-9)*

43. Some insurance companies give a discount of 15% on homeowner's or renter's insurance if smoke detectors are installed. Suppose you need 3 smoke detectors and they cost $14.99 each. The insurance bill is $250 a year without smoke detectors. *(Lesson 2-5)*
a. How much will you save each year by installing smoke detectors?
b. What will be the total savings after five years?

Life savers. *Smoke detectors are devices used to warn occupants of a building that a fire has started. The devices are inexpensive, easy to install, and save lives. But regardless of cost, every home should have at least one smoke detector installed.*

Exploration

44. Mathematics is a worldwide language, so mathematicians from different countries usually use the same symbols. Name some symbols outside mathematics that are used throughout the world.

45. What is Esperanto?

46. a. Ask a computer to evaluate the expressions of Questions 32–35. (Hint: For Question 32, you should type **?2*3 + 8** or **PRINT 2*3 + 8** and press **RETURN.**)

b. Does your computer follow the rules for order of operations stated in this lesson? If not, what rules does it seem to follow?

Artistic and natural patterns. *The Navajo blanket shown was woven with various instances of the zig-zag pattern. The same kind of repeating pattern can be found in waves.*

What Are Patterns?

A **pattern** is a general idea for which there are many examples. An example of a pattern is called an **instance.** Here are three instances of a pattern with percent.

$$5\% = 5 \times .01$$
$$43.2\% = 43.2 \times .01$$
$$78\% = 78 \times .01$$

In Lesson 2–4, this pattern was described using English words:

> The percent sign % means to multiply
> the number in front of it by .01.

But there is a simpler way to describe this pattern.

$$n\% = n \cdot .01$$

What Are Variables?

The letter *n* in the above equation is called a **variable.** *A variable is a symbol that can stand for any one of a set of numbers or other objects.* Here *n* can stand for any number. When *n* is 40, we write $n = 40$, and the instance becomes $40\% = 40 \cdot .01$.

Variables are usually letters. With variables, using × for multiplication could be confused with using the letter x. So the raised dot · is used instead.

Describing Number Patterns with Variables

Descriptions with variables have two major advantages over descriptions using words. They look like the instances. Also, they are often shorter than the verbal descriptions. You can see this in Example 1.

Example 1

Here are three instances of a pattern.

$$\frac{3}{3} = 1 \qquad \frac{657.2}{657.2} = 1 \qquad \frac{2/5}{2/5} = 1$$

a. Describe the pattern in words.
b. Describe the pattern with variables.

Solution

a. There is more than one way to write this in words. Here are two possible solutions.
  1. If a number is divided by itself, the quotient is equal to one.
  2. If the numerator and denominator of a fraction are the same number, the value of the fraction is 1.
b. First, write down everything that is the same in all three instances.

$$\frac{\quad}{\quad} = 1$$

Next, determine how many variables are needed. In this pattern, only one variable is needed because the numerator and denominator are the same in each instance.

$$\frac{r}{r} = 1$$

In Example 1, any other letter or symbol could be used in place of r.

Example 2

Describe the pattern with variables.

$$1.43 + 2.9 = 2.9 + 1.43$$
$$12 + 37 = 37 + 12$$
$$\frac{8}{3} + \frac{7}{5} = \frac{7}{5} + \frac{8}{3}$$

Solution

First, write everything that is the same in all the instances.

$$\underline{\quad} + \underline{\quad} = \underline{\quad} + \underline{\quad}$$

Next, determine how many variables are needed. Because each instance has two different numbers, the pattern requires two variables. If a represents the first number and b represents the second, then

$$a + b = b + a.$$

Because any letter may be used as a variable, there are many possible descriptions of a pattern. The sentence $a + b = b + a$ describes the same pattern as the sentence $y + z = z + y$.

A correct description of a pattern must work for all instances. The pattern in Example 2 works for all numbers. It works whether you use decimals or fractions. It is so important that it has a special name you may already know: the Commutative Property of Addition. You will study this property in the next chapter.

Describing Verbal Patterns with Variables

Patterns with words can also be described with variables.

Example 3

Describe the pattern using variables.

> One person has 2 eyes.
> Two people have 4 eyes in all.
> Three people have 6 eyes in all.
> Four people have 8 eyes in all.

Solution

First, rewrite the instances in a way that helps you see the pattern.

> 1 person has $1 \cdot 2$ eyes.
> 2 people have $2 \cdot 2$ eyes in all.
> 3 people have $3 \cdot 2$ eyes in all.
> 4 people have $4 \cdot 2$ eyes in all.

Second, write everything that is the same in all four instances.

> ___ people have ___ $\cdot$ 2 eyes in all.

Since the missing part in each instance is the same number, one variable is all that is needed.

> p people have $p \cdot 2$ eyes in all.

What Is Algebra?

Elementary algebra is the study of variables and the operations of arithmetic with variables. Mathematicians in many different ancient cultures, including Egyptian, Babylonian, Indian, and Chinese, solved problems that today would be solved using algebra. Greeks and Arabs first developed methods for solving problems we call algebra problems.

Algebra, the way it is done worldwide today, began in 1591. That was the year when François Viète (Fraw swah Vee yet), a French mathematician, first used letters of the alphabet to describe patterns. Viète's work quickly led to the invention of a great deal more mathematics. Within 100 years, the ideas behind almost all of elementary algebra and calculus had been discovered. (Notice how many symbols in the chart opening this chapter were invented in the 1600s.) For this reason, Viète is sometimes called the "father of algebra."

He knew the laws of mathematics. *François Viète was a lawyer by profession, but he devoted himself to mathematics. He made major contributions in the areas of algebra, analytic geometry, and trigonometry.*

Covering the Reading

1. What is a variable?

2. What use of variables is explained in this lesson?

3. Name two advantages of using variables to describe patterns.

In 4–7, a pattern is described with variables. Give two instances of the pattern.

4. $\frac{x}{x} = 1$

5. $n\% = n \cdot .01$

6. p people have $p \cdot 2$ eyes.

7. $x + y = y + x$

8. Today's algebra problems were solved long ago by what peoples?

9. What is elementary algebra?

10. Who is sometimes called the "father of algebra," and why?

11. *Multiple choice.* Algebra was developed about how many years ago?
    (a) 100   (b) 200   (c) 400   (d) 1700

Applying the Mathematics

In 12–14, give four instances of the pattern.

12. $12 + y = 5 + y + 7$

13. $6 \cdot a + 13 \cdot a = 19 \cdot a$

14. If your book is d days overdue, your fine will be $20 + d \cdot 5$ cents.

In 15–17, three instances of a general pattern are given. Describe the pattern using variables. Only one variable is needed for each description.

15. $10 \cdot 0 = 0$
     $8.9 \cdot 0 = 0$
     $\frac{15}{5} \cdot 0 = 0$

16. $5 \cdot 40 = 3 \cdot 40 + 2 \cdot 40$
     $5 \cdot \frac{3}{8} = 3 \cdot \frac{3}{8} + 2 \cdot \frac{3}{8}$
     $5 \cdot 0.2995 = 3 \cdot 0.2995 + 2 \cdot 0.2995$

17. In 3 years, we expect $3 \cdot 100$ more students and $3 \cdot 5$ more teachers. In 4 years, we expect $4 \cdot 100$ more students and $4 \cdot 5$ more teachers. In 1 year, we expect 100 more students and 5 more teachers.

18. Give three instances of the pattern $a \cdot b = b \cdot a$.

19. Is $2 + 2 = 2 + 2$ an instance of the pattern $a + b = b + a$? Why or why not?

Books on wheels. *In some areas, library books come to you. Traveling bookmobiles visit different neighborhoods, greatly enlarging the areas serviced by a library.*

20. Examine these four instances.

$$\frac{1}{3} + \frac{5}{3} = \frac{1 + 5}{3}$$

$$\frac{11}{3} + \frac{46}{3} = \frac{11 + 46}{3}$$

$$\frac{0}{3} + \frac{7}{3} = \frac{0 + 7}{3}$$

$$\frac{7}{4} + \frac{2}{4} = \frac{7 + 2}{4}$$

    a. Describe the pattern of the first three instances using two variables.

    b. Describe the pattern of all four instances using three variables.

21. Li noticed that $3 + .5$ is not an integer, $4 + .5$ is not an integer, and $7.8 + .5$ is not an integer.

    a. Describe a general pattern of these three instances.

    b. Find an instance where the general pattern is not true.

    c. Explain why the pattern is not always true.

22. Show that the pattern $x^2 > x$ is false by finding an instance that is not true.

Review

In 23–28, evaluate. *(Lessons 2-2, 2-5, 4-1)*

23. $25\% \times 60 + 40$ **24.** $7 \times 2 \times 8 - 7 \times 2$

25. $60 + 40 \div 4 + 4$ **26.** $12.5 - 11.5 \div 5$

27. $72 - 4 \cdot 3^2$ **28.** $170 - 5^3$

In 29 and 30, translate into mathematical symbols. Then evaluate. *(Lessons 1-1, 1-2)*

29. fifty divided by one ten-thousandth

30. the product of five hundred and five hundredths

31. 300 centimeters is how many meters? *(Lesson 3-4)*

Exploration

In 32–39, examine the sequence of numbers.
a. Find a pattern.
b. Describe the pattern you have found in words or with variables.
c. Write the next term according to your pattern.

32. 6, 12, 18, 24, 30, . . .   **33.** 1, 4, 9, 16, 25, . . .

34. 5, 6, 9, 10, 13, 14, . . .   **35.** $\frac{1}{3}, \frac{8}{3}, \frac{27}{3}, \frac{64}{3}, \frac{125}{3}, \ldots$

36. 2, 5, 8, 11, 14, . . .   **37.** 1, 3, 6, 10, 15, 21, . . .

38. 1, 1, 2, 3, 5, 8, 13, . . .   **39.** 4, 7, 13, 25, 49, . . .

Silent translation. *Sign language is a visual language for the hearing impaired. The American Sign Language is the fourth most common language in the U.S.*

Algebraic Expressions

Recall from Lesson 4-1 that $4 + 52$, $9 - 6 \cdot 7$, and $\frac{1}{6}$ are numerical expressions. You know how to translate English into numerical expressions.

English expression	numerical expression
the sum of three and five	$3 + 5$
the product of two tenths and fifty	$.2 \times 50$

If an expression contains a variable alone or with number and operation symbols, it is called an **algebraic expression.** Here are some algebraic expressions.

$$t \qquad 3 \cdot a^2 \qquad z + \frac{400.3}{5} \qquad m - n + m$$

English expressions can be translated into algebraic expressions.

English expression	algebraic expression
the sum of a number and five	$n + 5$
the product of length and width	$\ell \cdot w$

The value of an algebraic expression depends on what is substituted for the variables. You will evaluate algebraic expressions in the next lesson.

Words Leading to Algebraic Expressions with Addition or Subtraction

Many English expressions can translate into the same algebraic expression. Below are some common English expressions and their translations. Notice that in subtraction you must be careful about the order of the numbers.

English expression	algebraic expression
a number *plus* five the *sum* of a number and 5 a number *increased* by five five *more than* a number *add* five to a number	$a + 5$ or $5 + a$
a number *minus* eight *subtract* 8 from a number 8 *less than* a number a number *less* 8 a number *decreased by* 8	$h - 8$
eight *minus* a number *subtract* a number from 8 8 *less* a number 8 *decreased by* a number	$8 - n$

Often you have a choice of what letter to use for a number. We use the letter S in Example 1 because the word *salary* begins with that letter.

Example 1

A person's annual salary is S. It is increased by $700. What is the new salary?

Solution

"Increase" means "add to." So add $700 to S. The answer is $S + \$700$.

Example 2

Carmen is five years younger than her sister Anna. If Anna's age is A, what expression stands for Carmen's age?

Solution

Carmen's age is five less than Anna's.

Five less than A is $A - 5$. The answer is $A - 5$.

Words Leading to Algebraic Expressions with Multiplication or Division

Here are some English expressions for multiplication and division. In division, as in subtraction, you must be careful about the order of the numbers.

English expression	algebraic expression
two *times* a number the *product* of two and a number *twice* a number	$2 \cdot m$ or $m \cdot 2$
six *divided by* a number a number *divided into* six	$\frac{6}{u}$
a number *divided by* six six *divided into* a number	$\frac{u}{6}$

Some English expressions combine operations.

Example 3

Translate "five times a number, increased by 3."

Solution

Let n stand for the number.
Five times *n* is $5 \cdot n$.
$5 \cdot n$ increased by 3 is $5 \cdot n + 3$.

In Example 3, suppose there were no comma after the word *number*. The expression "five times a number increased by 3" would then be ambiguous. *Ambiguous* means the expression has more than one possible meaning. We would not know which to do first, to increase the number by 3, or to multiply the number by 5.

You can think of algebra as a language. As with any language, it is useful to be able to translate to and from other languages. Throughout this book, you will get a lot of practice in order to increase your ability to translate.

A country of many languages. *Nestled in the heart of Europe and bordering four other countries, Switzerland is a country with four national languages—German, French, Italian, and Romansh. The Swiss town shown is Spiez.*

Covering the Reading

1. What is the difference between a numerical expression and an algebraic expression?

In 2–11, let *n* stand for the number. Then translate the English into an algebraic expression.

2. twice the number

3. three more than the number

4. the number multiplied by four

5. the number less five

6. six less the number

7. seven less than the number

8. eight into the number

9. the number divided by nine

10. the number increased by ten

11. eleven decreased by the number

12. Translate into mathematics. Be careful. The answers are all different.
    a. six is less than a number
    b. six less than a number
    c. six less a number

13. A person's salary is currently *S* dollars a week. Write an expression for the new salary if:
    a. the person gets a raise of $50 a week.
    b. the salary is lowered by $12 a week.
    c. the salary is tripled.

14. What is the meaning of the word *ambiguous?*

15. Give an example of an English expression that is ambiguous.

In 16 and 17, give three possible English expressions for the algebraic expression.

16. $x + 10$

17. $2 - y$

18. Translate "a number times six, decreased by five" into an algebraic expression.

I love my new Cougar !

Applying the Mathematics

19. Tell why "fourteen less five plus three" is ambiguous.

In 20 and 21, translate into an algebraic expression. Use *C* to stand for the number. In working with a number between 0 and 1, the word *of* is often a signal to multiply.

20. half of the number

21. 6% of the number

22. Write two algebraic expressions for "a number times itself" using two different operations.

23. Much of what we know about mathematics in ancient Egypt comes from two documents written on papyrus, the *Ahmes Mathematical Papyrus* and the *Moscow Mathematical Papyrus*. Both documents were written around 1650 B.C. Problem 26 on the *Ahmes Papyrus* asks you to find a quantity such that when it is added to $\frac{1}{4}$ of itself, the result is 15. Translate the expression "a quantity added to $\frac{1}{4}$ of itself" into algebra.

24. "Trebled" means "multiplied by three."
a. What does "quintupled" mean?
b. What is the word for "multiplying by four"?

25. Why is the "quotient of 2 and 4" an ambiguous phrase?

Key to the past. *Pictured here is part of the Ahmes Papyrus, named after the scribe who copied it. It is sometimes called the Rhind Papyrus, named after Alexander Rhind of Scotland who purchased it in 1851.*

Review

26. Give three instances of this pattern. *(Lesson 4-2)*
$$7 \cdot x - x = 6 \cdot x$$

27. Three instances of a general pattern are given. Describe the pattern using one variable. *(Lesson 4-2)*
$$\frac{1 \text{ million}}{1} = 1 \text{ million}$$
$$\frac{10^2}{1} = 10^2$$
$$\frac{8.3}{1} = 8.3$$

28. Four instances of a general pattern are given. Describe the pattern using two variables. *(Lesson 4-2)*
$$4 + 5 + 12 - 5 = 4 + 12$$
$$\tfrac{1}{2} + 5 + \tfrac{1}{3} - 5 = \tfrac{1}{2} + \tfrac{1}{3}$$
$$1.7 + 5 + 6 - 5 = 1.7 + 6$$
$$0 + 5 + 0 - 5 = 0 + 0$$

In 29–31, evaluate each expression. Show your work. *(Lesson 4-1)*
29. $5^2 + 9^2$ **30.** $2090 - 4 \times 16^2$ **31.** $23 + .09 \times 11$

32. Find the volume of a cubical box with 10-foot edges. *(Lesson 3-9)*

33. a. Draw a triangle with three acute angles.
b. Draw a triangle with a right angle.
c. Draw a triangle with an obtuse angle. *(Lessons 3-6, 3-7)*

34. It is about 1110 km by air from Paris to Rome. In miles, about how far is it? *(Lesson 3-5)*

35. Measure this segment to the nearest eighth of an inch. *(Lesson 3-1)*

36. a. Give a word name for 10^{-5}.
   b. Give a decimal for 10^{-5}.
   c. Give a fraction for 10^{-5}. *(Lesson 2-8)*

37. According to the U.S. Bureau of the Census, the 1990 resident population of 248,709,873 was distributed approximately as follows:

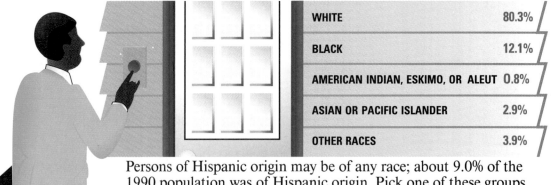

WHITE	**80.3%**
BLACK	**12.1%**
AMERICAN INDIAN, ESKIMO, OR ALEUT	**0.8%**
ASIAN OR PACIFIC ISLANDER	**2.9%**
OTHER RACES	**3.9%**

Persons of Hispanic origin may be of any race; about 9.0% of the 1990 population was of Hispanic origin. Pick one of these groups and estimate the number of people in the United States in that group in 1990. *(Lesson 2-5)*

38. Compare the two sides of each line listed below. Without calculating, decide if $<$, $=$, or $>$ makes each sentence true. Be ready to justify your choice. *(Previous course, Lesson 1-9)*

   a. $13 \times 17 \times 9 \underline{?} 17 \times 9 \times 13$
   b. $218 + 617 + 314 + 8 \underline{?} 617 + 314 + 8 + 218$
   c. $373 + 936 + 475 \underline{?} 327 + 934 + 306$
   d. $208 + 208 + 208 \underline{?} 3 \times 208$
   e. $42 + 43 + 44 \underline{?} 43 \times 3$
   f. $52 \times 25 \underline{?} 25 \times 52$
   g. $530 \div 8 \underline{?} 530 \div 10$
   h. $1840 \div 65 \underline{?} 1820 \div 65$
   i. $67 \times 73 \underline{?} 76 \times 73$

INPUT
1
2
3
4
5

OUTPUT
3
6
11
18
27

Exploration

39. This machine performs the same operations with each input number, resulting in the output numbers in the order given. If a number n is input into the machine, what is the output?

4-4

Evaluating Algebraic Expressions

Libraries use mathematics. *Library fines, book inventories, and budgets may be determined using algebraic expressions. Shown here is the main floor of Harper Memorial Library, one of several libraries at the University of Chicago.*

A library charges 20¢ if a book is not returned on time. Added to the fine is 5¢ for each day overdue. Now let *n* be the number of days overdue. Then the fine is

$$20 + 5 \cdot n \text{ cents.}$$

The fine can now be calculated for any number of days overdue. For a book 13 days overdue, just replace *n* by 13. The fine is

$$20 + 5 \cdot 13 \text{ cents.}$$

This computes to 85¢.

We say that 13 days is the **value of the variable** *n*. We can write *n* = 13. The quantity 85¢ is the **value of the expression.** We have evaluated the expression by letting *n* = 13 and finding the value of the resulting expression.

Example 1

Find the value of the expression $25 - 3 \cdot n$ when:
a. *n* = 2.   **b.**   *n* = 8.

Solution

a. Substitute 2 for *n*. Then evaluate the resulting arithmetic expression.

When n = 2, $25 - 3 \cdot n = 25 - 3 \cdot 2 = 19.$

b. Substitute 8 for *n*.

When n = 8, $25 - 3 \cdot n = 25 - 3 \cdot 8 = 1.$

We have been using a dot to stand for multiplication. When one of the numbers being multiplied is a variable, the dot is usually not used.

With · for multiplication	Without ·
$6 \cdot t$	$6t$
$27 - 2 \cdot r + r$	$27 - 2r + r$
$4 \cdot x + 3 \cdot y$	$4x + 3y$

The rules for order of operations apply even when the multiplication symbol is absent.

Example 2

Find the value of each expression if $r = 5$.
a. $6r$ **b.** $27 - 2r + r$ **c.** πr^2

Solution

a. If $r = 5$, $6r = 6 \cdot 5 = 30$.

b. Do the multiplication before the subtraction or addition.

$$
\begin{aligned}
\text{If } r = 5, \ 27 - 2r + r &= 27 - 2 \cdot 5 + 5 \\
&= 27 - 10 + 5 \\
&= 22
\end{aligned}
$$

c. Do the power before the multiplication by π.

$$
\begin{aligned}
\text{If } r = 5, \ \pi r^2 &= \pi \cdot 25 \\
&\approx 3.14 \cdot 25 \\
&\approx 78.5
\end{aligned}
$$

When an expression contains more than one variable, you need a value for each variable to get a numerical value for the expression.

Example 3

Evaluate $4x + 3y$ when $x = 10$ and $y = 7$.

Solution

Substitute 10 for x and 7 for y.

$$
\begin{aligned}
4x + 3y &= 4 \cdot 10 + 3 \cdot 7 \\
&= 40 + 21 \\
&= 61
\end{aligned}
$$

QUESTIONS

Covering the Reading

1. *Multiple choice.* If $n = 3$, then $5 \cdot n =$
   (a) 53   (b) 8   (c) 15   (d) none of these

2. Consider the expression $5 \cdot n$ from Question 1. Identify the:
  a. variable. **b.** value of the variable.
  c. expression. **d.** value of the expression.

3. What is the more common way of writing $5 \cdot n$?

4. Do the rules of order of operations apply to variables?

5. Suppose that a book is n days overdue and the fine is $20 + 5n$ cents. Calculate the fine for a book that is:
  a. 1 day overdue. **b.** 6 days overdue.
  c. 20 days overdue.

6. The area of a circle with radius r is πr^2. Which example in this lesson calculates the area of a circle with radius 5?

In 7–10, evaluate the expression when d is 5.
  7. $d + d$ **8.** $88 - 4d$

  9. $2 + 3d$ **10.** $d\%$

In 11–14, give the value of the expression when $m = 5$ and $x = 9$.
  11. $4m + 7x$ **12.** $2mx$

  13. $1.6x + m^3$ **14.** πx^2

Applying the Mathematics

15. Let A be an age between 1 and 7 years. A boy of age A weighs, on the average, about $17 + 5A$ pounds.
  a. What is the average weight for 6-year-old boys?
  b. What is the average weight for 2-year-old boys?
  c. For each additional year of age, by how much does the average weight change?

16. Suppose x is 4712 and y is 368.
  a. Evaluate $xy - yx$.
  b. Will the answer to part **a** change if the values of x and y are changed? Why or why not?

17. a. Evaluate $2v + 1$ when v is 1, 2, 3, 4, and 5.
  b. Your answers to part **a** should form a pattern. Describe the pattern in English.

In 18–20, an English expression is given.
a. Translate into an algebraic expression.
b. Evaluate that expression when the number has the value 10.

18. eight less than five times a number

19. the product of a number and 4, increased by nine

20. the third power of a number

21. *Multiple choice.* Which is the largest? *(Lesson 4-3)*
    (a) the sum of 10 and 1   (b) the product of 10 and 1
    (b) 10 divided by 1   (d) 10 to the first power

22. Four horses have 4 · 4 legs, 4 · 2 ears, and 4 tails. *(Lesson 4-2)*
    a. Six horses have 6 · 4 legs, __?__ · 2 ears, and __?__ tails.
    b. Eleven horses have __?__ legs, __?__ ears, and __?__ tails.
    c. *H* horses have __?__ legs, __?__ ears, and __?__ tails.

In 23 and 24, three instances of a pattern are given. Describe the pattern using one variable. *(Lesson 4-2)*

23. $5 + 0 = 5$
     $43.0 + 0 = 43.0$
     $\frac{1}{2} + 0 = \frac{1}{2}$

24. $1 \times 60\% = 60\%$
     $1 \times 2 = 2$
     $1 \times 1 = 1$

25. What is the metric prefix meaning $\frac{1}{1000}$? *(Lesson 3-4)*

26. Convert $\frac{2.4}{10.24}$ into a simple fraction in lowest terms. *(Lesson 2-6)*

27. *Multiple choice.* Which two of these refer to the same numbers?
    (Lessons 1-1, 1-8)
    (a) the whole numbers   (b) the natural numbers
    (c) the integers   (d) the positive integers

Horse history. *Fossils found in Europe and North America indicate that horses date back at least 54 million years. Around 10,000 years ago, horses mysteriously disappeared from North America. Horses were then reintroduced by Spaniards in the early 16th century.*

Exploration

28. A library decides to charge *m* cents for an overdue book and *A* more cents for every day the book is overdue. What will be the fine for a book that is *d* days overdue?

29. Computers can evaluate algebraic expressions. Here is a program that evaluates a particular expression.

```
10 PRINT "GIVE VALUE OF YOUR VARIABLE"
20 INPUT X
30 V = 30 * X − 12
40 PRINT "VALUE OF THE EXPRESSION IS " V
50 END
```

    a. What expression does the above program evaluate?
    b. What value does the computer give if you input 3.5 for *X*?
    c. Modify the program so that it evaluates the expression $25X + X^4$ and test your program when $X = 1$, $X = 2$, and $X = 17$. What values do you get?

LESSON
4-5

Parentheses

Winter wonderland. *The queen of the Quebec Winter Carnival is crowned in front of a genuine ice castle in Quebec, Canada. The annual carnival, attracting 500,000 tourists, is held in early February when the average temperature of 12°F, or -11°C, ensures the ice will remain firm.*

A Formula That Uses Parentheses

Here are the directions given in an almanac for converting Fahrenheit temperatures (the U.S. everyday temperature scale) to Celsius (the temperature scale used by all scientists and people almost everywhere outside the U.S.).

> To convert Fahrenheit to Celsius,
> subtract 32 degrees and multiply by 5, divide by 9.

With algebra, this description can be shortened. Let F be the Fahrenheit temperature. First subtract 32 from F, resulting in

$$F - 32.$$

Now this entire expression is to be multiplied by 5 and divided by 9. This is equivalent to multiplying by $\frac{5}{9}$. Parentheses are used to show this multiplication.

$$\frac{5}{9} \cdot (F - 32)$$

The parentheses signify that the subtraction of 32 is done first. If there were no parentheses, the multiplication would be done first.

Order of Operations Parentheses Rule
Work inside parentheses before doing anything else.

Example 1

What is the Celsius equivalent of a Fahrenheit temperature of 68°?

Solution

Using the information on page 197, the Celsius equivalent is
$$\frac{5}{9} \cdot (F - 32),$$
where $F = 68$. So substitute 68 for F.

$$\frac{5}{9} \cdot (F - 32) = \frac{5}{9} \cdot (68 - 32)$$
$$= \frac{5}{9} \cdot (36)$$
$$= \frac{5}{9} \cdot 36$$
$$= 20$$

68° Fahrenheit is equal to 20° Celsius.

In the Solution to Example 1 we removed the parentheses around 36. You can drop parentheses in an expression whenever doing so does not change the value of the expression.

As with other order-of-operation problems, it is useful to write the steps of calculations vertically as shown in the examples below.

Example 2

Simplify $7 + 9 \cdot (2 + 3)$.

Solution

Write the original expression.	$7 + 9 \cdot (2 + 3)$
Work inside ( ) first.	$= 7 + 9 \cdot (5)$
Drop the parentheses.	$= 7 + 9 \cdot 5$
Now evaluate as before.	$= 7 + 45$
	$= 52$

Work inside parentheses even before taking powers.

Example 3

Evaluate $6 + 5 \cdot (4x)^3$ when $x = 2$.

Solution

Write the original expression.	$6 + 5 \cdot (4x)^3$
Substitute 2 for x.	$= 6 + 5 \cdot (4 \cdot 2)^3$
Work inside ( ).	$= 6 + 5 \cdot (8)^3$
Powering comes next.	$= 6 + 5 \cdot 512$
Multiplication comes next.	$= 6 + 2560$
Addition comes last.	$= 2566$

The dot for multiplication is generally not used with parentheses. For the expression in Example 3, it is more common to see $6 + 5(4x)^3$. The 5 next to the parentheses signals multiplication.

Example 4

Evaluate $(y + 15)(11 - 2y)$ when $y = 3$.

Solution 1

Write the original expression.	$(y + 15)(11 - 2y)$
Substitute 3 for y everywhere y occurs.	$= (3 + 15)(11 - 2 \cdot 3)$
Work inside (). Follow order of operations.	$= (18)(11 - 6)$
	$= 18 \cdot 5$
	$= 90$

Using Parenthesis Keys on Calculators

Most scientific calculators have parenthesis keys ⟨ and ⟩. To use these keys, enter the parentheses where they appear in the problem.

Remember you may need to key in ⟨×⟩ even though the times sign is not written in the problem. Here is how Example 4 would be done with a calculator.

Solution 2

Key in: ⟨(⟩ 3 ⟨+⟩ 15 ⟨)⟩ ⟨×⟩ ⟨(⟩ 11 ⟨−⟩ 2 ⟨×⟩ 3 ⟨)⟩ ⟨=⟩.

The problem is not very difficult, but the calculator solution requires many key strokes. So unless the numbers are quite complicated, most people prefer not to use a calculator for this kind of problem.

Nested Parentheses

It is possible to have parentheses inside parentheses. These are called **nested parentheses.** With nested parentheses, work inside the innermost parentheses first.

Example 5

Simplify $300 - (40 + .6(30 - 8))$.

Solution

Write the original expression.	$300 - (40 + .6(30 - 8))$
Work in the innermost () first.	$= 300 - (40 + .6(22))$
Multiply before adding.	$= 300 - (40 + 13.2)$
Work inside the remaining ().	$= 300 - (53.2)$
	$= 246.8$

Some calculators enable you to nest parentheses; some do not. You should explore your calculator to see what it allows. (See Question 39.)

The given expression in Example 5 is too long for some calculators. With these calculators, you may have to evaluate the expression a little at a time.

Activity

Describe how to evaluate $300 - (40 + .6(30 - 8))$ on your calculator.

QUESTIONS

Covering the Reading

1. What is one reason for using parentheses?

In 2–5, *true or false*.

2. $5 + 4 \cdot 3 = (5 + 4) \cdot 3$ **3.** $5 + 4 \cdot 3 = 5 + (4 \cdot 3)$

4. $5(3) = 15$ **5.** $2 + (4) = 8$

6. *Multiple choice.* When evaluating an expression with parentheses, how should you begin?
(a) First do additions or subtractions, from left to right.
(b) First do multiplications or divisions, from left to right.
(c) First work inside parentheses.
(d) First take powers.
(e) Work from left to right regardless of the operation.

In 7–12, evaluate.

7. $4 + 3(7 + 9)$ **8.** $(12)(3 + 4)$

9. $10 + 20 \div (2 + 3 \cdot 6)$ **10.** $40 - (30 - 5)$

11. $(6 + 6)(6 - 6)$ **12.** $3(2 + 4)^5$

13. a. Multiplication occurs twice in the expression $2(5 + 4n)$. What are these places?
b. Suppose $n = 8$. Write the key sequence that will evaluate $2(5 + 4n)$ on your calculator.

14. Find the Celsius equivalent of 122° Fahrenheit.

In 15–20, $a = 2$, $b = 3$, and $c = 5$. Evaluate each expression.

15. $0.30(c - b)$ **16.** $(a + b)(7a + 2b)$

17. $(a + 100b) - (7a - 2b)$ **18.** $20a + 5(c - 3)$

19. $a^{(b + c)}$ **20.** $2(a + 2b)^3$

21. What are *nested* parentheses?

22. a. To simplify $1000 - (100 - (10 - 1))$, what should you do first?
b. Simplify $1000 - (100 - (10 - 1))$.

23. Give a key sequence for evaluating $300 - (40 + .6(30 - 8))$ on your calculator.

In 24 and 25, let $a = 10$ and $b = 31$. Which expression has the larger value?

24. $b - (a - 2)$ or $b - a - 2$ **25.** $a + 4(b + 3)$ or $(a + 4)b + 3$

In 26 and 27, when $n = 3$, do the two expressions have the same value?

26. $4(n + 2)$ and $4n + 8$ **27.** $33 - 7n$ and $(33 - 7)n$

28. *Multiple choice.* Begin with a number n. Add 5 to it. Multiply the sum by 4. What number results?
(a) $n + 5 \cdot 4$ (b) $4(n + 5)$
(c) $n + 5 + n \cdot 4$ (d) $n + 5 + (n + 5)4$

29. To convert a Celsius temperature C to its Fahrenheit equivalent, you can multiply the temperature by 9, divide the product by 5, and then add 32.
  a. *Multiple choice.* Which algebraic expression describes this rule?
  (i) $\dfrac{9(C + 32)}{5}$ (ii) $9C + \dfrac{32}{5}$
  (iii) $\dfrac{9C}{5} + 32$ (iv) $\dfrac{9C + 32}{5}$
  b. What is the Fahrenheit equivalent of 23° Celsius?

30. Three instances of a pattern are given. Describe the general pattern using three variables.

$$11(3 + 2) = 11 \cdot 3 + 11 \cdot 2$$
$$5(12 + 19.3) = 5 \cdot 12 + 5 \cdot 19.3$$
$$2(0 + 6) = 2 \cdot 0 + 2 \cdot 6$$

31. Evaluate $2(3 + n(n + 1) + 5)$ when $n = 14$.

In 32 and 33, insert parentheses to make the sentence true.

32. $16 - 15 - 9 = 10$ **33.** $16 - 8 - 4 - 2 = 14$

In 34 and 35, consider the expression $6T + E + 3F + 2S$. *(Lessons 4-3, 4-4)*

34. Evaluate this expression when $T = 3$, $E = 2$, $F = 1$, and $S = 0$.

35. This expression has something to do with scoring in football. The letters T, E, F, and S have been chosen carefully. With these hints,
  a. what precisely do T, E, F, and S stand for?
  b. what does the value of the expression tell you?

36. Three instances of a pattern are given. Describe the general pattern using one variable. *(Lesson 4-2)*
  $3^4 = 3 \cdot 3 \cdot 3 \cdot 3$ $6^4 = 6 \cdot 6 \cdot 6 \cdot 6$ $1^4 = 1 \cdot 1 \cdot 1 \cdot 1$

37. In many places in the Midwest, a township is a square, 6 miles on a side. What is the area of a township? *(Lesson 3-8)*

Safety! *The official is signalling a "safety," giving 2 points to the team on defense. Typically, a safety is scored when the ball carrier is tackled in his own end zone or when a blocked punt goes out of the end zone. The scoring team then gains possession of the ball.*

38. A person types a line in 5.7 seconds. Another person types it in one tenth of a second less time. What is the second person's time? *(Lesson 1-2)*

---

39. Use the following exercise to find out how many nested parentheses with operations inside your calculator will allow. Key in

$$100 - (1 - (2 - (3 - (4 \ldots \text{ and so on.}$$

When your calculator shows an error message, the last number keyed in is the calculator's limit. What is your calculator's limit?

40. This program finds values of the expression in Example 4 of this lesson.

```
10 PRINT "GIVE VALUE OF YOUR VARIABLE"
20 INPUT Y
30 V = (Y + 15) * (11 − 2 * Y)
40 PRINT "VALUE OF EXPRESSION IS " V
50 END
```

a. Run the program when $Y = 3$ to check that it gives the value of Example 4.
b. Run the program for at least five other values of Y.

41. Modify the program in Question 40 so that it gives the Celsius equivalent of any Fahrenheit temperature.

Inside out. *Just as we must simplify complicated expressions by working within innermost parentheses first, the animal must find a way out of the innermost section first.*

Brackets

Parentheses are the most common **grouping symbols. Brackets []** are grouping symbols sometimes used when there are nested parentheses. As grouping symbols, brackets and parentheses mean the same thing.

$$5[x + 2y(3 + z)] \text{ and } 5(x + 2y(3 + z)) \text{ are identical.}$$

Many people find brackets clearer than a second pair of parentheses. But some calculators and computer languages do not allow brackets.

Example 1

Simplify $2[4 + 6(3 \cdot 5 - 4)] - 3(30 - 3)$.

Solution

Write the original expression.	$2[4 + 6(3 \cdot 5 - 4)] - 3(30 - 3)$
Evaluate the innermost () first.	$= 2[4 + 6(15 - 4)] - 3(30 - 3)$
Now there are no nested groupings.	$= 2[4 + 6 \cdot 11] - 3(30 - 3)$
	$= 2[4 + 66] - 3(30 - 3)$
	$= 2 \cdot 70 - 3 \cdot 27$
Do multiplication before subtraction.	$= 140 - 81$
	$= 59$

The Fraction Bar

Another important grouping symbol is the *fraction bar*. You may not have realized that the fraction bar acts like parentheses. Here is how it works.

Suppose you want to calculate the average of 10, 20, and 36. The average is given by the expression

$$\frac{10 + 20 + 36}{3}.$$

If you write this fraction using the slash, /, like this,

$$10 + 20 + 36/3,$$

then, by order of operations, the division will be done first and only the 36 will be divided by 3. So you need to use parentheses.

$$\frac{10 + 20 + 36}{3} = (10 + 20 + 36)/3$$

In this way the slash, /, and the fraction bar, −, are different. (Of course, with something as simple as $\frac{1}{2}$ or 1/2, there is no difference.) A fraction bar always operates as if it has unwritten parentheses. Because the fraction bar is a grouping symbol, you *must* evaluate the numerator and denominator of a fraction separately before dividing.

Example 2

Simplify $\frac{4 + 9}{2 + 3}$.

Solution

Think $\frac{(4 + 9)}{(2 + 3)}$ and get $\frac{13}{5}$.

Example 3

Evaluate $\frac{4n + 1}{3n - 1}$ when $n = 11$.

Solution 1

Write the original expression.	$\frac{4n + 1}{3n - 1}$
Substitute.	$= \frac{4 \cdot 11 + 1}{3 \cdot 11 - 1}$
Work separately in the numerator and denominator.	$= \frac{44 + 1}{33 - 1}$
	$= \frac{45}{32}$

Solution 2

If you use a calculator to evaluate $\frac{4 \cdot 11 + 1}{3 \cdot 11 - 1}$, you must insert parentheses to ensure that the entire numerator is divided by the entire denominator. One possible key sequence is shown here.

$$(\ 4 \ \times \ 11 \ + \ 1 \ ) \ \div \ (\ 3 \ \times \ 11 \ - \ 1 \ ) \ =$$

A calculator displays 1.40625, the decimal equivalent to $\frac{45}{32}$.

At times, as in Example 3, using a calculator may be more complicated and time-consuming than evaluating with paper and pencil. Also, it loses the fraction. So in evaluating a fraction, it is not always efficient to use a calculator.

Parentheses in Fractions

Parentheses in the numerator or denominator of a fraction are like the innermost of nested parentheses. So work with them first.

Example 4

Evaluate $x + \dfrac{100(4 + 2x) - 25}{200 - x}$ when $x = 50$.

Solution

First substitute 50 for x wherever x appears.

$$50 + \frac{100(4 + 2 \cdot 50) - 25}{200 - 50}$$

Work inside the parentheses first.

$$= 50 + \frac{100(104) - 25}{200 - 50}$$
$$= 50 + \frac{10{,}400 - 25}{150}$$
$$= 50 + \frac{10{,}375}{150}$$
$$= 50 + 69.1\overline{6}$$
$$= 119.1\overline{6}$$

Summary of Rules of Order of Operations

1. First, do operations within parentheses or other grouping symbols. If there are nested grouping symbols, work within the innermost symbols first. Remember that fraction bars are grouping symbols and can be different from /.
2. Within grouping symbols or if there are no grouping symbols:
   A. First, take all powers.
   B. Second, do all multiplications or divisions in order, from left to right.
   C. Then do all additions or subtractions in order, from left to right.

QUESTIONS

Covering the Reading

1. What are the symbols [ ] called?

2. *True or false.* $2[x + 4]$ and $2(x + 4)$ mean the same thing.

3. When are the symbols [ ] usually used?

4. Name three different grouping symbols.

5. When there are grouping symbols within grouping symbols, what should you do first?

In 6–8, simplify. Show your work.

6. $3[2 + 4(5 - 2)]$

7. $39 - [20 \div 4 + 2(3 + 6)]$

8. $[(3 - 1)^3 + (5 - 1)^4]^2$

9. a. Write a key sequence for evaluating the expression of Question 6 on your calculator.

b. Which do you think is the better method for Question 6: paper and pencil, or calculator? Explain your choice.

10. *Multiple choice.* Written on one line, $\dfrac{20 + 2 \cdot 30}{6 + 4} =$

(a) $20 + 2 \cdot 30/6 + 4$.

(b) $20 + 2 \cdot 30/(6 + 4)$.

(c) $(20 + 2 \cdot 30)/6 + 4$.

(d) $(20 + 2 \cdot 30)/(6 + 4)$.

In 11 and 12, simplify. Show each step.

11. $\dfrac{50 + 40}{50 - 40}$

12. $\dfrac{560}{7(6 + 3 \cdot 4.5)}$

13. a. Write a key sequence for evaluating Question 12 on your calculator.

b. Which do you think is the better method for doing Question 12: paper and pencil, or calculator?

In 14 and 15, evaluate when $a = 5$ and $x = 4$. Show each step.

14. $\dfrac{a + 3x}{a + x}$

15. $\dfrac{5x - 2}{(x - 1)(x - 2)}$

Applying the Mathematics

In 16–19, recall that the **mean** or **average** of a collection of numbers is their sum divided by the number of numbers in the collection.

16. Write an expression for the mean of a, b, c, d, and e.

17. A bookcase has three shelves with 42, 37, and 28 books on them. What is the average number of books on a shelf of this bookcase?

18. A student scores 83, 91, 86, and 89 on 4 tests. What is the average?

19. Grades can range from 0 to 100 on tests. A student scores 85 and 90 on the first two tests.

a. What is the lowest the student can average for all 3 tests?

b. What is the highest the student can average for the 3 tests?

In 20 and 21, show each step in evaluating the expression.

20. $5[x + 2y(3 + 2z)]$ when $x = 1$, $y = 2$, and $z = 3$

21. $\dfrac{x + 3y}{z} + \dfrac{4y + z}{3x}$ when $x = 3$, $y = 2$, and $z = 1$

In 22 and 23, insert grouping symbols to make the equation true.

22. $3 + 5 \cdot 6 - 8 \cdot 2 = 80$

23. $3 \cdot 8 - 6/2 + 3 = 12$

24. Write the algebraic expression of Example 4 on one line.

25. Translate into mathematical symbols. *(Lesson 4-3)*
   a. the sum of thirty and twenty
   b. three more than a number
   c. a number decreased by one tenth

26. Here are four instances of a pattern. Describe the pattern using 1 variable. *(Lesson 4-2)*

$$2 \cdot 3 + 2 \cdot 3 = 4 \cdot 3$$
$$2(7.2) + 2(7.2) = 4(7.2)$$
$$2 \cdot \frac{3}{8} + 2 \cdot \frac{3}{8} = 4 \cdot \frac{3}{8}$$
$$2(6\%) + 2(6\%) = 4(6\%)$$

27. Suppose a formal garden is made up of squares as in the array below.

Each of the squares is 8 feet on a side. A bag of organic fertilizer covers 25 square yards and costs $6.95. How much will it cost to fertilize the entire garden? *(Lesson 3-8)*

28. According to the *New York Times,* June 26, 1992, about 4% of the 150 million cars on the road in the U.S. in 1992 had airbags. Estimate the number of U.S. cars with airbags in 1992. *(Lesson 2-5)*

29. The U.S. budget deficit in 1991 was about $268,729 million. (Source: *The World Almanac and Book of Facts* 1993.)

   a. Write this number as a decimal.

   b. Round this number to the nearest billion.

   c. Write the original number in scientific notation.
   (Lessons 1-4, 2-1, 2-3)

30. Arrange the numbers $7\frac{2}{3}$, $7\frac{3}{5}$, and 7.65 in correct order with the symbol $>$ between them. *(Lessons 1-6, 1-9)*

Exploration

31. The numerical expression $4 + \frac{4}{4} - 4$ uses four 4s and has a value of 1. Find numerical expressions using only four 4s that have the values of each integer from 2 to 10. You may use any operations and grouping symbols.

Formulas

Formulas in a can. *Due to the wide range of colors of paint that people desire, paint sellers often stock several base paints. Then they mix these colors according to exact formulas to create the desired colors.*

A Formula for the Area of a Rectangle

The rectangle drawn here has an actual length of about 6.3 cm, a width of about 2.5 cm, and area about 15.75 cm^2.

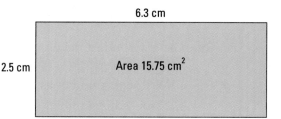

6.3 cm

2.5 cm   Area 15.75 cm^2

The length, width, and area are related by a simple pattern.

$$\text{Area} = \text{length times width}$$

Using the variables A for area, ℓ for length, and w for width,

$$A = \ell w.$$

(Remember that multiplication signs are usually not written between variables.) The sentence $A = \ell w$ is a **formula** for the area of a rectangle **in terms of** its length and width.

Formulas are very useful. For instance, the formula $A = \ell w$ works for *any* rectangle. So suppose a rectangular field has length of 110 ft and a width of 30 ft. Then its area can be calculated using the formula.

$$A = 110 \text{ ft} \cdot 30 \text{ ft}$$
$$= 3300 \text{ ft}^2$$

Units in Formulas

Remember, area is typically measured in *square* units. It is therefore important when using the formula $A = \ell w$ that the length and width are given in the *same* units. For instance, suppose a rectangle has width 3 inches and length 4 feet. You should first either change inches to feet or feet to inches to calculate the area. Then you will get a sensible unit for measuring area, square inches or square feet.

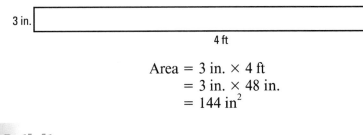

$$\text{Area} = 3 \text{ in.} \times 4 \text{ ft}$$
$$= 3 \text{ in.} \times 48 \text{ in.}$$
$$= 144 \text{ in}^2$$

Activity

Find the area of the front cover of your textbook.

Step 1: First, measure the length and width of the front cover to the nearest tenth of a centimeter.

Step 2: Use the area formula to find the area.

Repeat Steps 1 and 2 with measurements made to the nearest quarter inch.

Choosing Letters in Formulas

Letters in formulas are chosen carefully. Usually they are the first letters of the quantities they represent. That is why we use A for area, ℓ for length, and w for width. But be careful. In formulas, capital and small letters often stand for *different* things.

Example

The area A of the shaded region can be found by the formula $A = S^2 - s^2$. Find the area if $S = 1.75$ in. and $s = 1.25$ in.

Solution

Substitute 1.75 in. for S and 1.25 in. for s in the formula. The unit of the answer is square inches.

$$A = S^2 - s^2$$
$$= (1.75 \text{ in.})^2 - (1.25 \text{ in.})^2$$
$$= 3.0625 \text{ in}^2 - 1.5625 \text{ in}^2$$
$$= 1.5 \text{ in}^2$$

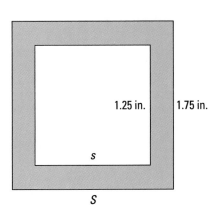

Writing Letters in Formulas

Because capital and small letters can mean different things, in algebra you should not change capital letters to small letters, or vice versa. And you need to learn to write small letters differently from capitals, not just smaller. Do this by writing the small letters with curves and hooks. Here are the most confusing letters, and one way to write them.

$$\text{CFIJKLMPSUVWXYZ}$$

$$\textit{c f i j k l m p s u v w x y z}$$

QUESTIONS

Covering the Reading

1. The formula $A = \ell w$ gives the __?__ of a rectangle in terms of its __?__ and __?__.

2. How do the letters in the formula of Question 1 make that question easier to answer?

3. What is the area of a rectangle that is 43 cm long and 0.1 cm wide?

4. a. What is the area of a rectangle that is 3 in. long and 4 ft wide?
 b. What does part **a** tell you about units in formulas?

In 5–7, use the formula $A = S^2 - s^2$ in the Example.

5. Do S and s stand for the same thing?

6. What do S and s stand for?

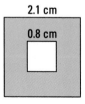

2.1 cm

0.8 cm

7. Give the area between the two squares drawn at the left.

8. In a formula, why should you not change a capital letter to a small letter?

9. a. Write the alphabet from A to Z in capital letters.
 b. Write the alphabet in small letters. Make sure your small letters look different from (not just smaller than) the capital letters.

10. Refer to the Activity in this lesson. What values do you get for the area of the front cover of this book?

11. A gardener wishes to surround a square garden by a walkway one meter wide. The garden is 5 meters on a side. How much ground will the gardener have to clear for the walkway?

12. One cube is inside another as shown below.

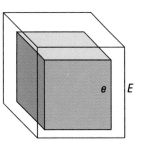

a. If the edges of the cubes have lengths E and e, what is the volume of the space between them?
b. If a cube with edge of 4″ is put inside a cube with edge of 5″, what is the volume between them?

In 13–16, use this information. When a team wins W games and loses L games, then its winning percentage P is given by the formula

$$P = \frac{W}{W + L}.$$

This number is always converted to a decimal. Then it is rounded to the nearest thousandth. Calculate the winning percentage for each team.

13. the 1993 NCAA women's basketball champion Texas Tech, who had a season record of 31 wins and 3 losses

14. the 1992 U.S. Olympic Men's basketball team, who won all 8 games they played

15. a volleyball team that wins 7 games and loses 7 games

16. a team that wins 3 games and loses 12

In 17 and 18, use the formula $T = t - \frac{2f}{1000}$. This formula gives the estimated air temperature T (in degrees Fahrenheit) at an altitude of f feet when the ground temperature is t. Suppose it is 75°F at ground level.

17. Find the temperature at an altitude of 1000 ft.

18. Find the temperature at an altitude of 3000 ft.

Pictured here are the Lady Raiders of Texas Tech University, the 1993 NCAA women's basketball champs.

19. Write a formula that expresses the relationship between a radius and a diameter of a circle.

20. *Multiple choice.* In a *Bill James Baseball Abstract,* the expression $\frac{2(HR \times SB)}{HR + SB}$ is called the Power/Speed number. A newspaper writer once incorrectly wrote the expression as $2[HR \times SB]/HR + SB$. What was wrong with the newspaper expression?
    (a) It used brackets instead of parentheses.
    (b) It should have used parentheses around $HR + SB$.
    (c) It used a $\times$ sign for multiplication.
    (d) It was written on one line.
    (e) It used HR and SB instead of single letters for variables.

Review

In 21–24, evaluate each expression. *(Lessons 4-5, 4-6)*

21. $9 + 5(3 + 2 \cdot 7)$

22. $3 + 4 \cdot 5^2$

23. $12 - 4x$ when $x = .5$

24. $8/(8/(2 + 6))$

25. Put into the correct order of operations. *(Lessons 4-1, 4-5)*
    (a) multiplications or divisions from left to right
    (b) powers
    (c) inside parentheses
    (d) additions or subtractions from left to right

In 26–29, put one of the symbols $>$, $=$, or $<$ in the blank.
(Lessons 1-7, 1-9, 2-2, 4-4)

26. $-5 \underline{\ ?\ } -5.1$

27. $4.09 \underline{\ ?\ } 4\frac{1}{11}$

28. $x + 14 \underline{\ ?\ } x - 14$ when $x = 20$.

29. $a^3 \underline{\ ?\ } a^2$ when $a = 0.7$.

30. One liter is how many cubic centimeters? *(Lesson 3-9)*

31. Convert 6 kilograms to grams. *(Lesson 3-5)*

32. Change $\frac{8}{4}, \frac{9}{4}, \frac{10}{4}, \frac{11}{4}$, and $\frac{12}{4}$ to decimals. *(Lesson 1-6)*

Exploration

33. Look in a newspaper to find the wins, losses, and winning percentage for a particular team in your area. Use the formula from Questions 13–16 to see if you get the same percentage as the one you find in the newspaper.

Relative Frequency

IN·CLASS
ACTIVITY

When you toss a coin there are two possible *outcomes,* "heads" and "tails." You expect heads to occur about 50% of the time and tails the other 50% of the time. However, this does not always happen. See what happens in your class.

The **frequency** of an outcome is the number of times the outcome occurs. If you toss the coin N times and an outcome occurs with frequency F, the **relative frequency** of the outcome is $\frac{F}{N}$.

First, work alone.

1 **a.** Toss a coin 10 times, writing down an H or T each time the coin is tossed.
   b. Record the frequency of heads.
   c. Determine the relative frequency of heads.
   d. Write the relative frequency as a percent.
   e. If you continue, do you think you will get about the same relative frequency of heads and tails in the long run? Why or why not?

2 **a.** Repeat step 1 four more times, for a total of 50 tosses.
   b. Calculate the total frequency of heads for the 50 tosses.
   c. Determine the relative frequency of heads for the 50 tosses.
   d. Write the relative frequency as a percent.
   e. Now do you think you will get about the same number of heads and tails in the long run? Why or why not?

Now work in a group of 3 or 4 students.

3 **a.** Combine your results with those of the others in your group.
   b. How many tosses were made altogether?
   c. How many of these were heads?
   d. Determine the relative frequency of heads for all the tosses made by those in your group.
   e. In the long run, what percent of tosses do you think will be heads?

Finally, work together as a class.

4 Combine the results of your group with all other groups in your class. Repeat step 3.

5 A coin is considered to be fair if in the long run the relative frequency of heads is $\frac{1}{2}$. Do you think the coins tossed in your class were fair? Explain why you think the way you do.

213

Rain expected today? *The probability of rain in this girl's neighborhood was 70% on the day this picture was taken. Weather forecasts are predictions of what might happen—and thus should be taken seriously.*

What Is Probability?

A **probability** is a number from 0 to 1 which tells you how likely something is to happen. A probability of 0 means that the event is impossible. A probability of $\frac{1}{5}$ means that in the long run it is expected the event will happen about 1 in 5 times. A probability of 70% means that the event is expected to happen about 70% of the time. A probability of 1, the highest probability possible, means that the event *must* happen.

The closer a probability is to 1, the more likely the event. Since 70% is greater than $\frac{1}{5}$, an event with a probability of 70% is more likely than one with a probability of $\frac{1}{5}$. This can be pictured on a number line.

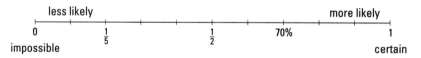

Probabilities can be written as fractions, decimals, or percents, or any other way that numbers are written.

Example 1

Suppose the weather bureau says there is a 1 in 3 chance of rain tomorrow. It reports a 40% precipitation probability the day after tomorrow. On which day does the weather bureau think rain is more likely?

Solution

The "1 in 3 chance of rain tomorrow" means the probability is $\frac{1}{3}$. To compare $\frac{1}{3}$ with 40%, change both numbers to decimals.

$\frac{1}{3} = 0.\overline{3}$ and 40% = 0.4, so 40% is larger, and rain is more likely the day after tomorrow.

Determining a Probability

There are three common ways that people determine probabilities.

(1) Guess. This is not a great method, but sometimes it is the only thing you can do. For instance, suppose you would like a ten-speed bicycle for your birthday. You tell your friend you think the probability of getting the bike is only about 25%. This means that you think there is a 1 in 4 chance (since $\frac{1}{4} = 25\%$) you will get the bike. You are guessing at the probability because you have no other way to determine it.

(2) Perform an experiment and choose a probability close to the relative frequency you found. Suppose you thought some of the coins tossed in the In-class Activity on page 213 might be unbalanced. If the coin tossing led to 1224 heads in 2500 tosses, the relative frequency is then $\frac{F}{N} = \frac{1224}{2500} = 48.96\%$. You could choose the probability of heads for these coins to be 49%. Others might choose 50%. A probability estimated from a relative frequency may be different for different people. But it is more likely to be close to what would happen in the long run than a guess would be. Also, the more times an experiment is repeated, the closer the relative frequency should be to the probability.

(3) Assume outcomes are equally likely. In many situations, people assume outcomes are equally likely. For instance, when you toss a coin, there are 2 possible outcomes. One of these is heads. If the outcomes are equally likely, the probability the event will occur is $\frac{1}{2}$.

Probabilities when Outcomes Are Assumed to Be Equally Likely

Consider tossing a dime and a quarter. Pictured here are the four possible outcomes.

Assume the outcomes are equally likely. Only one outcome has zero heads, so the probability of getting zero heads is $\frac{1}{4}$. There is also only one way to get two heads. So, the probability of getting 2 heads is $\frac{1}{4}$. Exactly one head can appear in two different ways: the dime can show heads and the quarter tails, or the quarter can show heads and the dime tails. So the probability of getting exactly one head is $\frac{2}{4} = \frac{1}{2}$.

When all outcomes of a situation like "tossing two coins" have the same probability, the outcomes are called **random**. The coins are said to be **fair,** or **unbiased.** In this lesson, dice and coins are assumed to be fair.

Any collection of outcomes from the same situation is called an **event.** For instance, when tossing a single die the event "an even number is tossed" is the collection of outcomes 2, 4, and 6. When outcomes are equally likely, there is a simple formula for finding the probability of an event.

Probability Formula for Equally Likely Outcomes
Suppose a situation has N equally likely possible outcomes and an event includes E of these. Let P be the probability that the event will occur. Then
$$P = \frac{E}{N}.$$

Example 2

Suppose a fair die is tossed once. What is the probability of tossing a 5 or a 6?

Solution

Determine the number of possible outcomes. In this situation there are 6, as shown at the left.

The event in this situation is "tossing a 5 or a 6." The event includes two of the six outcomes. So The probability is $\frac{2}{6}$, or $\frac{1}{3}$.

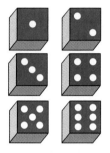

Outcomes are not always equally likely. Consider the probability that you will find $1,000,000 on your way home today. There are two outcomes: finding the money and not finding it. If the outcomes were equally likely, the probability of each would be $\frac{1}{2}$. But "finding $1,000,000 today" and "not finding $1,000,000 today" are not equally-likely outcomes. The probability of finding the money is clearly much less than the probability of not finding it. In this case, you cannot use $P = \frac{E}{N}$.

QUESTIONS

Covering the Reading

1. A six-sided die was tossed 100 times. The numbers that showed up are given here, along with the number of times each occurred.

1	16 times	3	18 times	5	17 times
2	19 times	4	12 times	6	18 times

   a. What is the frequency of the outcome "5 showed up"?
   b. What is the relative frequency of the outcome "5 showed up"?
   c. If the outcomes were equally likely, what would be the probability of each outcome?
   d. Do you think this die is fair? Explain your answer.

2. A probability can be no larger than __?__ and no smaller than __?__.

3. If an event is impossible, its probability is __?__.

4. What is the probability of a sure thing?

5. Suppose event A has a probability of $\frac{1}{2}$ and event B has a probability of $\frac{1}{3}$. Which event is more likely?

6. The weather bureau reports a 70% precipitation probability for tomorrow and a $\frac{3}{5}$ chance of thunderstorms the day after tomorrow. Which is thought to be more likely, precipitation tomorrow or thunderstorms the day after?

7. Identify three ways in which people determine probabilities.

8. Write the formula for finding the probability of an event in a situation with equally likely outcomes.

9. A fair die is tossed. What is the probability of getting a 1, 2, 5, or 6?

Applying the Mathematics

10. Toss a die 30 times. Record the number of times each number appears.
    a. What was your relative frequency of the event "tossing a 5 or a 6"?
    b. How does your relative frequency compare to the probability found in Example 2?

11. Explain the difference between relative frequency and probability.

In 12 and 13, imagine a situation where fifty raffle tickets are put into a hat. Each ticket has the same probability of being selected to win the raffle prize.

12. If one of these tickets is yours, what is the probability you will win the raffle?

13. If you and two friends have one ticket apiece, what is the probability that one of you will win the raffle?

14. A person says, "There is a negative probability that my uncle will leave me a million dollars." What is wrong with that statement?

15. A spinner in a game is pictured below. Assume all directions of the spinner are equally likely. Also assume radii that look perpendicular are.

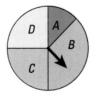

    a. If the angle which forms region A has measure 45°, what is the probability that the spinner will land in region A?
    b. Determine the probabilities of landing in the other regions.

16. What way of determining probability is being used in Question 15?

In 17 and 18, use this property: if the probability that an event will occur is *p*, then the probability that the event will not occur is $1 - p$.

17. Suppose the probability that your teacher will give a test next Friday is 80%. What is the probability that your teacher will not give a test next Friday?

18. You listen to a radio station for an hour. You estimate that $\frac{1}{10}$ is the probability that your favorite song will be played. What is the probability that your favorite song will not be played?

Review

19. Which rectangle has greater area?   *(Lesson 4-7)*

20. The area of a circle with radius *r* is $A = \pi r^2$. Find the area of a circle with radius 3 meters, to the nearest hundredth of a square meter. *(Lessons 1-3, 4-7)*

21. Suppose *y* = 3.2 and *x* = 1.5. Evaluate each expression.   *(Lesson 4-6)*
   a. $4x$ **b.** $4y - 10$ **c.** $[40 - (4y - 10)]/(4x)$

22. a. 6 miles + 1000 feet = _?_ feet
   b. 6 meters + 1000 centimeters = _?_ meters   *(Lessons 3-1, 3-4)*

23. The St. Gotthard Tunnel in Switzerland is the longest car tunnel in the world. Its length is about 16.4 km. About how many miles is this?   *(Lesson 3-4)*

24. According to one source, 60% of Americans don't exercise regularly. Using 250,000,000 for the U.S. population, estimate the number of people in the United States who do exercise regularly.   *(Lesson 2-5)*

Turn on the lights. *The St. Gotthard road tunnel, pictured above, was opened in 1980 as a companion to the St. Gotthard rail tunnel opened in 1882. Switzerland relies on tunnels to aid passage through its mountainous regions.*

Exploration

25. The **odds for** an event happening is expressed as the number of ways that the event can happen to the number of ways that the event cannot happen. The **odds against** an event is the number of ways the event cannot happen to the number of ways the event can happen. For instance, suppose a fair die is tossed once. The odds for tossing a six are 1 to 5. The odds against tossing a six are 5 to 1.

   a. Suppose Carol estimates that the odds for her passing the next math test are 10 to 1. What is Carol's estimate of the probability that she will pass?
   b. Find an example of odds in a newspaper or magazine article. Write a probability question based on the odds you found.

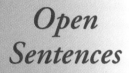

Open Sentences

Problem solving. *An entertainment committee may solve an equation to determine the number of tickets that must be sold to cover the costs of an event. The event might include the Jesse White Tumblers shown above.*

What Is an Open Sentence?

A sentence with an equal sign = is called an **equation.** Here are some equations.

$$27 = 9(4 - 1) \qquad 1 + 1 = 3 \qquad x + 7 = 50$$

The left equation is true. The middle equation is false. The right equation may be true or false, depending on what you substitute for *x*. If you substitute 57, the equation is false, because

$$57 + 7 \text{ does not equal } 50.$$

If you substitute 43 for *x,* the equation is true.

$$43 + 7 = 50$$

The equation $x + 7 = 50$ is an example of an *open sentence.* An **open sentence** is a sentence containing one or more variables. An open sentence can be true or false, depending on what you substitute for the variables. The *solution* to this open sentence is 43. A **solution** to an open sentence is a value of a variable that makes the sentence true.

Finding Solutions to Open Sentences

Sometimes you can find some solutions to open sentences in your head.

Example 1

Find the solution to the open sentence $40 - A = 38$.

Solution

Since $40 - 2 = 38$, the number 2 is the solution. You can also write A = 2.

If you are given a set of choices, you can find solutions by trying each choice to see if it works.

Example 2

Multiple choice. Which number is a solution to $30m = 6$?
(a) 24   (b) 0.5   (c) 5   (d) 0.2

Solution

The solution is the value that makes the sentence true. Substitute to find the solution.
(a) $30 \cdot 24 = 720$, so 24 is not a solution.
(b) $30 \cdot 0.5 = 15$, so 0.5 is not a solution.
(c) $30 \cdot 5 = 150$, so 5 is not a solution.
(d) $30 \cdot 0.2 = 6$, so 0.2 is a solution.
Choice (d) is the correct choice.

How Formulas Lead to Open Sentences

One of the most important skills in algebra is finding solutions to open sentences. This skill is important because open sentences occur often in situations where decisions have to be made.

Example 3

A store is to have an area of 10,000 square feet. But it can be only 80 feet wide. Let ℓ be the length of the store. The value of ℓ is the solution to what open sentence?

Solution

Drawing a picture can help. We know $A = \ell w$. The given information tells us that $A = 10,000$ and $w = 80$. Substitute 10,000 for A and 80 for w in the formula. ℓ is the solution to the open sentence $10,000 = \ell \cdot 80$.

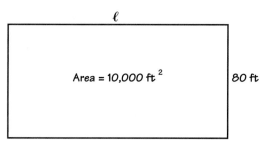

ℓ

Area = 10,000 ft^2 80 ft

The solution to $10,000 = \ell \cdot 80$ is not obvious. The variable ℓ is called the **unknown.** Finding values of the unknown (or unknowns) that make the sentence true is called **solving the sentence.** At this point, you are expected to be able to solve only very simple sentences. The sentences will be the kind you can solve in your head, or ones with a set of choices for solutions.

Example 4

Multiple choice.   If a basketball team loses 10 games, how many games must it win to have a winning percentage of .600?
(a) 6   (b) 15   (c) 16   (d) 25

Solution

From Questions 13–16 of Lesson 4-7, if a team wins W games and loses L games, then its winning percentage P is given by the formula $P = \dfrac{W}{W + L}$. Here $P = .600$ and $L = 10$. So the sentence to solve is $.600 = \dfrac{W}{W + 10}$. Now test each choice until you find the solution.

Activity

Finish Example 4.

QUESTIONS

Covering the Reading

1. What is an *equation?*

2. Give an example of an equation that is false.

3. What is an *open sentence?*

4. Give an example of an open sentence.

5. Define: solution to an open sentence.

6. What is meant by *solving* an open sentence?

7. *Multiple choice.*   The solution to $4x + 3 = 12$ is
   (a) 2.25.   (b) 0.
   (c) 2.5.   (d) 1.

8. *Multiple choice.*   A solution to $4n^2 = 64$ is
   (a) 1.   (b) 2.
   (c) 4.   (d) 8.

In 9–12, solve the sentence in your head.

9. $18 + A = 19$ 10. $2B = 10$

11. $C = 5 - .1$ 12. $4 = 3.5 + t$

13. Rhee Taylor wants a store with an area of 1000 square meters. The store can be only 40 meters wide. Rhee wants to know how many meters long the store should be.

   a. What equation can Rhee solve to find the length ℓ of the store?

   b. Which of these is the solution: 25 m, 960 m, 1040 m, or 4000 m?

14. What is the answer to the question of Example 4?

15. *Multiple choice.* If a volleyball team wins 12 games, how many games can it lose and still have a winning percentage of .750?

(a) 3   (b) 4   (c) 8   (d) 9

Applying the Mathematics

In 16–19, give one solution to each of these nonmathematical open sentences.

16. ___?___ is currently President of the United States.

17. In population, ___?___ is a bigger city than Detroit.

18. A trio has ___?___ members.

19. An octet has ___?___ members.

In 20–23, give at least one solution.

20. There are y millimeters in a meter.

21. x is a negative integer.

22. You move the decimal point two places to the right when multiplying by m.

23. The number one million, written as a decimal, is a 1 followed by z zeros.

24. *Multiple choice.* Let n be the number of days a book is overdue. Let F be the fine. Suppose $F = .20 + .05n$ dollars, the situation described in Lesson 4-4. When the fine F is $1.00, what number of days is the book overdue?

(a) 20   (b) 15
(c) 16   (d) 25

25. Suppose $y = 2x + 3$. When $x = 4$, what value of y is a solution?

In 26–31, find a solution to the sentence in your head.

26. $n + n = 16$ **27.** $m \cdot m = 16$

28. $z - 4 = 99$ **29.** $y \cdot 25 = 25$

30. $\frac{1}{2}w = \frac{5}{2}$ **31.** $p + 2\frac{1}{3} = 5\frac{1}{3}$

Architectural splendor. *The Renaissance Center in downtown Detroit, Michigan, houses a circular 73-story hotel and four 39-story office buildings.*

32. An organization has a raffle and sells n tickets. You have bought 5 of the tickets. What is the probability that you will win the raffle? *(Lesson 4-8)*

33. Slips of paper numbered 1 to 100 are put into a hat. A slip is taken from the hat. What is the probability that the number is a multiple of 3? *(Lesson 4-8)*

34. The sum of the integers from 1 to n is given by the expression $\frac{n(n + 1)}{2}$.
   a. Evaluate the expression when $n = 5$.
   b. Verify your answer to part **a** by adding the integers from 1 to 5.
   c. Find the sum of the integers from 1 to 100. *(Lessons 4-6, 4-7)*

35. Evaluate this rather complicated expression. *(Lesson 4-6)*
$$\frac{6(2 + 4^3)^2 - 3^2}{40 - 13 \cdot 3}$$

36. Lynette used her calculator to find the average of the three numbers 83, 91, and 89. The answer came out higher than any of the numbers. Explain why Lynette must have made a mistake. *(Lesson 4-6)*

37. Put the three quantities into one sentence with two inequality signs: 5 km, 500 m, 50,000 mm. *(Lessons 1-9, 3-4)*

38. In 1991, the world population was estimated by the Bureau of the Census as 5.423 billion. *(Lessons 1-4, 2-1, 2-3)*
   a. Round this number to the nearest ten million.
   b. Write your answer to part **a** in scientific notation.

Exploration

39. Estimate a solution to $(x + 1)(x + 2) = 100$
   a. between consecutive integers.
   b. between consecutive tenths.
   c. between consecutive hundredths.

Slow down! *It is unsafe to drive at a speed that* is greater than *the speed limit. The fine in many states for speeding* is greater than or equal to *$75*.

What Is an Inequality?

An **inequality** is a sentence with one of the following symbols.

$\ne$ is not equal to
$<$ is less than
$\le$ is less than or equal to
$>$ is greater than
$\ge$ is greater than or equal to

$5 > \text{-}7$ is an inequality that is true. So is $5 \ge \text{-}7$. An example of an inequality that is false is $3 < 2$. The inequality $50\% \ne 0.5$ is also false.

The inequality $x < 30$ is an open sentence, neither true nor false. The number 20 is a solution, because $20 < 30$. The number -14 is also a solution, since $\text{-}14 < 30$. However, 30 is not a solution, because $30 < 30$ is not true. Also, 554 is not a solution, because 554 is not less than 30.

Inequalities that are open sentences usually have many solutions. The solutions are often easier to graph than to list.

Graphing Simple Inequalities

Example 1

Graph all solutions to $y < 2$.

Solution

Any number less than 2 is a solution. So there are infinitely many of them. Among the solutions are 0, 1, 1.5, -5, -4.3, and -2,000,000.

Recall that to graph a single number means to darken a dot on the number line. To graph all of the numbers on a part of the number line, thicken that part of the line. So here we thicken the part of the line with numbers that are less than 2. The open circle around the tick mark for 2 indicates that 2 is not a solution. It is not true that $2 < 2$. The thickened arrowhead below indicates that the solutions go on forever in that direction. Put a y at the right of the number line to indicate the variable.

$$y < 2$$

Remember that $2 > y$ means the same as $y < 2$. So the graph of all solutions to $2 > y$ is identical to the graph of $y < 2$.

Example 2

Graph all solutions to $x \geq -1$.

Solution

Any number greater than or equal to -1 is a solution. Therefore, that part of the line is thickened. Since -1 is a solution, the circle at -1 is filled in. Here is the graph.

$$x \geq -1$$

Graphing Double Inequalities

On some interstate highways, a car's speed must be from 45 mph to 55 mph. The graph of all possible legal speeds s is the graph of solutions to the double inequality $45 \leq s \leq 55$. This can be read as: "s is greater than or equal to 45 and s is less than or equal to 55" or "s ranges from 45 to 55." Another way to write this is with $\geq$ signs. $55 \geq s \geq 45$. Some solutions are 48, 50.3, 51, 55, 45, and 49. Because both 45 and 55 are solutions, the circles at 45 and 55 are filled in.

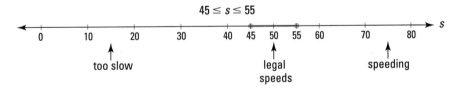

If 55 were not included, the sentence would be $45 \leq s < 55$. The graph would have an open circle at 55.

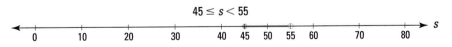

If neither 45 nor 55 were included, the sentence would be $45 < s < 55$. There would be open circles at both 45 and 55.

$45 < s < 55$

Example 3

Write an inequality for the following graph. Use m as the variable.

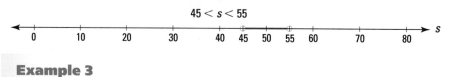

Solution

The solutions are the numbers from -4 to 5, including 5 but not including -4. One double inequality is $-4 < m \le 5$. Another correct answer is $5 \ge m > -4$.

QUESTIONS

Covering the Reading

1. What is an inequality?

In 2–5, give the meaning of the symbol.

2. $>$

3. $\ge$

4. $<$

5. $\le$

6. Write an inequality that means the same thing as $y > 5$.

In 7 and 8, is the sentence an open sentence?

7. $3y \ge 90$

8. $2 \le 9$

In 9–12, *true or false.*

9. $-2 < 1$

10. $3 < \frac{6}{2}$

11. $-5 \ge 5$

12. $6 \ne 5 + 1$

In 13–16, name one solution to the sentence.

13. $A > 5000$

14. $n \le -5$

15. $6\frac{1}{2} < d < 7\frac{1}{4}$

16. $55 \ge s \ge 45$

17. *Multiple choice.* The solutions to which sentences are graphed here?

(a) $x > 8$ (b) $x < 8$ (c) $x \ge 8$ (d) $x \le 8$

18. Write a sentence whose solutions are graphed below.

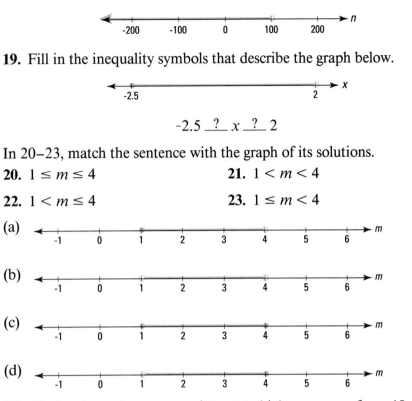

19. Fill in the inequality symbols that describe the graph below.

$$-2.5 \underline{\ ?\ } x \underline{\ ?\ } 2$$

In 20–23, match the sentence with the graph of its solutions.

20. $1 \le m \le 4$ **21.** $1 < m < 4$

22. $1 < m \le 4$ **23.** $1 \le m < 4$

(a)

(b)

(c)

(d)

24. The legal speeds s on some interstate highways range from 45 mph to 65 mph. Graph all legal speeds.

Applying the Mathematics

25. *Multiple choice.* Which sentence has the same solutions as $6 < a \le 9\frac{2}{3}$?

  (a) $9\frac{2}{3} < a \le 6$ (b) $9\frac{2}{3} > a \ge 6$

  (c) $9\frac{2}{3} \le a < 6$ (d) $9\frac{2}{3} \ge a > 6$

In 26–29, graph all the solutions to the sentence.

26. $w < \frac{40}{3}$ **27.** $-6 \le y$ **28.** $-2 < y < 3$ **29.** $-1 \ge z \ge -5$

In 30–32, a situation is given.

a. What inequality describes the situation?

b. Give three solutions to the inequality.

c. Graph all possible solutions.

30. The speed limit is 55 mph. A person is driving f miles per hour and is speeding.

31. A person earns d dollars a year. The amount d is less than $25,000 a year.

32. The area is A. Rounded up to the next hundred, A is 500.

33. Let P be a probability.
   a. Write an inequality showing all possible values of P.
   b. Graph the inequality.

Review

In 34–36, suppose it costs 30¢ for the first minute and 18¢ for each additional minute on a long-distance phone call. Then $c = .30 + .18(m - 1)$, where c is the total cost and m is the number of minutes talked.   *(Lessons 4-5, 4-9)*

34. Calculate c when $m = 6$.

35. What will it cost to talk for 10 minutes?

36. *Multiple choice.*   For $2.10, how long can you talk?

   (a) 7 minutes   (b) 9 minutes
   (c) 10 minutes   (d) 11 minutes

37. If the probability that an event will occur is x, what is the probability it will not occur?   *(Lesson 4-8)*

In 38 and 39, evaluate the expression when $a = 3$, $b = 5$, and $c = 7$. *(Lessons 4-4, 4-5, 4-6)*

38. $3(c + 10b - a^2)$

39. $a[a + b(b + c)]$

40. Give three instances of this pattern: There are $6n$ legs on n insects. *(Lesson 4-2)*

41. Three instances of a general pattern are given. Describe the pattern using two variables.   *(Lesson 4-2)*

$$\frac{31.4}{2} \cdot \frac{2}{31.4} = 1 \qquad \frac{7}{8} \cdot \frac{8}{7} = 1 \qquad \frac{100}{11} \cdot \frac{11}{100} = 1$$

Exploration

42. Describe a real-life situation that can lead to each inequality. The letter used is a hint to one possible description.
   a. $13 \le a \le 19$ **b.** $w > 250$ **c.** $d > 1.5$

ET, phone home.
Imagine ET's phone bill if he were charged by the kilometer for his very long-distance phone call!

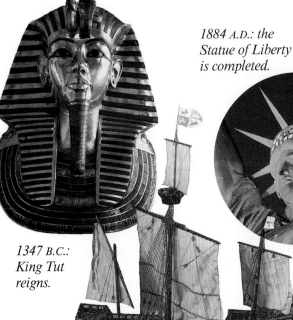 PROJECTS
4
CHAPTER FOUR

A project presents an opportunity for you to extend your knowledge of a topic related to the material of this chapter. You should allow more time for a project than you do for typical homework questions.

1 The Bricklayer's Formula

The formula $N = 7LH$ gives the number of bricks needed in a wall of length L feet and height H feet. Go with a tape measure to a brick building in your neighborhood. Consider a part of the wall with length and height that you should call L and H. How many bricks are in that part of the wall? Does the formula $N = 7LH$ work for that part of the wall? Try other parts of the wall. If the formula does not work, suggest a formula that will work for this wall.

2 A Parentheses Problem

Consider the subtraction problem $4 - 2 - 1$. Parentheses can be placed in two different places and the answer you get is different. $(4 - 2) - 1 = 1$ but $4 - (2 - 1) = 3$. How many different answers can you get by placing parentheses in $8 - 4 - 2 - 1$? Be careful—you can use more than one set of parentheses and you can put parentheses inside other parentheses. How many different answers can you get by placing parentheses in $16 - 8 - 4 - 2 - 1$? Write down any patterns you see in the answers you get, and use them to predict how many answers you might get for $32 - 16 - 8 - 4 - 2 - 1$.

3 Arithmetic Time Line

On page 175 dates are given for the invention of many of the symbols of arithmetic. Organize this information on a number line of dates (known as a time line). On this time line, put dates of at least ten other important events that happened around these times. Make your time line big enough so that it is easy for someone else to see and read. (You may need to use poster board.)

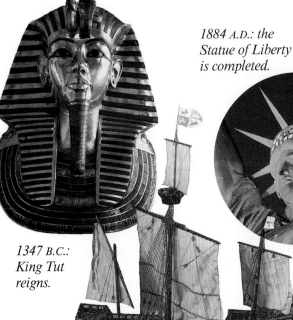

1347 B.C.: King Tut reigns.

1884 A.D.: the Statue of Liberty is completed.

1992 A.D.: Dr. Mae Jemison joins the Endeavour crew.

1964 A.D.: the Beatles tour the U.S.

2500 B.C.

1492 A.D.: Christopher Columbus sails to America.

2000 A.D.

▶

4 An Experiment with Dice

Test whether some dice you have are fair. Pick one die and toss it at least 100 times, recording the number that comes up each time. Count the number of times that 1, 2, 3, 4, 5, and 6 each come up. Compare the relative frequencies with the probability $\frac{1}{6}$ that these numbers would come up if your die was fair. Repeat this experiment with another die. Do you think your dice are fair? Explain why you think the way you do.

5 Using Inequality Symbols

Write a short true or fantasy story in which you use all of the inequality symbols ($\neq$, $<$, $\leq$, $>$, $\geq$). You might write about a place to visit, a sports event, or a type of vacation.

6 The Beginning of Algebra

Among the people who contributed to the beginnings of algebra were a Greek, Diophantus, and an Arab, al-Khowarizmi. Look in an encyclopedia or other reference book and write a short essay about each of these great mathematicians.

SUMMARY

A variable is a symbol that can stand for any one set of numbers or other objects. Usually a variable is a single letter. Algebra is the study of variables and operations on them. An algebraic expression is an expression with one or more variables in it. Algebraic expressions are often found in equations and inequalities.

Numerical and algebraic expressions may contain a number of operations. To avoid confusion, mathematicians have agreed upon rules for order of operations. To operate in an order different from that given by these rules, parentheses and other grouping symbols are used.

Introduced in this chapter are four uses of variables:

(1) Variables enable *patterns* to be described. For example, *n* people have 2*n* eyes. Descriptions with variables tend to be shorter and look like the instances of the pattern.

(2) Variables describe *properties* of numbers. For example, $a + b = b + a$.

(3) Variables are shorthand for quantities in *formulas*. For example, $A = \ell w$ is shorthand for area = length × width. $P = \frac{E}{N}$ gives the probability *P* that an event will occur if a situation has *N* equally likely possible outcomes, of which *E* are included in the event.

(4) Variables may be *unknowns*. For instance, suppose the area of a rectangle is 4 square meters and its length is 3 meters. Then the width *w* is the solution to the equation $4 = 3w$. The value of *w* can be found by solving this sentence.

VOCABULARY

You should be able to give a general description and a specific example of each of the following ideas.

Lesson 4-1
symbols $+$, $-$, $\times$, $/$, and so on, for operations
numerical expression
value
evaluating an expression
order of operations

Lesson 4-2
pattern
instance
variable
symbol · for multiplication
algebra

Lesson 4-3
algebraic expression

Lesson 4-4
value of a variable
value of an expression

Lesson 4-5
parentheses, ( )
nested parentheses

Lesson 4-6
grouping symbols
brackets, [ ]
mean, average

Lesson 4-7
formula
one variable in terms of others

Lesson 4-8
frequency, relative frequency
probability
outcome
event
random
fair, unbiased

Lesson 4-9
equation
open sentence
solution
unknown
solving an open sentence

Lesson 4-10
inequality symbols $\leq$, $\geq$, $\neq$

PROGRESS SELF-TEST

Take this test as you would take a test in class. You will need a ruler. Then check your work with the solutions in the Selected Answers section in the back of the book.

In 1–4, evaluate the expression.

1. $6 + 8 \cdot 7 + 9$ **2.** $(40-5)+(60-10)$

3. $75 - 50 - 3 - 1$ **4.** $5 + 3 \cdot 4^2$

5. Round to the nearest integer: $\dfrac{100 + 2 \cdot 5}{10 + 5}$.

In 6–8, evaluate when $a = 3$, $b = 4$, $x = 10$, and $y = 100$.

6. $x + 3y$ **7.** $(a + b)(b - a)$

8. $y + 5[y + 4(y + 3)]$

9. *Multiple choice.* Which is a solution to $(4x)^2 = 64$?
(a) 2   (b) 4   (c) 8   (d) 16

In 10–12, three instances of a pattern are given. Describe the pattern using variables.

10. Use one variable.
$$10 \cdot 5 = 6 \cdot 5 + 4 \cdot 5$$
$$10 \cdot 8.2 = 6 \cdot 8.2 + 4 \cdot 8.2$$
$$10 \cdot 0.04 = 6 \cdot 0.04 + 4 \cdot 0.04$$

11. Use two variables.
$$2 + 8 = 8 + 2$$
$$3.7 + 7.3 = 7.3 + 3.7$$
$$0 + 4 = 4 + 0$$

12. Use one variable.
In one year, we expect the town to grow by 200 people.
In two years, we expect the town to grow by $2 \cdot 200$ people.
In three years, we expect the town to grow by $3 \cdot 200$ people.

In 13 and 14, the formula $c = 23n + 6$ gives the cost of first-class postage in 1993; c is the cost in cents; n is the weight in ounces of the mail, rounded up to the nearest ounce.

13. If $n = 5$, find c.

14. Find the cost in dollars and cents of mailing a 9-oz letter first-class.

15. In the formula $p = s - c$, calculate p if $s = \$45$ and $c = \$22.37$.

In 16–19, translate into a numerical or algebraic expression or sentence.

16. the product of twelve and sixteen

17. forty is less than forty-seven

18. a number is greater than or equal to zero

19. a number is divided into nine

20. *Multiple choice.* Written on one line, $\dfrac{W}{W + L}$ is equal to:
(a) $W/W + L$ (b) $W/(W + L)$
(c) $(W/W) + L$ (d) $(W + L)/W$

21. *Multiple choice.* Most of the symbols we use for arithmetic operations were invented
(a) before 1 A.D.
(b) between 1 and 1000 A.D.
(c) between 1000 and 1800 A.D.
(d) since 1800 A.D.

22. Suppose a box landed face up 8 of the first 50 times it was dropped. What is the relative frequency that the box landed face up?

23. In the band, 5 boys and 7 girls play clarinet. If the director chooses 1 of these students at random to play a solo, what is the probability that the student will be a boy?

24. *Multiple choice.* A sentence that means the same as $2 < y$ is
(a) $2 \le y$.   (b) $y > 2$.
(c) $y \ge 2$.   (d) $y < 2$.

In 25 and 26, find a solution to the sentence.

25. $6x = 42$ **26.** $-5 < y < -4$

27. Why must units in formulas be consistent? Give an example to support your answer.

28. If $x = 7y$ and $y = 3$, what is the value of x?

29. Graph all solutions to $x < 12$.

30. Below are the solutions to what sentence?

In 31 and 32, give two instances of the pattern.

31. If your age is A years, your sister's age is $A - 5$ years.

32. $x + y - x = y$

CHAPTER REVIEW

Questions on SPUR Objectives

SPUR stands for **S**kills, **P**roperties, **U**ses, and **R**epresentations. The Chapter Review questions are grouped according to the SPUR Objectives for this chapter.

SKILLS DEAL WITH THE PROCEDURES USED TO GET ANSWERS.

Objective A. *Use order of operations to evaluate numerical expressions.* *(Lessons 4-1, 4-5, 4-6)*

In 1–14, evaluate the given expression.

1. $235 - 5 \times 4$

2. $32 \div 16 \div 8 \times 12$

3. $2 + 3^4$

4. $4 \times 2^3 + \frac{28}{56}$

5. $5 + 8 \times 3 + 2$

6. $100 - \frac{80}{5} - 1$

7. $1984 - (1947 - 1929)$ **8.** $40 - 30/(20 - 10/2)$

9. $6 + 8(12 + 7)$

10. $(6 + 3)(6 - 4)$

11. $3 + [2 + 4(6 - 3 \cdot 2)]$

12. $4[7 - 2(2 + 1)]$

13. $\frac{4 + 5 \cdot 2}{13 \cdot 5}$

14. $\frac{3^3}{3^2}$

Objective B. *Evaluate algebraic expressions given the values of all variables.* *(Lessons 4-4, 4-5, 4-6)*

15. If $x = 4$, then $6x = \underline{\ ?\ }$.

16. If $m = 7$, evaluate $3m + (m + 2)$.

17. Find the value of $2 + a + 11$ when $a = 5$.

18. Find the value of $3x^2$ when $x = 10$.

19. Evaluate $2(a + b - c)$ when $a = 11$, $b = 10$, and $c = 9$.

20. Find the value of $x^3 + 2^y$ when $x = 5$ and $y = 5$.

21. Evaluate $(3m + 5)(2m - 4)$ when $m = 6$.

22. Evaluate $(3m + 5) - (2m - 4)$ when $m = 6$.

23. Evaluate $\frac{3a + 2b}{2a + 4b}$ when $a = 1$ and $b = 2.5$.

24. Find the value of $x + [1 + x(2 + x)]$ when $x = 7$.

Objective C. *Find solutions to equations and inequalities involving simple arithmetic.* *(Lessons 4-9, 4-10)*

25. *Multiple choice.* Which of these is a solution to $3x + 11 = 26$?
(a) 15   (b) 5   (c) 45   (d) 37

26. *Multiple choice.* Which of these is a solution to $y > \text{-}5$?
(a) -4   (b) -5   (c) -6   (d) -7

27. Find a solution to $3x = 12$.

28. Find a solution to $100 - t = 99$.

29. What is a solution to $y + 8 = 10$?

30. What value of m works in $20 = m \cdot 4$?

PROPERTIES DEAL WITH THE PRINCIPLES BEHIND THE MATHEMATICS.

Objective D. *Know the correct order of operations.* *(Lessons 4-1, 4-5, 4-6)*

31. An expression contains only two operations, a powering and a multiplication. Which should you do first?

32. *True or false.* If an expression contains nested parentheses, you should work the outside parentheses first.

33. *Multiple choice.* Written on one line, $\frac{30 + 5}{30 - 5} =$
(a) $30 + 5/30 - 5$.   (b) $(30 + 5)/30 - 5$.
(c) $30 + 5/(30 - 5)$.   (d) $(30 + 5)/(30 - 5)$.

34. *Multiple choice.* In which expression can the grouping symbols be removed without changing its value?
(a) $10 - (7 - 2)$ (b) $(4 \cdot 87 \cdot 0) + 5$
(c) $10/(5 - 2^2)/2$ (d) $(9 \cdot 3)^2$

Objective E. *Given instances of a pattern, write a description of the pattern using variables.* *(Lesson 4-2)*

35. Three instances of a pattern are given. Describe the general pattern using one variable.

$$5 \cdot 12 + 9 \cdot 12 = 14 \cdot 12$$
$$5 \cdot 88 + 9 \cdot 88 = 14 \cdot 88$$
$$5 \cdot \pi + 9 \cdot \pi = 14 \cdot \pi$$

36. Three instances of a pattern are given. Describe the general pattern using three variables.

$$6 + 7 - 8 = 6 - 8 + 7$$
$$10.2 + 0.5 - 0.22 = 10.2 - 0.22 + 0.5$$
$$30\% + 10\% - 20\% = 30\% - 20\% + 10\%$$

37. Four instances of a pattern are given. Describe the general pattern using two variables.

$$\frac{1}{9} + \frac{5}{9} = \frac{1 + 5}{9} \qquad \frac{0}{9} + \frac{25}{9} = \frac{0 + 25}{9}$$
$$\frac{11}{9} + \frac{44}{9} = \frac{11 + 44}{9} \qquad \frac{9}{9} + \frac{9}{9} = \frac{9 + 9}{9}$$

Objective F. *Give instances of a pattern described with variables.* *(Lesson 4-2)*

38. Give two instances of the pattern $5(x + y) = 5x + 5y$.

39. Give three instances of the pattern $2 + A = 1 + A + 1$.

40. Give two instances of the pattern $ab - c = ba - c$.

USES DEAL WITH APPLICATIONS OF MATHEMATICS IN REAL SITUATIONS.

Objective G. *Given instances of a real-world pattern, write a description of the pattern using variables.* *(Lesson 4-2)*

41. Three instances of a pattern are given. Describe the general pattern using variables.

If the weight is 5 ounces, the postage is 5¢ + 5 · 20¢.
If the weight is 3 ounces, the postage is 5¢ + 3 · 20¢.
If the weight is 1 ounce, the postage is 5¢ + 1 · 20¢.

42. Four instances of a pattern are given. Describe the general pattern using variables.

One person has 10 fingers.
Two people have 2 · 10 fingers.
Seven people have 7 · 10 fingers.
One hundred people have 100 · 10 fingers.

Objective H. *Write a numerical or algebraic expression for an English expression involving arithmetic operations.* *(Lesson 4-3)*

In 43–46, translate into mathematical symbols.

43. the sum of eighteen and twenty-seven

44. fifteen less than one hundred thousand

45. the product of four and twenty, decreased by one

46. seven less six

In 47–50, translate into an algebraic expression.

47. seven more than twice a number

48. a number divided by six, the quotient decreased by three

49. A number is less than five.

50. the product of thirty-nine and a number

Objective I. *Calculate the value of a variable, given the values of other variables in a formula.* *(Lesson 4-7)*

51. The formula $I = 100m/c$ is sometimes used to measure a person's IQ. The IQ is I, mental age is m, and chronological age is c. What is I if $m = 7$ and $c = 5.5$?

52. The formula $F = 1.8C + 32$ relates Fahrenheit (F) and Celsius (C) temperature. If C is 10, what is F?

53. The formula $A = bh$ gives the area A of a parallelogram in terms of its base b and height h. What is the area of a parallelogram with base 1 foot and height 6 inches?

54. The formula $C = 0.6n + 4$ estimates the temperature C in degrees Celsius when n is the number of cricket chirps in 15 seconds. If a cricket chirps 25 times in 15 seconds, what is an estimate for the temperature?

Objective J. *Calculate probabilities and relative frequencies in a situation with known numbers of outcomes.* *(Lesson 4-8)*

55. What is the largest value that a probability can have?

56. If an experiment has 3 equally likely outcomes, what is the probability of each outcome?

57. A grab bag has 10 prizes. Two are calculators. If you choose a prize without looking, what is the probability you will select a calculator?

58. Fifty slips of paper numbered 1 to 50 are placed in a hat. A person picks a slip of paper out of the hat. What is the probability that the number on the slip ends in 4?

59. A coin was tossed 30 times and 16 times landed heads up. What was the relative frequency of *tails?*

60. A survey of 200 voters after an election indicated that 105 of them voted for the Republican candidate, 60 for the Democrat, and 35 for others. What is the relative frequency that a voter in this survey voted Democrat?

REPRESENTATIONS DEAL WITH PICTURES, GRAPHS, OR OBJECTS THAT ILLUSTRATE CONCEPTS.

Objective K. *Graph the solutions to any inequality of the form $x < a$ and similar inequalities, and identify such graphs.* *(Lesson 4-10)*

61. The solutions to what sentence are graphed here?

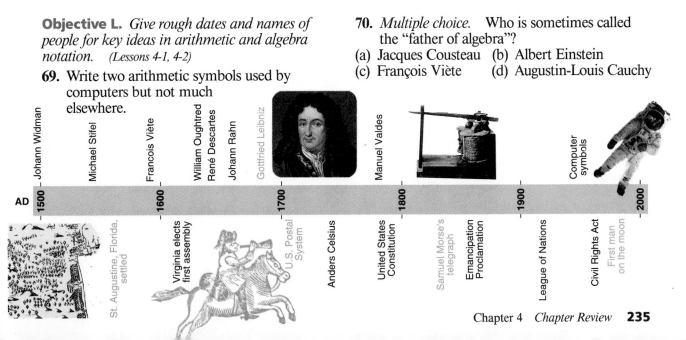

62. The solutions to what sentence are graphed here?

In 63–68, graph all solutions to the sentence on a number line.

63. $x < 24$

64. $y > 2$

65. $-4 \geq t$

66. $6 \leq d$

67. $3 < x < 7$

68. $-1 > y \geq -2$

CULTURE DEALS WITH THE PEOPLES AND THE HISTORY RELATED TO THE DEVELOPMENT OF MATHEMATICAL IDEAS.

Objective L. *Give rough dates and names of people for key ideas in arithmetic and algebra notation.* *(Lessons 4-1, 4-2)*

69. Write two arithmetic symbols used by computers but not much elsewhere.

70. *Multiple choice.* Who is sometimes called the "father of algebra"?
(a) Jacques Cousteau   (b) Albert Einstein
(c) François Viète   (d) Augustin-Louis Cauchy

PATTERNS LEADING TO ADDITION

Karen S. Peterson wrote an article in *USA Today* in 1989 about the amount of time that experts say adults should spend on various activities. The table below is taken from that article.

Just how much can you fit into a day?	Experts suggest
Exercise	30 min.
Personal grooming	45 min.
Time with children	4 hrs.
Reading newspapers	45 min.
Pets	50 min.
Housekeeping/chores	1–2 hrs.
Work	7–10+ hrs.
Commuting	$1\frac{1}{2}$ hrs.
Errands	up to 2 hrs.
Grocery shopping	(men) 17.88 min.
	(women) 22.25 min.
Cooking, eating dinner	1 hr.
Entertaining	1 hr.
Dental care	18 min.
General time with spouse	6 hrs. 50 min.
Volunteering	30 min.
Time with plants	10 min.
Time for you	1 hr.
Reading a book	15 min.
Spiritual development	15 min.
Sleep	7.5 hrs.

Karen Peterson concluded that a person would need a day of about 42 hours to do everything the experts say you're supposed to do. To arrive at this conclusion, she had to add whole numbers, fractions, and decimals. She also had to worry about units. Of course, she was being funny. But she was also trying to be serious. Some people think they are not perfect unless they try to do everything the experts say. She used addition to point out that this was impossible.

In this chapter, you will study many other uses of addition.

5-1

Models for Addition

Running water. *Niagara Falls, which lies on the U.S.-Canadian border, consists of two waterfalls—the American Falls in New York and the Horseshoe Falls in Ontario. The combined average water flow is about 212,200 cubic feet per second.*

Addition is important because adding gives answers in many actual situations. It is impossible to list all the uses of addition. So we give a general pattern that includes many of the uses. We call this general pattern a model for the operation. The two most important models for addition are called putting together and slide.

The Putting-Together Model for Addition

Here are three instances of the *Putting-Together Model for Addition.*

1. The two countries that share land borders with the United States are Canada and Mexico. In 1991, the population of Canada was estimated as 26,835,500 and the population of Mexico was estimated as 90,007,000. What was the total estimated population of these two countries in 1991?

2. Suppose a book is purchased for $4.95 and the tax is $0.30. What is the total cost?

3. Taken from a person's pay are 20% for taxes and 7.6% for social security. What percent is taken altogether?

The numbers put together may be large or small. They may be written as decimals, percents, or fractions. The general pattern is easy to describe using variables.

> **Putting-Together Model for Addition**
> Suppose count or measure x is put together with a count or measure y with the same units. If there is no overlap, then the result has count or measure $x + y$.

When the units are different and you want an answer in terms of a single unit, decide on an appropriate unit for your answer and then convert everything to that unit.

Example 1

2 meters + 46 centimeters

Solution

Either change everything to meters or change everything to centimeters.

Change to meters.	Change to centimeters.
2 meters + 46 centimeters	2 meters + 46 centimeters
= 2 meters + 0.46 meter	= 200 cm + 46 cm
= 2.46 meters	= 246 centimeters

The two answers, 2.46 meters and 246 centimeters, are equal.

Example 2

8 apples + 14 oranges

Solution

Apples cannot be changed to oranges, or vice versa. But they are both fruits.

$$8 \text{ apples} + 14 \text{ oranges}$$
$$= 8 \text{ pieces of fruit} + 14 \text{ pieces of fruit}$$
$$= 22 \text{ pieces of fruit}$$

Numbers to be added are called **addends.** If one addend is unknown, the putting-together model can still be used. For instance:

You weigh y and a cat weighs c. If you step on a scale together, the total weight is $y + c$.

This idea can be applied to weigh a cat that is too small to be weighed on an adult scale.

First weigh yourself on the scale. Suppose you weigh 107.5 lb. Together you and the cat must weigh $(c + 107.5)$ lb. Now you pick up the cat and step on the scale again. Suppose the total weight is 113.25 lb. Then the cat's weight is the solution to the equation $c + 107.5 = 113.25$.

If you do not know how to solve an equation of this type, you will learn how to do so in Lesson 5-8.

When there is overlap in the addends, the total cannot be found by just adding. For example, imagine that 50 pages of a book have pictures and 110 pages have tables. You cannot conclude that 160 pages have either pictures or tables. Some pages might have both pictures and tables. You will deal with this situation in Chapter 7.

The Slide Model for Addition

Recall that negative numbers are used when a situation has two opposite directions. Examples are deposits and withdrawals in a savings account, ups and downs of temperatures or weight, profits and losses in business, and gains and losses in football or other games. In these situations, you may need to add negative numbers. Here is a situation that leads to the addition problem 10 + -12.

Example 3

Flood waters rise 10 feet and then recede 12 feet. What is the end result?

Solution

Think of the waters sliding up and then sliding back down. Picture this with two vertical arrows, up for 10 and down for -12. Start the arrow for -12 where the arrow for 10 finished.

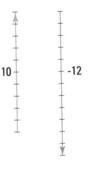

The arrow for -12 finishes 2 units below the starting position for the arrow for 10. In symbols, 10 + -12 = -2. The end result is that the waters went down 2 feet.

Unbelievably wet.
The boy and his grandfather are trying to cope with the Great Flood of 1993 along the banks of the Mississippi River. Due to spring and summer floods, the river was above flood stage for over $\frac{1}{3}$ of the year.

This example is an instance of the *Slide Model for Addition.* In the slide model, positive numbers are shifts or changes or slides in one direction. Negative numbers are slides in the opposite direction. The + sign means "followed by." The sum indicates the net result.

Slide Model for Addition
If a slide x is followed by a slide y, the result is a slide $x + y$.

Example 4

Tony spends $4 for dinner and then earns $7 for baby-sitting.
a. What is the net result of this?
b. What addition problem leads to this answer?

Solution

a. Tony has $3 more than he had before dinner.
b. -4 + 7 gives the net result. The -4 is for spending $4. The 7 is for earning $7.

The solution to Example 4 can be pictured on a horizontal number line.

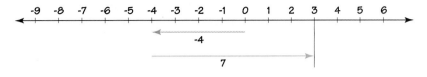

Start at 0. Think of ⁻4 as a slide 4 units to the left. Draw the arrow pointing left. Think of 7 as an arrow going 7 units to the right but starting at ⁻4. Where does the second arrow end? At 3, the sum.

Positive numbers are usually pictured as slides up or to the right. Negative numbers are usually pictured as slides down or to the left.

Example 5

a. Picture ⁻3 + ⁻2 on a horizontal number line.
b. What is the result?

Solution

a. Think of a slide 3 units to the left, followed by a slide 2 more units to the left.

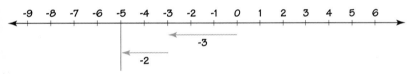

b. The result is a slide 5 units to the left. ⁻3 + ⁻2 = ⁻5

The slide model can also be used with positive numbers. If the temperature goes up 9° and then goes up 7° more, the result is an increase of 16°. Of course you know this. You have known some instances of both the putting-together and slide models for many years.

QUESTIONS

Covering the Reading

1. What is an *addend?*

2. State and give an example of the Putting-Together Model for Addition.

3. Mary gets on a scale and weighs *M* kg. Craig gets on a scale and it registers 60 kg. Together they get on the scale and the scale shows 108 kg. Write an equation relating *M*, 60, and 108.

4. Carla says you can add apples and oranges. Peter says you cannot. Who is right? Justify your answer.

5. Write 3 meters + 4 centimeters as one quantity.

6. State and give an example of the Slide Model for Addition.

In 7–10, what does the phrase mean in the slide model?

7. a positive number

8. a negative number

9. the + sign

10. the sum

11. a. Draw a picture of 5 + -6 using arrows and a number line.
   b. 5 + -6 = ?

12. a. Draw a picture of -3 + -8 using arrows and a number line.
   b. -3 + -8 = ?

In 13–16, a situation is given. **a.** What question related to the situation can be answered by addition? **b.** What is the answer to that addition? (Draw a picture if you need to.)

13. The temperature fell 5°, then fell 3° more.

14. A person gets a paycheck for $250, then buys a coat for $150.

15. Flood waters rise 15 feet, recede 19 feet, then rise 5 feet 4 inches.

16. In making a cake, a person puts in $\frac{1}{2}$ teaspoon of salt. Then the person forgets and puts in another $\frac{1}{2}$ teaspoon of salt.

17. Examine the table on page 237.
   a. Add up all the times experts suggest.
   b. Refer to your answer to part **a,** and explain why an adult cannot spend the suggested time on each activity.

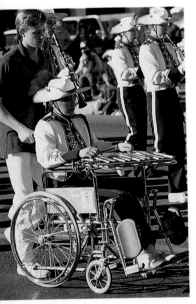

Applying the Mathematics

18. Joe needs to buy a new pair of basketball shoes. He estimates that it will take $\frac{1}{2}$ hour to bike to the mall and 20 minutes to purchase the shoes. How long should he allow for shopping?

19. Together Dan and Diane have $20. Together Diane and Donna have $15. How much do the three of them have in total?

20. Rosa has 2 brothers and 1 sister. Her sister Maria has 2 brothers and 1 sister. How many children are in the family?

21. Michelle has b brothers and s sisters. How many children are in the family?

22. Of the 25 students in Ms. Jones's class, 36% are on a school team, 40% are in the band or chorus, and 28% are in some other school activity.
   a. Is this possible?   **b.** Why or why not?

23. Last year in Springfield it rained or snowed on 140 days. The sun shone on *s* days. Can *s* be determined from this information? Why or why not?

24. Suppose the temperature was 58°F when you woke up at 7:00 A.M., then rose 23° from 7:00 A.M. to 2:00 P.M. and fell 8° from 2:00 P.M. to 6:00 P.M. What was the temperature at 6:00 P.M.?

25. What is the result of walking north 300 feet, then south 120 feet, and then north 40 feet?

26. Give an inequality relating the three numbers mentioned. The bench will support at most 250 pounds. Mike weighs *M* pounds. Nina weighs 112 pounds. The bench will hold both of them.

In 27–29, tell whether the sum is always, sometimes, or never positive.

27. Two negative numbers are added.

28. Two positive numbers are added.

29. A positive number and a negative number are added.

<hr>

Review

30. Tell whether the number is a solution to $0 \le x < 50$. Write *yes* or *no*. *(Lesson 4-10)*
  a. 35.2   **b.** -4   **c.** 50
  d. 0   **e.** 1/100   **f.** 60%

31. The percent *p* of discount on an item can be calculated by using the formula $p = 100(1 - n/g)$. In this formula, *g* is the original price, and *n* is the new price. Find the percent of discount on an item reduced from:
  a. $2 to $1.   **b.** $2.95 to $1.95.   *(Lesson 4-7)*

32. How many degrees are there in the given figure? *(Lesson 3-7)*
  a. a right angle   **b.** half a circle   **c.** an acute angle

33. Round 99.3% to the nearest whole number. (Hint: Watch out!) *(Lesson 2-4)*

34. *True or false.* $-10 \le -8$ *(Lesson 1-9)*

35. If climbing up 2 meters is represented by the number 2, what number will represent each event? *(Lesson 1-8)*
  a. climbing down 6 meters   **b.** staying at the same height

<hr>

Exploration

36. Ask two adults how much time they spent yesterday on the activities listed in the table on page 237. How do their times compare with the times suggested by experts?

37. When 1 cup of sugar is added to 1 cup of water, the result is not 2 cups of the mixture. Why not?

But I have only two hands! *Doing several things at once is one way to get everything done.*

Profits add up. *At some garage sales, children earn money selling homemade refreshments.*

Adding Zero

Suppose you have $25. If you earn $4, then you will have 25 + 4, or 29, dollars. If you spend $4, you will have 25 + -4, or 21 dollars. If you do nothing, you will have 25 + 0, or 25, dollars. When you add 0 to a number, the result is the original number. We say that adding 0 to a number keeps the identity of that number. So 0 is called the **additive identity.**

> **Additive Identity Property of Zero**
> For any number n, $n + 0 = n$.

For example, $-\frac{1}{2} + 0 = -\frac{1}{2}$, $77\% + 0 = 77\%$, and $0 + 3.469 = 3.469$.

Adding Opposites

Remember that 5 and -5 are called opposites. They are called opposites because if 5 stands for gaining 5 pounds, then -5 stands for its opposite, losing 5 pounds. Now suppose you gain 5 pounds and then lose 5 pounds. This is a slide model situation. So the result is found by adding 5 + -5.

You know the result! If you gain 5 pounds and then lose 5 pounds, the result is no change in weight. This verifies that 5 + -5 = 0.

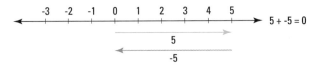

The same result occurs regardless of what number you use. The sum of a number and its opposite is 0. In symbols, $-n$ is the opposite of n. Some people call the general pattern the Property of Opposites.

> **Property of Opposites**
> For any number n, $n + \text{-}n = 0$.

The word "inverse" means opposite. Because of the property of opposites, the numbers n and $\text{-}n$ are called additive inverses of each other. For example, the additive inverse of 40 is -40. The additive inverse of -40 is 40. The additive inverse of -6.3 is 6.3. This can be written as -(-6.3) = 6.3. The parentheses are used so that it is clear that there are two dashes. Because $0 + 0 = 0$, the opposite of 0 is zero. That is, -0 = 0. Some people call $n + \text{-}n = 0$ the Additive Inverse Property.

The Opposite of an Opposite

The opposite of the opposite of any number is the number itself. We call this the Opposite of Opposites Property, or the Op-op Property for short. Your class may wish to make up a different name.

> **Opposite of Opposites (Op-op) Property**
> For any number n, $-(\text{-}n) = n$.

To verify the Opposite of Opposites Property on a calculator, display any number. Now press the +/− key twice. After the first pressing, you will get the opposite of the number. After the second pressing, the original number will appear.

Caution: When x is negative, $\text{-}x$ is positive. For instance, if $x = \text{-}3$, then $\text{-}x = \text{-}(\text{-}3) = 3$. For this reason, we read $\text{-}x$ as "the opposite of x," not as "negative x."

See	Read
-3	negative 3 or opposite of 3
-t	opposite of t

As important as zero and negative numbers are, it took a long time to discover them. The Mayas of Central America are the first known culture to have a symbol for zero, dating from about the year 300. Negative numbers were first seen as solutions to equations by some Western European mathematicians in the late 1400s.

Honoring the harvest.
The Mayas created an advanced civilization based on agriculture. Today, Mayas of Oaxaca, Mexico, enjoy the Guelaguetza, *honoring the corn harvest.*

QUESTIONS

Covering the Reading

1. Why is zero called the additive identity?

2. Add: $0 + 3 + 0 + 4$.

3. Another name for additive inverse is __?__ .

In 4–7, give the additive inverse of each number.

4. 70 **5.** -13 **6.** $-\frac{1}{2}$ **7.** $-x$

8. State the Property of Opposites.

9. Describe a real situation that illustrates $7 + -7 = 0$.

10. State the Opposite of Opposites Property.

In 11–14, an instance of what property is given?

11. $2 + -2 = 0$ **12.** $-9.4 = 0 + -9.4$

13. $7 = -(-7)$ **14.** $0 = 1 + -1$

In 15 and 16, a situation is given.
a. Translate the words into an equation involving addition.
b. An instance of what property is given?

15. Withdraw $25, then make no other withdrawal or other deposit, and you have decreased the amount in your account by $25.

16. Walk 40 meters east, then 40 meters west, and you are back where you started.

17. When does $-n$ stand for a positive number?

18. If $x = -10$, give the value of $-x$.

Applying the Mathematics

In 19–22, perform the additions.

19. $-51 + -9 + 51 + 2$ **20.** $x + 0 + -x$

21. $a + b + -b + b + -a$ **22.** $-\frac{8}{3} + -\frac{2}{7} + 0 + \frac{2}{7} + \frac{8}{3} + \frac{14}{11}$

In 23–26, simplify.

23. $-(-(-5))$ **24.** $-(-(-(-6)))$

25. $-(-(-x))$ **26.** $-(-(-7 + 1))$

27. Suppose you have entered the number 5 on your calculator. After you press the $\boxed{+/-}$ key 50 times, what number will be displayed?

28. Suppose you have entered the number -6 on your calculator. Then you press the $\boxed{+/-}$ key n times.
a. For what values of n will 6 be displayed?
b. For what values of n will -6 be displayed?

In 29–32, evaluate the expression given that $a = 4$ and $b = -5$.

29. $-a + -b$ **30.** $-b + 18$

31. $a + -b$ **32.** $a + b + -b + -a$

In 33 and 34 a situation is given.
a. What is the result for each situation?
b. What addition problem gets that result?   *(Lesson 5-1)*

33. The temperature is -11° and then goes up 2°.

34. A person is $150 in debt and takes out another loan of $100.

In 35 and 36, give an equation or inequality with addition to relate the three numbers.   *(Lesson 5-1)*

35. We were at an altitude of *t* meters. We went down 35 meters in altitude. Our altitude is now 60 meters below sea level.

36. We need $50 to buy an anniversary gift for our parents. You have *Y* dollars. I have $14.50. Together we have more than enough.

37. A rope *d* inches in diameter can lift a maximum of about *w* pounds, where $w = 5000d(d + 1)$.   *(Lesson 4-6)*

   a. About how many pounds can a rope of diameter 1″ lift?

   b. About how many pounds can a rope with a diameter of one-half inch lift?

   c. A rope has diameter $\frac{9}{16}$ inch. Can it lift 5,000 pounds?

38. Put in order from smallest to largest.   *(Lessons 2-2, 2-8)*
3.2×10^4 9.7×10^{-5} 5.1×10^7

Like a bird. *Hang gliders are generally launched from the tops of hills. Changes in altitude are affected by wind currents.*

39. Opposites are common outside of mathematics. You know that fast and slow are opposites, and so are big and little. Each of the eight animal names given here has a meaning other than as an animal. Identify the four pairs of opposites.

donkey	dove	hawk	mouse
tiger	elephant	bear	bull

5-3

Rules for Adding Positive and Negative Numbers

Take stock in this. *A stock exchange enables investors to buy and sell bonds or shares of stocks. The prices of stocks and bonds can go up or down, resulting in positive or negative changes.*

Adding Without Using Rules

Suppose you want to add -50 and 30. In Lesson 5-1, you learned two ways of doing this. One way is to think of a situation using the numbers -50 and 30.

> I spend $50 and I earn $30.
> What is the net result?

The net result is that you have $20 less than when you started.

A second way is to represent -50 and 30 with arrows on a number line.

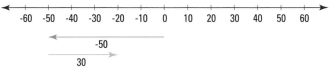

The bottom arrow ends under the -20 on the number line. Like the spending and earning idea, it shows that -50 + 30 = -20.

The Absolute Value of a Number

It isn't always easy to think of a situation for an addition problem, and it can be time consuming to draw a number line and arrows. So it helps to have a rule for adding positive and negative numbers. This rule is described using an idea called the *absolute value of a number.*

The **absolute value** of a number *n*, written |*n*|, is the distance of *n* from 0 on the number line. The absolute value of *n* is also the length of the arrow that describes *n*.

Consider the numbers -2, $-\frac{1}{3}$, 0, 3.8, and 5.

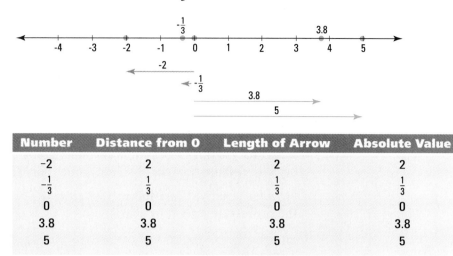

Number	Distance from 0	Length of Arrow	Absolute Value
-2	2	2	2
$-\frac{1}{3}$	$\frac{1}{3}$	$\frac{1}{3}$	$\frac{1}{3}$
0	0	0	0
3.8	3.8	3.8	3.8
5	5	5	5

Notice how simple it is to calculate absolute value.

The absolute value of a positive number is that number. The absolute value of 5 is 5. In mathematical symbols, this is written $|5| = 5$. The absolute value of 3.8 is 3.8, so $|3.8| = 3.8$. The absolute value of 4 million is 4 million.

The absolute value of a negative number is the opposite of that number. The absolute value of -2 is 2. In symbols, $|-2| = 2$. The absolute value of $-\frac{1}{3}$ is $\frac{1}{3}$, so $\left|-\frac{1}{3}\right| = \frac{1}{3}$. The absolute value of -4000 is 4000. The absolute value of zero is 0. $|0| = 0$.

A Rule for Adding Two Negative Numbers

Using the idea of absolute value, it is possible to state rules for adding positive and negative numbers. If the numbers are both positive, no rule is needed. You have been adding positive numbers since first grade. For instance, $7 + 4 = 11$.

Suppose the numbers are both negative, as in $-7 + -4$. How are they added? First, you can add 7 and 4. (You have added the absolute values of the numbers.) Now put the negative sign on. (You have taken the opposite.)

> To add two negative numbers, add their absolute values and then take the opposite.

A Rule for Adding a Positive and a Negative Number

The big advantage of the idea of absolute value comes when one addend is positive and one is negative.

Consider 13 + -40. You are 13 points ahead but you lose 40 points. How do you stand? If you don't know the answer, arrows can help.

Let's analyze this. The sum is certainly going to be negative. This is because -40 has a larger absolute value than 13.

How far negative? Just subtract 13 from 40. So the sum is -27. Here is the process described in words.

> To add a positive number and a negative number, take their absolute values. The sum is in the direction of the number with the larger absolute value. Subtract the smaller absolute value from the larger to find out how far the sum is in that direction.

Example 1

Add 5 and -8.9 using the above rule.

Solution

First take the absolute values. $|5| = 5$ and $|-8.9| = 8.9$. Since -8.9 has the larger absolute value, the sum is negative. Now subtract the smaller absolute value from the larger. $8.9 - 5 = 3.9$. So the sum is -3.9.

Example 2

$76 + -5 = ?$

Solution

The absolute values of the addends are 76 and 5. Since 76 has the larger absolute value, the sum is positive. Now subtract the absolute values.
$76 - 5 = 71$. So the sum is 71.

You may be able to find sums of positive and negative numbers without using the above rule. As long as you do not make errors, this is fine.

Activity

The $\boxed{\pm}$ or $\boxed{+/-}$ key on a calculator changes a number to its opposite. The following key sequence can be used to add -3 and -2. Should the answer be positive or negative? Answer in your mind before going on.
Key in 3 $\boxed{\pm}$ $\boxed{+}$ 2 $\boxed{\pm}$ $\boxed{=}$ on your calculator.

Enter the addition problem -263 + 159 on your calculator. Is the result positive or negative?

QUESTIONS

1. *Multiple choice.* Which are ways of finding the sum of positive and negative numbers?
   (a) Think of a situation fitting the slide model for addition.
   (b) Use arrows and a number line.
   (c) Take absolute values and use a rule.
   (d) All of the above are valid ways.

In 2–5, give the absolute value of the number.

2. -58 3. 4.01 4. 0 5. -11

In 6–9, simplify.

6. $|12|$ 7. $|-20|$ 8. $|0.0032|$ 9. $|0|$

In 10–12, an addition problem is given.
a. Without adding, tell whether the sum is positive or negative.
b. Add.
c. Check using a calculator. Show the key sequence you used.

10. -40 + 41 11. -7.3 + -0.8 12. 6 + 7

In 13–18, *true or false.*

13. The sum of two positive numbers is always positive.

14. The sum of two negative numbers is always negative.

15. The sum of a negative number and a positive number is always negative.

16. The absolute value of a number is always positive.

17. The numbers 50 and -50 have the same absolute value.

18. The absolute value of a negative number is the opposite of that number.

In 19–22, find the sum.

19. 3 + -6 20. -10 + -4 21. -1.7 + -.85 22. -473 + 2920

In 23–32, simplify.

23. $-2\frac{1}{3} + 1\frac{2}{3}$ 24. $5 + -\frac{3}{4}$

25. -3 + 8 + -7 + -9 + 7 26. 11 + -86 + -11 + -7 + 105

27. $-|-2|$ 28. $-\left|\frac{15}{2}\right|$

29. $|-0.74| + -|-0.74|$ 30. $|3| + |-3|$

31. $|3| - |-3|$ 32. $-|2.5| + |-6.8|$

33. *True or false.*
    a. When x is positive, $|x| = x$.
    b. When x is negative, $|x| = -x$.
    c. When $x = 0$, $|x| = x$.

In 34 and 35, the absolute value sign occurs outside an expression. When this happens, $|\ \ |$ is acting as a grouping symbol as well as indicating absolute value. So work within the absolute value sign first.

34. What is the value of $|x + y|$ when $x = -5$ and $y = 4$?

35. *Multiple choice.* Which is *not* true?
    (a) $|10 + 32| = |10| + |32|$ (b) $|10 + -32| = |10| + |-32|$
    (c) $|-10 + -32| = |-10| + |-32|$

36. Can the absolute value of a number ever be negative? Explain.

In 37 and 38, use this information. Values of exports are considered as positive quantities. Imports are considered as negative quantities. The total of exports and imports is called the *balance of trade.*

37. In 1991, the United States exported 421.730 billion dollars of goods and imported 487.129 billion dollars of goods. What was the balance of trade for the U.S. in 1991?

38. U.S. exports to Algeria in 1991 were 726.7 million dollars. The imports were 2,102.6 million dollars. What was the balance of trade with Algeria?

Shown above is Ghardaia Market in Algiers, Algeria. The U.S. is the biggest buyer of Algerian goods, purchasing nearly half the country's oil production. In return, Algeria imports machinery, raw materials, and most of its food.

Review

39. Evaluate $-h + 13$ for the given value of h.   *(Lesson 5-2)*
    a. $h = 13$ **b.** $h = -2$ **c.** $h = 0$

40. Translate into mathematical symbols. The sum of a number and its additive inverse is the additive identity.   *(Lessons 4-3, 5-2)*

41. Suppose a person is $150 in debt and pays off $50.
    a. What addition problem gives the result for this situation?
    b. What is that result?   *(Lesson 5-1)*

In 42 and 43, use the drawing at the left.

42. a. Without measuring, tell whether $\angle B$ is acute, obtuse, or right.
    b. Measure $\angle B$ to the nearest degree.   *(Lessons 3-6, 3-7)*

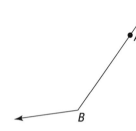

43. Measure the distance from A to B to the nearest $\frac{1}{8}$ inch.   *(Lesson 3-1)*

Exploration

44. Find three numbers that satisfy all of the following conditions.
    (1) The sum of the numbers is negative.
    (2) The sum of the numbers is greater than -1.
    (3) The sum of the absolute values of the numbers is larger than 1,000,000.

*Combining
Turns*

IN·CLASS
ACTIVITY

Recall that a circle has 360°. If you turn all the way around, you have made one **full turn** or one **revolution.** A full turn has **magnitude** 360°.

Pictured below are two **quarter turns.** Each is $\frac{1}{4}$ of a revolution, or 90°.

Notice that the turns are in different directions. In the turn at left, think of standing at point *O* (the center of the turn) and looking at point *P.* Then turn so that you are looking at point *Q.* You have turned **counterclockwise,** the opposite of the way clock hands usually move. At right, imagine yourself at point *B.* Look first at point *A,* then turn to look at point *C.* You have turned 90° **clockwise.**

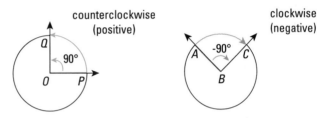

In most of mathematics, the counterclockwise direction is considered positive and the clockwise direction is negative. The turn with center *O* above is a turn with the positive magnitude 90°. The turn with center *B* is a turn of -90°.

1 Face the front of the room. Imagine that you are standing at the center of a circle on the floor. Turn 90° clockwise. If your entire class is doing this together, wait for signals from your teacher. What is in front of you?

2 Turn another 90° clockwise. What is in front of you?

3 Turn 45° counterclockwise. What is in front of you?

4 Begin again facing the front of the room. Your teacher will select an object in the room. Estimate how far (clockwise or counterclockwise) you have to turn to face the object directly.

5 Your teacher may select other objects. For each object, follow the directions of step 4.

6 Select 4 objects in the room. For each object, give the magnitude of a clockwise and a counterclockwise turn you would make to face each object. What is true about the magnitudes?

253

Combining Turns

Spirals, leaps, and spins. *In the 1992 Winter Olympics, Russians Natalia Mishkutienok and Artur Dmitriev of the Unified Team performed their turns and lifts with grace and precision. They won the gold medal in pair skating.*

Examples of Turns

In the picture below, the six small angles at M are drawn to have equal measures. Each angle has measure 60° because each is $\frac{1}{6}$ of 360°. Think of standing at M and facing point S. If you now turn clockwise to face point U, you have turned -120°. If you turn from facing S to face point X, you have turned 60°.

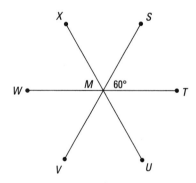

For any clockwise turn, there is a counterclockwise turn that ends in the same place. For instance, the -120° turn from S to U above can also be achieved by going 240° counterclockwise. So a turn of -120° is considered identical to a turn of 240°.

You can use protractors to help measure magnitudes of turns. But a protractor cannot tell you the direction of the turn.

Following One Turn by Another

Turns are discussed in this chapter because addition is used in finding results of combining two turns. Pictured below is a 35° turn followed by a 90° turn. The result is a 125° turn. In general, if one turn is followed by another, the magnitude of the result is found by adding the magnitudes of the turns.

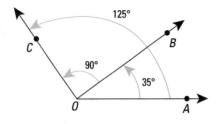

This addition property holds even if the turns are in opposite directions, but you then need to add positive and negative numbers. Shown below is a 60° turn followed by a -90° turn. The result is a -30° turn, or in other words, a turn of 30° clockwise.

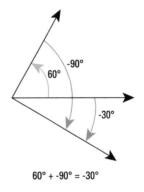

$$60° + {-90°} = {-30°}$$

Fundamental Property of Turns
If a turn of magnitude *x* is followed by a turn of magnitude *y*, the result is a turn of magnitude *x* + *y*.

Tops, figure skaters, gears, the earth, pinwheels, and phonograph records turn. All of these objects may turn many revolutions. So it is possible to have turns with magnitudes over 360°. For example, a dancer who spins $1\frac{1}{2}$ times around has spun 540°, since the dancer has made a full turn of 360° followed by a half turn of 180°.

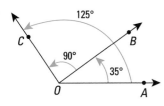

Adding Measures of Adjacent Angles

In the drawing at the left, ray $\overrightarrow{OB}$ is in the *interior* of $\angle AOC$. This ray is a side of two angles, *AOB* and *BOC*, that have no common interior points. The angles are called **adjacent angles.** The absolute values of the magnitudes of the turns are the measures of these angles. You can see that

$$\mathrm{m}\angle AOB + \mathrm{m}\angle BOC = \mathrm{m}\angle AOC.$$
$$35° \quad + \quad 90° \quad = \quad 125°$$

This is called the *Angle Addition Property.*

Angle Addition Property

If $\overrightarrow{OB}$ is in the interior of $\angle AOC$, then $\mathrm{m}\angle AOB + \mathrm{m}\angle BOC = \mathrm{m}\angle AOC$.

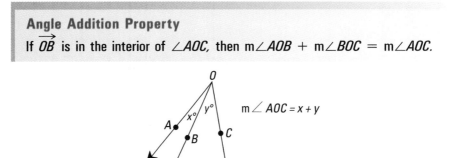

QUESTIONS

Covering the Reading

In 1–4, give the number of degrees in the turn.

1. a full turn

2. one revolution

3. a half turn

4. a quarter turn

5. In the drawing above, is the 35° turn clockwise or counterclockwise?

6. _?_ turns have positive magnitudes.

7. _?_ turns have negative magnitudes.

In 8–11, all small angles at *O* have the same measure. Give the magnitude of each turn around point *O*.

8. the counterclockwise turn from *D* to *B*.

9. the counterclockwise turn from *C* to *F*.

10. the clockwise turn from *D* to *E*.

11. the clockwise turn from *A* to *F*.

12. In the drawing of Questions 8–11, what is $\mathrm{m}\angle BOF$?

13. A turn of -80° ends in the same place as a turn of what positive magnitude?

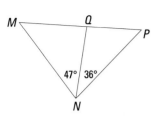

In 14–16, give the result of

14. a 90° turn followed by a 35° turn.

15. a 90° turn followed by a -35° turn.

16. a -5° turn followed by a -6° turn.

17. State the Fundamental Property of Turns.

18. a. In the drawing at the left, how are ∠MNQ and ∠QNP related?
   b. Find m∠MNP.

Applying the Mathematics

19. An airplane pilot is flying along the line shown toward point *B*.
When the plane reaches point *A*, the pilot changes course toward
point *C*. Use a protractor to determine the magnitude of the turn
that is needed.

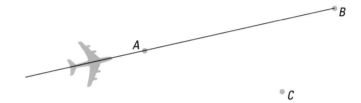

In 20 and 21, the highlighted spokes divide the wheel into equal
sections. What is the measure of a central angle?

20. (rear wheel)   **21.** (front wheel)

In 22–24, a Ferris wheel is pictured. The spokes are equally spaced.

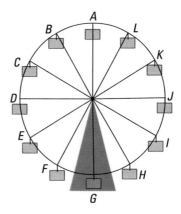

22. a. What magnitude turn will bring seat *J* to the position of seat *L*?
b. Where will seat *B* wind up on this turn?

23. a. What magnitude turn will bring seat *K* to the position of seat *G*?
b. What magnitude turn will bring seat *K* back to its original position?

24. If seat *C* is moved to the position of seat *G*, what seat will be moved to the top?

25. If $\overrightarrow{BC}$ and $\overrightarrow{BD}$ are perpendicular, how many degrees are in $\angle ABC$?

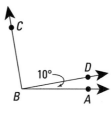

26. How many degrees does the hour hand of a clock turn in the given amount of time?
a. one hour   **b.** 10 minutes

27. How many degrees are in 6 revolutions?

28. Describe a situation involving turns that is not mentioned in this lesson.

Review

In 29–34, let $a = -5$, $b = -6$, and $c = 20$. Find the value of the expression.   *(Lessons 4-1, 5-1, 5-2, 5-3)*

29. $a + b$ **30.** $a + -b$ **31.** $|b + c|$

32. $|b| + |c|$ **33.** $-a + -b + -c$ **34.** $-(a + b + c)$

35. Consider this situation. It took Ralph 2 hours to do homework last night. He spent 45 minutes on math and *m* minutes on the rest. Give an equation that relates all the quantities in the situation. *(Lesson 5-1)*

36. The first Ferris wheel was erected in 1893 by G.W.G. Ferris for the World's Columbian Exposition in Chicago. It was huge. The wheel and cars weighed 2100 tons; the levers and machinery weighed 2200 tons; and a full load of passengers weighed 150 tons. What was the total weight of the wheel and machinery with a full load of passengers? *(Lesson 5-1)*

37. If *Y* stands for a year and time is measured in years, what stands for two years before *Y*? *(Lesson 4-3)*

38. Four instances of a general pattern are given. Describe the general pattern using two variables. *(Lesson 4-2)*

$$3 + 4 + 3 = 4 + 2 \cdot 3$$
$$-2 + 9.1 + -2 = 9.1 + 2 \cdot -2$$
$$7 + 0 + 7 = 0 + 2 \cdot 7$$
$$\frac{2}{7} + \frac{3}{7} + \frac{2}{7} = \frac{3}{7} + 2 \cdot \frac{2}{7}$$

39. How many meters are in 6 kilometers? *(Lesson 3-4)*

40. Write 63.2 million
   a. as a decimal.   **b.** in scientific notation. *(Lessons 2-1, 2-3)*

Exploration

41. Here is a brainteaser from *Quantum,* a mathematics and science magazine for students.

Starting from the inside, the above spiral of 35 toothpicks is wound clockwise. Move four toothpicks to rewind it counterclockwise.

A big wheel. *George Washington Gale Ferris built his wheel after his design won a contest held by promoters of the Exposition. The wheel was 264 ft high and had 36 seats. But each "seat" held 60 people and was about the size of a bus!*

Adding Positive and Negative Fractions

Using fractions. *Fractions are commonplace in food science or cooking classes.*

Adding Fractions with the Same Denominator

Nanette put $\frac{1}{3}$ cup of milk in a casserole. She forgot and then put in another $\frac{1}{3}$ cup of milk. No, no, Nanette! To find the total amount of milk Nanette put in the casserole, think of thirds as units. By the Putting-Together Model for Addition:

$$1 \text{ third} + 1 \text{ third} = 2 \text{ thirds}$$
$$\frac{1}{3} \quad + \quad \frac{1}{3} \quad = \quad \frac{2}{3}$$

Nanette put $\frac{2}{3}$ cup milk in the casserole.

Milton jogs laps on a quarter-mile track. Two days ago he ran 3 laps or 3/4 mile. Yesterday he ran 9 laps. Today he ran 10 laps. My, my, Milton. The total amount he ran is:

$$3 \text{ laps} + 9 \text{ laps} + 10 \text{ laps} = 22 \text{ laps}$$
$$\frac{3}{4} \quad + \quad \frac{9}{4} \quad + \quad \frac{10}{4} \quad = \quad \frac{22}{4}$$

Milton ran $\frac{22}{4}$ miles altogether.

The general pattern is easy to see. To add fractions with the same denominator, add the numerators and keep the denominator the same. Here is a description of the pattern, using variables:

Adding Fractions Property

For all numbers *a*, *b*, and *c*, with $c \neq 0$, $\frac{a}{c} + \frac{b}{c} = \frac{a+b}{c}$.

Example 1

Simplify $\frac{3}{x} + \frac{5}{x}$ and check your answer.

Solution

Using the Adding Fractions Property,

$$\frac{3}{x} + \frac{5}{x} = \frac{3+5}{x} = \frac{8}{x}.$$

Check

This should work for every value of x other than zero. So to check, we substitute some value for x. We pick 4, because fourths are terminating decimals and are easier to check than repeating decimals.

Does $\frac{3}{4} + \frac{5}{4} = \frac{8}{4}$? Check by rewriting the fractions as decimals.
Does $0.75 + 1.25 = 2$? Yes.

One good turn deserves another. *With just a few minutes between classes, students must make accurate clockwise and counterclockwise turns to quickly open their combination locks and grab their books.*

Example 2

A quarter turn counterclockwise is followed by a half turn clockwise. What is the result?

Solution 1

Draw a picture like the one below. Think of standing at O, facing P. After the quarter turn counterclockwise you will be facing Q. Then, after the half turn clockwise you will be facing R. The result is a quarter turn clockwise.

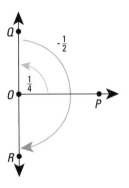

Solution 2

Use the Adding Fractions Property.

quarter turn counterclockwise + half turn clockwise

$= \frac{1}{4} + -\frac{1}{2}$	Translate to fractions.
$= \frac{1}{4} + -\frac{2}{4}$	Change to common denominator.
$= \frac{1 + -2}{4}$	Use the Adding Fractions Property. $(-\frac{2}{4} = \frac{-2}{4})$
$= -\frac{1}{4}$	Add 1 and -2.
= quarter turn clockwise	Translate back to words.

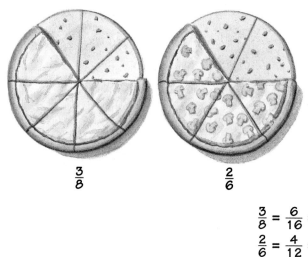

$$\frac{3}{8}$$

$$\frac{2}{6}$$

In Example 2, the denominators are not the same. Here is another example of adding fractions with different denominators.

Bill ate 3 of 8 pieces of a personal spinach pizza and 2 of 6 pieces of a mushroom pizza the same size. How much pizza did he eat in all? Use the Equal Fractions Property to find equal fractions with the same denominator. One way to start is to write

$$\frac{3}{8} = \frac{6}{16} = \frac{9}{24} = \frac{12}{32} = \cdots$$
$$\frac{2}{6} = \frac{4}{12} = \frac{6}{18} = \frac{8}{24} = \cdots$$

Stop, as we have, when two fractions have the same denominator. The number 24 is a **common denominator.**

$$\frac{3}{8} + \frac{2}{6} = \frac{9}{24} + \frac{8}{24}$$
$$= \frac{17}{24}$$

Bill ate the equivalent of $\frac{17}{24}$ of a whole pizza.

Notice that the common denominator 24 is a multiple of both 8 and 6. Other common multiples would be 48, 72, 96, and so on. Since 24 is the smallest common integer multiple of 8 and 6, it is called the **least common multiple (LCM)** of these numbers. You could use any common multiple as the common denominator. However, when the least common multiple is the common denominator, the numbers are smaller than with any other multiple.

Example 3

Add $\frac{5}{6} + \text{-}\frac{1}{2} + \frac{7}{3}$, and check.

Solution

You could find a common denominator by finding the product of the denominators: $6 \cdot 2 \cdot 3 = 36$. But that is much bigger than is needed. Since 6 is a multiple of the other denominators, 6 can be a common denominator.

$$\frac{5}{6} + \text{-}\frac{1}{2} + \frac{7}{3}$$

$$= \frac{5}{6} + \text{-}\frac{3}{6} + \frac{14}{6} \qquad \text{Since } \frac{1}{2} = \frac{1 \cdot 3}{2 \cdot 3} = \frac{3}{6} \text{ and } \frac{7}{3} = \frac{7 \cdot 2}{3 \cdot 2} = \frac{14}{6}.$$

$$= \frac{5 + \text{-}3 + 14}{6} \qquad \text{Adding Fractions Property}$$

$$= \frac{16}{6}$$

$$= \frac{8}{3}$$

▶

▶ **Check**

You can check using decimals even if the decimals repeat.

$$\frac{5}{6} \approx 0.833333$$

$$-\frac{1}{2} = -0.5$$

$$\frac{7}{3} \approx 2.333333$$

So the sum of the fractions is about
0.833333 + -0.5 + 2.333333 = 2.666666.
Since $\frac{8}{3}$ = 2.666666 . . . , the answer checks.

Using a Fraction Calculator

There are calculators that can add fractions directly. On those calculators you do not have to find a common denominator. You just enter the fractions, add, and the answer is given as a fraction.

Changing Mixed Numbers to Simple Fractions

When adding an integer and a fraction, use the denominator of the fraction as the common denominator. With this process, you can change any mixed number to a simple fraction.

Example 4

Write $2\frac{3}{5}$ as a simple fraction.

Solution

Think of $2\frac{3}{5}$ as $2 + \frac{3}{5}$.

$$2 + \frac{3}{5} = \frac{2}{1} + \frac{3}{5}$$

Use 5 as common denominator.

$$= \frac{10}{5} + \frac{3}{5}$$

$$= \frac{13}{5}$$

To do Example 4, many people use a shortcut to find the numerator of the simple fraction: multiply 5 by 2, and then add 3.

QUESTIONS

Covering the Reading

1. On Monday, Mary ran 7 laps. On Tuesday, she ran 9 laps. If each lap is a quarter mile, how far in miles did Mary run?

2. Sam used $\frac{5}{2}$ cups of flour in one recipe and $\frac{3}{2}$ cups of flour in another. How much flour did he use altogether?

3. A $\frac{3}{4}$ clockwise turn is followed by a full turn counterclockwise.
   a. Draw a picture of this.
   b. What is the result?

4. Terri ate $\frac{1}{3}$ of one pizza and $\frac{1}{4}$ of another the same size.
   a. What problem with fractions determines the total that Terri ate?
   b. Name five fractions equal to $\frac{1}{3}$.
   c. Name five fractions equal to $\frac{1}{4}$.
   d. What is the total amount of pizza that Terri ate?

5. a. Find three common multiples of 5 and 7.
   b. What is the least common multiple of 5 and 7?
   c. Evaluate $\frac{2}{7} + \frac{3}{5}$. Show your work.

6. a. Write $\frac{4}{5} + \frac{3}{10}$ as a single fraction.
   b. Check by using decimals.

In 7–14, write the sum as a simple fraction. Show each step in your work.

7. $\frac{50}{11} + \frac{5}{11}$ **8.** $\frac{13}{x} + \frac{4}{x}$

9. $\frac{3}{z} + -\frac{9}{z}$ **10.** $\frac{-4}{9} + \frac{1}{3}$

11. $\frac{8}{9} + \frac{1}{15}$ **12.** $\frac{5}{8} + \frac{2}{5}$

13. $\frac{11}{7} + \frac{6}{7} + \frac{-15}{7}$ **14.** $\frac{2}{3} + \frac{5}{3} + \frac{-5}{6}$

15. a. Write $4 + \frac{2}{5}$ as a simple fraction.
   b. Write $4 + \frac{2}{5}$ as a mixed number.

Applying the Mathematics

16. a. Divide to write $\frac{15}{4}$ as a mixed number.
   b. Draw a circle representation of $\frac{15}{4}$.

17. In 1896, Fannie Merritt Farmer (1857–1915) published the first cookbook that used standard measuring units such as teaspoons and cups. (Earlier cookbooks used such unclear terms as "a handful" or "a glass.") One of the relationships used in her cookbook is
1 teaspoon $= \frac{1}{3}$ tablespoon.
   a. Add $\frac{1}{3}$ tablespoon to $\frac{2}{3}$ tablespoon.
   b. Convert the tablespoons to teaspoons in part **a** and add the corresponding quantities.
   c. Are your answers to parts **a** and **b** equal quantities? Why or why not?

18. In music, two eighth notes take the same time as a quarter note. What addition of fractions explains this?

19. In music, what amount of time is taken by a sixteenth note followed by an eighth note?

Bon appetit. *The* Fannie Farmer Cook Book *is the most famous American cookbook ever published. Over 3 million copies have been sold. Many of the recipes were tested and revised at Miss Farmer's Boston Cooking School.*

In 20–23, write the sum as a simple fraction. Show all your steps in writing out each solution.

20. $4\frac{3}{8} + 2\frac{1}{5}$

21. $4\frac{3}{8} + 2\frac{2}{3}$

22. $\frac{6}{7} + 3\frac{7}{12}$

23. $9.75 + \frac{1}{4} + 3$

24. Because of a rainstorm, the water level in a swimming pool rose by $1\frac{1}{2}''$. The following day, it was $2\frac{1}{4}''$ lower.

    a. What was the total change in the water level?

    b. If the deep end of the pool was originally 9 ft deep, how deep was the water at the end of the second day?

Review

25. A clockwise turn of 52° is followed by a counterclockwise turn of 120°. What is the result?   *(Lesson 5-4)*

In 26–29, simplify.   *(Lessons 5-2, 5-3)*

26. $3 + {}^-2 + {}^-1$

27. $-(-({}^-4.7))$

28. ${}^-6 + {}^-8 + {}^-10$

29. ${}^-6 + 8 + {}^-10$

30. Minnie has 5 brothers and 2 sisters. Her brother Dennis has 4 brothers and 3 sisters. How many children are in the family?   *(Lesson 5-1)*

In 31 and 32, a sentence is given. **a.** Name one solution. **b.** Graph all solutions.   *(Lesson 4-10)*

31. $x < 3$

32. $-2 \geq y > {}^-6.5$

33. The dinosaur known as *Brachiosaurus,* weighed over 30,000 kg.
    a. Write this number in scientific notation.
    b. About how much is this weight in pounds?
    (Lessons 2-3, 3-5)

The biggest known dinosaur. *Shown here is a friendly Brachiosaurus as depicted in the movie* Jurassic Park.

Exploration

34. a. List all the factors of 28 and 49. Then list all the factors of the least common multiple of 28 and 49.

    b. List all the factors of 30 and 40. Then list all the factors of the least common multiple of 30 and 40.

    c. Suppose you know the factors of two numbers. Describe how to find the factors of the least common multiple of the numbers.

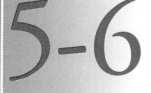

LESSON
5-6

Adding Probabilities

Family fun. *You are on Pennsylvania Avenue on the game board. It is your turn to toss the dice. Do you know the probability of landing on Boardwalk (5 spaces away)?*

Certain situations lead to the adding of probabilities.

Example 1

Consider the spinner shown here. If all directions of the spinner are equally likely, what is the probability that the spinner will land either in region *A* or in region *B*?

Solution

The circle has a total of 360°.

The probability of landing in A is $\frac{90}{360}$.

The probability of landing in B is $\frac{120}{360}$.

The probability of landing in region A or in region B is the sum of these two probabilities.

$$\frac{90}{360} + \frac{120}{360} = \frac{210}{360}$$

$$= \frac{7 \cdot 30}{12 \cdot 30} = \frac{7}{12}.$$

The probability that the spinner will land either in region A or in region B is $\frac{7}{12}$.

Check

$\frac{7}{12}$ is a little more than $\frac{6}{12}$, which is $\frac{1}{2}$. Do regions *A* and *B* cover a little more than half the circle? It seems so.

Probabilities of Mutually Exclusive Events

Addition works in Example 1 because the events "landing in region A" and "landing in region B" do not overlap. The spinner is not able to land in both regions at the same time. Two events which cannot occur at the same time are called **mutually exclusive.** This name comes from the fact that the occurrence of one event "excludes" the occurrence of the other event.

Example 2

Two normal dice are tossed at the same time. Assume the dice are fair.
Let C = one or more of the dice show a 1.
Let D = the sum of the numbers on the dice is 10.
a. Are C and D mutually exclusive events?
b. What is the probability of C or D occurring?

Solution

A display of all 36 possible outcomes from tossing two dice is helpful.

The outcomes in event C are outlined in blue. There are 11 outcomes in event C.

The outcomes in event D are outlined in orange. There are three possible ways of rolling a sum of 10, so event D has three outcomes.
a. Since C and D have no outcomes in common, they are mutually exclusive.
b. There are 14 outcomes in events C or D. Since the dice are fair, each outcome has probability $\frac{1}{36}$. The probability that C or D will occur is $\frac{14}{36}$.

The general pattern is quite simple.

> **Probability of A or B**
> If A and B are mutually exclusive events, the probability of the event A or B is the probability of A plus the probability of B.

Addition of probabilities does not work if the events are not mutually exclusive.

Example 3

Three of eight patients in the pediatrician's waiting room have sore throats and half have a fever. What is wrong with the following reasoning?

"$\frac{3}{8} + \frac{1}{2} = \frac{3}{8} + \frac{4}{8} = \frac{7}{8}$, so $\frac{7}{8}$ of the patients have either a sore throat or a fever."

Solution

Since it is possible that a patient has both a sore throat and a fever, the two events are not mutually exclusive. Because of the overlap, the putting-together model does not apply.

Probabilities of Complementary Events

For any event E, there is a corresponding event consisting of all the outcomes that are not in E. This event is called the **complement** of the event E, and is written as ***not E.*** You can see that an event and its complement are always mutually exclusive.

For instance, in Example 2, you might wish to know the probability that neither die shows a 1. This is the event *not C*. By counting, you can see that 25 outcomes do not show a 1, so the probability of *not C* is $\frac{25}{36}$.

The events E and *not E* never have any overlap and together include all possible outcomes. So the sum of their probabilities is 1. In the case of the events C and *not C* discussed above, $\frac{11}{36} + \frac{25}{36} = \frac{36}{36} = 1$.

Probabilities of Complements

The sum of the probabilities of any event E and its complement *not E* is 1.

For instance, if the probability of rain today is 30%, then the probability of no rain has to be 70%, since 30% + 70% = 1.

QUESTIONS

Covering the Reading

In 1–4, suppose the spinner in Example 1 is spun once.

1. **a.** What is the probability that the spinner will land in region C?
   b. What is the probability that the spinner will land in region D?
   c. What is the probability that the spinner will land in region C or region D?

2. What is the probability that the spinner lands in region A or region C?

3. What is the probability that the spinner will not land in region B?

LESSON 5-7

The Commutative and Associative Properties

Which way is up? *Switch them around and call them Curly, Moe, and Larry, or Moe, Curly, and Larry, or Moe, Larry, and Curly. But no matter how mixed up they become, they are still The Three Stooges.*

The Commutative Property of Addition

When two things have been put together, it does not make any difference which came first. Buying two items in different order results in the same total cost. Likewise, the order of slides or turns makes no difference. In football, gaining 3 yards and then losing 4 yards gives the same end result as first losing 4 yards and then gaining 3: $3 + -4 = -4 + 3$.

The general pattern was first given a name by François Servois in 1814. He used the French word *commutatif,* which means "switchable." The English name is the *commutative property.*

> **Commutative Property of Addition**
> For any numbers a and b, $a + b = b + a$.

The Associative Property of Addition

Think of spending $5, earning $14, and then spending $20. The end result is given by the addition $-5 + 14 + -20$. From order of operations, you know you should work from left to right. But does it matter in which order the additions are done? Follow this experiment. By using parentheses, we can change which addition is to be done first.

Doing the left addition first is shown by:
$$(-5 + 14) + -20$$
$$= 9 + -20$$
$$= -11$$

Doing the right addition first is shown by:
$$-5 + (14 + -20)$$
$$= -5 + -6$$
$$= -11$$

The sums are equal.

$$(-5 + 14) + -20 = -5 + (14 + -20)$$

The genius of Sir William Rowan Hamilton was identified before he was 3. During his career, he made major contributions in mathematics and physics.

On the left-hand side, the 14 is associated with the -5 first. On the right-hand side, the 14 is associated with the -20. For this reason, in 1835, the Irish mathematician Sir William Rowan Hamilton called the general pattern the *associative property*.

Associative Property of Addition
For any numbers a, b, and c, $(a + b) + c = a + (b + c)$.

Distinguishing Between the Properties

Both the commutative property and the associative property have to do with changing order. The commutative property says you can change the order of the *numbers* being added. The associative property says you can change the order of the *operations*.

Example 1

Which property is demonstrated by $3 + (18 + -12) = 3 + (-12 + 18)$?

Solution

All that has been changed is the order of numbers within parentheses. So this is an instance of the **Commutative Property of Addition.**

Example 2

Below is an instance of which property of addition?
$$150 + -73 + -22 + 8 = 150 + (-73 + -22) + 8$$

Solution

The order of the numbers has not been changed. On the left-hand side of the equation, we would add the 150 and the -73 first. On the right-hand side, we would add the -73 to the -22 first. So the order of additions has been changed. This can be done because of the **Associative Property of Addition.**

Applying the Properties

Because addition is both commutative and associative, when an expression involves *only* addition:

(1)   addends can be put in any order before adding them;

(2)   parentheses can be removed or put in whenever you wish;

(3)   you can speak of adding three or more numbers.

Example 3

Consider the following checking account, opened with $150 on October 24. How much was in the account after the deposit of November 6?

DATE	DEPOSIT	WITHDRAWAL (check)
10-24	$150.00	
10-27		$70.00
10-31	100.00	
11-01		40.00
11-03		60.00
11-06	50.00	

Solution

In order, the deposits and withdrawals lead to the expression

$$150 + \text{-}70 + 100 + \text{-}40 + \text{-}60 + 50.$$

These numbers are all added, so they may be added in any order. It is easier to add positives to positives and negatives to negatives, so group all deposits together and all withdrawals together. Then find the positive total and the negative total.

$$= (150 + 100 + 50) + (\text{-}70 + \text{-}40 + \text{-}60)$$
$$= 300 + \text{-}170$$
$$= 130$$

There was $130 in the account after the deposit of November 6.

QUESTIONS

Covering the Reading

1. Give an example of the Commutative Property of Addition.

2. Give an example of the Associative Property of Addition.

3. In an expression involving only additions, you can change the order of the additions. Which property implies this?

In 4–7, *multiple choice.*   Tell which property is illustrated.

(a) only the Commutative Property of Addition
(b) only the Associative Property of Addition
(c) both the commutative and associative properties
(d) neither the commutative nor the associative property

4. $\frac{2}{3} + \frac{3}{4} = \frac{3}{4} + \frac{2}{3}$ **5.** $8 + (36 + \text{-}24) + \text{-}16 = (8 + 36) + (\text{-}24 + \text{-}16)$

6. $3(4 + 9) = 3(9 + 4)$ **7.** $1 + 2 + 3 = 3 + 2 + 1$

8. In what century were the names commutative and associative first used? (Remember that the years 1901–2000 constitute the 20th century.)

9. Give a real-world situation that could lead someone to add many positive and negative numbers.

10. On November 1, a person had $400 in a checking account. Here are the transactions for the next two weeks.

DATE	DEPOSIT	WITHDRAWAL (check)
11-03	$102.00	
11-05		$35.00
11-08		75.00
11-11	40.00	
11-12		200.00

a. What addition can you do to calculate the amount in the account at the end of the day on November 12?
b. How much was in the account at that time?

Applying the Mathematics

11. What does a negative answer to a problem like Question 10b mean?

12. Harry and Kerry start at the same place. Harry walks 300 meters east and then 500 meters west. Kerry walks 500 meters west and then 300 meters east. Do they end at the same place? Explain.

In 13 and 14, simplify.

13. $17 + \text{-}1 + \text{-}4$

14. $0 + \text{-}3 + \text{-}2 + 4 + \text{-}6 + 1$

In 15 and 16, add the given numbers.

15. $99, \text{-}46, 12, \text{-}99, 46, \text{-}12$

16. $-\frac{3}{8}, -\frac{3}{8}, \frac{40}{3}, -\frac{3}{8}, -\frac{40}{3}, \frac{3}{8}$

17. A family keeps a weekly budget. During a five-week period, they are $12.50 over, $6.30 under, $21 over, $7.05 under, and $9.90 under. How are they doing?

18. A robot turns 50°, -75°, 120°, -103°, and 17°. What is the total turn?

19. The daily low temperatures last week were -1°, 5°, 6°, 4°, 3°, -4°, and -6°. Find the mean of these temperatures.

In 20 and 21, tell whether the equation illustrates the Commutative Property of Addition, the Associative Property of Addition, or both.

20. $\text{-}3x + (4 + 3x) = \text{-}3x + (3x + 4)$

21. $8x + \text{-}5y + 5y = 8x + (\text{-}5y + 5y)$

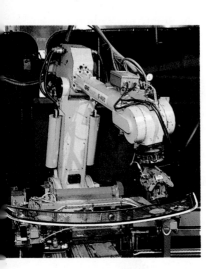

What do robots do?
Robots were first introduced as toys. Today, robots are indispensable in industry. The robot pictured above is applying adhesive to a new car window on a production line.

Review

22. Suppose an airplane flight has a 15% probability of being more than 10 minutes early, a 60% probability of being within 10 minutes of its scheduled arrival time, and a 25% probability of being over 10 minutes late. The plane is due at 8:10 P.M.
a. What is the probability that the plane will land before 8:20 P.M.?
b. What is the probability that the plane will land after 8:00 P.M.?
c. Are the events in parts **a** and **b** mutually exclusive?
d. Are the events in parts **a** and **b** complements? *(Lesson 5-6)*

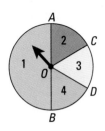

23. In the spinner at the left, $\overline{AB}$ is a diameter of the circle.
m∠AOC = m∠COD = m∠BOD.
a. What is m∠AOC?   **b.** What is m∠AOD?
c. Name all acute angles.   **d.** Name all right angles.
e. Name all obtuse angles in the picture.
f. If all positions of the pointer are equally likely, what is the probability the pointer will stop on an odd number?
(Lessons 3-7, 4-8, 5-4, 5-6)

24. To open a lock, you turn it $2\frac{1}{2}$ turns clockwise and $1\frac{3}{4}$ turns counterclockwise. What is the result of these two turns? *(Lessons 5-4, 5-5)*

25. Refer to the picture of the Ferris wheel on page 258. If seat *B* is turned to the position of seat *G*, what seat will be moved to the top? *(Lesson 5-4)*

26. How many degrees does the second hand of a watch turn in 10 seconds? *(Lesson 5-4)*

27. Give an instance of the Property of Opposites. *(Lesson 5-2)*

28. Another term for "additive inverse" is __?__. *(Lesson 5-2)*

29. The temperature is $t°$ and goes down 3°. What is the end result? *(Lesson 5-1)*

30. Morrie and Lori together have 7 tickets to a concert. Corey and Lori together have 5 tickets. How many tickets do Corey, Lori, and Morrie have altogether? *(Lesson 5-1)*

31. Simplify: $-5 + 6(3.7 + 1.3/2) - 4$. *(Lesson 4-5)*

32. At many professional athletic contests, fans are asked to guess the day's attendance from five choices displayed on the scoreboard. Assuming you guess randomly:
a. What is the probability of picking the right number?
b. Write the answer to part **a** as a percent, a decimal, and a fraction.
c. If the correct attendance figure was 38,833, about how many people will guess the correct number? Round this answer for reasonable accuracy. *(Lessons 1-4, 2-6, 4-8)*

33. How many feet are in 6.2 miles? *(Lesson 3-2)*

34. State the Multiplication Property of Equality. *(Lesson 3-2)*

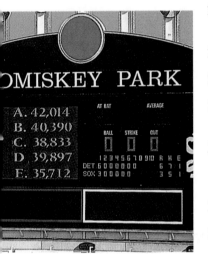

What's your best guess?
The fans at a baseball game are being asked to guess the attendance.

Exploration

---

35. a. Add 203,275 + -89,635 + 265,164 on a simple or scientific calculator.
b. Turn the calculator 180°. What word does the display spell?
c. Find two negative numbers and a positive number whose sum on the calculator, when turned 180°, is ShOE.
d. Make up a problem to spell a word on the calculator display.

Examples of Equations of the Form $x + a = b$

Early in elementary school, you could see how to fill in the blank in

$$\underline{\quad} + 3 = 4.$$

In algebra, usually the blank is replaced by a letter. If the sentence involves simple numbers, you should be able to find the solution mentally. For instance, in

$$x + 35 = 48$$

you should know that $x = 13$.

But some equations involve complicated numbers. Other equations involve only variables and no specific numbers at all. It helps to have a systematic way of solving these equations. On one side of both $\underline{\quad} + 3 = 4$ and $x + 35 = 48$, a number is being added to an (unknown) variable. On the other side is a single number. These equations are *equations of the form $x + a = b$*. Think of x as the unknown. Think of a and b as known numbers.

A Useful Property for Solving Equations

One property of addition is particularly helpful in solving equations. Here is an example. Begin with equal numbers.

$$\tfrac{1}{4} = 0.25$$

If the same number is added to both sides, then the sums will be equal. For instance, if 3 is added to both sides:

$$3 + \tfrac{1}{4} = 3 + 0.25.$$

That is, $3\tfrac{1}{4} = 3.25$.

Of course, you could add *any* number to both sides. The sums would still be equal. This property is a special form of the Substitution Principle first mentioned in Lesson 2-5.

Addition Property of Equality
If $a = b$, then $a + c = b + c$.

Using the Addition Property of Equality

The Addition Property of Equality says: If two numbers are equal, then you can add anything you want to both numbers and the sums will be equal. It is similar to the Multiplication Property of Equality. Here is how this property is used in solving equations.

Example 1

Solve $x + -8 = 57$.

Solution

Add 8 to both sides. (You can do this because of the Addition Property of Equality.) Because addition is associative, you do not have to worry about grouping.

Original equation	$x + -8 = 57$
Use Addition Property of Equality	$x + -8 + 8 = 57 + 8$
Property of Opposites	$x + 0 = 57 + 8$
Additive Identity Property of Zero and simplification	$x = 65$

Check

Substitute 65 for x in the original equation. Does $65 + -8 = 57$? Yes. So 65 is the solution.

The key is to know what number to add to both sides. In Example 1, we added 8 to both sides. This caused the left side to simplify to just x.

You may be able to solve equations like these in your head. Still, you must learn the general strategy. More complicated problems require it.

In Example 1, after adding the same number to both sides, all you have to do is simplify the sides to find the solution. So there is only one step to remember.

To solve an equation of the form $x + a = b$, add $-a$ to both sides and simplify.

Solving $x + a = b$ can be pictured with a balance scale. Think of x as the unknown weight of a sack. Various objects have been placed on the two pans of the scale. Because the scale is not tipped one way or the other, the total weights of the two sides are equal.

$$x + a = b$$

Clearly one way to find x is to remove a weight equal to a from both pans. This is like adding $-a$ to each side. The result is that $x = b + -a$.

In Example 2, no equation is given. An equation has to be found from the situation.

Example 2

The temperature was 4° this afternoon and now is -6°. By how much has it changed?

Solution

Let c be the change in temperature. (If c is positive, the temperature has gone up. If c is negative, the temperature has gone down.)

Then $4 + c = -6$.

Since 4 is added to c, we add -4 to both sides. This ensures that c will wind up alone on its side of the equation.

$$4 + c = -6$$
$$-4 + 4 + c = -4 + -6$$
$$0 + c = -10$$
$$c = -10$$

The change is -10°.

Check

To check, look at the original question. If the temperature was 4° and it changed -10°, is it now -6°? Since the answer is yes, -10° is the correct solution.

More Comments on Solving $x + a = b$

When solving an equation, it is important to organize your work carefully. Write each step *underneath* the previous step. Some teachers like students to name properties. Here the equation $x + 43 = -18$ is solved with the properties named where they are used.

```
        x + 43 = -18            (original equation)
x + 43 + -43 = -18 + -43        Addition Property of Equality
      x + 0 = -18 + -43         Property of Opposites
          x = -18 + -43         Additive Identity Property of Zero
          x = -61               (arithmetic computation)
```

Check

Substitute -61 for x in the original equation.
Does -61 + 43 = -18? Yes. So -61 is the solution.

In solving an equation, the unknown can be on either side. (The sack can be on either side of the scale.) And since addition is commutative, the unknown may be the first or second addend on that side. So all four equations below have the same solution.

$$x + 43 = -18 \qquad 43 + x = -18 \qquad -18 = x + 43 \qquad -18 = 43 + x$$

QUESTIONS

Covering the Reading

1. Give the solution to $8 + \underline{\ ?\ } = 13$.

2. In algebra, the blank of Question 1 is usually replaced by what?

3. Begin with the true equation $\frac{4}{5} = 0.8$.
   a. Is it true that $6 + \frac{4}{5} = 6 + 0.8$?
   b. Is it true that $-1 + \frac{4}{5} = -1 + 0.8$?
   c. Is it true that $17.43 + \frac{4}{5} = 17.43 + 0.8$?
   d. Parts **a** to **c** are instances of what property?

4. Here are steps in the solution of the equation $3.28 = A + -5$. Give the reason for each step.
$$3.28 = A + -5$$
   a. $3.28 + 5 = A + -5 + 5$
   b. $3.28 + 5 = A + 0$
   c. $3.28 + 5 = A$
   d. $8.28 = A$

5. *Multiple choice.* Which equation does not have the same solution as the others?
   (a) $13 + x = -6$ (b) $-6 + x = 13$ (c) $x + 13 = -6$ (d) $-6 = x + 13$

In 6–11, an equation is given. **a.** To solve the equation, what number should you add to each side of the equation? **b.** Find the solution. **c.** Check your answer.

6. $x + 86 = 230$ **7.** $-12 + y = 7$ **8.** $-5.9 = A + -3.2$

9. $60 = z + \frac{-22}{3}$ **10.** $431 + B = -812$ **11.** $C + \frac{1}{4} = \frac{2}{5}$

12. The temperature was $-15°$ yesterday and is $-20°$ today. Let c be the change in the temperature.
   a. What equation can be solved to find c?
   b. Solve that equation.
   c. Check your answer.

13. Suppose the temperature is two degrees below zero. By how much must it change to become three degrees above zero?

Applying the Mathematics

14. If $a = b$, the Addition Property of Equality says that $a + c = b + c$. But it is also true that $c + a = c + b$. Why?

In 15 and 16, first simplify one side of the equation. Then solve and check.

15. $A + 43 + -5 = 120$ **16.** $-35 = 16 + d + 5$

17. A family's income I satisfies the formula $I = F + M + C$, where F is the amount the father earns, M is the amount the mother earns, and C is the amount the children earn. If $I = 40{,}325$, $F = 18{,}800$, and $M = 20{,}500$, how much did the children earn?

18. Roberto had $48.83. His mother gave him some money for his birthday. He then had $63.33.
   a. Write an equation to determine how much money Roberto got for his birthday.
   b. Solve and check.

19. On Monday, Vito's savings account showed a balance of $103.52. He deposited $35 into the account Tuesday. Wednesday he withdrew $12.50. Thursday he asked the bank to tell him how much was in the account. They said that $130.05 was in the account. What happened is that the bank paid Vito some interest. How much interest?

20. George is G years old. Wilma is W years old.
   a. What will be George's age 10 years from now?
   b. What will be Wilma's age 10 years from now?
   c. If George and Wilma are the same age, how are G and W related?
   d. Translate into mathematics: If George and Wilma are the same age now, then they will be the same age 10 years from now.
   e. Part **d** of this question is an instance of what property?

21. Solve for b: $a + b = c$.

22. Which property is used in getting to each step? *(Lessons 5-2, 5-7)*

$(5 + 7) + {}^-5$
a. $= 5 + (7 + {}^-5)$
b. $= 5 + ({}^-5 + 7)$
c. $= (5 + {}^-5) + 7$
d. $= 0 + 7$
e. $= 7$

23. You make a random choice from a hat filled with red, white, and blue chips. The probability of choosing a red chip is $\frac{2}{3}$. The probability of choosing a white chip is $\frac{1}{4}$.
a. What is the probability of choosing a red or white chip?
b. Describe the complement of the event in part **a**.
c. Find the probability of the event you described in part **b**. *(Lesson 5-6)*

24. If a $210\frac{1}{2}^{\circ}$ counterclockwise turn is followed by a $150\frac{1}{8}^{\circ}$ counterclockwise turn, what results? *(Lessons 5-4, 5-5)*

25. Consider the inequality $x < {}^-3$.
a. Name a value of x that is a solution.
b. Name a value of x that is not a solution. *(Lesson 4-10)*

26. Simplify: $(4 \cdot 3 - 2 \cdot 1)(4 \cdot 3 + 2 \cdot 1)$. *(Lesson 4-5)*

In 27–29, translate into mathematics. *(Lesson 4-3)*

27. triple a number n

28. five less than a number t

29. a number B divided by twice a second number C

In 30 and 31, translate into English words. *(Lesson 3-6)*
30. $\overrightarrow{AB}$

31. $\mathrm{m}\angle LNP = 35°$

32. Draw a line segment with length 73 mm. *(Lessons 3-1, 3-3)*

33. How many zeros follow the 1 in the decimal form of 10^{30}? *(Lesson 2-2)*

34. Approximate $\frac{7}{16}$ to the nearest hundredth. *(Lessons 1-4, 1-6)*

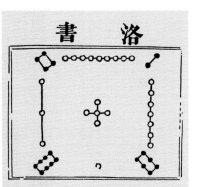

Where did it come from? *The earliest known magic squares came from China. According to tradition, the Emperor Yu received a magic square called the* Lo shu *from a divine tortoise in about* 2900 B.C.

35. A magic square is a square array of numbers in which the sum along every row, column or main diagonal is the same. Copy and complete the square at the right, such that

(1) Each row, column, and diagonal adds to 2.

(2) Each integer from $^-7$ to 8 is used exactly once.

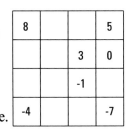

8			5
		3	0
		-1	
-4			-7

IN-CLASS
ACTIVITY

With this activity, you will make a special hexagon. Start with an $8\frac{1}{2}''$ by $11''$ or $8''$ by $10\frac{1}{2}''$ piece of paper. Your teacher may guide you through the steps, so wait for instructions.

1 Hold your paper with the long side horizontal. Fold in half along the dotted line as shown.

2 With the folded edge to the left, again fold the paper in half along the dotted line.

3 With the folds at the top and left, starting at the upper left, label the corners *J, K, L,* and *M*. (This rectangle has $\frac{1}{4}$ the area of the original sheet of paper.)

4 Pick up corner *K*. Move $\overline{JK}$ so that the measure of ∠*KJM* is equal to the measure of ∠*KJN*. Be as careful as you can here.

5 Pick up the corner *M* and fold it over $\overline{JK}$ to meet $\overline{JN}$. (If you have made m∠*KJM* = m∠*KJN* in Step 4, this will work.)

6 Cut through all the sheets of paper along $\overline{PM}$.

7 Unfold △*JPM*. You now should have a hexagon to use for Questions 20–24 of Lesson 5-9. Label the vertices of your hexagon *H, E, X, A, G, N*.

Hexagonal structures. *The hexagon is the polygonal shape of the Navajo hogan. Most hogans are log-and-mud structures built to face the rising sun.*

Naming Lines and Line Segments

Let *A* and *B* be any points. There is only one line containing both of them. The line goes on forever. To show this, when we draw a picture of a line, we put arrows at both ends. This line is written $\overleftrightarrow{AB}$ with the two-sided arrow above the letters.

The points on $\overleftrightarrow{AB}$ that are between *A* and *B*, together with the points *A* and *B* themselves, make up a **line segment** or **segment,** written as $\overline{AB}$. *A* and *B* are the **endpoints** of $\overline{AB}$.

What Is a Polygon?

A **polygon** is a union of segments in which each segment intersects two others, one at each of its endpoints. The segments are called the **sides** of the polygon. Two sides meet at a **vertex** of the polygon. (The plural of vertex is **vertices.**) The number of sides of a polygon equals the number of its vertices. Polygons are classified by that number.

Number of sides	Number of vertices	Type of polygon	Example
3	3	triangle	
4	4	quadrilateral	
5	5	pentagon	
6	6	hexagon	
7	7	heptagon	
8	8	octagon	
9	9	nonagon	
10	10	decagon	

Polygons with more than 10 sides do not always have special names. A polygon with 11 sides is called an 11-gon. A polygon with 42 sides is called a 42-gon. In general, a polygon with *n* sides is called an ***n*-gon.**

Naming Polygons

Polygons are named by giving their vertices in order around the figure. The polygon below is *WTUV*. You could also call it *TUVW, WVUT,* or five other names. *WTVU* is not a correct name for this polygon because the vertices *W, T, V,* and *U* are not in order.

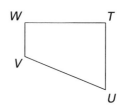

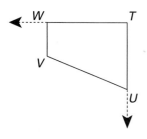

Extend two sides of a polygon drawn from a particular vertex to be rays with the vertex as the endpoint. These rays form an **angle of the polygon.** Every polygon has as many angles as it has sides or vertices. In polygon *WTUV*, angles *W* and *T* look like right angles. Angle *U* is acute. Angle *V* is obtuse.

QUESTIONS

Covering the Reading

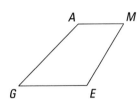

1. The line through points A and B is written __?__.

2. The line segment with endpoints C and D consists of what points?

3. The line segment with endpoints F and E is written __?__.

4. Refer to the polygon pictured at the left.
   a. Name this polygon.
   b. Name its sides.
   c. Name its vertices.
   d. Name its angles.
   e. What type of polygon is this?

In 5–12, what name is given to a polygon with the given number of sides?

5. 5   6. 6   7. 7   8. 8

9. 9   10. n 11. 11   12. 1994

13. How many angles does a 12-gon have?

14. A decagon has __?__ sides, __?__ angles, and __?__ vertices.

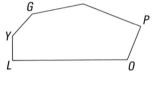

15. Refer to the figure at the left.
    a. Give three possible names for this polygon.
    b. What type of polygon is this?

Applying the Mathematics

16. A square is what type of polygon mentioned in this lesson?

17. Temples in the Bahá'í religion have 9 sides. Suppose you view a Bahá'í temple from directly overhead. The outline of the temple will have what shape?

18. A famous building just outside of Washington, DC, is pictured below. A clue to its name is given by its shape. What is its name?

19. A **diagonal** of a polygon is a segment that connects two vertices of the polygon but is not a side of the polygon. For example, $\overline{AC}$ and $\overline{BD}$ are the diagonals of quadrilateral *ABCD* drawn below.

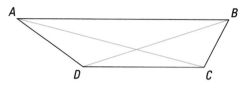

a. Draw a pentagon *PQRST* and name its diagonals.
b. How many diagonals does it have?

In 20–24, use the hexagon you made in the In-class Activity preceding this lesson.

20. Measure $\angle H$, $\angle E$, and $\angle X$. Compare your measures with those of a neighbor. Make a conjecture about the measures of $\angle A$, $\angle G$, and $\angle N$. Measure the angles. How do they compare to the measures of angles *H*, *E*, *X*? Can you make a conjecture about the angles of *HEXAGN*? What is the approximate sum of the measures of angles *H, E, X, A, G,* and *N*?

21. Measure $\overline{HE}$, $\overline{EX}$, and $\overline{XA}$ to the nearest centimeter. Compare your measures with those of a neighbor. Make a conjecture about the lengths of $\overline{AG}$, $\overline{GN}$, and $\overline{NH}$. Do you see a pattern? If so, what is it?

22. $\overline{HX}$ is a diagonal. List the other diagonals of *HEXAGN*. Measure their lengths to the nearest centimeter.

23. Explain why $\overline{XA}$ is not a diagonal.

24. Write a paragraph describing some characteristics of *HEXAGN*.

Review

25. *Multiple choice.* Which equation does not have the same solution as the others? *(Lesson 5-8)*
(a) $23 = 60 + y$ (b) $y + 60 = 23$
(c) $23 = y + 60$ (d) $y + 23 = 60$

In 26–29, solve and check. *(Lesson 5-8)*

26. $x + 12 = -10$ **27.** $300 = 172 + (45 + w)$

28. $-1 + -2 + m + -3 + -4 = -100$ **29.** $d + -1.3 = 6.8$

30. Given are steps in solving the general equation $a + x = b$ for *x*. Give the reason why each step is correct. *(Lessons 5-2, 5-8)*

$a + x = b$
a. $-a + a + x = -a + b$
b. $0 + x = -a + b$
c. $x = -a + b$

31. A sofa has length *s* inches. If the sofa were 3 inches longer, it would be 8 feet long. *(Lessons 4-3, 5-8)*
 a. Give an equation of the form $a + x = b$ that involves the three quantities mentioned in the previous two sentences.
 b. Solve that equation for *s*.

32. Find the area of the six-pointed star below using the following steps.
 a. Trace the figure below onto a piece of paper.
 b. Cut the figure into the 5 pieces shown.
 c. Rearrange the pieces and measure to find the area.

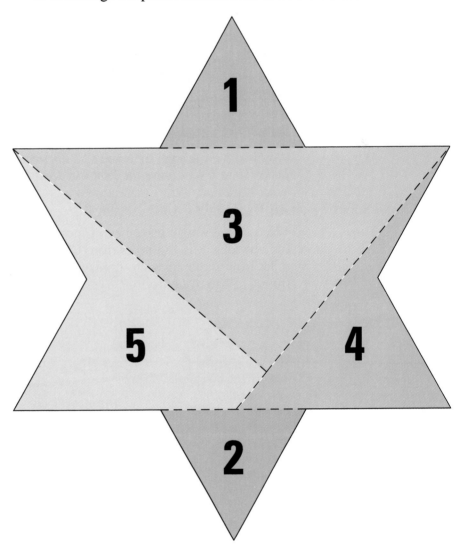

Winning by a length or a nose? *Hundreds of thousands of people travel to Louisville each year to see the Kentucky Derby at Churchill Downs.*

An Example of Adding Lengths

The quickest way to drive from Chicago to Louisville is to go through Indianapolis. It is 185 miles from Chicago to Indianapolis. It is 114 miles from Indianapolis to Louisville. Along this route, how far is it from Chicago to Louisville?

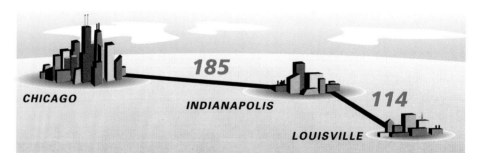

The answer is easy to find. Just add.

$$185 + 114 = 299$$

It is 299 driving miles from Chicago to Louisville.

The general pattern is an instance of the Putting-Together Model for Addition. If you have two lengths x and y, the total length is $x + y$.

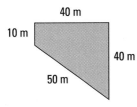

40 m

10 m

40 m

50 m

What Is Meant by Perimeter?

There are many situations in which more than two lengths are put together. Imagine walking around the building outlined at the left. You will have walked $40 + 40 + 50 + 10$ meters. This is the idea behind *perimeter*.

> The **perimeter** of a polygon is the sum of the lengths of the sides of the polygon.

Let the sides of a pentagon have lengths v, w, x, y, and z. If the perimeter is p, then $p = v + w + x + y + z$.

Now suppose $v = 12$, $w = 19$, $x = 15$, $z = 22$, and the perimeter $p = 82$. The situation is pictured below. What is the value of y? You can solve an equation to find out.

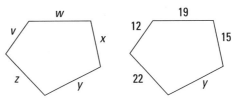

Begin with the general formula.	$p = v + w + x + y + z$
Substitute.	$82 = 12 + 19 + 15 + y + 22$
Simplify.	$82 = 68 + y$
Add -68 to both sides.	$-68 + 82 = -68 + 68 + y$
Simplify.	$14 = y$

Naming the Distance Between Two Points

Recall that if A and B are points, we use the following symbols.

$\overleftrightarrow{AB}$ line through A and B

$\overline{AB}$ segment with endpoints A and B

$\overrightarrow{AB}$ ray with endpoint A and containing B

There is one other related symbol. The symbol **AB** (with nothing over it) stands for the *length* of the segment AB. (It makes no sense to multiply points. So when A and B are points, putting the letters next to each other does not mean multiplication.) Here are some examples of this notation. The perimeter of triangle PQR is $PQ + QR + RP$.

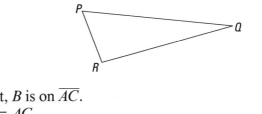

At the right, B is on $\overline{AC}$.

$AB + BC = AC$

$12 + 24 = AC$

$36 = AC$

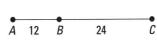

A   12   B   24   C

Covering the Reading

1. On Route 1 along the California coast, San Luis Obispo is located between San Francisco and Los Angeles. A map shows that San Francisco is 227 miles from San Luis Obispo. Los Angeles is 195 miles from San Luis Obispo. By this route, how far is it from San Francisco to Los Angeles?

2. The situation in Question 1 is an instance of what model for addition?

In 3–6, *P* and *Q* are points.
a. Give the meaning of the symbol.
b. Draw a picture (if possible) of the figure the symbol stands for.

3. $\overleftrightarrow{PQ}$

4. $\overrightarrow{PQ}$

5. $\overline{PQ}$

6. PQ

In 7–10, use the drawing. *B* is on $\overline{AC}$.

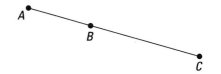

7. If $AB = 5$ and $BC = 3$, what is AC?

8. If $AB = x$ and $BC = y$, what is AC?

9. If $AB = x$ and $BC = 3$, what is AC?

10. If $AB = 14.3$ cm, $BC = t$ cm, and $AC = v$ cm, how are 14.3, t, and v related?

11. What is the perimeter of a polygon?

In 12 and 13, give the perimeter of the polygon.

12.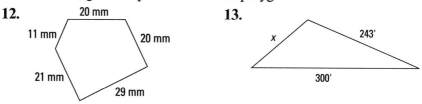

13.

14. Suppose the perimeter of the triangle in Question 13 is 689 feet.
a. What equation can you solve to find x?
b. Solve the equation.

In 15 and 16,
a. Write an equation that will help answer the question.
b. Solve the equation.

15. Two sides of a pentagon have length 9 and two sides have length 5. The perimeter of the pentagon is 30. What is the length of the fifth side?

16. A race is to be 10 km long. The organizers want the runners to run along a bicycle path that is 2800 meters long and a final leg that is 650 meters long. How long must the rest of the course be?

17. Two sides of a triangle ABC have the same length, and $AB = 40$. The perimeter of triangle ABC is 100. What are the possible lengths for $\overline{BC}$ and $\overline{AC}$?

How much is enough?
In Leo Tolstoy's short story, How Much Land Does a Man Need?, *a farmer pays 1,000 rubles for as much land as he wants. The catch is that he must walk the perimeter of the land in one day.*

18. *Multiple choice.* Only one of these equations is not true. Which one is it?
(a) $\overleftrightarrow{AB} = \overleftrightarrow{BA}$
(b) $\overrightarrow{AB} = \overrightarrow{BA}$
(c) $\overline{AB} = \overline{BA}$
(d) $AB = BA$

19. A square has one side of length 5 m.
a. Is this enough information to find its perimeter?
b. If so, find it. If not, tell why there is not enough information.

A —————————— B C

20. Use the drawing at the left. B is on $\overline{AC}$. If $AC = 1$ ft and $BC = 1$ in., what is AB?

21. Draw a polygon called *NICE* and put $\overrightarrow{CN}$ and $\overline{IE}$ on your drawing. *(Lesson 5-9)*

22. The temperature was -22° yesterday and 10° today. Let t be the change in the temperature.
a. What equation can be used to find t?
b. Solve the equation.
c. Check your answer. *(Lesson 5-8)*

23. Suppose the probability of rain is $\frac{1}{50}$, and the probability of no rain is x.

    a. What equation can be solved to find x?

    b. Solve this equation. *(Lessons 5-6, 5-8)*

24. Solve $-0.3 + A = 6.3$. *(Lesson 5-8)*

25. The sum of 23.6 and some number is 40.05. Find the number. *(Lesson 5-8)*

26. Perform the intended operations. *(Lesson 5-2)*

    a. $-6 + -(-6)$ **b.** $-(-x) + -(-y)$

In 27 and 28, graph the solutions to the inequality. *(Lesson 4-10)*

27. $1 \le x < 5$ **28.** $y \ge -\frac{3}{2}$

29. Ernestine is E years old. Alfredo is A years old. *(Lesson 4-3)*

    a. How old was Ernestine 4 years ago?

    b. How old was Alfredo 4 years ago?

    c. Translate into mathematics: If Ernestine and Alfredo are the same age now, then they were the same age four years ago.

    d. Part **c** of this question is an instance of what property?

In 30–33, use the drawing.

30. Measure the lengths of all sides of the quadrilateral to the nearest millimeter. *(Lesson 3-1)*

31. Which angles of the heptagon seem to be obtuse? *(Lessons 3-7, 5-9)*

32. Which angles of the pentagon seem to be acute? *(Lessons 3-7, 5-9)*

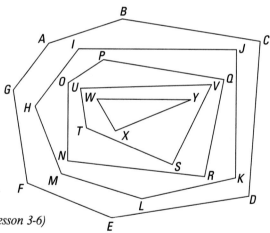

33. Measure angle *TSV*. *(Lesson 3-6)*

Exploration

34. Find the length of sides in all triangles satisfying *all* these conditions.

    (1) The perimeter is 40.

    (2) No side has length under 10.

    (3) All sides have different lengths.

    (4) The lengths of all sides are integers.

        (Hint: Organize your work.)

PROJECTS
5
CHAPTER FIVE

A project presents an opportunity for you to extend your knowledge of a topic related to the material of this chapter. You should allow more time for a project than you do for typical homework questions.

1 Polygons in Architecture

In Lesson 5-9 there is a picture of the Pentagon building outside of Washington DC, a building that received its name from its shape. But a building does not have to have the shape or name of a polygon in order to have polygons in it. Take photographs or draw pictures of buildings near where you live and identify various kinds of polygons on them.

2 Fractions that Add to 1

The fractions $\frac{1}{2}$, $\frac{1}{3}$, and $\frac{1}{6}$ are *unit fractions* (their numerators are 1) and their sum is 1. Find four different unit fractions whose sum is 1. Then find five different unit fractions whose sum is 1. Try to find a general procedure that could enable a person to find any number of different unit fractions whose sum is 1.

3 The Mayas

The Mayas, who are the first known people to have a symbol for zero, developed their own number system. Find out about this number system and write an essay on what you find.

(continued)

4 A Probability Experiment

Extend the Lesson 5-6 Exploration by using a computer or calculator to generate at least 500 random numbers. Assign those numbers in equal probabilities to the eight outcomes possible when tossing three

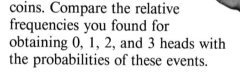

coins. Compare the relative frequencies you found for obtaining 0, 1, 2, and 3 heads with the probabilities of these events.

5 Turns in Sport

In gymnastics, diving, throwing the discus, and some other sports, a competitor may do a number of turns while competing. Select a sport and analyze some of these turns mathematically, giving the magnitude and direction of each turn.

6 Financial Log

Keep track of the money you receive and the money you spend—to the penny—over a week-long period in an organized table. Consider all receipts as positive and expenses as negative, and explain how addition of positive and negative numbers can indicate how you are doing.

SUMMARY

A large variety of situations lead to addition. These situations are of two types: putting together or slide. Some situations can be interpreted as either of these types.

In a putting-together situation, a count or measure x is put together with a count or measure y. If x and y have the same units and there is no overlap, the result has count or measure $x + y$. Probabilities of mutually exclusive events can be added to find the probability that either one or the other event will occur. When the lengths of the sides of a polygon are put together, the result is the perimeter of the polygon. In a putting-together situation, x and y must be positive or zero.

In slide situations, a slide x is followed by a slide y. The result is a slide $x + y$. There may be changes forward or back, up or down, in or out, clockwise or counterclockwise. One direction is positive, the other negative. So in a slide situation, x and y can be positive, zero, or negative.

From these situations, you can see many properties of addition. Zero is the additive identity. Every number has an opposite or additive inverse. Addition is both commutative and associative.

Suppose you know one addend a and the sum b and would like to find the other addend. Then you are trying to find x in the equation $x + a = b$. This sentence can be solved using the Addition Property of Equality and the other properties of addition.

VOCABULARY

You should be able to give a general description and a specific example of each of the following ideas.

Lesson 5-1
Putting-Together Model for
    Addition
addend
Slide Model for Addition

Lesson 5-2
additive identity
Additive Identity Property
    of Zero
additive inverse
Property of Opposites
Opposite of Opposites
    (Op-Op) Property
$\pm$ or $+/-$

Lesson 5-3
absolute value, $|\ |$

Lesson 5-4
revolution, full turn,
    quarter turn
clockwise, counterclockwise
magnitude of turn
Fundamental Property
    of Turns
adjacent angles
Angle Addition Property

Lesson 5-5
Adding Fractions Property
common denominator
least common multiple
    (LCM)

Lesson 5-6
mutually exclusive
complement of an event
probability of A or B
probabilities of complements

Lesson 5-7
Commutative Property
    of Addition
Associative Property of Addition

Lesson 5-8
Addition Property of Equality
equation of the form $x + a = b$

Lesson 5-9
line segment, segment, $\overline{AB}$,
    endpoints, $\overleftrightarrow{AB}$
polygon, side, vertex (vertices),
angle of polygon
triangle, quadrilateral, pentagon,
    hexagon, . . . , n-gon
diagonal of a polygon

Lesson 5-10
perimeter
$\overline{AB}$ for length $\overline{AB}$

PROGRESS SELF-TEST

Take this test as you would take a test in class. Then check your work with the solutions in the Selected Answers section in the back of the book.

In 1–7, simplify.

1. $3 + -10$

2. $-460 + -250$

3. $-9.8 + -(-1)$

4. $x + y + -x + 4$

5. $|-8|$

6. $|-2 + 1| + |0|$

7. $-6 + 42 + -11 + 16 + -12$

8. Evaluate $|-A + 8|$ when $A = -3$.

In 9–11, solve.

9. $x + 43 = 31$

10. $-2.5 + y = -1.2$

11. $8 = -2 + z + -5$

In 12–15, write as a single fraction in lowest terms.

12. $\frac{53}{12} + \frac{11}{12}$

13. $\frac{5}{x} + \frac{10}{x}$

14. $\frac{17}{9} + -\frac{8}{3}$

15. $\frac{1}{4} + \frac{3}{8} + \frac{2}{16}$

In 16 and 17, Sally was 20 points behind. Now she is 150 points ahead. Let c be the change in Sally's status.

16. What equation can be solved to find c?

17. Solve that equation.

18. $(2 + 3) + 4 = 2 + (3 + 4)$ is an instance of what property?

19. Give an instance of the Addition Property of Zero.

20. A polygon with 6 sides is called a __?__.

21. Give an example of two mutually exclusive events and tell why they are mutually exclusive.

22. A pentagon has two sides of length 3 cm and three sides of length 4 cm. What is its perimeter?

23. *Multiple choice.* If L and K are points, which symbol stands for a number?
    (a) LK (b) $\overline{LK}$ (c) $\overrightarrow{LK}$ (d) $\overleftrightarrow{LK}$

24. Ms. A's class has m students. Mr. B's class has n students. Together there are 50 students in the classes. How are m, n, and 50 related?

25. Anita and Ajay will play a game of chess. The probability that Anita will win is 50%. The probability that Ajay will win is 45%. What is the probability of a draw?

26. An iron bar is 3 cm longer than 5 meters. In meters, how long is the bar?

27. Is $-5.498765432101 + 5.498765432102$ positive, negative, or zero?

In 28 and 29, use the figure below. A is on $\overline{MP}$.

28. If $MA = 16$ and $AP = 8$, what is MP?

M ———————————— A ———— P

29. If $MA = 2.3$ and $MP = 3$, what is PA?

30. Picture the addition problem $-3 + 2$ on a number line and give the sum.

31. What is the probability of getting a number less than 3 on one toss of a fair die?

32. What is the result when a 50° clockwise turn is followed by a 250° counterclockwise turn?

In 33 and 34, use the figure below. Assume all small angles with vertex O have the same measure.

33. What is $m\angle VOW$?

34. If you are standing at O facing U and turn to X, what is the magnitude of your turn?

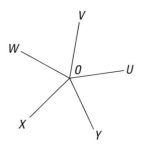

CHAPTER REVIEW

Questions on SPUR Objectives

SPUR stands for **S**kills, **P**roperties, **U**ses, and **R**epresentations. The Chapter Review questions are grouped according to the **SPUR** Objectives for this chapter.

SKILLS DEAL WITH THE PROCEDURES USED TO GET ANSWERS.

Objective A. *Add positive and negative numbers.* *(Lessons 5-1, 5-3, 5-5)*

In 1–10, perform the addition.

1. $-16 + 4$
2. $-7 + -8 + -9$
3. $7 + -2.4 + 5$
4. $-31 + 32$
5. $\frac{6}{11} + 2\frac{5}{11}$
6. $\frac{12}{17} + -\frac{12}{17}$
7. $6 + -\frac{8}{9}$
8. $\frac{2}{3} + \frac{6}{7}$
9. $\frac{1}{2} + \frac{1}{3} + -\frac{1}{4}$
10. $\frac{40}{c} + \frac{-10}{c}$

Objective B. *Calculate absolute value.*
(Lesson 5-3)

In 11–16, simplify.

11. $|-12|$
12. $|4 + -9 + -2|$
13. $|0| + |3| + |-5|$
14. $-|7| + |4|$
15. $-|-8 + 7|$
16. $|x + y|$, when $x = 40$ and $y = -40$

Objective C. *Apply properties of addition to simplify expressions.* *(Lessons 5-2, 5-7)*

In 17–22, simplify.

17. $-(-(-17))$
18. $-(-4) + 3$
19. $-40 + 0$
20. $(86 + -14) + (-86 + 14)$
21. $-(-(0 + \frac{2}{7}))$
22. $\frac{11}{4} + y + -\frac{11}{4}$
23. When $a = -42$, find $-a + 6$.
24. If $b = \frac{3}{5}$, find the value of $b + -b$.

Objective D. *Solve equations of the form* $x + a = b$. *(Lesson 5-8)*

In 25–34, solve.

25. $x + -32 = -12$
26. $6.3 = t + 2.9$
27. $\frac{10}{3} + y = \frac{1}{3}$
28. $0 + a = 4 + 1$
29. $3 + c + -5 = 36$
30. $-8 = 14 + (d + -6)$
31. $7034 = v + 1112$
32. $312.9 = 163.4 + b$
33. $-1 + e = \frac{1}{6}$
34. $-\frac{11}{5} + w = -\frac{4}{5}$

Objective E. *Find the perimeter of a polygon.* *(Lesson 5-10)*

35. What is the perimeter of a square in which one side has length 3?

36. If $x = 23$, what is the perimeter of the polygon *ABCDE*?

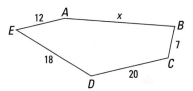

37. Measure the sides of polygon *GHIJ* to find its perimeter to the nearest centimeter.

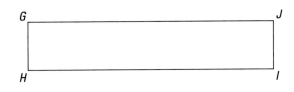

38. An octagon has 3 sides of length 7 and 5 sides of length 6. What is its perimeter?

39. For polygon *ABCDE* of Question 36, if the perimeter is 82, what equation can be solved to find x?

PROPERTIES DEAL WITH THE PRINCIPLES BEHIND THE MATHEMATICS.

Objective F. *Identify the following properties of addition: Commutative Property of Addition, Associative Property of Addition, Additive Identity Property of Zero, Addition Property of Equality, Property of Opposites, Opposite of Opposites Property.* (*Lessons 5-2, 5-7, 5-8*)

In 40–43, an instance of what property of addition is given?

40. 3.53 meters + 6.74 meters = 6.74 meters + 3.53 meters

41. Since $30\% = \frac{3}{10}$, it is also true that $\frac{1}{2} + 30\% = \frac{1}{2} + \frac{3}{10}$.

42. $^-941 + 941 = 0$

43. $(1 + 2) + 3 = (2 + 1) + 3$

Objective G. *Tell whether events are mutually exclusive or not.* (*Lesson 5-6*)

In 44–48, what events are mutually exclusive?

44. It is raining outside. It is not raining outside.

45. I am a student. I have a job.

46. Mathematics is Beth's favorite subject. English is Gilbert's favorite subject.

47. Mathematics is Greg's favorite subject. English is Greg's favorite subject.

48. If the probability of one event is $\frac{1}{2}$ and of another is $\frac{3}{4}$, explain why they cannot be mutually exclusive.

Objective H. *Identify parts and give names of polygons.* (*Lesson 5-9*)

49. Which is not a correct name for the polygon below? *LAKE, LEAK, KALE, ELAK*

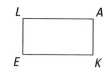

50. Name the two diagonals of the polygon in Question 49.

51. The polygon in Question 49 has __?__ vertices and __?__ sides.

52. Name a side of the polygon *ABCDE*.

USES DEAL WITH APPLICATIONS OF MATHEMATICS IN REAL SITUATIONS.

Objective I. *Use the Putting-Together Model for Addition to form sentences involving addition.* (*Lessons 5-1, 5-8, 5-10*)

53. Bob rode his bicycle a mile and a half to school. Then he rode $\frac{3}{5}$ of a mile to his friend's house. Altogether he rode *M* miles. What equation connects these distances?

54. You have read *x* books this year. A friend has read *y* books. Together you have read 16 books. What equation connects *x*, *y*, and 16?

55. In the Johannson family, Dad earned *D* dollars last year. Mom earned *M* dollars. The children earned *C* dollars. The total family income was *T* dollars. What equation relates *D*, *M*, *C*, and *T*?

In 56–58, use the figure shown. *B* is on $\overline{AC}$.

56. What equation connects *AB*, *BC*, and *AC*?

57. If *AB* = *x* and *BC* = 3, what is *AC*?

58. Let *AC* = 10.4 and *BC* = 7.8.
   a. What equation can be used to find *AB*?
   b. What is *AB*?

Objective J. *Use the Slide Model for Addition to form sentences involving addition.* (*Lessons 5-1, 5-8*)

59. Charyl's stock rose $\frac{3}{8}$ of a point on one day and fell $\frac{1}{4}$ point the next day.
   a. What addition gives the total change in Charyl's stock?
   b. What is the total change?

60. Bernie gained 5 pounds one week, lost 7 the next, and lost 3 the next.
  a. What addition gives the total change in Bernie's weight?
  b. What is the total change?

61. A scuba diving team was 60 feet below sea level then came up 25 feet.
  a. What addition tells where the team wound up?
  b. Where did they wind up?

62. The temperature was -3° and changed c° to reach -10°. What is c?

Objective K. *Calculate the probability of mutually exclusive events or complements of events.* *(Lesson 5-6)*

63. From prior experience, we estimate that when Frank comes up to bat in a baseball game, he has a 12% probability of hitting a single, a 6% probability of hitting a double, a 2% probability of hitting a triple, and a 4% probability of hitting a home run. These are the only possible hits. What is the probability Frank will get a hit when he bats?

64. If Lee works hard, he has a $\frac{1}{3}$ probability of getting an A, a $\frac{1}{2}$ probability of getting a B,

and a $\frac{1}{6}$ probability of getting a C. If Lee works hard, will he pass?

In 65 and 66, assume the spinner below is spun once. All angles at the center have equal measure. Assume all positions of the arrow are equally likely.

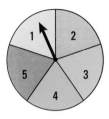

65. What is the probability that the spinner will land in an even-numbered region?

66. What is the probability that the spinner will land in an odd-numbered region?

67. Explain why the events in Questions 65 and 66 are complements.

68. A weighted quarter is tossed. Let E = heads will occur. Suppose the probability of E is $\frac{3}{5}$.
  a. Describe the event that is the complement of E.
  b. Give the probability of that event.

REPRESENTATIONS DEAL WITH PICTURES, GRAPHS, OR OBJECTS THAT ILLUSTRATE CONCEPTS.

Objective L. *Calculate magnitudes of turns given angle measures or revolutions.* *(Lesson 5-4)*

In 69–71, all small angles at O have the same measure. Think of turns around the point O.

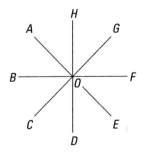

69. What is the magnitude of the turn from F to A?

70. What is the magnitude of the turn from E to C?

71. Suppose you are facing H and turn 315° clockwise and then 45° counterclockwise.
  a. What point will you then be facing?
  b. How much will you have turned?

72. If you turn 90° counterclockwise, and then turn 40° clockwise, what is the result?

Objective M. *Picture addition of positive and negative numbers using arrows on a number line.* *(Lesson 5-1)*

73. Picture the addition -5 + 8 on a number line and give the sum.

74. Picture the addition 17 + -4 on a number line and give the sum.

75. Picture the addition -3.5 + -6.5 on a number line and give the sum.

CHAPTER 6

PROBLEM-SOLVING STRATEGIES

McGURK'S MOB

To solve a problem is to find a way where no way is known off-hand, to find a way out of a difficulty, to find a way around an obstacle, to attain a desired end that is not immediately attainable, by appropriate means.

—George Polya (1887–1985)

Solving problems is human nature itself.

—Polya

The person who makes no mistakes usually makes nothing.

—Anonymous

Amazing Maize Maze. *When solving problems, you need not feel as if you are lost in a maze. Strategies can help you find solutions. Shown here is the* Amazing Maize Maze, *carved from a cornfield in Annville, Pennsylvania.*

What Is an Algorithm?

In Chapter 5, you learned how to solve any equation of the form $x + a = b$ for x. The method was to add $-a$ to each side and then simplify. This method is an example of an *algorithm*. An **algorithm** is a sequence of steps that leads to a desired result.

Not all algorithms are short. The algorithm called *long division* can involve many steps.

$$
\begin{array}{r}
.75 \\
34\overline{)25.50} \\
\underline{23\ 8} \\
1\ 70 \\
\underline{1\ 70}
\end{array}
$$

You know another algorithm for dividing decimals. It is the calculator algorithm. It is much easier to describe the algorithm for division on a calculator than the long division algorithm. This key sequence describes the algorithm.

Key sequence: 25.50 ÷ 34 =
Display: 25.50 25.50 34. 0.75

There is another way to think about algorithms. An algorithm is something that a computer can be programmed to do.

How Do Problems Differ from Exercises?

An **exercise** is a question that you know how to answer. For you, adding whole numbers is an exercise because you know an algorithm for addition. A **problem** for you is a question you do not know how to answer. It is a question for which you have no algorithm. Many people think that if they do not have an algorithm, then they cannot solve a problem. But that isn't true. By following some advice, almost anyone can become a better problem solver.

General Advice for Solving Problems

In this chapter you will learn many strategies for solving problems. But you must do three things.

1. Take your time. Few people solve problems fast. (If a person solves a problem fast, then it may not have been a problem for that person!)

2. Don't give up. You will never solve a problem if you do not try. Do something! (In the cartoon on page 301, Rick and the man give up too soon.)

3. Be flexible. If at first you don't succeed, try another way. And if the second way does not work, try a third way.

George Polya was a mathematician at Stanford University who was famous for his writing about solving problems. He once wrote:

> Solving problems is a practical skill like, let us say, swimming. We acquire any practical skill by imitation and practice. Trying to swim, you imitate what other people do with their hands and feet to keep their heads above water, and, finally, you learn to swim by practicing swimming. Trying to solve problems, you have to observe and to imitate what other people do when solving problems and, finally, you learn to do problems by doing them.

How Do Good Problem Solvers Solve Problems?

In a famous book called *How to Solve It,* Polya described what good problem solvers do.

- They *read the problem carefully.* They try to understand every word. They make sure they know what is asked for. They reread. They make certain that they are using the correct information. They look up words they do not know.

- They *devise a plan.* They even plan their guesses. They arrange information in tables. They draw pictures. They compare the problem to other problems they know. They decide to try something.

- They *carry out the plan.* They attempt to solve. They work with care. They write things down so they can read them later. If the attempt does not work, they go back to read the problem again.

- They *check work.* In fact, they check their work at every step. They do not check by repeating what they did. They check by estimating or by trying to find another way of doing the problem.

These are the kinds of strategies you will study in this chapter. You will also learn about some important problems and some fun problems.

QUESTIONS

Covering the Reading

1. What is an algorithm?

2. Give an example of an algorithm.

3. What is a *problem?* Give an example of a question that for you is a problem.

4. What is an *exercise?* Give an example of a question that for you is an exercise.

In 5–7, what problem-solving advice is suggested by the traffic sign?

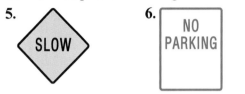

5. SLOW

6. NO PARKING

7. DETOUR

8. Who was George Polya?

George Polya (1887–1985)

9. What famous book did Polya write?

In 10–12, tell what the person should have done.

10. Monty tried one way to do the problem. When that did not work, he tried the same way again.

11. Priscilla said to herself, "I've never seen any problem like that before. I won't be able to do it."

12. Adam wrote down an answer very quickly and went on to the next problem.

13. What four things do successful problem solvers do when solving a problem?

14. Suppose you want to devise a plan to solve a problem. What are some of the things you can do?

Applying the Mathematics

In 15 and 16, refer to the algorithm pictured below.

$$
\begin{array}{r}
3\,4.2 \\
5.6\,7 \\
\hline
2\,3\,9\,4 \\
2\,0\,5\,2 \\
1\,7\,1\,0 \\
\hline
1\,9\,3.9\,1\,4
\end{array}
$$

15. What question does this algorithm answer?

16. Write the calculator algorithm that answers the same question.

17. Consider the letters *L, U, N, C, H*. Make two-letter monograms with these letters, such as *LU, UL, LL*. How many two-letter monograms are there? (Hint: Make an organized list.)

18. You can buy stickers for the digits from 0 through 9 at most hardware stores. They can be used for house addresses and room numbers. Suppose you buy stickers for all the digits in all the integers from 1 through 100.
   a. What digit will be used most?
   b. What digit will be used least?

19. Trains leave Union Station every hour from 7 A.M. through 7 P.M. bound for Kroy. How many trains is that in a day?

20. Suppose in Question 19, trains left every *half* hour, from 7 A.M. through 7 P.M. How many trains is that in a day?

21. Theo wants to buy a newspaper from a newspaper vending machine. The newspaper costs $0.55. The machine takes nickels, dimes, and quarters. He has 4 dimes, 6 nickels, and 3 quarters. List all the different ways Theo can buy the newspaper.

Review

22. Simplify $\frac{2}{3} + \frac{5}{7} - \frac{1}{14}$.   *(Lessons 5-3, 5-5)*

23. Find the value of $3 + 4a$ when $a = 2.5$.   *(Lesson 4-4)*

24. A cube has how many edges?   *(Lesson 3-8)*

25. What is the volume of a cube with an edge of length 5 centimeters? *(Lesson 3-8)*

26. Here are changes in populations for the ten largest cities in the United States from 1980 to 1990.

New York	250,925
Los Angeles	518,548
Chicago	-221,346
Houston	35,415
Philadelphia	-102,633
San Diego	235,011
Detroit	-175,365
Dallas	102,799
Phoenix	193,699
San Antonio	150,053

a. Which cities lost in population from 1980 to 1990?

b. What was the total change in population for these ten cities combined? *(Lessons 1-8, 5-7)*

27. How many feet are in 5 miles? *(Lesson 3-2)*

28. One-sixth is equal to ___?___ percent. *(Lesson 2-6)*

29. Of the 5.4 billion people on Earth in 1991, about 1% lived in Egypt. How many people is this? *(Lessons 2-1, 2-5)*

30. Give the fraction in lowest terms equal to the given percent.
a. 10% b. 75%
c. 25% d. $66\frac{2}{3}\%$ *(Lesson 2-4)*

Exploration

31. In this addition problem each letter stands for a different digit. The sum is 10440. Find the digits.

```
      U
    C A N
      D O
  T H I S
1 0 4 4 0
```

LESSON

6-2

Read Carefully

A sign of the times? *Drivers at the intersection shown above need to read carefully and obey only those signs that apply to them. The photo above was taken at an intersection in Muncie, Indiana.*

General Advice About Reading

To solve a problem, you must understand it. So, if the problem is written down, you must *read it carefully.* Here are three things to do to get the necessary information from reading.

---

1. Know the meaning of all words and symbols in the problem.
2. Sort the information into what is needed and what is not needed.
3. Determine if there is enough information to solve the problem.

Example 1

List the ten smallest positive composite integers.

Strategy

To answer this question, you must know the meaning of "positive composite integer." In this book, we have already discussed the positive integers. So the new word here is "composite." Our dictionary has many definitions of "composite," but only one of "composite number." Here is that definition: "Any number exactly divisible by one or more numbers other than itself and 1; opposed to *prime number.*" So a composite number is a positive integer that is not prime nor equal to 1. Now the problem can be solved.

Solution

The positive integers are 1, 2, 3, 4, ... Those that are prime are 2, 3, 5, 7, 11, 13, 17, and so on. The ten smallest composites are the others: 4, 6, 8, 9, 10, 12, 14, 15, 16, 18.

Finding the Meanings of Words and Symbols

If you do not know the meaning of a word, look it up in a dictionary. Your school may have a mathematics dictionary, one that specializes in mathematical terms. Some mathematics books have glossaries or indexes in the back that can help you locate words.

Some terms have more than one meaning even in mathematics. The word *divisor* can mean *the number divided by* in a division problem. (In $12 \div 5 = 2.4$, 5 is the divisor.) But *divisor* also means *a number that divides another number with a zero remainder*. For example, 7 is a divisor of 21. In this situation *divisor* has the same meaning as *factor*.

Some symbols have more than one meaning. The dash (-) can mean subtraction. It can also mean "the opposite of." However, in phone numbers, like 555-1212, it has no meaning other than to separate the number to make it easier to remember. You must look at the situation to determine which meaning is correct in a given problem.

Sorting Information into What Is Needed

Here is an example of a problem with too much information. Can you tell what information is not needed?

Example 2

Last year the Williams family decided that they should try to read more. So they kept track of the books they read. Mrs. Williams read 20 books. Mr. Williams read 16 books. Their son Jed read 12 books. Their daughter Josie read 14 books, and their daughter Julie read 7 books. How many books did these children of Mr. and Mrs. Williams read altogether?

Solution

Did you see that the problem asks only about the *children?* The information about the books read by Mr. and Mrs. Williams is not needed. The children read 12 + 14 + 7 books, for a total of 33 books.

Sometimes too little information is given to answer a question.

Example 3

Read Example 2 again. How many children do the Williamses have?

Solution

Do you think 3? Reread the problem. Nowhere does it say that these are the only children. The Williamses have at least 3 children.

QUESTIONS

1. What three things should you do when you read a problem?

2. Can a word have more than one mathematical meaning?

3. Give an example of a mathematical symbol that has more than one meaning.

4. Which of the following numbers is not a composite integer?
   6   7   8   9   10

5. Which of the following numbers is not a prime number?
   11   13   15   17   19

6. List the ten smallest positive prime numbers.

7. Which of the following numbers is a divisor of 91?
   3   5   7   9   11

8. Which of the following numbers is a factor of 91?
   13   15   17   19   21

In 9–11, refer to Example 2.

9. How many books did Mr. and Mrs. Williams read altogether?

10. What word seems to hint that the Williamses have more than three children?

11. Of the three Williams children named, who is the youngest?

12. Name two places to look to find a definition of a mathematical term.

Applying the Mathematics

In 13–15, use this definition. A *natural number* is one of the numbers 1, 2, 3, 4, ...

13. *Multiple choice.* Which of (a) to (c) is the same as natural number?
    (a) whole number   (c) positive integer
    (b) integer   (d) none of these

14. Name all natural numbers that are solutions to $10 > x \geq 7$.

15. How many natural numbers are solutions of $v < 40$?

16. List all the divisors of 36.

17. In the fraction $\frac{24}{35}$, which number is the divisor: 24 or 35?

18. Explain why 7^5 is not a prime number.

In 19–21, you may need to look in a dictionary or other source.

19. Draw a regular hexagon.

20. Give an example of a perfect number.

21. Give an example of a pair of twin primes.

In 22 and 23, use the following information. The class in room 25 had 23 students last year and has 27 students this year. The class in room 24 had 25 students last year and has 26 students this year. There are 22 students this year and there were 28 students last year in room 23.

22. What was the total number of students in these classrooms last year?

23. In which year were there more students in these classrooms?

In 24 and 25, use the following information. The class in room a had b students last year and has c students this year. The class in room d had e students last year and has f students this year. There are g students this year and there were h students last year in room i.

24. What was the total number of students in these classrooms last year?

25. In which year were there more students in these classrooms?

Review

26. What are Polya's four steps in problem solving? *(Lesson 6-1)*

27. How many diagonals does a hexagon have? *(Lesson 5-9)*

28. Solve for x: $x + 3 + y = y + 8$. *(Lesson 5-8)*

29. In a Scrabble® tournament, a person played six different people with the following results: won by 12, lost by 30, won by 65, won by 47, lost by 3, and lost by 91. What was the total point difference between this person and the other six people combined? *(Lesson 5-3)*

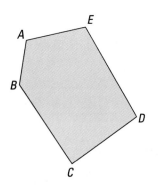

30. Simplify $-(-(-(-y)))$. *(Lesson 5-3)*

31. Simplify $5 + -(-(-3 + 4 \cdot 2))$. *(Lessons 4-5, 5-2)*

32. Name two segments in the drawing at the left that look perpendicular. *(Lesson 3-7)*

Exploration

33. Ronald has a truckload of fewer than 5000 apples. He sold exactly 20% of them to Mary. Next he sold exactly 20% of the remaining apples to Gregory. Next he sold exactly 20% of those remaining to Carolyn. He sold exactly 20% of those remaining to Joy. Then he sold exactly 20% of those remaining to Anne. If Ronald cannot sell a fraction of an apple, how many apples did he begin with? How many does he have now? Explain your answers.

Draw a Picture

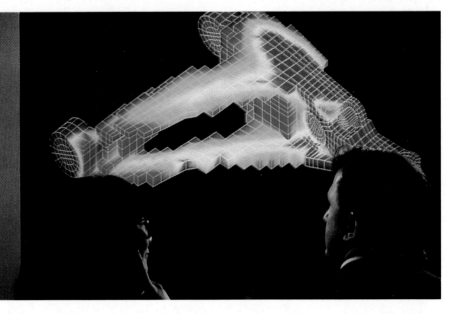

Picture perfect. *Designers of automobiles can create, test, and modify plans using computer-aided design programs. This picture from Ford Motor Company shows a computerized image of the lower control arm of an automobile.*

Pictures for Geometry Problems

You can understand every word in a problem yet still not be able to solve it immediately. One useful strategy is to *draw a picture.* This is particularly helpful when the problem involves a geometric figure.

Activity 1

How many diagonals does a heptagon have?

Strategy

One strategy is obvious. Draw a heptagon and draw its diagonals. Count the diagonals as you draw them.

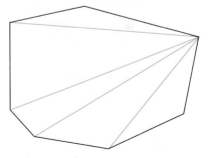

Trace the heptagon above and draw the remaining diagonals, counting as you draw. How many diagonals does a heptagon have?

Suppose you did not remember that a heptagon has 7 sides, or suppose you forget what a diagonal is. Then you could look in the glossary of this book. Or you could look in a dictionary.

Pictures for Problems That Are Not Geometric

Even when a problem is not geometric, a drawing can still help. Example 1 uses the heptagon drawing in a way that may surprise you.

Example 1

Seven teams are to play each other in a tournament. How many games are needed?

Solution

Name the teams *A, B, C, D, E, F,* and *G.* Use these letters to name points in a drawing. When *A* plays *D,* draw the segment $\overline{AD}$. So each segment between two points represents a different game.

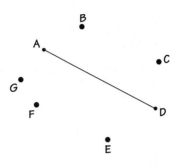

The number of games played is the number of segments that can be drawn. But the segments are the sides and diagonals of a heptagon. A heptagon has 7 sides. Your work in Activity 1 should have shown that there are 14 diagonals in a heptagon. So there are 21 segments altogether. *Twenty-one games are needed for the tournament.*

Activity 2

List each of the 21 games needed for the tournament. (Identify each game by a segment.)

Drawing a picture is a very important problem-solving idea. You may never have thought of picturing teams as we did, but there are many types of problems for which drawing a picture is a helpful, but not obvious, way to find a solution.

QUESTIONS

Covering the Reading

1. How many diagonals does a heptagon have?

2. Seven teams *A, B, C, D, E, F,* and *G* are to play each other in a tournament.
   a. How many games are needed?
   b. List the games that are needed.

3. How many diagonals does a quadrilateral have?

4. Four teams are to play each other in a volleyball tournament. How many games are needed?

Applying the Mathematics

5. Eight teams are to play each other two times in a season. How many games are needed?

6. Have you heard this problem? If you haven't, watch out. A snail wants to climb out of a hole that is 10 feet deep. The snail climbs up 2 feet each day and falls back 1 foot each night. How many days will it take the snail to climb out of the hole?

In 7–10, drawing a picture will help.

7. A square dog pen is 10 feet on a side. There is a post in each corner. There are posts every 2 feet on each side. How many posts are there in all?

8. Bill is older than Wanda and younger than Jill. Jill is older than Chris and younger than Pete. Chris is older than Wanda. Bill is younger than Pete. Chris is older than Bill. Who is youngest?

9. Amy is driving along Interstate 55 in Illinois from Collinsville to Joliet. She will pass through Litchfield and then through Atlanta. The distance from Collinsville to Litchfield is half the distance from Litchfield to Atlanta. The distance from Atlanta to Joliet is three times the distance from Collinsville to Litchfield. The distance from Atlanta to Joliet is 114 miles. How far is it from Collinsville to Joliet?

10. The Sherman family has a pool 30 feet long and 25 feet wide. There is a walkway 4.5 feet wide around the pool.
    a. What is the perimeter of the pool?
    b. What is the perimeter of the outside edge of the walkway?

11. Name two places to look if you do not know the meaning of a mathematical term.   *(Lesson 6-2)*

12. List all the natural number factors of 40.   *(Lesson 6-2)*

13. List the natural numbers between 8.4 and 4.2.   *(Lesson 6-2)*

14. List all even prime numbers. Explain your answer.   *(Lesson 6-2)*

In 15 and 16, suppose that Sarah has S dollars and Dana has D dollars. Translate into English.   *(Lesson 5-8)*

15. $S = D$

16. If $S = D$, then $S + 2 = D + 2$.

17. When $a = {}^-3$ and $b = 6$, what is the value of ${}^-(a + b)$?
(Lessons 4-5, 5-3)

18. What is the absolute value of ${}^-30$?   *(Lesson 5-3)*

19. Find the value of $6x^4$ when $x = 3$.   *(Lesson 4-4)*

20. 10 pounds is about how many kilograms?   *(Lesson 3-5)*

21. Which is larger, 45 centimeters or 800 millimeters?   *(Lesson 3-4)*

22. What number will you get if you enter this key sequence?
20 $\boxed{y^x}$ 3 $\boxed{\times}$ 2 $\boxed{\div}$ $\boxed{(}$ 5 $\boxed{\div}$ 4 $\boxed{)}$ $\boxed{=}$ *(Lessons 1-5, 2-2)*

23. Round 56.831 to the nearest hundredth.   *(Lesson 1-4)*

24. 27 small cubes are arranged to form one big $3 \times 3 \times 3$ cube. Then the entire big cube is dipped in paint. How many small cubes will now be painted on exactly two sides?

25. One local soccer league has 5 teams and another has 6 teams. Each team in the first league is to play each team in the second league once. How many games are needed?

LESSON
6-4

Trial and Error

Try, try again. *The strategy of trial and error is necessary in order to solve a jigsaw puzzle. If a piece does not fit, don't give up—try another piece.*

Trial and error is a problem-solving strategy everyone uses at one time or another. In trial and error, you try an answer. (The word *trial* comes from *try*.) If the answer is in error, you try something else. You keep trying until you get a correct answer.

Trial and error is a particularly good strategy if a question has only a few possible answers.

Example 1

Which of the numbers 4, 5, and 6 is a solution to $(n + 3)(n - 2) = 36$?

Strategy

This type of equation is not discussed in this book. You probably do not know how to solve it. You have no algorithm. So it is a problem for you. However, since there are only three choices for a value of *n*, trial and error is a suitable strategy.

Solution

Try the numbers one at a time.
Try 4. Does $(4 + 3)(4 - 2) = 36$? Does $7 \cdot 2 = 36$? No, $14 \neq 36$.
Try 5. Does $(5 + 3)(5 - 2) = 36$? Does $8 \cdot 3 = 36$? No, $24 \neq 36$, but 24 is closer than 14.
Try 6. Does $(6 + 3)(6 - 2) = 36$? Does $9 \cdot 4 = 36$? Yes.
So 6 is a solution.

Organizing the Trials

Trial and error works best if the trials are organized. So try possible answers in some logical order and write down your results, noting even the errors. A wrong answer or patterns in wrong answers may lead you to a right answer.

Example 2

A polygon has exactly 9 diagonals. How many sides does the polygon have?

Strategy

Combine two strategies. Draw pictures of various polygons with their diagonals, starting with a triangle. Keep adding a side until a polygon with 9 diagonals is found. Count the diagonals as you draw them.

Solution

A triangle has 0 diagonals.

A quadrilateral has 2 diagonals.

A pentagon has 5 diagonals.
A polygon with 9 diagonals has 6 sides.

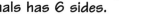

A hexagon has 9 diagonals.

Using Technology

Computers and calculators are able to do a lot of calculations very quickly. When there are many possible solutions or complicated calculations, it may be worth the effort to program a computer to do a trial-and-error search. The FOR-NEXT-STEP command sets up a loop to test several values for a variable. FOR assigns a beginning value and an ending value to a variable. NEXT increases the value of the variable by the amount indicated in STEP. If the increase is 1, the STEP part of the command is not needed.

Example 3

Estimate, to the nearest tenth, the positive value of x that satisfies $x^2 + x = 10$.

Solution

First try to estimate x.

Let x = 1: $1^2 + 1 = 2$, so 1 is too small.
Let x = 2: $2^2 + 2 = 6$, so 2 is too small.
Let x = 3: $3^2 + 3 = 12$, so 3 is too big.
Thus 2 < x < 3.

▶

▶ Finding x to the nearest tenth requires a more difficult calculation, so a computer program is appropriate.

The following program evaluates and prints $x^2 + x$ for 2.0, 2.1, 2.2, 2.3, . . . , 3.0.

```
10 FOR X = 2 TO 3 STEP .1
20 LET Y = X ^ 2 + X
30 PRINT X, Y
40 NEXT X
50 END
```

When this program is run, the computer will print a table with two columns of numbers. The left column is x. The right column is $x^2 + x$.

2	6
2.1	6.51
2.2	7.04
2.3	7.59
2.4	8.16
2.5	8.75
2.6	9.36
2.7	9.99
2.8	10.64
2.9	11.31
3	12

Read the columns to see that $x^2 + x$ is closest to 10 when $x \approx 2.7$. In fact, since $2.71^2 + 2.71 > 10$, $2.7 < x < 2.71$.

To the nearest tenth, the value of x that satisfies $x^2 + x = 10$ is 2.7.

Some graphing calculators have the capability to make tables like the one in Example 3.

QUESTIONS

Covering the Reading

1. Describe the problem-solving strategy called "trial and error."

2. When is trial and error a useful strategy?

3. When using trial and error, what can you do to make the strategy work well?

In 4–7, which of the numbers 1, 2, 3, 4, or 5 is a solution?

4. $(x + 7)(x + 2)(x + 3) = 300$ 5. $3y - 2 + 5y = 30$

6. $1 + \frac{4}{3} = 2$ 7. $11 - x = 7 + x$

8. What polygons have 5 diagonals?

9. A polygon has 14 diagonals. How many sides does it have?

10. Can a polygon have exactly 10 diagonals? Explain your answer.

11. In Example 3, how do you know the computer should check numbers between 2 and 3?

12. Use the information in Example 3 to estimate to the nearest tenth the value of x that satisfies $x^2 + x = 7$.

Applying the Mathematics

In 13–17, trial and error is a useful strategy.

13. Choose one number from each row so that the sum of the numbers is 300.

Row 1:	147	152	128
Row 2:	132	103	118
Row 3:	63	35	41

14. Joel was 13 years old in a year in the 1980s when he noticed that his age was a factor of the year. In what year was Joel born?

15. What two positive integers whose sum is 14 have the smallest product?

16.
```
10 FOR N = 1 TO 100
20 IF (N+3) * (N−2) = 696 THEN PRINT N
30 NEXT N
40 END
```

  a. What does this program find?
  b. Modify the program so that it finds all integer solutions between 1 and 50 to the equation $x(x + 40) = 329$.
  c. Modify the program so that it finds all integer solutions between -100 and 100 to $n(n + 18) = \dfrac{25(3n + 28)}{n}$.

17. Write a computer program to find a positive integer solution to $x^3 + x^2 + x = 2954$. Be sure the program will print the answer. If you have access to a computer, run your program.

18. Describe a situation where you have used trial and error to solve a problem outside of math class.

Review

19. Kenneth is older than Ali and Bill. Carla is younger than Ali and older than Bill. Don is older than Carla but younger than Kenneth. Don is Ali's older brother. If the ages of these people are 11, 12, 13, 14, and 15, who is the 13-year-old? *(Lesson 6-3)*

20. How many sides are in a dodecagon?   *(Lesson 6-2)*

21. List all composite numbers between 11 and 23.   *(Lesson 6-2)*

22. Give an example of the Associative Property of Addition.   *(Lesson 5-7)*

23. Add: $\frac{1}{10} + -\frac{2}{5} + \frac{7}{18}$. Write your answer as a simple fraction.
    (Lessons 5-1, 5-3, 5-5)

24. Evaluate $2xy + \frac{x}{z}$ when $x = 3$, $y = 4$, and $z = 1.5$.   *(Lesson 4-4)*

25. To the nearest tenth of a foot, how many feet are in 50 inches?
    (Lesson 3-2)

26. Consider the number 5.843%.   *(Lessons 1-2, 1-4, 2-4)*
    a. Convert this number to a decimal.
    b. Round the number in part a to the nearest thousandth.
    c. Between what two integers is this number?

27. Repeat Question 26 for the number $\frac{99}{70}$.   *(Lessons 1-4, 1-6)*

Exploration

28. Put one of the digits 0, 1, 2, 3, 4, 5, 6, 7, 8, 9, in each circle, using each digit only once. This will form two five-digit numbers.

    a. Make the sum as large as possible. (For example, 53,812 and 64,097 use all the digits, but their sum of 117,909 is easy to beat.)

    b. Make the difference the smallest possible positive number.

Using a Table to Solve a Problem

Vadim thought he knew everything about diagonals and polygons. But this question had him stumped:

How many diagonals does a 13-gon have?

He read the question carefully. It was easy to understand. He drew a picture of a 13-gon.

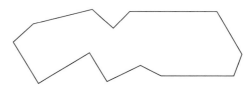

He started drawing diagonals and quickly gave up this approach. He thought about trial and error. He even thought about guessing. He had used all of the strategies of the last three lessons. What could he do?

Sometimes it helps to *make a table.* You saw one table in Example 3 of Lesson 6-4. A **table** is an arrangement with rows and columns. Making a table is just one way of organizing what you know. Vadim's friend Rose recalled the following information from Example 2 of Lesson 6-4. She wrote down a table with two columns.

Number of sides	Number of diagonals
3	0
4	2
5	5
6	9
7	14

She noticed that the number of diagonals increases by 2, then 3, then 4, then 5. She figured that if the pattern continued, then these would be the next three rows of the table.

8	20
9	27
10	35

She checked her work by drawing an octagon. It did have 20 diagonals. So she continued the pattern and answered Vadim's question. (You are asked to do this yourself in the questions following this lesson.)

Reasons for Making a Table

In solving problems, there are three reasons for making a table.

1. A table organizes your thoughts.
2. It helps you find patterns.
3. It can help you to make a *generalization.*

Using a Table to Make a Generalization

A **generalization** is a statement that is true about many instances. The following story is about making a generalization.

Francie lives in Phoenix. She has a telephone in her room. She pays for all long-distance calls she makes. One Sunday she called her cousin Meg in San Diego. As they talked, Francie forgot about the time. They talked for at least an hour! She called her long-distance operator for rates and found that it cost $.25 for the first minute and $.18 for each additional minute. She had a problem.

What is the cost of a one-hour phone call, if the first minute costs $.25 and each additional minute costs $.18?

Francie began by making a table.

Number of minutes talked	Cost of phone call
1	$.25
2	.43
3	.61
4	.79
5	.97

It wasn't working at all. Yet the pattern was simple. You add .18 to get the next cost. But she wanted to know the cost for 60 minutes. She did not want to add .18 all those times. She might make a mistake, even with a calculator.

George helped Francie by rewriting the costs to show a different pattern.

Number of minutes talked	Cost of phone call
1	$.25
2	.25 + .18
3	.25 + 2 · .18
4	.25 + 3 · .18
5	.25 + 4 · .18

"Look across the rows," he said. "See what stays the same and what changes." George noted that in the right column, the only number that changes is the number multiplied by .18. It is always one less than the number in the left column. So,

60	.25 + 59 · .18

"Now you can calculate the cost." Francie did the calculations.

$$.25 + 59 · .18 = 10.87$$

The cost of her phone call was over $10.87!

George and Francie then realized that the table makes it possible to write a formula—a general pattern—for the cost.

m	.25 + (m − 1) · .18

If Francie spoke for m minutes, it would cost $.25 + (m − 1) · .18$ dollars, not including tax.

The formula enables you to find the cost for any number of minutes easily. It is a generalization made from the pattern the instances formed in the table.

QUESTIONS

Covering the Reading

In 1–3, use information given in this lesson.

1. How many diagonals does a 10-gon have?

2. How many diagonals does an 11-gon have?

3. Answer the question Vadim could not answer.

In 4–7, refer to the story with Francie in this lesson.

4. If Francie had talked to her cousin Meg for only 5 minutes, what would have been the cost of the phone call?

5. If Francie had talked to her cousin Meg for 30 minutes, what would have been the cost of the call?

6. In the story about Francie, what generalization did George and Francie make?

7. For a two-hour phone call to Meg, what would it cost Francie?

8. a. What is a *generalization?*
   b. Give an example.

9. Erich wants to call his cousin Inge in Germany. The cost for a late-night call is $1.45 for the first minute and 81¢ for each additional minute.

   a. Make a table indicating the cost for calls lasting 1, 2, 3, 4, and 5 minutes.

   b. What is the cost of a 25-minute call?

   c. What is the cost of a call lasting *m* minutes?

Applying the Mathematics

10. Make a table to help answer this question: In a 50-gon, all the diagonals from *one* particular vertex are drawn. How many triangles are formed?

11. In 1993, it cost 29¢ to mail a one-ounce first-class letter. It cost 23¢ for each additional ounce or part of an ounce up to 11 ounces.

   a. What was the cost to mail a letter weighing 4 ounces?

   b. What was the cost to mail a letter weighing 9.5 ounces?

   c. What was the cost to mail a letter weighing *w* ounces, if *w* is a whole number?

12. Calculate enough of the sums

$$1 + 2$$
$$1 + 2 + 4$$
$$1 + 2 + 4 + 8$$
$$\cdot$$
$$\cdot$$
$$\cdot$$

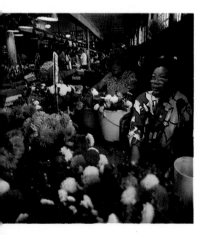

The Pacific Northwest is known for its flowers. Shown here is Farmers' Market in Seattle, Washington.

so that you see a pattern in the sums and can give the answer to

$$2^0 + 2^1 + 2^2 + 2^3 + \ldots + 2^{20}$$

without adding. (You are allowed to use a calculator, but not the $+$ key on it.)

Review

13. To drive from Seattle to Chehalis, you must first drive from Seattle to Tacoma, then from Tacoma to Olympia, and finally from Olympia to Chehalis. It is 4 miles farther from Seattle to Tacoma than from Tacoma to Olympia. And it is 4 miles farther from Tacoma to Olympia than from Olympia to Chehalis. If it is 84 miles from Seattle to Chehalis, how far is it from Tacoma to Olympia? *(Lessons 6-3, 6-4)*

14. $x^y = 343$ and x and y are integers between 1 and 10. Find x and y.
(Lessons 2-2, 6-4)

15. 10 FOR X = 10 TO 30 STEP .5
20 IF X ^ 4 + X ^ 2 = 360900.3125 THEN PRINT "X =";X
30 NEXT X
40 END

    a. What problem does this program solve?
    b. How many values for x will it test?
    c. If you have a computer, run the program and record the output.
    (Lesson 6-4)

16. One type of polygon is the *undecagon*. How many sides does an undecagon have? *(Lesson 6-2)*

17. Mabel was 5 pounds overweight. Now she is 3 pounds underweight. Let c be the change in her weight.
    a. Write an equation relating 5, 3, and c.
    b. Find c. *(Lesson 5-8)*

18. List all the factors of 54. *(Lesson 1-10)*

Exploration

---

19. Find the number of squares on a regular 8×8 checkerboard. (Include squares of all possible sizes.)

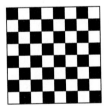

*Use a
Spreadsheet*

Spreading the workload. *Many companies use computer spreadsheets to help them keep track of their inventory. This woman is checking to see if the store's inventory matches the computer records.*

What Is a Spreadsheet?

A **spreadsheet** is a table, usually with many rows and columns. In 1978, the first electronic spreadsheet, VisiCalc, was created by Dan Bricklin and Bob Frankston. Before this time, spreadsheets were used primarily by accountants and other people to keep track of income and expenses for companies. Now there are spreadsheet programs available for every computer and some calculators can deal with them. They are used by many people to organize, explore, and analyze information. Spreadsheets have become all-purpose tools for storing information and solving problems.

	A	B	C	D
1				
2				
3				
4				
5	Gym Shorts			
6			24	
7				

Spreadsheets are made up of **columns** and **rows.** Columns are vertical and labeled by capital letters. Rows are horizontal and are labeled by numbers. The intersection of each column and row is a **cell.** The label Gym Shorts is located in cell A5 which is in column A and row 5. In cell C6 is the number 24.

Entering Information into a Spreadsheet

To use a spreadsheet, you must know how to enter information into its cells. Here is a sample spreadsheet for a school supply store.

	A	B	C	D	E	F	G
1		January	Sales				
2							
3	Item	Unit Price	Number Sold	Total Sales			
4							
5	Gym Shorts	6.00	15				
6	Shirts	8.00	24				
7	Sweatshirts	12.00	32				
8	Tablets	.60	38				
9	Pencils	.10	123				
10	Folders	.50	21				
11							
12	Monthly Sales						

Activity 1

Create the spreadsheet shown above.

Step 1

Turn on your computer and load in your spreadsheet package. The cell which is highlighted or in a different color is called the **active cell**. It is the cell that takes the information you type. You can change the active cell by pressing the arrow keys: → for right, ← for left, ↑ for up, and ↓ for down. With some machines, you can change cells by using a mouse.

Step 2

Use the arrow keys or mouse to make cell B1 the active cell.
Type: January
Make C1 the active cell.
Type: Sales
Make A3 the active cell.
Continue typing until you have entered all of columns A, B, and C. If you make a mistake, use the arrow keys to go back, and then type over your mistakes.

Cells can hold formulas as well as numbers and words. Each formula indicates how a number in a particular place in the spreadsheet is computed from other numbers in the spreadsheet. When a number is changed, the computer uses the formula to change the other numbers in the spreadsheet automatically.

Activity 2

Use the spreadsheet you created in Activity 1 to calculate total sales.

	A	B	C	D
1		January	Sales	
2				
3	Item	Unit Price	Number Sold	Total Sales
4				
5	Gym Shorts	6.00	15	=B5*C5

Step 1

Each number in column D, Total Sales, is the product of a value in column B and a value in column C. For instance, the Total Sales for Gym Shorts is the product of the number in cell B5 and the number in cell C5. So in cell D5, enter the formula = B5 * C5. Now press the enter key.

	A	B	C	D
1		January	Sales	
2				
3	Item	Unit Price	Number Sold	Total Sales
4				
5	Gym Shorts	6.00	15	90.00

You should see 90, or 90.00, displayed in cell D5.

Note: Different spreadsheets have different methods for entering formulas. Some choices are: an equal sign (=) in front of the formula, parentheses, or an arithmetic operator. You will need to know what your spreadsheet requires.

Step 2

Fill in cells D6 through D10. One way to do this is to type = B6 * C6 in cell D6, = B7 * C7 in cell D7, and so on. However, most spreadsheets have a way of copying formulas from one cell to another. The command may be called COPY, FILL, or REPLICATE. Use this command to copy cell D5. Now paste it onto cells D6, D7, D8, D9, and D10. You should see the correct values for Total Sales appear in the appropriate cells.

Row 12 of the spreadsheet can now be completed.

Activity 3

Find the Monthly Sales Total for the data you entered in Activity 1.

Activate cell D12 and enter the formula needed. What were the total sales?

Saving and Updating a Spreadsheet

You should save your work on a **data disk.** Sometimes, you will also want a printed copy of the table you have produced. The commands used for saving and printing vary with the computer and with the spreadsheet package being used. Make sure you know how to save and print with the package you use.

Once formulas are in a spreadsheet, any change in the cells used in the formula will result in automatic recalculation. For example, suppose 27 shirts rather than 24 shirts had been sold. If 27 is typed in cell C6, cell D6 will automatically change to 216. To update the spreadsheet for February, you could change cell B1 to February and put the February sales figures in column C. If any unit prices change, then update column B. The totals in column D for February will be calculated automatically.

Problem Solving with a Spreadsheet

The next activity shows the power of a spreadsheet to solve mathematical problems. Here is a question from Lesson 6-5.

Activity 4

Use a spreadsheet to determine the cost of a one-hour phone call, if the first minute costs $.25 and each additional minute costs $.18.

Step 1
Create the following skeleton for the spreadsheet.

	A	B
1	Minutes	Cost
2	1	.25
3		
4		
5		
6		

Step 2
To enter the consecutive integers 2, 3, 4, 5, . . . in the cells of column A, first type $= A2 + 1$ in cell A3. Then copy that cell onto cells A4 to A61.

Step 3
To increase each consecutive value in column B by .18, type $= B2 + .18$ in cell B3. Then copy that cell onto cells B4 through B61. The answer to the question of the activity should be found in row 61.

QUESTIONS

1. What is a spreadsheet?

2. Explain two uses of a spreadsheet.

In 3–5, what is displayed in the given cell of the spreadsheet in Activity 1?

3. C5   4. A10   5. B8

In 6 and 7, refer to the spreadsheet in Activity 1.

6. Give the heading for row 7.

7. Give the heading for column B.

8. What is meant by the active cell?

9. How do you change the active cell?

In 10 and 11, refer to the spreadsheet in Activity 2.

10. **a.** What is the formula for cell D8?
    b. What values appear in the other cells of column D?

11. What happens within the spreadsheet if the value in cell B7 is changed to 15?

12. Formulas are entered with a symbol typed in front of them. What symbol do you need to use on your computer?

13. **a.** What formula did you type for Activity 3?
    b. What is the result?

14. When were electronic spreadsheets first introduced?

In 15 and 16, refer to Activity 4.

15. What formulas are entered in cells A6 and B6?

16. What formula is in row 51?

Applying the Mathematics

In 17–20, use the spreadsheet for Nautilus Subs shown below.

	A	B	C	D	E	F	G
1	Sandwich	Quantity	Small	Quantity	Large	Item Total	
2							
3	Tuna Salad		3.00		4.35		
4	Ham & Cheese		3.25		4.65		
5	Turkey		3.75		5.65		
6	Roast Beef		4.00		5.95		
7	Italian		4.75		6.75		
8							
9	Beverages						
10							
11	Soda		.60		.75		
12	Milk		.40		.55		
13	Coffee		.50		.65		
14							
15					Subtotal:		
16					Tax:		
17							
18					Total:		

17. Fill in the blanks to give a formula for cell F3.
= ___ * C3 + ___ * ___

18. What is a formula for cell F15?

19. If the tax is 5%, what is a formula for cell F16?

20. Nancy and Chuck plan to stop at Nautilus Subs for dinner.
a. If they order a large turkey sandwich, a small tuna salad sandwich, two large sodas, and one small milk, what entries will the cashier make in the spreadsheet?
b. What is the formula in cell F18?
c. What will be shown in that cell?

21. **a.** Create a spreadsheet that lists times and costs for a phone call to Nigeria if the cost is $1.19 for the first minute and 79¢ for each additional minute.
b. What will it cost for a 27-minute phone call?

In 22 and 23, use the spreadsheet shown below.

	A	B	C	D	E
1	X				
2	2	4	8	16	32
3	3	9	27	81	243
4	5	25	125	625	3125

22. Give a heading for each column.
   a. B   **b.** C   **c.** D   **d.** E

23. Give the formula that is in each indicated cell.
   a. B2   **b.** C3   **c.** D4   **d.** E2

24. Consider the school supply store spreadsheet in this lesson. Suppose that there is an 8% increase in the price of all clothing sold at the store. What spreadsheet cells need to change value and what are the new prices?

Review

25. Admission to Video-Rama, which includes one game token, costs $4.75. Additional game tokens cost $.35 each.
   a. How much is one admission with three tokens?
   b. How much is one admission with 4 tokens?
   c. What is the cost of one admission with t tokens? *(Lesson 6-5)*

26. Find all sets of three integers which satisfy all of the following:
   Condition 1: The numbers are different.
   Condition 2: The sum of the numbers is 0.
   Condition 3: No number is greater than 4.
   Condition 4: No number is less than -4. *(Lesson 6-4)*

27. Hidden inside the box at the left are two numbers between 10 and 99. The difference between the numbers is 54. The sum of the digits in each number is 10. What are the two numbers? *(Lesson 6-4)*

In 28–31, write answers in lowest terms. *(Lesson 5-5)*

28. $\frac{5}{6} + \frac{1}{6}$

29. $2\frac{3}{4} + 1\frac{2}{3}$

30. $-\frac{1}{3} + \frac{1}{4} + \frac{1}{5}$

31. $5 + 2\frac{1}{4}$

Exploration

32. Spreadsheets contain many built-in commands and functions. Explore the commands that enable you to obtain quickly the sum and the mean of numbers in a row or column. Write a short summary of what you find.

Special Cases and Simpler Numbers

An Example of Testing a Special Case

Elaine was not sure whether the pattern

$$x(y + z) = xy + xz$$

is true for all numbers x, y, and z. So she tried some numbers. She substituted 10 for x, 3 for y, and 2 for z. She wrote:

Does 10(3 + 2) = 10 · 3 + 10 · 2?

Following the rules for order of operations, she simplified. She wrote:

Does 10(5) = 30 + 20?
Yes.

Elaine found that the equation is true for this *special case*. A **special case** is an instance of a pattern used for some definite purpose. Elaine was using a strategy called *testing a special case*.

Special cases help to test whether a property or generalization is true.

Example 1

Is it true that $-(a + b) = -a + -b$ for any numbers a and b?

Strategy
Work with a special case. Let $a = 3$ and $b = 2$.
Does -(3 + 2) = -3 + -2?
Does   -(5) = -5? Yes.

This case may be too special. Both numbers are positive. So try another special case, this time substituting a negative number. Let $a = -4$ and $b = 7$. Now substitute.

▶

▶

$$\text{Does } -(-4 + 7) = -(-4) + -7?$$
$$\text{Does } -(3) = 4 + -7? \text{ Again, yes.} \quad \text{But it wasn't as obvious this time.}$$

You might want to try other special cases.

Solution
We have evidence that the property is true.
We still *do* not know for certain that it is true.

Caution:
Even if several special cases of a pattern are true, the pattern may not always be true.

Trying a special case can show you that a pattern is not always true.

Example 2

Is x^2 always greater than or equal to x?

Strategy
Try several special cases.

Solution
Try $x = 6$. $6^2 = 36$, which is greater than 6.
Try $x = 13$. $13^2 = 169$, which is greater than 13.

So you might think $x^2 \geq x$ is always true.

But now try $x = 0.5$.
$0.5^2 = 0.25$, which is less than 0.5.
x^2 is not always greater than or equal to x.

Using Special Cases to Make Choices

Suppose you know that only one of a small number of possibilities is true. Then working with a special case can help you pick the right one.

Example 3

Multiple choice. In an n-gon, how many diagonals can be drawn from one vertex?

(a) n (b) $n - 2$ (c) $n - 3$ (d) $\frac{n}{2}$

Strategy
Pick a special case where n is a small number that is easy to work with. We want a polygon with diagonals, so we need $n \geq 4$. Reword the question for the special case when n is 4.

▶

Solution

Special case

Multiple choice. In a 4-gon, how many diagonals can be drawn from one vertex?

(a) 4   (b) 4 − 2   (c) 4 − 3   (d) $\frac{4}{2}$

Draw a picture to answer the question for the special case.

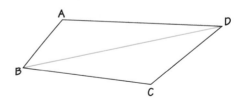

There is only one diagonal from 1 vertex in a quadrilateral (a 4-gon). Choice (c) is correct for the quadrilateral. So choice (c) n − 3 is correct for an n-gon.

Check

Try another special case, perhaps a pentagon. Are there 5 − 3 diagonals from each vertex in a 5-gon? Yes.

Solving Problems by Trying Simpler Numbers

A strategy similar to trying a special case is called *try simpler numbers*. The idea is that complicated numbers or variables may make it hard to devise a plan, so substitute simpler numbers. Simpler numbers often make it easier to see the steps required to solve the problem. Then you can use these same steps with the complicated numbers or variables.

Example 4

Steve can bike one mile in 8 minutes. If he can keep up this rate, how far will he ride in one hour?

Strategy

The numbers are complicated. There are minutes and hours—that makes things more difficult. Modify the problem so that the numbers are simpler and you can answer the question. Then look for a pattern.

Modified problem

Steve can bike one mile in 10 minutes. If he can keep up this rate, how far will he bike in 60 minutes (1 hour)?

Solution to Modified Problem

The answer to this new problem is 6 miles. (Do you see why?)

Solution

The answer 6 to the modified problem is 60 divided by 10. This suggests that the original problem can be solved by division. Dividing the corresponding numbers of the original problem, the number of miles is 60/8.

Steve can bike 60 ÷ 8, or 7.5 miles in an hour.

A long tour. *The Tour de France is a famous bicycle race that lasts about 24 days. The participants travel about 4000 km, or about 2500 miles, throughout Europe.*

Example 5

Therese can bike one mile in t minutes. At this rate, how far can she bike in u minutes?

Strategy

This is similar to Example 4 except that variables are used instead of numbers. Replace the numbers with variables to find the answer.

Solution

When u was 60 and t was 10, the answer was $\frac{60}{10}$.

Therese can bike u/t miles.

QUESTIONS

Covering the Reading

1. What is a special case of a pattern?

In 2–4, consider the pattern $-(a + b) = -a + -b$.

2. Give an instance of this pattern.

3. Let $a = -5$ and $b = -11$. Is this special case true?

4. Let $a = 4.8$ and $b = 3.25$. Is this special case true?

5. *True or false.* If more than two special cases of a pattern are true, then the pattern is true.

6. Consider the pattern $x^2 \geq x$.
   a. Give two instances that are true.
   b. Give an instance that is false.
   c. What do parts **a** and **b** tell you about special cases?

7. *True or false.* When a special case of a pattern is not true, then the general pattern is not true.

In 8–12, consider the following pattern: In an n-gon, $n - 3$ diagonals can be drawn from one vertex.

8. a. When $n = 4$, to what kind of polygon does the pattern refer?
   b. Is the pattern true for that kind of polygon?

9. a. When $n = 7$, to what kind of polygon does the pattern refer?
   b. Is the pattern true for that kind of polygon?

10. a. What value of n refers to a hexagon?
    b. Is the pattern true for hexagons?

11. a. What value of n refers to a triangle?
    b. Is the pattern true for triangles?

12. Show, by drawing, that this pattern is true for an octagon.

In 13 and 14, use the formula from Example 5.

13. Buzz can bike one mile in 7 minutes. At this rate, how far can he bike in an hour?

14. Kristin can bike one mile in *M* minutes. At this rate, how far can she bike in 30 minutes?

15. *True or false.* A simpler number is always a smaller number.

Applying the Mathematics

Electrical vehicle technology holds great promise for the future. This electric van, being tested by utility and delivery companies, is expected to be technically and commercially viable by the year 2000.

16. Tell how many miles per gallon the car is getting.
    a. Isaac drove 250 miles on 10 gallons of gas.
    b. Judy drove 250 miles on 11.2 gallons of gas.
    c. Ken drove *m* miles on 11.2 gallons of gas.
    d. Louise drove *m* miles on *g* gallons of gas.

17. Consider the pattern $-(-m + 9) = m + -9$.
    a. Test a special case with a positive number.
    b. Test a special case with a negative number.
    c. Decide whether the pattern is possibly true or definitely not always true.

18. Which is easier, showing a pattern is true or showing a pattern is false? Explain your answer.

19. *Multiple choice.* Use special cases to help you select. The sum of all the whole numbers from 1 to *n* is
    (a) $n + 1$.
    (b) $n + 2$.
    (c) $\frac{n(n + 1)}{2}$.
    (d) n^2.

20. *Multiple choice.* Use special cases to help you select.
    For all whole numbers *w, n,* and *d,* where $d \neq 0$, $w + \frac{n}{d} =$
    (a) $\frac{dw + n}{d}$.
    (b) wn/d.
    (c) $\frac{w + n}{w + d}$.
    (d) $\frac{w + n}{d}$.

In 21–25, if you cannot answer the question, try a special case or use simpler numbers.

21. A roll of paper towels originally had *R* sheets. If *Z* sheets are used, how many sheets remain?

22. There are 8 boys and 7 girls at a party. A photographer wants to take a picture of each boy with each girl. How many pictures are required?

23. A coat costs $49.95. You give the clerk *G* dollars. How much change should you receive?

24. How should you move the decimal point in order to divide a decimal by .001?

25. *Multiple choice.* The sum of the measures of the four angles of any quadrilateral is
(a) 180°.   (b) 360°.   (c) 540°.   (d) 720°.

Review

In 26–28, use the following spreadsheet.   *(Lessons 4-6, 6-6)*

	A	B	C	D
1		Student	Grades	
2	Name	Test 1	Test 2	Average
3	John	86	78	
4	Paul	55	80	
5	George	97	94	
6	James	23	18	
7	Average			=(B11+C11)/2

26. Write the formula that calculates John's test average, and name the cell in which the formula is entered.

27. What formula should be entered in cell B7 to calculate the group's average on Test 1?

28. a. In which cell will the formula used to calculate the group's average on Test 2 be entered?
b. What formula should be entered there?

29. A downtown parking lot charges $1.00 for the first hour and $.50 for each additional hour or part of an hour.   *(Lesson 6-5)*
a. What will it cost to park from 9 A.M. to 4:45 P.M.?
b. At this rate, what does it cost for h hours of parking, when h is a whole number?

30. The number 17 is a divisor of x, and $925 < x < 950$. Find x.
(Lesson 6-4)

31. Virginia's garden is in the shape of a triangle. Each side of the triangle is 12 feet long. There is a stake at each corner. There are stakes every 3 feet on each side. How many stakes are there in all?   *(Lesson 6-3)*

32. Evaluate a^{b+c} when $a = 5$, $b = 2$, and $c = 4$.   *(Lessons 2-2, 4-4)*

33. Would an adult be more likely to weigh 10 kg, 70 kg, or 170 kg?
(Lesson 3-3)

Exploration

34. Put the digits 1, 2, 3, 4, 5, 6, 7, 8, and 9 in the circles at the left to make the largest product possible.

A project presents an opportunity for you to extend your knowledge of a topic related to the material of this chapter. You should allow more time for a project than you do for typical homework questions.

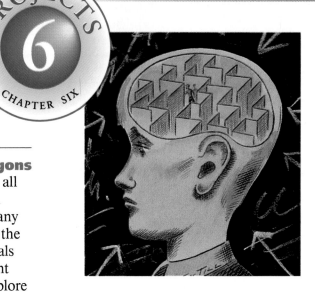

1 Diagonals of Regular Polygons

A *regular polygon* is one in which all sides have the same length and all angles have the same measure. Many of the diagonals of regular polygons are of the same length. For instance, the nine diagonals of a regular hexagon have only two different lengths: 3 are longer and 6 are shorter. Explore the diagonals of regular polygons. How many different lengths are there in the diagonals of a regular *n*-gon?

2 Spreadsheets

Explore one of the best-selling spreadsheets, such as Microsoft Excel® or Lotus 1-2-3®. Write an essay describing what these spreadsheets can do.

3 Problem Solving

Being able to solve problems is important in almost every walk of life. Books have been written for people who want to solve problems in their work, in their personal lives, or in various subjects. Find a book that talks about solving problems. Summarize the main advice given in that book.

4 Cryptarithms

Question 31 in Lesson 6-1 is a type of problem known as a *cryptarithm*. Cryptarithms are normally solved with a good amount of trial and error combined with a knowledge of arithmetic. Sometimes there is more than one solution. One of the

most famous of these problems was (according to a story) sent by a boy in college to his father:

```
  S E N D
+ M O R E
M O N E Y
```

His father responded:

```
  U S E
+ L E S S
S O N N Y
```

Solve these cryptarithms. That is, find out what digit each letter stands for in these addition problems. In each problem, each different letter stands for a different digit. The same letter may stand for different digits in the two problems.

5 Special Cases that Work, but Generalizations that Do Not

3 is an odd number and 3 is prime. 5 is an odd number and 5 is prime. 7 is an odd number and 7 is prime. A person might be tempted to make the generalization that all odd numbers are prime. But of course they are not. Make up *ten* examples of situations in which some special cases have a certain property, but the generalization is not true. Not all of your examples have to be mathematical.

6 Special Numbers

Find the meaning of each of the following terms. Give at least two examples of each. Explain why the term is appropriate for these numbers.

abundant number   amicable numbers
deficient number   relatively prime numbers

SUMMARY

If you have an algorithm for answering a question, it is an exercise for you. If not, then the question is a problem for you. Problems are frustrating unless you know strategies that can help you approach them. This chapter discusses six general strategies. They are:

1. Read the problem carefully. Look up definitions for any unfamiliar words. Determine if enough information is given. Sort information into what is needed and what is not needed.

2. Draw an accurate picture.

3. Use trial and error. Use earlier trials to help in choosing better later trials.

4. Make a table to organize your thoughts. Use the table to find patterns. Use the table to make generalizations. A spreadsheet may be a good way to make your table.

5. Use special cases to help decide whether a pattern is true.

6. Try simpler numbers. Generalize to more complicated numbers or variables.

These strategies can help in understanding a problem, in devising a plan, in solving, and in checking. Good problem solvers take their time. They do not give up. They are flexible. And they continually check their work.

VOCABULARY

You should be able to give a general description and a specific example of each of the following ideas.

Lesson 6-1
algorithm, exercise, problem

Lesson 6-2
prime number, composite number
divisor, factor
natural number

Lesson 6-4
trial and error

Lesson 6-5
table
generalization

Lesson 6-6
spreadsheet
row, cell, column
active cell
replicate
data disk

Lesson 6-7
special case
testing a special case
try simpler numbers

PROGRESS SELF-TEST

Take this test as you would take a test in class. Then check your work with the solutions in the Selected Answers section in the back of the book.

1. Name two places to find the meaning of the term "trapezoid."

2. How many diagonals does an octagon have?

3. There are six players in a singles tennis tournament. Each player will compete with each of the others. How many matches will be played?

4. What two positive integers whose product is 24 will have the smallest sum?

5. *True or false.* If you know an algorithm for multiplying fractions, then multiplying fractions is a problem for you.

6. List the prime numbers between 50 and 60.

7. Consider the sentence $|x| + |y| = |x + y|$. Find values of x and y that make this sentence false.

8. What is an advantage of using an electronic spreadsheet over a handwritten table?

9. How much will 1.62 pounds of hamburger cost if the price is $1.49 per pound?

10. *True or false.* If a special case of a pattern is true, then the general pattern may still be true.

11. What integer between 5 and 10 is a solution to $(x - 3) + (x - 4) = 9$?

12. Donna has some disks numbered 1, 2, 3, 4, and so on. She arranges them in order in a circle, equally spaced. If disk 3 is directly across from disk 10, how many disks are in the circle?

13. When all the possible diagonals are drawn, an n-gon has $n + 3$ diagonals. What is n?

14. How many decimal places and in which direction should you move a decimal point in order to divide by .01?

15. Brianna was given $1000 by her parents to spend during her first year of college. She decided to spend $25 a week. Make a table to show how much she will have left after 1, 2, 3, and 4 weeks.

16. In the situation of Question 15, how much will Brianna have after 31 weeks?

17. Auni carefully read a problem, prepared a table to help solve the problem, and worked until she reached a conclusion. She wrote her answer and went on to the next problem. What step in good problem solving did Auni omit?

18. Name two situations in which an electronic spreadsheet would be helpful.

CHAPTER REVIEW

Questions on SPUR Objectives

SPUR stands for **S**kills, **P**roperties, **U**ses, and **R**epresentations. The Chapter Review questions are grouped according to the SPUR Objectives for this chapter.

SKILLS DEAL WITH THE PROCEDURES USED TO GET ANSWERS.

Objective A. *Know the general strategies used by good problem solvers.*   *(Lesson 6-1)*

1. *Multiple choice.*   Which advice should you follow to become a better problem solver?
   (a) Check work by answering the question the same way you did it.
   (b) Be flexible.
   (c) Skip over words you don't understand as long as you can write an equation.
   (d) none of (a) through (c)

2. If you can apply an algorithm to a problem, then the problem becomes an __?__.

3. Nancy multiplied 1487×309 on her calculator. What would be a good way for her to check her answer?

Objective B. *Determine solutions to sentences by trial and error.*   *(Lesson 6-4)*

4. Which integer between 10 and 20 is a solution to $3x + 15 = 66$?

5. *Multiple choice.*   Which number is not a solution to $n^2 \geq n$?
   (a) 0   (b) 0.5
   (c) 1   (d) 2

6. Choose one number from each row so that the sum of the numbers is 255.

88	84	9
69	79	76
108	104	102

7. What number between 1000 and 1050 has 37 as a factor?

PROPERTIES DEAL WITH THE PRINCIPLES BEHIND THE MATHEMATICS.

Objective C. *Determine whether a number is prime or composite.*   *(Lesson 6-2)*

8. Explain why 2.3×10^4 is composite.

9. Is 49 prime or composite? Explain your answer.

10. Is 47 prime or composite? Explain your answer.

11. List all composite numbers n with $20 < n < 30$.

Objective D. *Find the meaning of unknown words.*   *(Lesson 6-2)*

12. What is a tetrahedron?

13. What is a perfect number?

Objective E. *Make a table to find patterns and make generalizations.*   *(Lesson 6-5)*

14. Mandy is saving to buy a present for her parents' anniversary. She has $10 now and will add $5 a week.

   a. How much will she save in 12 weeks?

   b. How much will she save in w weeks?

15. Consider 2, 4, 8, 16, 32, . . . (the positive integer powers of 2). Make a table listing all the factors of these numbers. How many factors does 256 have?

Objective F. *Work with a special case to determine whether a pattern is true.* *(Lesson 6-7)*

16. Is there any *n*-gon with *n* + 2 diagonals?

17. To divide a decimal by 1 million, you can move the decimal point __?__ places to the __?__.

Objective G. *Use special cases to determine that a property is false or to give evidence that it is true.* *(Lesson 6-7)*

18. Let *a* = 5 and *b* = -4 to test whether $2a + b = a + (b + a)$. Is the property false or do you have more evidence that it is true?

19. Show that $5x + 5y$ is not always equal to $10xy$ by choosing a special case.

USES DEAL WITH APPLICATIONS OF MATHEMATICS IN REAL SITUATIONS.

Objective H. *Use simpler numbers to answer a question requiring only one operation.* *(Lesson 6-7)*

20. If you fly 430 miles in 2.5 hours, how fast have you gone?

21. If you buy 7.3 gallons of gas at a cost of $1.19 per gallon, what is your total cost (to the penny)?

Objective I. *Use drawings to solve real problems.* *(Lesson 6-3)*

22. Nine teams are to play each other in a tournament. How many games are needed?

23. Five hockey teams are to play each other two times in a season. How many games are needed?

24. Bill is older than Becky. Becky is younger than Bob. Bob is older than Barbara. Barbara is older than Bill. Who is second oldest?

25. Interstate 25 runs north and south through Wyoming and Colorado. Denver is 70 miles from Colorado Springs and 112 miles from Pueblo. Cheyenne is 171 miles from Colorado Springs and 101 miles from Denver. Pueblo is 42 miles from Colorado Springs. Cheyenne is north of Denver. Which of these four cities, all on Interstate 25, is farthest south?

REPRESENTATIONS DEAL WITH PICTURES, GRAPHS, OR OBJECTS THAT ILLUSTRATE CONCEPTS.

Objective J. *Use a spreadsheet to answer questions in real situations.* *(Lesson 6-6)*

26. Jenny scored 82 on test 1, 84 on test 2, and 60 on test 3. Tom scored 93 on test 1, 85 on test 2, and 90 on test 3. David scored 75 on test 1, 100 on test 2, and 86 on test 3. Sara scored 99 on test 1, 100 on test 2, and 87 on test 3. Their teacher needs to know the average grade for each student and the average grade for each test. Design a spreadsheet to show this information.

27. A plumber charges $18.75 for a service call and $11 for every 15 minutes of labor. Design a spreadsheet that will show the costs for the plumber's time from 0 minutes to 6 hours, in 15-minute increments.

Objective K. *Draw a diagram to aid in solving geometric problems.* *(Lesson 6-3)*

28. How many diagonals does a hexagon have?

29. How many diagonals does a decagon have?

30. All the diagonals of a pentagon are drawn. Into how many sections is the interior divided?

31. In polygon *ABCDEFGHI* all diagonals from *A* are drawn. How many triangles are formed?

PATTERNS LEADING TO SUBTRACTION

Here is a breakdown of the resident population, in 1990, of each state and its gain or loss since 1980, through estimated net migration (moving, births, and deaths).

State	1990 Population	Change Since 1980	State	1990 Population	Change Since 1980
Alabama	4,040,587	146,562	Montana	799,065	12,375
Alaska	550,043	148,192	Nebraska	1,578,385	8,560
Arizona	3,665,228	948,682	Nevada	1,201,833	401,325
Arkansas	2,350,725	64,368	New Hampshire	1,109,252	188,642
California	29,760,021	6,092,257	New Jersey	7,730,188	365,177
Colorado	3,294,394	404,659	New Mexico	1,515,069	211,767
Connecticut	3,287,116	179,552	New York	17,990,455	432,290
Delaware	666,168	71,830	North Carolina	6,628,637	748,542
Florida	12,937,926	3,190,965	North Dakota	638,800	-13,917
Georgia	6,478,216	1,015,234	Ohio	10,847,115	49,512
Hawaii	1,108,229	143,538	Oklahoma	3,145,585	120,098
Idaho	1,006,749	62,622	Oregon	2,842,321	209,165
Illinois	11,430,602	3,193	Pennsylvania	11,881,643	16,923
Indiana	5,544,159	53,945	Rhode Island	1,003,464	56,310
Iowa	2,776,755	-137,053	South Carolina	3,486,703	365,974
Kansas	2,477,574	113,338	South Dakota	696,004	5,236
Kentucky	3,685,296	24,972	Tennessee	4,877,185	286,162
Louisiana	4,219,973	13,857	Texas	16,986,510	2,760,997
Maine	1,227,928	102,885	Utah	1,722,850	261,813
Maryland	4,781,468	564,535	Vermont	562,758	51,302
Massachusetts	6,016,425	279,332	Virginia	6,187,358	840,561
Michigan	9,295,297	33,253	Washington	4,866,692	734,339
Minnesota	4,375,099	299,129	West Virginia	1,793,477	-156,709
Mississippi	2,573,216	52,446	Wisconsin	4,891,769	186,127
Missouri	5,117,073	200,307	Wyoming	453,588	-15,969
Other areas					
D.C.	606,900	-31,532	Guam	133,152	27,173
Puerto Rico	3,522,037	325,517	American		
Virgin Islands	101,809	6,218	Samoa	46,773	14,476

Source: Bureau of the Census, U.S. Dept. of Commerce, 1980 and 1990 Censuses

You can determine the population of your state in 1980 from this table. You must subtract the change since 1980 from the 1990 population for your state. This subtraction may involve positive or negative numbers. It is one of many situations in which subtraction is the operation to use.

In this chapter, you will study many other situations that lead to subtraction. These naturally lead to equations that involve subtraction.

The cutting edge. *The technician is programming a high-speed saw to ensure that as little lumber as possible is wasted when cuts are made. Even the sawdust is recycled.*

Subtraction as Taking Away

Suppose you walk into a store with $10 and spend $2.56. The amount you have left is found by subtraction: $10 − $2.56 = $7.44. Recall that a model for an operation is a general pattern that includes many of the uses of the operation. This subtraction is an instance of the *Take-Away Model for Subtraction*.

> **Take-Away Model for Subtraction**
> If a quantity y is taken away from an original quantity x with the same units, the quantity left is $x - y$.

Here are other examples of the many different situations that use the take-away model.

Example 1

A piece of wood 32.5 centimeters long is cut from a board of original length 3 meters long. How long is the remaining piece?

Solution

Draw a picture. Units must be consistent, so change 3 meters to centimeters.

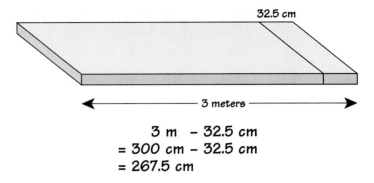

$$3 \text{ m} - 32.5 \text{ cm}$$
$$= 300 \text{ cm} - 32.5 \text{ cm}$$
$$= 267.5 \text{ cm}$$

The remaining piece is 267.5 cm long.

Example 1 can be generalized. If the units are the same, a piece C units long cut from a board L units long leaves a piece $L - C$ units long.

Recall from Lesson 5-4, that the angles ABC and CBD drawn below are adjacent angles. The sum of their measures is the measure of angle ABD. If you know m$\angle ABD$ and one of the smaller angles, you can find the other by subtraction. You can think of this as "taking away" one of the smaller angles from the larger.

An old saw. *This saying was probably based upon a comment by noted American author and naturalist, Henry David Thoreau (1817–1862). He said: "They (wood stumps) warmed me twice—once while I was splitting, and again when they were on the fire."*

Example 2

In the drawing below, what is the measure of $\angle ABC$?

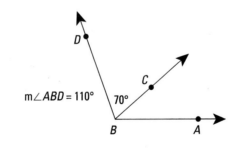

Solution
To find m$\angle ABC$, subtract the measure of $\angle CBD$ from the measure of $\angle ABD$.

$$\text{m}\angle ABC = \text{m}\angle ABD - \text{m}\angle CBD$$
$$= 110° - 70°$$
$$= 40°$$

Subtraction as Comparison

The first number in a subtraction problem is called the **minuend**. The second number is the **subtrahend**. The answer is called the **difference**.

$$
\begin{array}{rl}
436.2 & \text{minuend} \\
-98.5 & \text{subtrahend} \\
\hline
337.7 & \text{difference}
\end{array}
$$

The term "difference" comes from a second model of subtraction, the *comparison model.*

> **Comparison Model for Subtraction**
> $x - y$ is how much more x is than y.

The comparison model for subtraction is commonly used. Here is an example comparing lengths.

Example 3

Penelope is 164 cm tall. Her boyfriend, Ulysses, is 180.5 cm tall. How much taller is Ulysses than Penelope?

Solution

"How much taller" is asking for a comparison, so subtract.
180.5 – 164 = 16.5, so Ulysses is 16.5 cm taller.

The next example compares capacities of $3\frac{2}{3}$ cups and $1\frac{3}{4}$ cups. In it the mixed number $3\frac{2}{3}$ is changed first to $3\frac{8}{12}$ and then to $2\frac{20}{12}$. This can be done because $3\frac{8}{12} = 2 + 1\frac{8}{12} = 2\frac{20}{12}$.

Example 4

A recipe requires $3\frac{2}{3}$ cups of flour. You have $1\frac{3}{4}$ cups. How much more flour do you need?

Solution 1

You need to compare what you have to what you need. So the answer is given by the subtraction $3\frac{2}{3} - 1\frac{3}{4}$. Here is the usual paper-and-pencil algorithm to do this. As with addition, this method requires finding equal fractions with the same denominator.

$$
\begin{array}{r}
3\frac{2}{3} = 3\frac{8}{12} = 2\frac{20}{12} \\
- \ 1\frac{3}{4} = 1\frac{9}{12} = 1\frac{9}{12} \\
\hline
1\frac{11}{12}
\end{array}
$$

So you need $1\frac{11}{12}$ cups more.

Solution 2

Change both $3\frac{2}{3}$ and $1\frac{3}{4}$ to improper fractions with the same denominator.

$$
\begin{array}{r}
3\frac{2}{3} = 3\frac{8}{12} = \frac{44}{12} \\
- \ 1\frac{3}{4} = 1\frac{9}{12} = \frac{21}{12} \\
\hline
\frac{23}{12} = 1\frac{11}{12}
\end{array}
$$

So you need $1\frac{11}{12}$ cups more.

▶

▶ **Solution 3**

With a calculator that does fractions, you need only key in $3\frac{2}{3}$, then ⊖,
then $1\frac{3}{4}$, then ⊜. The answer $1\frac{11}{12}$ should appear.

QUESTIONS

Covering the Reading

1. **a.** There are 320 passenger seats in one wide-body jet plane. A flight
      attendant counts 4 vacant seats. How many passengers are on
      board?
   b. There are S passenger seats in a wide-body jet plane. A flight
      attendant counts V vacant seats. How many passengers are on
      board?

2. **a.** Hungry Hans ate 2 of the dozen rolls his mother prepared for
      dinner. How many rolls are left for the others at the table?
   b. Hungry Heloise ate A of the dozen rolls her mother prepared for
      dinner. How many rolls are left for the others at the table?

3. Questions 1–2 are instances of what model for subtraction?

In 4–6, use the picture at the right.

4. Angles ABD and CBD are __?__ angles.

5. If m$\angle ABC = 100°$ and m$\angle DBA = 84°$,
   what is m$\angle DBC$?

6. m$\angle ABC -$ m$\angle DBC =$ m$\angle$__?__.

7. Consider the subtraction fact $12 - 8 = 4$. Identify each.
   a. difference   **b.** minuend   **c.** subtrahend

8. State the Comparison Model for Subtraction.

9. Jim weighs 150 pounds but wants to get his weight down to 144
   pounds. How much does he need to lose?

10. Nina has $240 saved for a stereo system that costs $395. How much
    more money does she need?

11. Lori needs $2\frac{3}{4}$ cups of flour for a cake. She has $2\frac{2}{3}$ cups.
    a. Compare these two quantities by subtraction.
    b. Does Lori have not enough, just enough, or more than enough
       flour for the cake?

12. **a.** To subtract $13\frac{3}{8} - 6\frac{7}{8}$ without using a calculator, how would you
       rewrite $13\frac{3}{8}$?
    b. Do this subtraction.

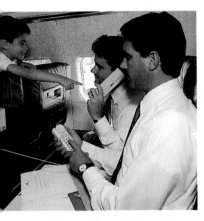

The modern skies. *Many
jet planes include high-
tech conveniences for the
passengers, such as
telephones, computers,
and fax machines.*

13. The head of a diver is x meters below a diving board that is 10 meters above the water in a pool. How far above the water is the diver's head?

10 m

x

In 14–16, use the diagram below of towns A, B, C, and D along a highway.

A   B   C   D

First female Nobel Prize recipient. *Marie Curie received the 1903 Nobel Prize in physics and the 1911 Nobel Prize in chemistry.*

14. Suppose $AD = 10$ km and $AC = 6$ km.
    a. What other distance can be found?
    b. What is that distance?

15. a. The difference $AD - CD$ is the distance between what two towns?
    b. $AD - AB - CD = $ ___?___

16. Suppose CD is 40% of AD. Also suppose AB is 25% of AD. Then BC is what percent of AD?

17. Bill's savings account has $510.75 in it. How much will be left after the given withdrawal?
    a. a withdrawal of $40   b. a withdrawal of W dollars

18. The famous scientist Marie Curie was born in 1867 and died in 1934. Using only this information, what was her age when she died? (Watch out. There are two possible answers.)

19. Use the table on page 345. Determine the 1980 population of each of these places.
    a. New York
    b. Wyoming
    c. the state or area where you live

20. The Roman poet, Livy, was born in 59 B.C. and died in 17 A.D. From this information, what was his age when he died? (Watch out again. There was no year 0. The year 1 A.D. followed 1 B.C.)

In 21–23, use the two squares pictured at the left.

21. What is the area of the shaded region?

22. Find the area of the shaded region if $a = 8$ and $b = 10$.

23. Find the area of the shaded region if $a = 4\frac{1}{2}$ and $b = 6.7$.

24. Try simpler numbers if you cannot get the answer right away.

    a. How many integers are between 100 and 1000, not including 100 or 1000?

    b. How many integers are between two integers I and J, not including I or J?

Review

25. Try positive and negative numbers to see whether it is true that $-(a + b) = -b + -a$. *(Lesson 6-7)*

26. Name one advantage of using a computer spreadsheet. *(Lesson 6-6)*

27. What is the last digit of 7^{1000}? *(Lesson 6-5)*

28. Solve $-3\frac{1}{2} + x = 4.2$. *(Lesson 5-8)*

29. Give the additive inverse of each number.

    a. 5   **b.** -3.4   **c.** 0 *(Lesson 5-2)*

30. *True or false.* $\frac{5}{4} + \frac{5}{4} = \frac{10}{8}$ *(Lesson 5-5)*

31. Evaluate $3p + q^4$ if $p = 4$ and $q = 5$. *(Lessons 2-2, 4-4)*

32. How many grams are in a pound? *(Lesson 3-5)*

Exploration

33. If the same number is added to both the minuend and the subtrahend, their difference is not changed. For instance, to subtract

$$\begin{array}{r} 4307 \\ -2998 \end{array}$$

you can add 2 to the minuend and subtrahend to get

$$\begin{array}{r} 4309 \\ -3000 \end{array}$$

which is much easier. The answer to both questions is 1309. For each of the subtraction questions below, find a number to add to make the subtraction easier. Then do the subtraction.

    a. $\begin{array}{r} 136 \\ -97 \end{array}$ **b.** $\begin{array}{r} 4905 \\ -1996 \end{array}$ **c.** $\begin{array}{r} 1117 \\ -989 \end{array}$

The Slide Model for Subtraction

Sliding with snowboards. *When temperatures slide down and snow falls, people can slide downhill. These provide images for subtraction.*

Subtraction as Sliding Down

Suppose the temperature is 50° and drops 12°. This situation can be pictured on a number line. Start at 50° and slide 12° to the left.

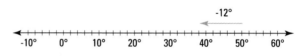

The resulting temperature is 38°.

The answer could also be found by subtracting $50 - 12$. This subtraction is not take-away or comparison, but a third model for subtraction called the *slide model*.

> **Slide Model for Subtraction**
> If a quantity x is decreased by an amount y, the resulting quantity is $x - y$.

In slide situations, you usually can slide up or down. Results of sliding up are found by addition. Results of sliding down can be found either by adding negative numbers or by subtracting.

	By subtraction	By addition	Answer
A person who weighs 60 kg loses 4 kg. What is the resulting weight?	60 kg − 4 kg	60 kg + ⁻4 kg	56 kg
The temperature is ⁻17° and falls 20°. What is the resulting temperature?	⁻17° − 20°	⁻17° + ⁻20°	⁻37°

These examples show a basic relationship between subtraction and addition. Its formal name is the *Algebraic Definition of Subtraction.* It is also called the *Adding Opposites Property of Subtraction,* or the *Add-Opp Property* for short.

Algebraic Definition of Subtraction (Add-Opp Property)
For any numbers x and y,
$$x - y = x + \text{-}y.$$
In words, subtracting y is the same as adding the opposite of y.

An Algorithm for Subtraction

The Algebraic Definition of Subtraction allows any subtraction to be converted to an addition. This is helpful because you already know how to add both positive and negative numbers.

Example 1

Simplify $\text{-}5 - 2$.

Solution 1

Use the slide model. Start at $\text{-}5$ and slide down 2.

The result is $\text{-}7$. So $\text{-}5 - 2 = \text{-}7$.

Solution 2

Use the Algebraic Definition of Subtraction. $\text{-}5 - 2 = \text{-}5 + \text{-}2$
(Instead of subtracting 2, add $\text{-}2$.) $= \text{-}7$

Example 2

Perform each subtraction by converting it to an addition using the Algebraic Definition of Subtraction.
a. $40 - 50.79$ **b.** $\text{-}9 - \text{-}11$ **c.** $\frac{3}{4} - \frac{7}{8}$ **d.** $x - \text{-}y$

Solution

	Given		Convert to Addition		Result
a.	$40 - 50.79$	$=$	$40 + \text{-}50.79$	$=$	$\text{-}10.79$
b.	$\text{-}9 - \text{-}11$	$=$	$\text{-}9 + 11$	$=$	2
c.	$\frac{3}{4} - \frac{7}{8}$	$=$	$\frac{3}{4} + \text{-}\frac{7}{8} = \frac{6}{8} + \text{-}\frac{7}{8}$	$=$	$\text{-}\frac{1}{8}$
d.	$x - \text{-}y$	$=$	$x + y$	$=$	$x + y$

A major reason for converting subtractions to additions is that addition has the commutative and associative properties. You do not have to worry about order of additions. This is particularly nice if there are more than two numbers involved in the subtraction.

Example 3

Simplify $-5 - -3 + 8 + -2 - 7 - 4$.

Solution

Start with the original expression.	$-5 - -3 + 8 + -2 - 7 - 4$
Convert all subtractions to addition.	$= -5 + 3 + 8 + -2 + -7 + -4$
Rearrange so all negatives are together.	$= -5 + -2 + -7 + -4 + 3 + 8$
Add negatives together and positives together.	$= -18 + 11$
Add the negative total to the positive total.	$= -7$

Uses of the − Sign

You have now learned three uses of the − sign. Each use has a different English word.

where − sign is found	example	in English
between numbers or variables	$2 - 5$	2 *minus* 5
in front of a positive number	-3	*negative* 3
in front of a variable or negative number	$-x$ $-(-4)$	*opposite of* x *opposite of negative* 4

For example, $-3 - -y$ is read "negative three minus the opposite of y."

Comparison with Negative Numbers

Special types of the Comparison Model for Subtraction can lead to subtracting negative numbers.

change = later value − earlier value

Example 4

The temperature was $-3°$ earlier today. Now it is $-17°$. By how much has it changed?

Solution

To find the change, start the subtraction with the later value.

$$-17 - -3 = -17 + 3 = -14$$

So the temperature has gone down 14°.

In questions like that in Example 4, many people get confused. They do not know which number is the later value. This should not cause you to worry, because in comparing numbers, you can subtract in either order. The number $x - y$ is always the opposite of $y - x$. (You are asked to explore this in the Questions.)

For instance, in Example 4 suppose you subtracted -17 from -3.

$$-3 - -17 = -3 + 17 = 14$$

This tells you that the answer to Example 4 is either 14 or -14. Since the temperature went from -3 to -17, it went down. So the correct answer must be the negative one of these, -14.

QUESTIONS

Covering the Reading

1. To picture the subtraction $3 - 4$ on the number line, you can start at __?__ and draw an arrow __?__ units long pointing to the __?__.

2. State the Slide Model for Subtraction.

In 3 and 4, a question is given.
a. Write a subtraction problem that will answer the question.
b. Write an addition problem that will answer the question.
c. Answer the question.

3. The temperature is 74°F and is supposed to drop 20° by this evening. What is the expected temperature this evening?

4. The temperature is -4°C and is supposed to drop 10° by morning. What is the expected morning temperature?

5. State the Algebraic Definition of Subtraction (Add-Opp Property)
a. in symbols; b. in words.

Too cold to melt. *At -4°C, snow will remain on the ground indefinitely.*

6. $5 - -8 = 5 + \underline{}$.

7. $-x - -y = \underline{} + \underline{}$.

In 8–15, simplify.

8. $2 - 5$

9. $83 - 100$

10. $-8 - 45$

11. $-1 - 1$

12. $3 - -7$

13. $0 - -41$

14. $-\dfrac{9}{5} - -\dfrac{6}{5}$

15. $m - -2$

16. Give two reasons why it is useful to be able to convert subtractions to additions.

17. Consider the expression $-43 - -x$.
a. Which of the three dashes (left, center, or right) is read as "minus"?
b. Which dash is read "opposite of"?
c. Which dash is read "negative"?

18. Translate into English: $-A - -4$.

In 19 and 20, simplify.

19. $-4 - 8 + 6 - 7 + -5 - -3$ **20.** $12 - 24 + -36 - -48 + 60$

In 21 and 22, calculate the change in temperature from yesterday to today.

21. yesterday, 27°; today, 13°   **22.** yesterday, -9°; today, -8°

23. Let $x = 46$ and $y = -16$. Give the value of each expression.
   a. $x - y$ **b.** $y - x$ **c.** $(x - y) + (y - x)$

Applying the Mathematics

24. The formula $p = s - c$ connects profit p, selling price s, and cost c.
   a. Calculate p when $s = \$49.95$ and $c = \$30.27$.
   b. Calculate p when $s = \$49.95$ and $c = \$56.52$.
   c. Your answer to part **b** should be a negative number. What does a negative profit indicate?

25. Calculate $a - b + c - d$ when $a = -1$, $b = -2$, $c = -3$, and $d = -4$.

26. Give a key sequence for your calculator that does the subtraction $3 - -4$.

In 27–29, use trial and error, or test special cases.

27. *Multiple choice.* If $9 = 7 - x$, then $x =$
   (a) 2.   (b) 16.   (c) -2.   (d) -16.

28. a. For all numbers x, y, and z, is it true that
   $(x - y) - z = x - (y - z)$?
   b. Does subtraction have the associative property?

29. a. In order for subtraction to be commutative, what relationship between any two numbers a and b would have to be true?
   b. Is subtraction commutative?

30. Here are populations of Portland, Oregon, according to the last four censuses.

1960	372,676
1970	379,967
1980	366,383
1990	437,319

   a. Calculate the change from each census to the next. You should get two positive numbers and one negative number.
   b. Add the three numbers you got in part **a**.
   c. What does the sum in part **b** mean?

31. Do the addition and subtraction and write your answer as a mixed number:
$$-3\tfrac{1}{2} + 2\tfrac{2}{3} - 6\tfrac{1}{6}$$

32. Daniel weighed $6\frac{15}{16}$ pounds at birth and $21\frac{1}{2}$ pounds on his first birthday. Kiera weighed $7\frac{1}{4}$ pounds at birth and $20\frac{1}{2}$ pounds at one year. How much more weight did Daniel gain in his first year than Kiera gained in hers? *(Lesson 7-1)*

33. A symphony is 43 minutes long. If an orchestra has been playing the symphony for m minutes, how many minutes remain? *(Lesson 7-1)*

Maestro, if you please.
Michael Morgan is shown here conducting the Chicago Youth Symphony Orchestra.

34. Find $m\angle HDA$ if $m\angle HDR = 136°$ and $m\angle ADR = 54°$. *(Lesson 7-1)*

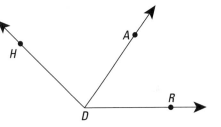

35. In a single elimination tournament a team plays until it is beaten. Eight teams play a single elimination tournament. How many games must be played in this tournament? *(Lesson 6-3)*

36. What property justifies adding the same number to both sides of an equation? *(Lesson 5-8)*

37. How many degrees does the minute hand of a watch turn in 1 minute? *(Lesson 5-4)*

38. Use a spreadsheet to evaluate $x - y$ and $y - x$ for at least twenty pairs of values of x and y. Use positives, negatives, fractions, and decimal values for each variable. What relationship do you find between $x - y$ and $y - x$?

Hawaiian beef. *The Parker Ranch, located on the island of Hawaii, covers 262,000 acres and is one of the largest ranches in the country.* Paniolos *(cowboys) are shown here rounding up the herd and collecting strays.*

Consider this problem.

> From a large herd of cattle, cowhands drove away 230 steers. There were 575 steers left in the herd. How large was the original herd?

One way to answer this question is to solve an equation. Let S be the number of steers in the herd. Then, by the Take-Away Model for Subtraction, $S - 230 = 575$.

The equation $S - 230 = 575$ is an equation of the form $x - a = b$, where x is the unknown. To solve this equation, just convert the subtraction to addition using the Algebraic Definition of Subtraction. Then solve the resulting equation as you did in Chapter 5.

Example 1

Solve $S - 230 = 575$ to answer the question at the top of this page.

Solution

Start with the original equation.	$S - 230 = 575$
Convert to addition.	$S + \text{-}230 = 575$
Add 230 to both sides.	$S + \text{-}230 + 230 = 575 + 230$
Simplify.	$S + 0 = 805$
	$S = 805$

The original herd had 805 steers.

Check

If there were 805 steers in the original herd and 230 were driven away, would 575 remain? Yes.

The situation in Example 2 leads to another equation of the form $x - a = b$. This time b is negative. Still, the same process used in Example 1 works. Convert the subtraction to addition and solve the resulting equation.

Example 2

A group of divers paused on a natural plateau below the surface of the sea. After descending 20 meters more, they found themselves 53 meters below sea level. What is the elevation of the plateau?

Solution

Let E be the elevation of the plateau (in meters). From the Slide Model for Subtraction, $E - 20 = -53$. Solve this equation.

Original equation	$E - 20 = -53$
Convert to addition.	$E + -20 = -53$
Add 20 to both sides.	$E + -20 + 20 = -53 + 20$
Simplify.	$E + 0 = -33$
	$E = -33$

The plateau is 33 meters below sea level.

The process used in Examples 1 and 2 will work with any equation of the form $x - a = b$. We show this by using the general properties of addition and subtraction on this equation. Since the general properties are true for all numbers, this process will work regardless of the values of a and b.

What is done	What is written	Why it can be done
Original equation	$x - a = b$	
Change to addition.	$x + -a = b$	Algebraic Definition of Subtraction
Add a to each side.	$(x + -a) + a = b + a$	Addition Property of Equality
Regroup.	$x + (-a + a) = b + a$	Associative Property of Addition
Add $-a$ and a.	$x + 0 = b + a$	Property of Opposites
Add 0 and x.	$x = b + a$	Additive Identity Property of Zero

Experienced solvers do not write in all the steps. In Examples 1 and 2, we combined the 3rd and 4th lines. You should begin by writing all (or almost all) of the lines. After a while, you may skip steps. Your teacher may have advice for you to follow.

Scuba diving. *The word* scuba *stands for* S*elf-* C*ontained* U*nderwater* B*reathing* A*pparatus. This scuba diver is swimming in an underwater cave in Australia.*

QUESTIONS

Covering the Reading

1. What is the name of the property that enables $x - a$ to be replaced by $x + -a$?

2. What is the only difference between solving an equation of the form $x - a = b$ and solving an equation of the form $x + a = b$?

3. Is -42 the solution to $x - 13 = -29$?

4. Is $y = -1$ the solution to $6 = y - -7$?

In 5–10, an equation is given.
a. Convert the equation to one with only addition in it.
b. Solve.
c. Check.

5. $a - 6 = 9$

6. $c - 12.5 = 3$

7. $x - 14 = -2$

8. $73 = y - 28$

9. $B - -5 = 6$

10. $3.01 = e - 9.2$

11. *Multiple choice.* If $x - a = b$, then $x =$
(a) $a - b$.   (b) $b - a$.   (c) $b + a$.   (d) $b + -a$.

In 12 and 13, a situation is given.
a. Write an equation involving subtraction describing the situation.
b. Solve the equation.
c. Answer the question.

12. After descending 75 meters, Antoine Clymer is at an elevation of 3,980 meters on the mountain. At what elevation was he before his descent?

13. Greta Lowenz spent $14.39 of her weekly allowance on fancy bows. She has $12.61 left. What is her weekly allowance?

Applying the Mathematics

The dancers shown here were among the many performers at the Festival of Philippine Arts and Culture held in Los Angeles, California.

14. The formula $p = s - c$ relates profit, selling price, and cost. Solve for the selling price s in terms of the profit and the cost.

15. Scott started with b dollars in his savings account. He withdrew $60. Two days later he withdrew $30, leaving a balance of $237.61. How much was in Scott's account to start with?

16. A fish starts at a depth of d feet below the surface of a pond. While searching for food it ascends 4 feet, then descends 12 feet. Its final depth is 15 feet below the surface. What was the fish's initial depth?

17. Solve for s: $s - 1\frac{5}{8} = \frac{1}{2}$.

Review

In 18–20, calculate. *(Lesson 7-2)*

18. $2 - 3 - 10$

19. $-8 - 9$

20. $-\frac{3}{8} - -\frac{2}{3}$

21. The number of Americans of Filipino descent living in the United States in 1980 was 774,652. In 1990, the number of Filipinos in the United States was 1,406,770. What was the change in the Filipino population in the United States from 1980 to 1990? *(Lesson 7-2)*

22. The highest temperature ever recorded in the United States was 134°F, in Death Valley, California, on July 10, 1913. The lowest temperature ever recorded in the U.S. was -80°F, in Prospect Creek, Alaska, on January 23, 1971.
a. To the nearest year, how many years separate the two dates?
b. What is the difference between the temperatures? *(Lessons 7-1, 7-2)*

23. Barry had $312 in his checking account and made out a check for $400. By how much was he overdrawn? *(Lessons 7-1, 7-2)*

24. A person was born in this century in the year *B* and died in the year *D*. What are the possible ages of this person at the time of death? (Use special cases if you cannot answer quickly.) *(Lessons 6-6, 7-1)*

25. *B* is on $\overline{AC}$. If *AC* = 1 meter and *AB* = 30 cm, what is *BC*? *(Lessons 3-4, 7-1)*

26. a. Find *x* if -*x* = -17.   **b.** Find *x* if -*x* = 3.   *(Lesson 5-2)*

27. Measure ∠*V* to the nearest degree.   *(Lesson 3-5)*

In 28–33, rewrite as a decimal.

28. sixty-four millionths   *(Lesson 2-8)*   **29.** 150%   *(Lesson 2-5)*

30. 6.34×10^6 *(Lesson 2-3)*   **31.** five trillion   *(Lesson 2-1)*

32. $\frac{6}{11}$ *(Lesson 1-8)*   **33.** $\frac{3}{4}$ *(Lesson 1-8)*

34. On personal checks, amounts must be written in English words. Write $4009 as you would need to for a personal check.   *(Lesson 1-1)*

Exploration

35. a. Replace each letter with a digit to make a true subtraction. Different letters stand for different digits. Each letter stands for the same digit wherever it occurs.

$$\begin{array}{r} \text{TWO} \\ -\text{ONE} \\ \hline \text{ONE} \end{array}$$

b. There is more than one solution. Find as many solutions as you can.

This person in Kotzebue, Alaska, is ice fishing—a popular sport in areas where heavy freezing occurs.

Solving
$a - x = b$

Georgetown. *Georgetown, a neighborhood in Washington, D.C., has many fine examples of well-preserved, early American architecture. Many of the buildings are 100–200 years old.*

A Situation Leading to $a - x = b$

Here is a portion of the table from the beginning of this chapter.

State	1990 Population	Change Since 1980	State	1990 Population	Change Since 1980
Michigan	9,295,297	33,253	Washington	4,866,692	734,339
Minnesota	4,375,099	299,129	West Virginia	1,793,477	−156,709
Mississippi	2,573,216	52,446	Wisconsin	4,891,769	186,127
Missouri	5,117,073	200,307	Wyoming	453,588	−15,969
Other areas					
D.C.	606,900	−31,532	American		
Puerto Rico	3,522,037	325,517	Samoa	46,773	14,476
Virgin Islands	101,809	6,218	Guam	133,152	27,173

Source: Bureau of the Census, U.S. Dept. of Commerce, 1980 and 1990 Censuses

As you know, from this information you can obtain the 1980 population of each listed state or other area. The general formula comes from the Comparison Model for Subtraction.

$$1990 \text{ population} - 1980 \text{ population} = \text{change}$$

For instance, to find the 1980 population of the District of Columbia (D.C.), let x be the 1980 population. Substitute the known values for the 1990 population and the change.

$$606,900 - x = \text{-}31,532$$

The equation to be solved is of the form $a - x = b$. The numbers a and b in this situation have many digits, so we begin by considering a simpler example. Notice that we use almost the same algorithm we used to solve $x - a = b$ in Lesson 7-3. But because the unknown in $a - x = b$ is subtracted, an extra step is needed.

Example 1

Solve $3 - x = 20$.

Solution 1

Original equation	$3 - x = 20$
Convert to addition.	$3 + -x = 20$
Add -3 to both sides.	$-3 + 3 + -x = -3 + 20$
Simplify.	$-x = 17$
To solve, you need to know x, not $-x$.	
Just take the opposite of both sides.	$-(-x) = -17$
Simplify.	$x = -17$

Check

Substitute in the original equation.
Does $3 - -17 = 20$? Yes.

Solution 2

Another way to solve $3 - x = 20$ is to add x to both sides.

Original equation	$3 - x = 20$
Convert to addition.	$3 + -x = 20$
Add x to both sides.	$3 + -x + x = 20 + x$
Simplify.	$3 = 20 + x$

This type of equation you have solved many times before.

Add -20 to both sides.	$-20 + 3 = -20 + 20 + x$
Simplify.	$-17 = x$

Example 2

Solve the equation $606{,}900 - x = -31{,}532$ to determine the population of the District of Columbia in 1980.

Solution

We use the strategy of Solution 1 from Example 1.

Original equation	$606{,}900 - x = -31{,}532$
Convert to addition.	$606{,}900 + -x = -31{,}532$
Add $-606{,}900$ to each side.	$-606{,}900 + 606{,}900 + -x = -606{,}900 + -31{,}532$
Simplify.	$-x = -638{,}432$
Take the opposite of each side.	$-(-x) = 638{,}432$
Simplify again.	$x = 638{,}432$

The population of the District of Columbia in 1980 was 638,432.

What Are Equivalent Equations?

The equations $x = 638{,}432$ and $606{,}900 - x = -31{,}532$ are **equivalent equations** because they have the same solutions. When you write down the steps to solve an equation, each step should be an equation equivalent to the equations in the previous steps.

Example 3

Multiple choice.   Which equation is not equivalent to the others?
(a) $x + 3 = 8$ (b) $2x = 10$ (c) $-1 = 4 - x$ (d) $x = 5$
(e) $x - 0.4 = 4.96$

Solution
Equations (a), (b), (c), and (d) all have the single solution 5. The solution to equation (e) is 5.36. *Equation (e) is not equivalent to the others.*

Of all equations with solution 5, the simplest is $x = 5$. The idea in solving equations is to find a sentence of the form $x = \underline{\quad}$ that is equivalent to the original equation.

What Are Equivalent Formulas?

Equivalent formulas are like equivalent equations. Here is an example. You have seen the formula $p = s - c$. (Profit on an item equals its selling price minus the cost of obtaining the item.) If the selling price is \$49.95 and the cost is \$30.27, then the profit is \$49.95 − \$30.27, or \$19.68.

If you solve the formula for s, you will get $s = c + p$. (Selling price equals cost plus profit.) The same three numbers work in the formula $s = c + p$.

$$p = s - c \qquad\qquad s = c + p$$
$$19.68 = 49.95 - 30.27 \qquad\qquad 49.95 = 30.27 + 19.68$$

The formulas $p = s - c$ and $s = c + p$ are **equivalent formulas** because the same numbers work in both of them. When you take a formula and solve for a variable in it, you always get an equivalent formula.

QUESTIONS

Covering the Reading

1. Consider the equation $3 - x = 20$. To solve this equation, you can convert the subtraction to an addition. What sentence results?

2. If $-x = 30$, then $x = \underline{\ ?\ }$.

3. If $-y = -\frac{1}{2}$, then $y = \underline{\ ?\ }$.

In 4–6, consider $-5 = 14 - t$.

4. To solve this equation using the algorithm of Example 1, Solution 1, what should be added to both sides?

5. To solve this equation using the algorithm of Example 1, Solution 2, what should be added to both sides?

6. Solve the equation using the algorithm you like better.

In 7–12, solve and check.

7. $300 - x = -2$ **8.** $61 = 180 - y$ **9.** $-45 = 45 - z$

10. $m - 3.3 = 1$ **11.** $A - 57 = -110$ **12.** $\frac{2}{3} - B = \frac{88}{9}$

13. Use the table in this lesson. Solve an equation of the form $a - x = b$ to find the 1980 population of West Virginia.

14. When are two sentences equivalent?

In 15–17, *multiple choice.*   Find the sentence that is not equivalent to the others.

15. (a) $x = 5$ (b) $5 - x = 10$ (c) $x = -5$

16. (a) $y + 1 = 4$ (b) $4 + 1 = y$ (c) $1 + y = 4$

17. (a) $a + \frac{2}{3} = b$ (b) $a - b = \frac{2}{3}$ (c) $b - a = \frac{2}{3}$

18. *Multiple choice.*   Which formula is not equivalent to the others?
  (a) $s = p + c$ (b) $p = s - c$ (c) $p = c - s$

Applying the Mathematics

19. Solve for c: $p = s - c$.

In 20–22, the question can lead to an equation of the form $a - x = b$.
a. Give that equation. **b.** Solve the equation. **c.** Answer the question.

20. There were 3500 tickets available for a concert. Only 212 are left. How many tickets have been sold?

21. The temperature was $14°$ just 6 hours ago. Now it is $3°$ below zero. How much has it decreased?

22. The Himalayan mountain climbers pitched camp at 22,500 feet yesterday. Today they pitched camp at 20,250 feet. How far did they come down the mountain?

In 23–26, solve. You will have to simplify first.

23. $40 - x + 20 = 180$ **24.** $-6 = -1 - y - 5$

25. $13 - 5 \cdot 2 = 9 - K - -7$ **26.** $\frac{6}{5} = \frac{2}{3} - -A$

27. *Multiple choice.*   Which sentence is equivalent to $A = \ell w$? (Hint: If you do not know the answer quickly, test a special case.)
  (a) $A\ell = w$ (b) $Aw = \ell$
  (c) $\frac{A}{\ell} = w$ (d) none of these

28. *Multiple choice.*   Which of (a) to (c) is not equal to $-b + a$?
  (a) $a + -b$ (b) $a - b$
  (c) $b + -a$ (d) All equal $-b + a$.

29. Give a reason for each step in this detailed solution of $a - x = b$ for x.

Step 1: $\qquad a + \text{-}x = b$
Step 2: $(a + \text{-}x) + x = b + x$
Step 3: $a + (\text{-}x + x) = b + x$
Step 4: $\qquad a + 0 = b + x$
Step 5: $\qquad\quad a = b + x$
Step 6: $\quad \text{-}b + a = \text{-}b + (b + x)$
Step 7: $\quad \text{-}b + a = (\text{-}b + b) + x$
Step 8: $\quad \text{-}b + a = 0 + x$
Step 9: $\quad \text{-}b + a = x$

Review

30. Solve $x - 11 = \text{-}11$. *(Lesson 7-3)*

31. Solve $8 = y - 40$. *(Lesson 7-3)*

32. In Montreal, Quebec, the average high temperature in January is -6°C. The average low temperature is -15°C. What is the average difference in the temperatures on a January day in Montreal? *(Lesson 7-2)*

33. Booker T. Washington was born in 1856 and died in 1915. What was his age when he died? *(Lesson 7-1)*

34. An *n*-gon has exactly 77 diagonals. What is *n*? *(Lesson 6-5)*

35. Solve for *d* in terms of *c*: $c + d = 90$. *(Lesson 5-8)*

36. 4 kilograms + 25 grams + 43 milligrams equals how many grams? *(Lessons 3-4, 5-1)*

37. a. Draw two perpendicular lines.
   b. What kind of angles are formed? *(Lesson 3-7)*

Washington and his sons.
Booker T. Washington was a leader and an educator. In 1881, he founded Tuskegee Normal and Industrial Institute, now called Tuskegee University, and served as its principal until 1915.

Exploration

38. Suppose $a = 5$, $b = 4$, $c = 3$, $d = 2$, and $e = 1$. Consider the expression $a - b - c - d - e$, in which all the dashes are for subtraction. Now put grouping symbols wherever you want. For instance, you could consider $(a - b) - (c - d) - e$; this gives the values $(5 - 4) - (3 - 2) - 1$, which simplifies to -1. Or consider $a - [b - (c - d)] - e$, which is $5 - [4 - (3 - 2)] - 1$, which simplifies to 1. How many different values of the expression are possible?

Counting and Probability with Overlap

Getting a kick out of soccer. *Although soccer has been popular around the world for decades, it did not gain popularity in the U.S. until the 1970s. Now it is one of the fastest growing team sports in the country.*

Last fall at Central Middle School, 43 boys and 54 girls played on a soccer team. Since there is no overlap between boys and girls, it is easy to find how many students played soccer. Use the Putting-Together Model for Addition to get $43 + 54 = 97$.

You can draw a picture, called a **Venn diagram,** to illustrate this situation. Separate circles represent quantities with no overlap.

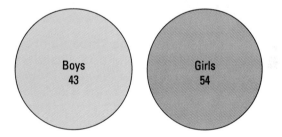

Counting When There Is Overlap

Situations with overlap are more complicated.

Example 1

37 girls from Central Middle School played basketball in the winter. If 9 girls played both soccer and basketball, how many were on at least one of these teams?

Solution

The 54 girls who played soccer overlap with the 37 who played basketball. On the next page, a Venn diagram with intersecting circles for the overlap is drawn to help answer the question.

▶

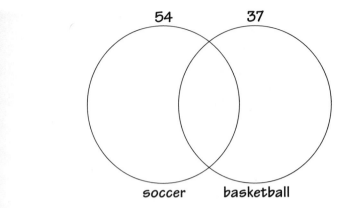

Nine girls belong in the overlap of the circles. Since 54 girls played soccer, 54 − 9 = 45 girls played only soccer. Likewise, 37 − 9 = 28 girls played only basketball. So 45 and 28 are in the remaining parts of the circles.

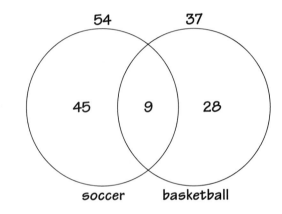

Now there are two ways to answer the question.

1. Add the three numbers: 45 + 9 + 28 = 82.
2. Add the soccer and basketball totals and subtract the overlap because it has been counted twice:
   54 + 37 − 9 = 82.

So 82 girls played at least one of these sports.

The solution to Example 1 illustrates the combination of putting-together with take-away. The following general principle holds for any counts or other measure.

Putting-Together with Overlap Model
If a quantity x is put together with a quantity y, and there is overlap z, the result is the quantity $x + y - z$.

When you know the value of $x + y - z$ and two of the variables in it, you can find the value of the third variable by solving an equation.

Example 2

The Central Middle School spring concert included both the orchestra and the band. The orchestra has 85 musicians and the band has 75. If 128 students played in the concert, how many played in both the band and the orchestra?

Solution

Think: $x = 85$, $y = 75$, and $x + y - z = 128$, where z is the overlap.

Start with the equation.	$x + y - z = 128$
Substitute.	$85 + 75 - z = 128$
Simplify.	$160 + \text{-}z = 128$
Add $\text{-}160$ to both sides.	$160 + \text{-}160 + \text{-}z = 128 + \text{-}160$
Simplify.	$\text{-}z = \text{-}32$
Take the opposite of both sides.	$z = 32$

32 students played in both the orchestra and the band.

Probability when There Is Overlap

When two events overlap, the probability that one of them occurs can be calculated using this model.

Example 3

Use the information given on the preceding page. If there are 175 girls in Central Middle School, what is the probability that a randomly selected girl in the school played soccer or basketball?

Solution 1

From the given information, the probability a girl played soccer is $\frac{54}{175}$; basketball is $\frac{37}{175}$; the probability a girl played both is $\frac{9}{175}$. So the probability a girl played one of the sports can be found using the Putting-Together with Overlap model. It is

$$\frac{54}{175} + \frac{37}{175} - \frac{9}{175} = \frac{82}{175}.$$

The probability a girl played at least one of the sports is $\frac{82}{175}$.

Solution 2

Use the solution to Example 1. 82 girls played at least one of the sports. There are 175 girls in all.

The probability that a girl played soccer or basketball is $\frac{82}{175}$.

All that jazz. *Jazz is a style of improvisational music influenced by early African-American spirituals and folk music, and by ragtime music. New Orleans was the home of the first jazz musicians in the early 1900s.*

Example 4

If you toss a pair of fair dice once, what is the probability that they will show doubles or a sum greater than seven?

Solution

Examine this diagram; it shows the 36 equally likely outcomes from tossing a pair of dice.

Six outcomes, outlined with rectangles, are doubles.

So the probability of doubles is $\frac{6}{36}$.

Fifteen outcomes, outlined with ellipses, have a sum greater than 7. So the probability that the sum is greater than 7 is $\frac{15}{36}$.

Three outcomes, 4 and 4, 5 and 5, and 6 and 6, have doubles and a sum greater than 7. So the probability of doubles and a sum greater than 7 is $\frac{3}{36}$. This is the probability of the overlap.

So, the probability of doubles or a sum greater than 7 is

$$\frac{6}{36} + \frac{15}{36} - \frac{3}{36} = \frac{18}{36} = \frac{1}{2}.$$

Check

Count the outcomes that are outlined with rectangles or ellipses. There are 18 of them, out of the 36.

QUESTIONS

Covering the Reading

1. Give an example of a situation with overlap.

In 2 and 3, draw a Venn diagram to illustrate the situation.

2. During 5th period at South Junior High, 23 seventh graders take English, 20 seventh graders take math, and 24 seventh graders are in science.

3. Samantha is in 5th-period English with 22 other ninth graders. Her 6th-period math class has 25 students with six students, including Samantha, in both these classes.

4. In the expression $x + y - z$ for putting-together with overlap, which variable stands for the number in the overlap?

In 5 and 6, refer to the information at the beginning of the lesson.

5. If 65 boys at Central Middle School play basketball and 20 play both soccer and basketball, draw a Venn diagram to represent the soccer and basketball players.

6. If there are 200 boys at Central Middle School, what is the probability a randomly selected boy plays soccer or basketball?

7. Draw a Venn diagram to check the answer to Example 2.

In 8 and 9, consider the situation of Example 4.

8. List the outcomes that overlap.

9. Find the probability that a pair of dice will show an even sum or a sum greater than eight.

Applying the Mathematics

10. Explain how the putting-together model can be considered to be a special case of putting-together with overlap.

11. The Venn diagram below shows the top five scorers at a school for the first 3 math contests of the year.

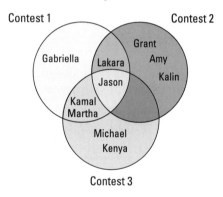

a. List the top 5 scoring students in contest 1.
b. Who finished in the top 5 in at least two contests?
c. Who finished in the top 5 in all three contests?
d. Which two contests had the least overlap?

12. Suppose a school has 600 students. Math teachers gave A's to 20% of the students. English teachers gave A's to 15% of the students. 8% received A's in both math and English.
a. What percent received an A in at least one of the two subjects?
b. How many students received an A in at least one of the two subjects?

13. In the Venn diagram below, if the total number of items in A and B is 37, how many items are in B?

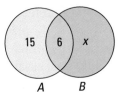

15 6 x

A B

14. In the Venn diagram below, if a total of 35 items are in R, S, and T, how many are in both S and T?

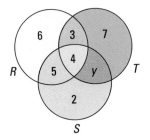

6   3   7

4

R 5 y T

2

S

15. A whole number x such that $1 \le x \le 100$ is chosen at random. Give the probability of each event.
  a. x is divisible by 2.
  b. x is divisible by 3.
  c. x is divisible by 2 or 3.

The Carson House, built in Eureka, CA, between 1884 and 1886 is an example of Victorian architecture. It was so named because the style was popular during Queen Victoria's reign.

Review

In 16 and 17, a situation is given. **a.** Write an equation involving subtraction to answer the question. **b.** Solve the equation. *(Lessons 7-3, 7-4)*

16. The hot air balloon fell 125 feet from its highest point. The new altitude was 630 ft. How high had it flown?

17. In the 1992 Summer Olympics all-around gymnastics competition, a Ukrainian gymnast, Tatiana Gutsu, defeated an American gymnast, Shannon Miller, by twelve hundredths of a point. If Miller's total score was 39.725, what was Gutsu's score?

18. Write a simpler equation equivalent to $40 = A + 12$. *(Lesson 7-4)*

19. a. Which of these famous English monarchs had the longest reign?

King Henry II, who reigned from 1154 to 1189
King Henry VIII, 1509–1547
Queen Elizabeth I, 1558–1603
Queen Victoria, 1837–1901

  b. Queen Elizabeth II ascended to the throne in 1952. In what year could she become the longest reigning English monarch?
  (Lessons 7-1, 7-3)

Queen Victoria

20. Evaluate $3.\overline{3} + 6\frac{2}{3} - \frac{1}{2} - -\frac{5}{6}$. *(Lessons 1-7, 5-5, 7-2)*

21. *Multiple choice.* Which of these numbers could be the sum of two consecutive integers? *(Lessons 6-5, 6-7)*
(a) 1992   (b) 1993   (c) 1994   (d) 2000

22. Use the figure below. If $m\angle BIT = 2°$ and $m\angle GIT = 3°$, what is $m\angle BIG$? *(Lesson 5-4)*

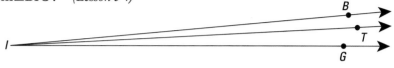

In 23–24, use the graph below. *(Lessons 2-5, 2-7, 3-6)*

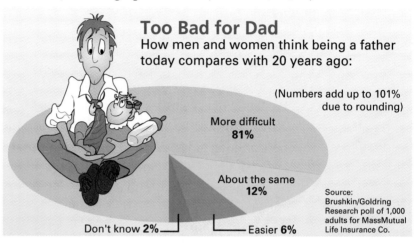

Too Bad for Dad
How men and women think being a father today compares with 20 years ago:

(Numbers add up to 101% due to rounding)

More difficult **81%**

About the same **12%**

Don't know **2%**

Easier **6%**

Source: Brushkin/Goldring Research poll of 1,000 adults for MassMutual Life Insurance Co.

23. Estimate the number of adults who said being a dad today is more difficult.

24. a. Without a protractor, estimate the number of degrees in the central angle for "About the same."
   b. Use a protractor to measure the angle in part **a.** You will have to extend the sides of the angle.
   c. What should the number of degrees in the central angle be?
   d. Is the graph accurate? Justify your answer.

25. Which holds more, a half-gallon watering can or a 2-liter watering can? *(Lesson 3-5)*

Exploration

26. Toss a pair of dice some multiple of 36 times and record the outcomes.
   a. What is your relative frequency of a double?
   b. What is your relative frequency of a sum greater than 7?
   c. What is your relative frequency of a double or a sum greater than 7?
   d. How do your answers for parts **a, b,** and **c** compare to the probability of each outcome?

IN-CLASS

ACTIVITY

In the figure below, $\overrightarrow{BA}$ and $\overrightarrow{BC}$ are called **opposite rays** because they have the same endpoint and together they form a line. Angles DBC and DBA are adjacent angles. Angle ABC is called a **straight angle.**

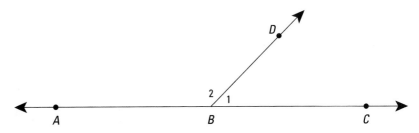

We name the angles 1 and 2 to make reading easier. These angles are called a **linear pair** because they have a common side, $\overrightarrow{BD}$, and their noncommon sides, $\overrightarrow{BA}$ and $\overrightarrow{BC}$, are opposite rays.

1 Measure angles 1 and 2 above and record your results.
m∠1 = _?_   m∠2 = _?_

2 If the ray $\overrightarrow{BE}$ opposite to $\overrightarrow{BD}$ is drawn, angles 3 and 4 are formed. Label ∠ABE angle 3 and ∠CBE angle 4. Measure angles 3 and 4 and record your results.

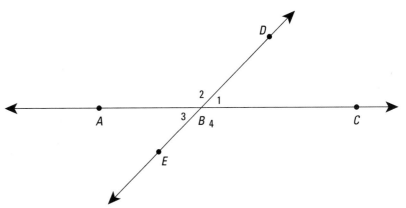

3 Angles 1 and 4 form a linear pair. What is the sum of your measurements of these angles? Angles 2 and 3 form a linear pair. What is the sum of their measures?

4 *Draw a conclusion.* What is the sum of the measures of the angles in a linear pair?

LESSON
7-6

Angles and Lines

Florida bridge. *Note the variety of angles in the Sunshine State Skyway Bridge, located in Tampa Bay, Florida. This cable-stayed bridge, completed in 1987, spans 1200 feet.*

What Are Vertical Angles?

Angles 1 and 3 on the preceding page are called *vertical angles.* Two angles are **vertical angles** when their sides are opposite rays. Vertical angles always have the same measure, so you should have found that $m\angle 1 = m\angle 3$. Angles 2 and 4 are vertical angles, so you should have found that $m\angle 2 = m\angle 4$.

In general, when two lines intersect, four angles are formed. Two of these have one measure, call it x. The other two each have measure y. The sum of x and y is 180°. That is, $x + y = 180$.

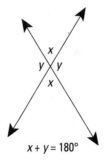

$$x + y = 180°$$

All of this is related to subtraction. We solve the equation $x + y = 180$ for y.

Original equation	$x + y = 180$
Addition Property of Equality	$-x + x + y = 180 + -x$
Property of Opposites	$0 + y = 180 + -x$
Additive Identity Property of Zero	$y = 180 + -x$
Add-Opp Property of Subtraction	$y = 180 - x$

So if you know x, the measure of one angle formed when two lines intersect, the measures of the other three angles are either x or $180° - x$. For instance, if $x = 110°$, then $y = 180° - x = 70°$. In the drawing below, one angle measures 110°. The other measures 70°.

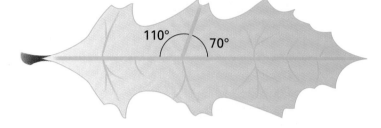

What Are Supplementary Angles?

Two lines intersecting is a common situation. So angles whose measures add to 180° appear often. Such angles are called **supplementary angles.** Supplementary angles do not have to form a linear pair. For instance, any two angles with measures 53° and 127° are supplementary no matter where the angles are located because $53 + 127 = 180$.

When two lines intersect to form right angles as in the diagram on the left, all four angles have measure 90°. So any two of the angles are supplementary. Recall that the lines are called perpendicular.

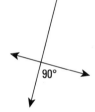

A Situation Leading to Perpendicular Segments

Right angles often occur where streets cross. Here is a sketch of part of a map of Chicago. Drexel and Ellis Avenues go north and south from 71st to 72nd streets. Then they bend so that they intersect South Chicago Avenue at right angles.

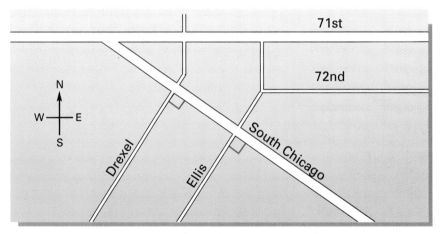

Suppose Drexel did not bend. Then the situation would be as pictured at the top of the next page. A driver going south on Drexel could easily see a car traveling northwest (from A to B) on South Chicago. But the driver could not easily see a car going southeast (from B to A) on South Chicago. Because the streets are perpendicular, a driver can see both directions equally well and the intersection is safer.

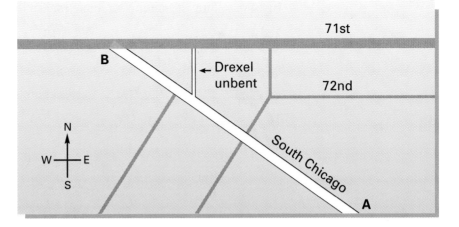

QUESTIONS

Covering the Reading

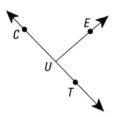

In 1–7, refer to the drawing at the left. *C, U,* and *T* are points on the same line.

1. Angle *CUT* is sometimes called a __?__ angle.

2. $\overrightarrow{UC}$ and $\overrightarrow{UT}$ are __?__ rays.

3. ∠*CUE* and ∠*EUT* form a __?__.

4. m∠*CUE* + m∠*EUT* = __?__ degrees.

5. If m∠*CUE* = 88°, then m∠*EUT* = __?__.

6. If m∠*CUE* = 90°, then $\overleftrightarrow{CT}$ and $\overleftrightarrow{UE}$ are __?__.

7. Angles *CUE* and *EUT* are __?__ angles.

8. In the In-class Activity on page 374, what did you find for the measures of angles 1, 2, 3, and 4?

In 9–13, $\overleftrightarrow{AB}$ and $\overleftrightarrow{CD}$ intersect at *E.*

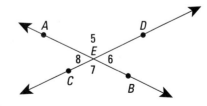

9. Name all pairs of vertical angles.

10. Name all linear pairs.

11. Name all pairs of supplementary angles.

12. If m∠5 = 125°, what are the measures of the other angles?

13. Measure angles 5, 6, 7, and 8 (to the nearest degree) with a protractor.

In 14–16, find the unknown angle measures without using a protractor.

14.

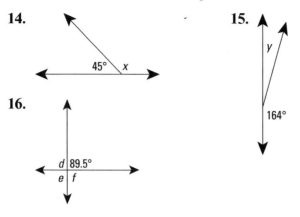

15.

16.

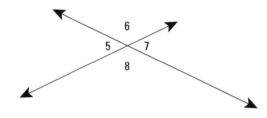

17. Two lines intersect below. If m∠5 = *t*, find the measures of the other three angles.

In 18–22, use the map below. Assume all the streets are two-way.

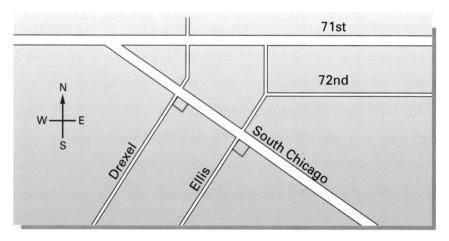

18. *True or false.* Drexel and South Chicago intersect at right angles.

19. *True or false.* Ellis and 71st intersect at right angles.

20. How many street intersections are pictured on the map?

21. At which intersection is the best visibility of oncoming traffic?

22. At which intersection is the worst visibility of oncoming traffic?

23. Entrance ramps onto expressways usually form very small angles with the expressway. Perpendicular ramps would allow better visibility. Explain why entrance ramps are not perpendicular to expressways.

In 24 and 25, name all linear pairs shown in the figure.

24.

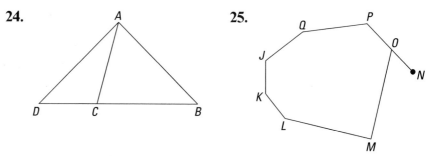

25.

26. If one angle of a linear pair is an acute angle, then the other angle is obtuse.

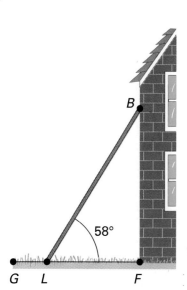

In 26 and 27, a statement is given.
a. Tell whether the statement is true or false.
b. Explain why you answered part **a** as you did.

26. If one angle of a linear pair is an acute angle, then the other angle is obtuse.

27. If one of two vertical angles is acute, the other is obtuse.

28. A firefighter places a ladder against a building so that m∠*BLF* = 58°, as in the diagram at left. What is m∠*BLG*?

In 29 and 30, suppose *x* and *y* are measures of two angles. How are *x* and *y* related in the given situation?

29. The angles are vertical angles.

30. The angles are supplementary.

31. An angle has measure 40°. A supplement to this angle has what measure?

32. In the figure below, $\overrightarrow{UO}$ and $\overrightarrow{UT}$ are opposite rays. Find the measure of ∠*RUN*.

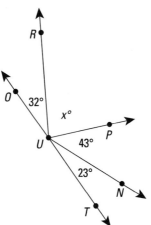

33. In Casimir Ewell's 5th-period math class, every student wore blue jeans or sneakers. 13 students wore blue jeans. 16 students wore sneakers. 5 students wore both.
   a. Draw a Venn diagram to illustrate this situation.
   b. How many students are in Mr. Ewell's class? *(Lesson 7-5)*

In 34–37, solve.

34. $180 - x = 23$ *(Lesson 7-4)*

35. $90 - y = 31$ *(Lesson 7-4)*

36. $17 - (2 - 6) + A = 5 - 2(5 + 6)$ *(Lessons 5-8, 7-2)*

37. $4c = 1200$ *(Lesson 4-9)*

38. Suppose the sign in the cartoon is true. How much would you pay for a down jacket normally selling for $129.95? *(Lesson 2-5)*

HERMAN®

© 1984 Universal Press Syndicate 8-1

"Salesman of the week gets to go to Hawaii."

39. In this lesson you learned about supplementary angles. What are **complementary** angles? Draw at least two examples.

Train lines. *At a train depot, many lines meet.*

Angles and Intersecting Lines

Two intersecting lines form four angles, as pictured below. Suppose ∠1 has measure x. Then angle 3, vertical to ∠1, has measure x. The other angles have measure $180 - x$.

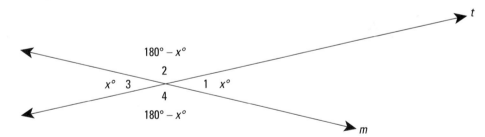

For example, if $x = 25$, then the original angle has measure 25°. The other angles measure either 25° or $180° - 25°$, which is 155°.

Angles Formed by Parallel Lines and a Transversal

Two different lines in a plane are **parallel** if they have no points in common. At the top of the next page, we draw a line n parallel to line m. This can be done easily by aligning line m with one edge of a ruler and drawing line n along the opposite edge. Line n intersects line t to form angles 5, 6, 7, and 8. Because parallel lines go in the same direction, the measures of the four new angles equal the measures of the four angles that were already there.

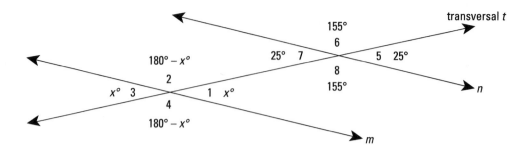

The line that is not parallel to the others is called a **transversal.** The eight angles have names that tell how they are related to the two parallel lines and the transversal. The interior angles are between the parallel lines. The exterior angles are outside the parallel lines. Alternate interior or alternate exterior angles are on opposite sides of the transversal.

Name	Angles in above figure	Measures
Corresponding angles	1 and 5, 2 and 6, 3 and 7, 4 and 8	equal
Interior angles	1, 2, 7, 8	
Alternate interior angles	2 and 8, 1 and 7	equal
Exterior angles	3, 4, 5, 6	
Alternate exterior angles	3 and 5, 4 and 6	equal

Symbols for Angles

In work with angles, it helps to use special symbols to indicate when angles have the same measure. Use a single arc when two angles have the same measure. For a second pair of equal angles, use a double arc, and so on.

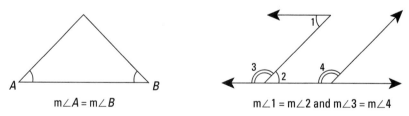

Symbols for Parallelism

Arrows in the middle of two lines that look parallel mean that they are parallel. If lines m and n are parallel, you can write $m // n$. The symbol $//$ means *is parallel to*. The drawing at the top of this page is repeated at the top of the next page. The symbols for parallel lines and for angles of equal measure are included. Also identified are the interior and exterior angles.

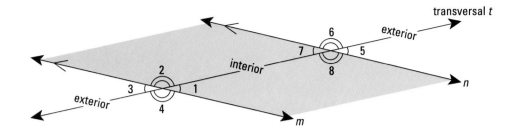

Symbols for Perpendicularity

There are also symbols for perpendicularity. In a drawing, the symbol ⌐
shows that lines or line segments are perpendicular. In writing, ⊥ means
is perpendicular to.

In drawings, put the ⌐ symbol on only one of the angles at the point of
intersection.

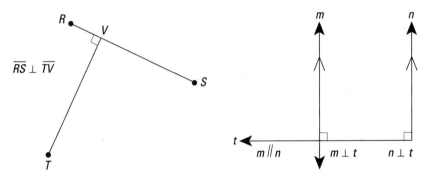

Notice how these symbols shorten the amount of writing. In the figure
above at the right, each parallelism and perpendicularity is indicated
twice.

QUESTIONS

Covering the Reading

1. When are two lines parallel?

2. Draw two parallel vertical lines and a transversal.

In 3–8, *m* // *n*. Give the measure of
the indicated angle.

3. ∠1   4. ∠2

5. ∠3   6. ∠4

7. ∠5   8. ∠6

9. In the drawing of Questions 3–8,
   which line is the transversal?

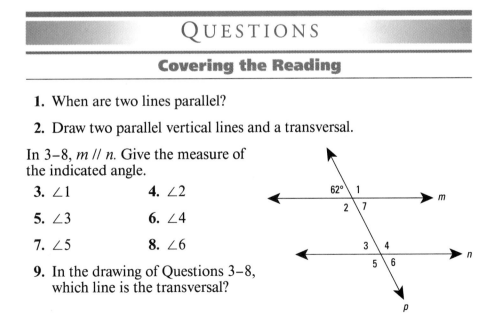

In 10–18, use the drawing below. Lines *r* and *s* are parallel.

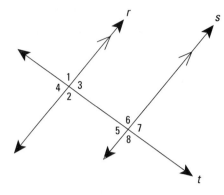

10. Name all pairs of corresponding angles.

11. Name the interior angles.

12. a. Explain why ∠3 and ∠5 are alternate interior angles.
b. Name the other pair of alternate interior angles.

13. Name two pairs of alternate exterior angles.

14. Do the corresponding angles 1 and 6 have the same measure?

15. Do the alternate interior angles 3 and 5 have the same measure?

16. If m∠3 = 84°, give the measure of all other angles.

17. If m∠5 = *x*, give the measure of all other angles.

18. If ∠6 is a right angle, which other angles are right angles?

In 19–22, use the drawing below. Assume segments that look perpendicular are perpendicular, and segments that look parallel are parallel.

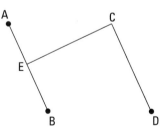

19. Copy the drawing and put in the symbols indicating parallel and perpendicular segments.

20. What symbol completes the statement? $\overline{AB}$ _?_ $\overline{CE}$.

21. What symbol completes the statement? $\overline{AB}$ _?_ $\overline{CD}$.

22. Name all right angles.

In 23 and 24, find the measure of each numbered angle.

23. **24.**

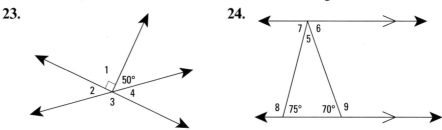

25. Find the sum of the measures of angles 1, 2, 3, and 4 in the drawing below.

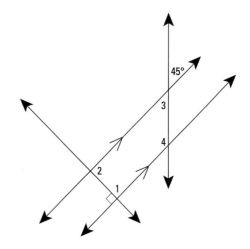

26. Line *m* is parallel to line *n*. Line *m* is perpendicular to line *t*.
   a. Make a drawing of this situation.
   b. Describe how *n* and *t* are related.

In 27–29, draw an accurate example and identify on your drawing.

27. a linear pair **28.** vertical angles

29. supplementary angles that are not a linear pair *(Lesson 7-6)*

30. Along I-70 in Kansas, it is 179 miles from Salina to Kansas City, 110 miles from Salina to Topeka, and 150 miles from Abilene to Kansas City. How far is it from Abilene to Topeka? *(Lesson 7-5)*

Kansas became a state and adopted its official seal in 1861. The 34 stars on the seal indicate Kansas was the 34th state. The farmer and cabin symbolize the anticipated prosperity through agriculture.

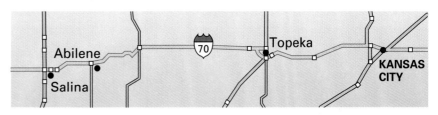

31. Solve $m - \frac{8}{3} = {}^-3$.   *(Lesson 7-3)*

32. Subtract $2\frac{6}{7}$ from $5\frac{2}{3}$.   *(Lesson 7-1)*

33. a. What is the area of the shaded region between the rectangles below?
   b. Describe at least two real situations where one might have to solve a problem similar to part **a.**   *(Lesson 7-1)*

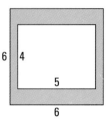

6 | 4
5
6

34. Solve for c: $a + b + c = 180$.   *(Lesson 5-8)*

35. A savings account contains C dollars. How much is in the account after the indicated transactions?
   a. $100 is withdrawn, and $30.45 is deposited.
   b. W dollars are withdrawn (from the original C dollars), and D dollars are deposited.   *(Lesson 5-3)*

36. Three instances of a pattern are given. Describe the pattern using two variables.   *(Lesson 4-2)*
$$\frac{73 \cdot 5}{5} = 73 \qquad \frac{8/7 \cdot 6}{6} = \frac{8}{7} \qquad \frac{{}^-4.02 \cdot 43}{43} = {}^-4.02$$

Exploration

37. The photograph below pictures balusters (the parallel vertical supports) and a banister (the transversal) on stairs. Give two other places where it is common to find parallel lines and transversals.

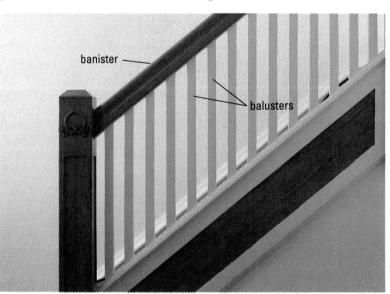

banister —
balusters

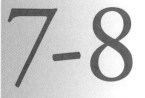

A field of quadrilaterals. *The rectangular panels shown above are part of the AIDS Quilt, a hand-sewn tribute to the tens of thousands of AIDS victims. The quilt was displayed in Washington, D.C., to raise funds for AIDS-related services.*

Here is a drawing of two parallel lines *m* and *n* and a transversal *t*. The pattern of angle measures for vertical angles and linear pairs is shown. Other patterns become clear as more parallel lines are added.

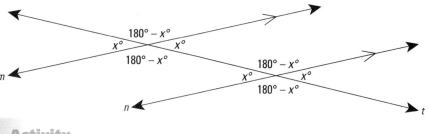

Activity

Step 1
Draw two parallel lines and a transversal as shown above.

Step 2
Choose and label any one angle as $x°$. Label its vertical angle $x°$, and the remaining two angles of the linear pairs $180° - x°$. Label the remaining four angles, using either alternate interior or corresponding angles.

Step 3
Draw another line parallel to the transversal. Label the newly formed angles appropriately. How many angles are now labeled?

Step 4
Trace the interior quadrilateral formed by the lines. Label the four vertices *T, E, A,* and *M* in consecutive order.

Step 5
Measure each angle of *TEAM* to the nearest degree and record your results. Measure the length of each side to the nearest millimeter and record your results.

What Is a Parallelogram?

Quadrilateral *TEAM* that you traced in step 4 of the Activity is a *parallelogram.* A **parallelogram** is a quadrilateral with two pairs of parallel sides. Our parallelogram *TEAM* is shown below. In parallelogram *TEAM,* $\overline{TE}$ and $\overline{MA}$ are called **opposite sides.** $\overline{TM}$ and $\overline{EA}$ are also opposite sides. Angles *A* and *T* are **opposite angles.** Angles *E* and *M* are also opposite angles.

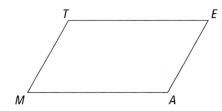

Properties of Parallelograms

From the parallel lines, we know that m∠*A* = m∠*T* and m∠*E* = m∠*M*. So opposite angles have the same measure. Also, opposite sides have the same length. So *TE* = *MA* and *TM* = *EA*. This is true in any parallelogram.

> **Properties of Parallelograms**
> In a parallelogram, opposite sides have the same length; opposite angles have the same measure.

For example, *ABCD* is a parallelogram with side and angle measures indicated. The angle measures have been rounded to the nearest degree. The side lengths are to the nearest millimeter.

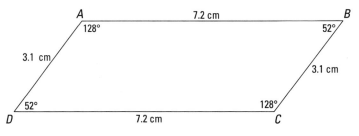

Special Types of Parallelograms

Many common figures are parallelograms. Among these are the rectangle, rhombus, and square. A **rectangle** is a parallelogram with a right angle. All angles of a rectangle have the same measure—they are all right angles. A **rhombus,** or **diamond,** is a parallelogram with all sides the same length. A **square** is a rectangle with the same length and width. Therefore, you can also think of a square as a special type of rhombus. The diagram at the top of the next page summarizes these relationships.

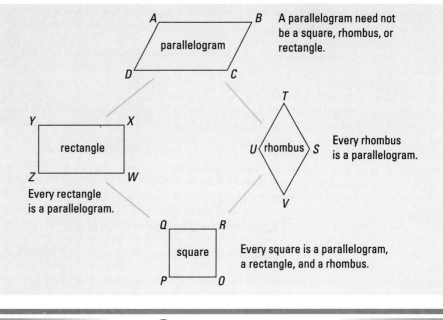

A parallelogram need not be a square, rhombus, or rectangle.

Every rectangle is a parallelogram.

Every rhombus is a parallelogram.

Every square is a parallelogram, a rectangle, and a rhombus.

QUESTIONS

Covering the Reading

1. If two parallel lines intersect one other line, how many angles are then formed?

2. Suppose two parallel lines intersect two other parallel lines.
   a. How many angles are formed?
   b. If one of the angles has measure x, then all angles have either measure __?__ or __?__.
   c. If one of the angles has measure 30°, then all angles have either measure __?__ or __?__.

In Questions 3–5, refer to parallelogram *TEAM* that you traced in step 4 of the Activity in this lesson.

3. To the nearest degree, what is the measure of each angle in *TEAM*?

4. To the nearest millimeter, what is the measure of each side of *TEAM*?

5. Does *TEAM* appear to be a rectangle, rhombus, or square?

In 6 and 7, refer to quadrilateral *ABCD*.

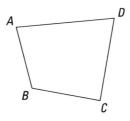

6. Name a pair of opposite sides.   7. Name a pair of opposite angles.

In 8–11, refer to parallelogram *MNPQ*.

8. ∠*N* and _?_ have the same measure.

9. $\overline{PQ}$ and _?_ have the same length.

10. If m∠*Q* = 70°, then m∠*M* = _?_.

11. If *PQ* = 5 and *PN* = 2.8, then *MN* = _?_.

12. What must a parallelogram have in order to be a rhombus?

In 13–16, consider the figures *A* through *H*.

Building blocks to a child's development. *Toys often are designed to help children develop their coordination or to increase their awareness of geometric shapes.*

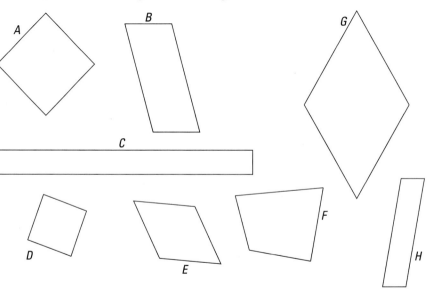

13. Which seem to be rectangles?

14. Which seem to be rhombuses?

15. Which seem to be squares?

16. Which seem to be parallelograms?

In 17–20, consider quadrilaterals, parallelograms, rhombuses, rectangles, and squares.

17. Which have all four sides equal in length?

18. Which have all four angles equal in measure?

19. Which have both pairs of opposite angles equal in measure?

20. Which have both pairs of opposite sides equal in length?

21. In a collection of 50 parallelograms, 35 are rectangles and 26 are rhombuses. Every figure is a rectangle or rhombus or both. How many squares are in the collection?

In 22 and 23, give the perimeter of each figure.

22. a rhombus with one side having length 10

23. a parallelogram with one side having length 10

24. One angle of a rhombus has measure 35°.
 a. Draw such a rhombus.
 b. What are the measures of its other angles?

25. In the figure below, *ABCD* and *AEFG* are parallelograms and $\overline{AB}$ and $\overline{AE}$ are perpendicular. m∠*ABC* = 147°. *G, A, D,* and *H* are on the same line. Find the measure of each indicated angle.
 a. ∠*BAG*   **b.** ∠*GAE*   **c.** ∠*AGF*   **d.** ∠*CDH*

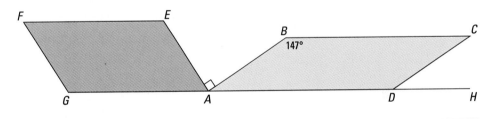

26. If $x + y = 180$, then $y =$ __?__.   *(Lesson 7-6)*

In 27–31, use the figure at the left.   *(Lesson 7-6)*

27. Angles *CBG* and __?__ form a linear pair.

28. Angles *ABD* and __?__ are vertical angles.

29. Suppose m∠*FBC* = 30°.
 a. What other angle measures can be found?
 b. Find them.

30. Suppose m∠*GBA* = 126°.
 a. What other angle measures can be found?
 b. Find them.

31. $\overline{BG}$ __?__ $\overline{DF}$

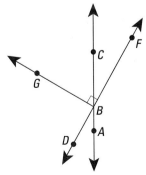

32. If you toss a pair of dice once, what is the probability of getting either doubles or a sum of seven?   *(Lesson 7-5)*

33. Here are census figures for Milwaukee, Wisconsin.

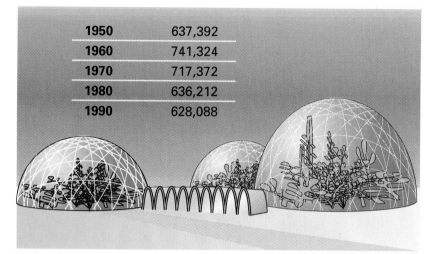

1950	637,392
1960	741,324
1970	717,372
1980	636,212
1990	628,088

Mitchell Conservatory, Milwaukee, Wisconsin

a. Calculate the change in population from each census to the next. You should get one positive and three negative numbers.
b. Add the four changes you find in part **a.**
c. What does the sum in part **b** mean?   *(Lessons 7-2, 7-4)*

Exploration

34. Rectangles are easy to find. Most windows, chalkboards, and doors are shaped like rectangles. Give an everyday example of something shaped like each figure.
a. a square
b. a parallelogram that is not a rectangle
c. a rhombus that is not a square

Kayo has been studying triangles. Read this cartoon to see what he considers awesome.

Activity

Draw a triangle on a large sheet of paper. Repeat what Kayo's grandfather did. Do you get the same results?

The famous property of triangles illustrated by the cartoon and the activity is the *Triangle-Sum Property.*

> **Triangle-Sum Property**
> In any triangle, the sum of the measures of the angles is 180°.

Finding the Measure of the Third Angle of a Triangle

Example 1

Suppose two angles of a triangle have measures 57° and 85°. What is the measure of the third angle?

Solution

First draw a picture. Then let x be the measure of the third angle.

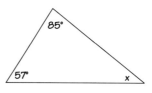

Use the Triangle-Sum Property.
$$57° + 85° + x = 180°$$
$$142° + x = 180°$$

You know how to solve this equation.
$$-142 + 142 + x = -142 + 180$$
$$x = 38$$
So the third angle has measure 38°.

Check

Add the measures. Does 38 + 57 + 85 = 180? Yes.

You have now studied many different ways to find measures of angles. Examples 2 and 3 use some of these ways.

Example 2

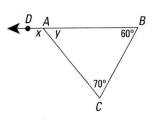

Find the measure of angle *DAC* in the drawing at the left.

Solution

Let x be the measure of ∠*DAC*. Let y be the measure of ∠*BAC*.
Use the Triangle-Sum Property to find y.

$$60° + 70° + y = 180°$$

Simplify. $\quad\quad\quad\quad\quad\quad 130° + y = 180°$

Solve for y. $\quad\quad -130° + 130° + y = -130° + 180°$

$$y = 50°$$

Now use the fact that angles *DAC* and *BAC* form a linear pair.

$$x + y = 180°$$

Substitute for y. $\quad\quad x + 50° = 180°$

Solve for x. $\quad\quad\quad\quad x = 130$

So m∠DAC = 130°.

Finding the Measures of Angles Given Triangles and Parallel Lines

Example 3

Use the information given in the drawing. Explain the steps needed to find m∠9.

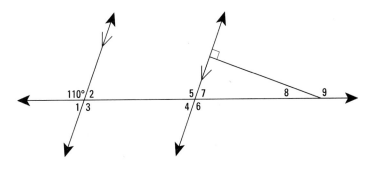

Solution

The drawing is complicated. Examine it carefully. You may wish to make an identical drawing and write in angle measures as they are found.

The arrows indicate that two of the lines are parallel. Angle 5 and the 110° angle are corresponding angles. So m∠5 = 110°. ∠7 forms a linear pair with ∠5, so they are supplementary. Thus m∠7 = 70°. The triangle has a 90° angle created by the perpendicular lines. But it also has ∠7, a 70° angle. The Triangle-Sum Property says that the measures of the three angles add to 180°. This forces m∠8 to be 20°. Finally, ∠9 and ∠8 form a linear pair, so m∠9 = 160°.

Covering the Reading

1. In the Activity, did you get the same results as Kayo's grandfather?

2. *Multiple choice.* Which is true?
(a) In some but not all triangles, the sum of the measures of the angles is 180°.
(b) In all triangles, the sum of the measures of the angles is 180°.
(c) The sum of the measures of the angles of a triangle can be any number from 180° to 360°.

3. What is the Triangle-Sum Property?

4. Why did the corners of the triangles in the cartoon fit together to make a straight line?

In 5–8, two angles of a triangle have the given measures. Find the measure of the third angle.

5. $30°, 60°$ **6.** $117°, 62°$ **7.** $1°, 2°$ **8.** $x°, 140° - x°$

Applying the Mathematics

9. Find $m\angle ABC$.

10. Find $m\angle XYZ$.

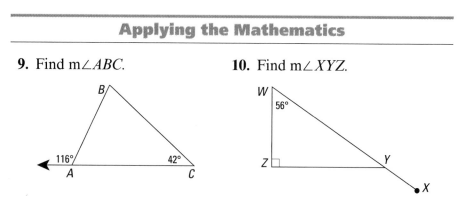

11. Find the measures of angles 1 through 8.

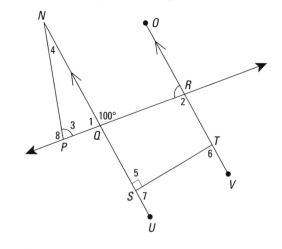

In 12 and 13, *multiple choice.* Use these choices.
  (a) complementary
  (b) supplementary
  (c) neither complementary nor supplementary

12. Which describes angles 5 and 7 of Question 11?

13. Which describes $\angle W$ and $\angle ZYW$ of Question 10?

14. What is the sum of the measures of the four angles of a rectangle?

15. Explain why a triangle cannot have two obtuse angles.

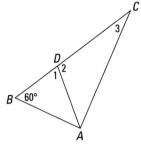

16. In the figure at the left, $\overline{BA} \perp \overline{AC}$. Angle BAC is bisected (split into two equal parts) by $\overline{AD}$. $m\angle B = 60°$. Find the measures of angles 1, 2, and 3.

17. *Multiple choice.* Two angles of a triangle have measures x and y. The third angle must have what measure?
  (a) $180 - x + y$ (b) $180 + x + y$
  (c) $180 + x - y$ (d) $180 - x - y$

18. In the figure at the left, $\overline{EH}$ and $\overline{GI}$ intersect at F. Find the measures of as many angles in the figure as you can.

Review

19. a. *True or false.* A square is a special type of rectangle.
  b. *True or false.* A square is a special type of rhombus.   *(Lesson 7-8)*

In 20–23, use the drawing below. What is the meaning of each symbol? *(Lesson 7-7)*

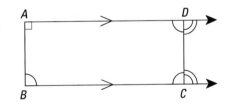

20. the arrows on segments $\overline{AD}$ and $\overline{BC}$

21. the sign that looks like a backward L by angle A

22. the single arcs by points B and C

23. the double arcs by points C and D

In 24 and 25, a mathematical sentence is given. **a.** Translate the sentence into an English sentence. **b.** Draw a picture.   *(Lesson 7-7)*
24. $\overline{AB} \mathbin{/\mkern-5mu/} \overrightarrow{CD}$ **25.** $\overleftrightarrow{EF} \perp \overline{GH}$

26. Solve $-5 = 15.4 - x$.   *(Lesson 7-4)*

27. Solve $x - 3 = -3$.   *(Lesson 7-3)*

28. Four people went to a health club. Their weights on February 1 and March 1 are shown below. **a.** How much did each person's weight change? **b.** Who gained the most? **c.** Who lost the most? *(Lesson 7-1)*

	February 1	March 1
Richard	65.3 kg	62.8 kg
Marlene	53.4 kg	54.3 kg
Evelyn	58.6 kg	55.1 kg
Daniel	71.1 kg	72.0 kg

29. Evaluate $a - b + c - d$ when $a = 0$, $b = -10$, $c = -100$, and $d = -1000$. *(Lesson 7-2)*

30. Evaluate $a - b + c - d$ when $a = -43$, $b = 2$, $c = 5$, and $d = 11$. *(Lesson 7-2)*

Exploration

31. Here is a quadrilateral that is not a parallelogram. **a.** Measure the four angles. **b.** What is the sum of the measures?

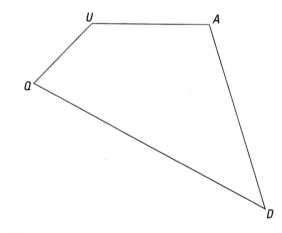

32. *Multiple choice.* What is the sum of the measures of the angles of an n-gon?
(a) $180n$ (b) $180(n + 1)$ (c) $180(n - 2)$ (d) $180(n + 3)$

33. Use computer software that can draw and measure to draw any size or shape of triangle. Then, use the measure tool to measure each of the angles of the triangle and find the sum of the angles.

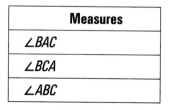

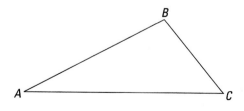

Measures
∠BAC
∠BCA
∠ABC

A project presents an opportunity for you to extend your knowledge of a topic related to the material of this chapter. You should allow more time for a project than you do for typical homework questions.

1 Population Changes

Obtain census information from the last two censuses for at least 25 towns and communities in your geographic area. Make a table like the one at the beginning of the chapter. Identify the places that are growing the fastest.

2 Probabilities with Playing Cards

Find a standard playing card deck (bridge deck) of 52 cards. Draw an array of the 52 cards similar to that found for dice in Example 4 of Lesson 7-5. Suppose a card is picked at random. Give the probability of each of the following events and explain how you found each answer.

a. The card is an ace.
b. The card is a diamond.
c. The card is the nine of hearts.
d. The card is a nine or a heart.
Make up three other questions of your own and answer them.

3 Angles of Chairs

To measure the angle between a chair leg and the floor, take the smallest of all the angles between the chair leg and lines on the floor drawn through the point of contact with the chair leg. Find ten chairs of different styles and measure the angles between their front legs and the floor. Then measure the angles between their back legs and the floor. Summarize your results.

4 A Large Triangle

What is the sum of the measures of the angles in a very large triangle? Pick three points outdoors that are at least 20 meters apart. These points will be the vertices of your triangle. Sight each point from the other two, and, as accurately as you can, measure the angles of the triangle determined by the three points. Add the measures to see what you get. Then repeat this process with another very large triangle.

5 Automatic Drawing Programs

There is computer software that will draw geometric figures of various kinds. Among this software are *GeoExplorer, Cabri, The Geometer's Sketchpad,* and *The Geometric Supposer.* Learn how to use this software to draw the various special quadrilaterals of Lesson 7-8. Print out examples of what you have drawn, and explain what you had to do in order to instruct the computer to draw the figures.

6 When $x - y$ Is Small

Suppose $x - y = z$ and z is a small positive fraction, say $z = \frac{1}{100}$. What fractions could x and y be?

a. Find a pair of fractions (in lowest terms) whose denominators are not 100, but whose difference is $\frac{1}{100}$.

b. Then find a pair of fractions (in lowest terms) whose denominators are not 1000, but whose difference is $\frac{1}{1000}$.

c. Then find a pair of fractions (in lowest terms) whose denominators are not 10,000, but whose difference is $\frac{1}{10,000}$.

d. Explain how you could continue this process and get fractions that are closer to each other than any number someone might give.

SUMMARY

Subtraction can arise from take-away, slide, comparison, or overlap situations. In take-away situations, $x - y$ stands for the amount left after y has been taken away from x. In the linear pair below, $m\angle BDA$ can be thought of as the amount left after x is taken from 180°. So $m\angle BDA = 180° - x°$.
Measures of linear pairs add to 180°. The angles are supplementary.

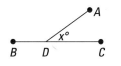

In comparison situations, $x - y$ is how much more x is than y. The word *difference* for the answer comes from this kind of subtraction.

In slide situations, $x - y$ is the result after x has been decreased by y. Earlier this was described as an addition slide situation $x + -y$. Consequently $x - y = x + -y$ always.

Overlap situations arise when two sets have elements in common. This can be pictured by a Venn diagram.

If a quantity x is put together with a quantity y with overlap z, the result is the quantity $x + y - z$.

All of these situations can lead to equations that simplify to $x - a = b$ or $a - x = b$.

Another situation that can lead to equations involving subtraction is the Triangle-Sum Property: In any triangle, the sum of the measures of the angles is 180°.

Two intersecting lines, like $\overleftrightarrow{AB}$ and $\overleftrightarrow{AD}$ below, form two pairs of vertical angles. Adding the line $\overleftrightarrow{DC}$ parallel to $\overleftrightarrow{AB}$ forms eight angles. Adding the line $\overleftrightarrow{BC}$ parallel to $\overleftrightarrow{AD}$ forms the parallelogram $ABCD$. Each of the angles formed has measure $x°$ or $180° - x°$. Special kinds of parallelograms are rectangles, rhombuses, and squares, so these ideas have many uses.

VOCABULARY

You should be able to give a general description and a specific example of each of the following ideas.

Lesson 7-1
Take-Away Model for
  Subtraction
minuend, subtrahend,
  difference
Comparison Model for
  Subtraction

Lesson 7-2
Slide Model for Subtraction
Algebraic Definition (Add-Opp
  Property) of Subtraction

Lesson 7-4
equivalent equations
equivalent formulas

Lesson 7-5
Venn diagram
Putting-Together with Overlap

Lesson 7-6
straight angle, opposite rays
linear pair, vertical angles
supplementary angles
complementary angles

Lesson 7-7
parallel lines, transversal
corresponding angles, //, ⊥
interior angles, exterior angles
alternate interior angles
alternate exterior angles

Lesson 7-8
parallelogram
opposite angles, opposite sides
rectangle, rhombus, square

Lesson 7-9
Triangle-Sum Property

PROGRESS SELF-TEST

Take this test as you would take a test in class. Then check your work with the solutions in the Selected Answers section in the back of the book.

1. Simplify $5 - (-5)$.

2. Picture the subtraction $-6 - 22$ on a number line and give the result.

3. Simplify $\frac{3}{4} - \frac{5}{6}$.

4. Evaluate $5 - x + 2 - y$ when $x = 13$ and $y = -11$.

5. Convert all subtractions to additions: $x - y - -5$.

6. Valleyview H.S. scored V points against Newtown H.S. Newtown scored N points and lost by L points. How are V, N, and L related?

7. Ray is Z inches tall. Fay is 67 inches tall. How much taller is Ray than Fay?

8. The outer square is 8 meters on a side. The inner square is 4 meters on a side. Find the area of the shaded region.

9. After dropping $7°$, the temperature is now $-3°$.
   a. Solving what equation will tell you what the temperature was before it dropped?
   b. Solve that equation.

In 10–12, solve.
10. $y - 14 = -24$
11. $-50 = 37 - x$
12. $g - 3.2 = -2$
13. Solve for a: $c - a = b$.
14. *Multiple choice.* Which formula is not equivalent to the others?
   (a) $180 = x + y$ (b) $180 - y = x$
   (c) $x - y = 180$ (d) $180 - x = y$
15. Suppose 50 people are drinking coffee on an airplane. If 15 drink it black (with nothing added), 25 add cream, and 20 add sugar, how many drink it with both cream and sugar?

In 16 and 17, use the figure at the right, in which $m\angle ABD = 25°$.
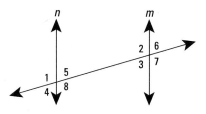
16. $m\angle CBD =$ ___?___
17. $m\angle ABE =$ ___?___

In 18–20, use the figure below. $m \mathbin{/\mkern-3mu/} n$
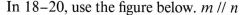

18. If $m\angle 5 = 74°$, then $m\angle 2 =$ ___?___ $°$.
19. Which angles have measures equal to the measure of angle 6?
20. Which angles are supplementary to angle 5?
21. Two angles of a triangle have measures $55°$ and $4°$. What is the measure of the third angle?
22. Why can't a triangle have three right angles?
23. What other angle in the figure below has the same measure as angle E?
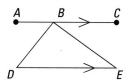

24. In parallelogram $WXYZ$, if $m\angle W = 50°$, what is $m\angle X$?

25. Of parallelograms, rectangles, rhombuses, and quadrilaterals, which have all sides equal in length?

CHAPTER REVIEW

Questions on SPUR Objectives

SPUR stands for **S**kills, **P**roperties, **U**ses, and **R**epresentations. The Chapter Review questions are grouped according to the SPUR Objectives for this chapter.

SKILLS DEAL WITH THE PROCEDURES USED TO GET ANSWERS.

Objective A: *Subtract any numbers written as decimals or fractions.* *(Lesson 7-2)*

In 1–6, simplify.

1. $-14 - 14$
2. $\frac{1}{2} - \frac{5}{2}$
3. $8.6 - 9.3$
4. $-9 - -2$
5. $\frac{2}{3} - (-\frac{4}{5})$
6. $11 - (-10)$
7. Evaluate $x - y - z$ when $x = 10.5$, $y = 3.8$, and $z = -7$.
8. Evaluate $a - (b - c)$ when $a = -2$, $b = -3$, and $c = \frac{1}{4}$.

Objective B: *Solve sentences of the form $x - a = b$ and $a - x = b$.* *(Lessons 7-3, 7-4)*

In 9–16, solve.

9. $x - 64 = 8$
10. $6 = y - \frac{1}{5}$
11. $-4.2 = V - -3$
12. $2 + m - 5 = 4$
13. $200 - b = 3$
14. $-28 = 28 - z$
15. $223 - x = 215$
16. $\frac{4}{9} - y = \frac{5}{18}$
17. Solve for c: $e = c - 45$.
18. Solve for y: $180 - y = x$.

Objective C: *Find measures of angles in figures with linear pairs, vertical angles, or perpendicular lines.* *(Lesson 7-6)*

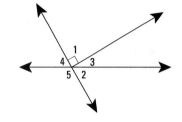

In 19–22, use the figure above to determine the measures.

19. $m\angle 1 = $ __?__ degrees
20. If $m\angle 2 = 60°$, then $m\angle 4 = $ __?__.
21. If $m\angle 4 = x°$, $m\angle 5 = $ __?__.
22. If $m\angle 5 = 125°$, then $m\angle 3 = $ __?__.

Objective D: *Find measures of angles in figures with parallel lines and transversals.* *(Lesson 7-7)*

In 23–26, use the figure below.

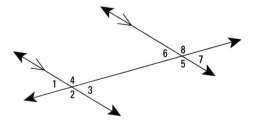

23. Which angles have the same measure as $\angle 6$?
24. If $m\angle 3 = 43°$, then $m\angle 7 = $ __?__.
25. If $m\angle 2 = y°$, then $m\angle 7 = $ __?__.
26. If $m\angle 8 = 135°$, then $m\angle 1 = $ __?__.

Objective E: *Use the Triangle-Sum Property to find measures of angles.* *(Lesson 7-9)*

In 27 and 28, use the figure below.

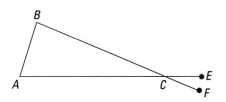

27. If $\overline{AB} \perp \overline{BC}$ and $m\angle ECF = 40°$, find $m\angle A$.
28. If $\overline{AB} \perp \overline{BC}$ and $m\angle A = 72°$, find $m\angle ECB$.
29. Two angles of a triangle measure 118° and 24°. What is the measure of the third angle?

30. Two angles of a triangle measure $y°$ and $150° - y°$. What is the measure of the third angle?

Objective F: *Find measures of angles and sides in special quadrilaterals without measuring.* *(Lesson 7-8)*

31. A rhombus has one side of length 2.5 cm. Is this enough to find its perimeter?
32. Each angle of a rectangle has measure ___?___.
33. In parallelogram *ABCD*, if *DC* = 4 and *BC* = 6, find *AB*.
34. In parallelogram *ABCD*, if $m\angle D = 105°$, find $m\angle A$.

PROPERTIES DEAL WITH THE PRINCIPLES BEHIND THE MATHEMATICS.

Objective G: *Apply the properties of subtraction.* *(Lessons 7-2, 7-4)*

35. According to the Algebraic Definition of Subtraction, $7 - 3 = 7 +$ ___?___.
36. If $967 - 432 = 535$, then $432 - 967 =$ ___?___.
37. Which sentence is not equivalent to the others?
    $a - c = b$　$a - b = c$　$a + c = b$　$b + c = a$
38. Which sentence is not equivalent to the others?
    $x - 8 = 3$　$3 - 8 = x$　$-x + 3 = -8$　$-8 + x = 3$

Objective H: *Know relationships among angles formed by intersecting lines, or by two parallel lines and a transversal.* *(Lessons 7-6, 7-7)*

39. Angles 1 and 2 form a linear pair. If $m\angle 1 = 40°$, what is $m\angle 2$?
40. Angles 1 and 2 are vertical angles. If $m\angle 1 = 40°$, what is $m\angle 2$?
41. Angles 1 and 2 are supplementary. If $m\angle 1 = x°$, what is $m\angle 2$?

In 42–45, use the figure below.

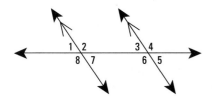

42. Angles 2 and ___?___ are corresponding angles.
43. Name the exterior angles.
44. Angles 3 and ___?___ are alternate interior angles.
45. Name four angles supplementary to angle 8.

Objective I: *Apply the definitions of parallelogram, rectangle, rhombus, and square to determine properties of these figures.* *(Lesson 7-8)*

In 46–49, consider parallelograms, rectangles, rhombuses, and squares.

46. In which of these figures are all sides equal in length?
47. In which of these figures are all angles equal in measure?
48. In which of these figures are both pairs of opposite sides equal in length?
49. In which of these figures are both pairs of opposite angles equal in measure?

Objective J: *Explain consequences of the Triangle-Sum Property.* *(Lesson 7-9)*

50. Why can't a triangle have two 100° angles?
51. Can a triangle have three acute angles? Explain your answer.

USES DEAL WITH APPLICATIONS OF MATHEMATICS IN REAL SITUATIONS.

Objective K: *Use the Take-Away Model for Subtraction to form sentences involving subtraction.* *(Lessons 7-1, 7-3, 7-4)*

52. A one-hour TV program allows $9\frac{1}{2}$ minutes for commercials. How much time is there for the program itself?

53. A 2000 square-foot house was built on a lot of area A square feet. Landscaping used the remaining 3500 square feet. How big was the original lot?

54. Anne must keep $100 in a bank account. The account had x dollars in it. Then Anne withdrew y dollars. There was just enough left in the account. How are x, y, and 100 related?

55. On the line below, $AE = 50$, $BC = 12$, and $CE = 17$. What is AB?

A   B   C   D   E

Objective L: *Use the Slide Model for Subtraction to form sentences involving subtraction.* *(Lessons 7-2, 7-3, 7-4)*

56. O'Hare Airport in Chicago is often 5°F colder than downtown Chicago. Suppose the record low recorded downtown for a day is -13°F. What is the possible low temperature at O'Hare?

57. *Multiple choice.* A person weighs 70 kg and goes on a diet, losing x kg. The resulting weight is y kg. Then $y =$
(a) $x - 70$.   (b) $70 - x$.
(c) $x - {-70}$.   (d) $-70 - x$.

58. Actor Michael Landon died in 1991 at age 55. When might he have been born?

Objective M: *Use the Comparison Model for Subtraction to form sentences involving subtraction.* *(Lessons 7-1, 7-3, 7-4)*

59. The number of 16- to 19-year-olds working in 1960 was about 4.1 million. By 1987 the number was about 6.6 million. How many more teens were working in 1987?

60. Yvette believes that her team will win its next game by 12 points. But they lose by 1 point. How far off was Yvette?

61. An airline fare of F dollars is reduced by R dollars. The lower fare is L dollars. How are F, R, and L related?

Objective N: *Use the Putting-Together with Overlap Model to solve sentences involving subtraction.* *(Lesson 7-5)*

62. Of 12 hot dogs, 6 have catsup, 8 have mustard, and 5 have both catsup and mustard. How many hot dogs have neither catsup nor mustard?

63. At International High, 80% of the students speak English. 90% of the students speak a language other than English. What percentage of students speak English and another language?

64. On a baseball team, 8 people can bat right-handed and 4 people can bat left-handed. Two players are "switch hitters" (can bat either right- or left-handed). How many people are on the team?

REPRESENTATIONS DEAL WITH PICTURES, GRAPHS, OR OBJECTS THAT ILLUSTRATE CONCEPTS.

Objective O: *Picture subtraction of positive and negative numbers on a number line.* *(Lesson 7-2)*

65. Picture the subtraction $-5 - 3$ on a number line and give the result.

66. Picture the subtraction $8 - 10$ on a number line and give the result.

Objective P: *Use Venn diagrams to describe or determine overlap.* *(Lesson 7-5)*

In 67 and 68, draw Venn diagrams.

67. 37 people are in the orchestra, 32 are in choir, and 12 are in both.

68. Suppose out of all car owners in a given place, 60% own American-made cars and 50% own foreign cars.

DISPLAYS

Everything that is visual is a *display* of some sort. Even a written word is a display of a thing, an idea, or a sound. A picture is a display that has been made by someone or something. Every culture has realized the value of pictures.

A picture shows me at a glance what it takes dozens of pages of a book to expound.
—Ivan Turgenev, *Fathers and Sons*

One picture is worth more than a thousand words.
—traditional Chinese proverb

Many people who study the learning of reading believe that the longer a sentence is, the more difficult it is to understand. Here is a display of the lengths of the first 36 sentences in the famous story *Alice's Adventures in Wonderland* by Lewis Carroll. From this information alone, do you think it would be easy or difficult to read this story?

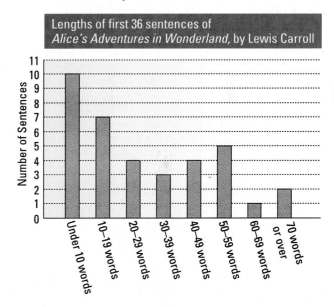

Lengths of first 36 sentences of *Alice's Adventures in Wonderland,* by Lewis Carroll

Displays in mathematics usually refer to diagrams, graphs, or tables. In this chapter, you will study stem-and-leaf displays, bar graphs, and coordinate graphs. In Lesson 8-1 you will see a stem-and-leaf display of the same data pictured in this bar graph.

*Graphs
and Other
Displays*

Graphic illustration. *During his campaign for the U.S. Presidency in 1992, Ross Perot demonstrated how the use of graphs can enhance a presentation.*

Why Use a Graph?

When looking through a newspaper or magazine, you are likely to see graphs of many kinds. People use graphs to display information for many reasons. Here are four.

Graphs can show a great deal of information in a small space.

Graphs are sometimes easier to understand than tables or prose writing.

Graphs can show trends visually.

Graphs can be startling and can be used to sway a reader.

Reading a Graph

When reading a graph, you should follow certain steps to make sure you understand it. First, read the title of the graph and any descriptive text. Then study the scales, if any, to be sure you know what is being measured. Are the numbers in percents, millions, dollars, people, or some other unit? Where does the scale begin? Are the intervals on the scale uniform? All this information can help you decide the purpose of the graph.

Example 1

Examine the line graph below.
a. What does this graph describe?
b. What does the horizontal scale represent?
c. What does the vertical scale represent?
d. What trend is shown by the graph?

Source: Nielsen Media Research, ABC, CBS, NBC

Solution
a. This graph is about the amount of television coverage of the Democratic National Convention from 1960 to 1992.
b. The horizontal scale represents the years these conventions were held. It is divided into four-year intervals, since presidential elections occur every four years.
c. The vertical scale represents the total number of hours of television coverage by ABC, CBS, and NBC.
d. The amount of coverage has been decreasing.

Notice how the graph of Example 1 visually displays the decrease in television time for the Democratic National Convention. However, the artist has taken liberties with the graph. The graph drops off at the far right, almost to 0. But the graph has no data past 1992. This is a misleading aspect of the graph.

There is another important reason for using graphs and other displays. You can sometimes gain information from a display that would be difficult to see without it. One type of display that is used to explore data for patterns is called a *stem-and-leaf display.*

What Is a Stem-and-Leaf Display?

A **stem-and-leaf display** begins with a set of numbers. First, each number is split between two specific decimal places. For instance, 135 might be split as 1|35 or 13|5. The digits to the left of the vertical line form the **stem.** The digits to the right form the **leaf.** The key to the display is that the stem is written only once, while the leaves are listed every time they appear. For instance, if both 135 and 136 appear and are split as 13|5 and 13|6, then you would write 13|5 6. If you made the split 1|35 and 1|36, then you would write 1|35 36.

Example 2

The number of words in the first 36 sentences in *Alice's Adventures in Wonderland* (popularly known as *Alice in Wonderland*), by Lewis Carroll, are as follows: 57, 55, 139, 21, 43, 34, 52, 55, 5, 13, 8, 17, 5, 3, 21, 11, 100, 14, 15, 40, 30, 25, 3, 12, 9, 4, 10, 3, 8, 23, 7, 52, 61, 43, 37, 40.

a. Describe this set of numbers with a stem-and-leaf display.

b. What does the display show?

Solution

a. The natural place to split the numbers is between the units place and the tens place. Think of a one-digit number as having a tens digit of 0. Since the largest number is 139, which will be split as 13|9, the largest possible stem is 13. List the possible stems in a vertical column.

```
 0
 1
 2
 3
 4
 5
 6
 7
 8
 9
10
11
12
13
```

Now put the leaves into the diagram. For 57, a 7 goes in the 5 row to the right of the bar. For 55, a 5 goes in the same row. Then, put a 9 to the right of the 13. Then, put a 1 to the right of the 2. Proceed in this manner until all 36 units digits have been entered. Here is the finished display with its title.

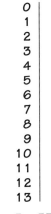

Lengths of the First
36 Sentences of
<u>Alice's Adventures
in Wonderland</u>
by Lewis Carroll

```
 0 | 5 8 5 3 3 9 4 3 8 7
 1 | 3 7 1 4 5 2 0
 2 | 1 1 5 3
 3 | 4 0 7
 4 | 3 0 3 0
 5 | 7 5 2 5 2
 6 | 1
 7 |
 8 |
 9 |
10 | 0
11 |
12 |
13 | 9
```

b. The display shows that shorter sentences are more common. There are 10 sentences with fewer than 10 words. However, there are 14 sentences with more than 30 words. The sentence with 139 words is much longer than any other sentence, though there is another with 100 words.

Stem-and-leaf displays were developed in the 1970s by John Tukey, a professor of statistics at Princeton University. Because of its newness, and because this kind of display is used to explore data, it is not used often to report data. So stem-and-leaf displays are rarely used in newspapers or magazines. But they are helpful in the study of some numerical information.

Why Is a Stem-and-Leaf Display Useful?

Notice that the stem-and-leaf display resembles the bar graph on page 407, turned 90° clockwise. But the stem-and-leaf display shows all the original values. So you can gather more information about the sentence lengths.

For instance, the **mode** of a collection of objects is the object that appears most often. The stem-and-leaf display shows three 3s by the stem 0. No other leaf appears as often in a single stem, so the mode is 3. More of these sentences are 3 words long than any other length.

From the stem-and-leaf display, you can also identify the **range** of the data. This is the difference of the highest and lowest numbers, $139 - 3$, or 136. You can also identify the **median,** the middle number if the list were in numerical order. Since there are 36 numbers in all, the middle of the list lies between the 18th and 19th numbers. The 18th number in order is 21; the 19th number is 21. The median sentence length is also 21 words, the mean of the numbers 21 and 21. If there were an odd number of numbers in the list, the median would be the number in the middle and no mean is needed. The range, median, and mode are important *statistics*. A **statistic** is a number that is used to describe a set of numbers.

Example 3

A student scores 89, 72, 99, 93, and 81 on five tests. Give the range, median, and mode of this set of numbers.

Solution

The range is the difference of the largest and smallest numbers.

$$99 - 72 = 27$$

The range is 27.

The median is the middle number if the numbers are in numerical order. So order the numbers: 72, 81, 89, 93, 99. The middle number is 89, the median.

No number appears more often than the others. Therefore, This set of numbers has no mode.

Covering the Reading

1. Give four reasons for using graphs and other displays.

2. Name at least two steps to follow when reading a graph.

3. What is a *statistic?*

In 4–7, use the graph of Example 1.

4. What are the missing values on the horizontal scale?

5. **a.** In what year was the most TV coverage of the Democratic National Convention?
   b. What was the amount of TV coverage in that year?

6. Explain the purpose of the graph.

7. Which of the four reasons for graphing do you think apply to this graph?

8. The next twenty-five sentences of *Alice's Adventures in Wonderland* have the following lengths: 45, 119, 34, 55, 11, 11, 24, 83, 23, 108, 44, 6, 8, 37, 43, 8, 30, 102, 22, 65, 14, 12, 37, 45, 88.
   a. Make a stem-and-leaf display of these numbers.
   b. What is the range of these numbers?
   c. What is the median?

In 9 and 10, a collection of numbers is given.
a. Give the median.
b. Give the range.

9. 10, 9, 9, 10, 10, 7, 9, and 8; the numbers of lessons in the first eight chapters of this book

10. 9,400; 6,900; 3,800; 17,300; 11,700; 3,300; 5,400; the areas (in thousands of square miles) of the seven continents

In 11–14, give the mode of each collection. (It is possible for a collection to have more than one mode.)

11. the collection of Question 8

12. the collection of Question 9

13. 1, 2, 2, 3, 2, 4, 2, 4, 3, 4; the numbers of integer factors of the numbers from one to ten

14. the first 50 decimal places of π (See page 12.)

15. a. Put the information in the graph of Example 1 into a table.
   b. Which do you prefer, the graph or the table, and why?

In 16–18, use the graph below.

16. Give the interval on each scale.

17. Estimate the percent of electoral votes California cast in 1960.

18. Why do you think the graph is on the arm of a strongman flexing his bicep?

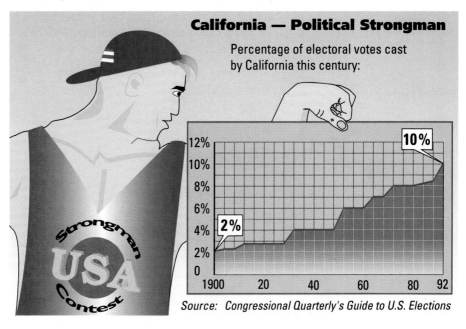

California — Political Strongman

Percentage of electoral votes cast by California this century:

Source: *Congressional Quarterly's Guide to U.S. Elections*

Review

19. Given: m∠*A* = m∠*T* = 58°. Find m∠*H*.   *(Lesson 7-9)*

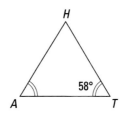

20. Use the figure below. Lines ℓ and *m* intersect at *P*. Find the measures of angles 1, 2, 3, 4, and 5.   *(Lesson 7-6)*

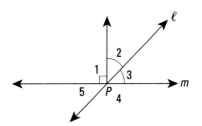

21. Try at least two special cases with positive and negative numbers to decide whether the pattern is probably true or definitely false for all numbers: $a + -b + -c = -(b + -a + c)$. *(Lesson 6-7)*

22. Evaluate each. When possible, use answers to one part to obtain answers to another. *(Lessons 5-3, 5-5, 7-2)*

   a. $6\frac{4}{5} - 3\frac{2}{3}$ **b.** $-6\frac{4}{5} - 3\frac{2}{3}$

   c. $6\frac{4}{5} + 3\frac{2}{3}$ **d.** $-6\frac{4}{5} + 3\frac{2}{3}$

   e. $-6\frac{4}{5} - -3\frac{2}{3}$

23. An age-guesser at a carnival gives a prize if the guess misses your actual age by more than two years. Let G be the guess. If a person is 26 years old, describe the values of G that will give that person a prize. *(Lessons 4-9, 7-2)*

24. Steve has h brothers and 4 sisters. There are c children in the family. Give a mathematical sentence relating these numbers. *(Lesson 5-1)*

25. The segment below represents 1 unit.

1 unit

   a. Draw a segment that represents $\frac{1}{3}$ unit.

   b. Draw a segment that represents $2\frac{3}{4}$ units.

   c. Draw a segment that represents 0.6 unit. *(Lesson 3-1)*

Exploration

26. Find a graph in a newspaper or magazine. Copy the graph or cut it out. Explain the purpose of the graph.

8-2

Bar Graphs

Family origins. *This Hispanic family is looking at its family tree. The U.S. Census Bureau has predicted that Hispanics will comprise 14% of the population and be the nation's largest minority group by the year 2010.*

According to the 1990 U.S. census, there were about 22,354,000 people of Hispanic origin living in the U.S. This number includes 13,496,000 people of Mexican ancestry, 2,728,000 of Puerto Rican ancestry, 1,044,000 of Cuban ancestry, and 5,086,000 of other Hispanic origin. The people of "other Hispanic origin" have origins from other Spanish-speaking countries of the Caribbean, Central or South America, or from Spain, or are persons identifying themselves as Spanish, Latino, Hispanic, Spanish-American, and so on.

The preceding paragraph shows numerical information in *prose* writing. **Prose** uses words and numerals but no pictures. Prose allows a person to insert opinions and extra information. It is the usual way you are taught to write.

Making a Bar Graph

Numbers in paragraphs are not always easy to follow. Many people prefer to see numbers displayed. One common display is the *bar graph.* In a **bar graph,** the lengths of bars correspond to the numbers that are represented. For instance, a bar that represents 5,000,000 is twice as long as a bar that represents 2,500,000.

Example 1 shows the steps necessary to display the above data in a bar graph.

Example 1

Display the numerical information on page 415 about persons of Hispanic origin in the United States in a bar graph.

Solution

Step 1: Every bar graph is based on a number line. With a ruler, draw a number line with a *uniform scale*. A **uniform scale** is one on which numbers are equally spaced so that each interval represents the same value. The interval of the scale below was chosen so that all of the numbers will fit on the graph.

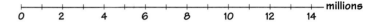

Step 2: Graph each of the numbers.

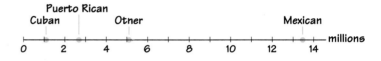

Step 3: Draw a segment from 0 to each number. Each segment is a *bar* of the bar graph. Then raise each bar above the number line.

Step 4: Identify the bars and write each bar's length by it. Finally, label the entire graph so that someone else will know what you have graphed.

U.S. Population of Hispanic Origin, 1990

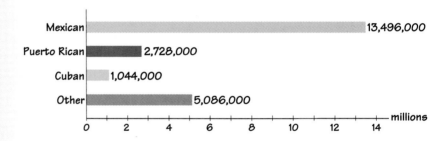

What Is a Double Bar Graph?

A single bar graph can take up more space than prose. To save space, two related bar graphs can be combined into one. The bars are different colors so that bars of the same color can be easily compared.

Example 2

The information below comes from a survey of the actual and desired ways of spending one's day. The times are averages of the responses of 1,400 adults.

Activity	Real schedule	What people desire
Work/commute	10 hr	8 hr, 18 min
Time with family	1 hr, 54 min	2 hr, 54 min
Physical activity	36 min	1 hr, 6 min

Display this information in a *double bar graph*. ▶

► **Solution**

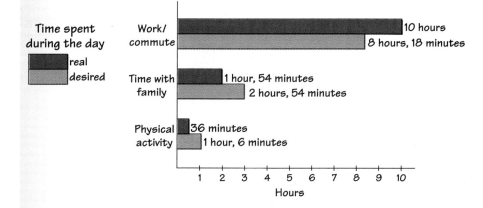

Each row of the table leads to two bars of the graph. To distinguish the bars, different colors are used.

Creating a Bar Graph from a Spreadsheet

Most spreadsheet programs have the capability of creating bar graphs. To create the bar graph, you first enter data in two rows or two columns. Then you direct the program to convert the information into a graph. For instance, the table below left lists the percent changes in mean gasoline prices from 1978 to 1992. The information for each year is the percent change from the previous year. A negative percent means that the price declined by that percent. The same information is displayed below in a vertical bar graph.

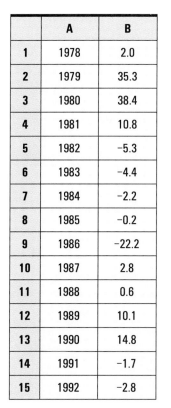

	A	B
1	1978	2.0
2	1979	35.3
3	1980	38.4
4	1981	10.8
5	1982	−5.3
6	1983	−4.4
7	1984	−2.2
8	1985	−0.2
9	1986	−22.2
10	1987	2.8
11	1988	0.6
12	1989	10.1
13	1990	14.8
14	1991	−1.7
15	1992	−2.8

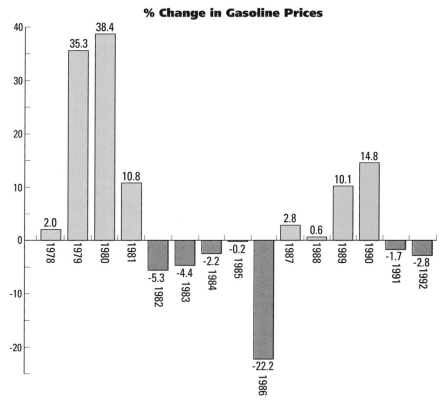

Covering the Reading

In 1–3 refer to Example 1.

1. What is the interval of the scale of the bar graph?

2. Are the bars on this graph horizontal or vertical?

3. Identify some information that is in the prose but not in the bar graph.

In 4–6, use the bar graph of Example 2.

4. What is the interval of the scale on this graph?

5. a. How much time does the average person spend working or commuting?
b. How much is desired?

6. Why is this graph called a double bar graph?

In 7–10, use the bar graph of changes in gasoline prices.

7. What is the interval of the vertical scale?

8. On this graph, are the bars vertical or horizontal?

9. a. In which year from 1978 to 1992 did prices increase the most?
b. What was the change that year?

10. What does the -5.3 by 1982 mean?

Applying the Mathematics

In 11–14, a scale on a number line is given.
a. Is the scale uniform?
b. If the scale is uniform, what is its interval? If the scale is not uniform, where is it not uniform?

11.

| 0 | 2 | 4 | 6 | 8 | 10 | 12 | 14 | 16 | 18 |

12.

| 4 | 7 | 10 | 13 | 16 | 19 | 22 | 25 | 28 | 31 |

13.

| 0 | 1.0 | 1.05 | 1.1 | 1.15 | 1.2 | 1.25 | 1.3 | 1.35 | 1.4 |

14.

| 0 | 2 | 4 | | 6 | | 8 | | 10 | |

In 15 and 16, suppose a bar graph is to display the numbers 3.39, 3.4, 3.391, and 3.294.

15. Which number will have the longest bar?

16. Which two bars will differ the most in length?

In 17 and 18, some information is given. To put the information into a bar graph, what might be a good interval to use for the scale of the graph?

17. annual average unemployment in the United States: 1986, 6.9%; 1987, 6.1%; 1988, 5.4%; 1989, 5.2%; 1990, 5.4%; 1991, 6.6%; 1992, 7.3%

18. popular vote in the Presidential election of 1860: Abraham Lincoln, 1,866,352 votes; Stephen A. Douglas, 1,375,157 votes; John C. Breckenridge, 845,763 votes; John Bell, 589,581 votes

19. Here are the number of miles of coastline of those states bordering on the Gulf of Mexico: Alabama, 53; Florida, 770; Louisiana, 397; Mississippi, 44; Texas, 367.
  a. If you wanted to display this information in a bar graph, what might be a good interval to use?
  b. Draw a bar graph of this information.

20. As of January 1993, the record high and low Fahrenheit temperatures in selected states were: Alaska, 100° and -80°; California, 134° and -45°; and Hawaii, 100° and 12°. Display this information in a double bar graph.

Review

21. Give two reasons for graphs. *(Lesson 8-1)*

22. Which of the states in Question 20 has had the greatest range of temperatures? What is that range? *(Lessons 7-2, 8-1)*

23. a. Solve the equation $105 + 45 + x = 180$.
  b. Write a problem about triangles that can lead to that equation. *(Lessons 5-8, 7-9)*

24. How many degrees are there in the turn with the given magnitude?
  a. half a revolution
  b. one third of a revolution
  c. one fifth of a revolution *(Lesson 3-6)*

In Questions 25 and 26, refer to the circle graph below. It shows the percent of land use of various types in the U.S. in 1987, excluding Alaska and the District of Columbia.

25. What percent of the land was developed (used for buildings, roads, and so on)? *(Lesson 2-7)*

26. What percent of land was not Federal land? *(Lesson 2-7)*

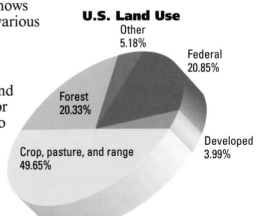

U.S. Land Use
Other 5.18%
Federal 20.85%
Forest 20.33%
Developed 3.99%
Crop, pasture, and range 49.65%

Source: U.S. Department of Agriculture

27. Use the information from this lesson. The mean price of a gallon of gas was about $1.06 in 1989. What was the mean price in 1990? What was the mean price in 1991? *(Lessons 2-5, 2-6)*

28. Use the information of Question 17. If about 125,000,000 people in the U.S. could work in 1992, how many people were unemployed? *(Lesson 2-5)*

29. Round the numbers of Question 18 to the nearest hundred thousand. *(Lesson 1-4)*

Exploration

30. Give the name and approximate population of the largest city in each of the following: Asia, Africa, Europe, North America, Australia, Central or South America. Display this information in a bar graph by hand or by using a spreadsheet program, if one is available.

31. Newspapers and magazines use color and imagination to create graphs that will grab the readers' attention. The graph at the right below is far more attractive than the bar graph at the left.

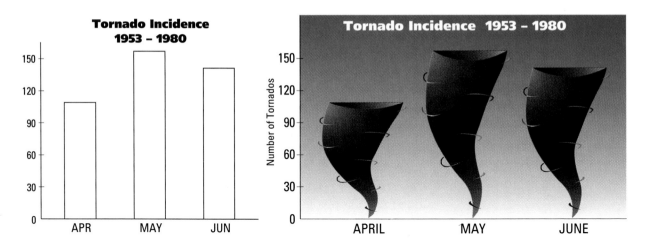

Using the following information, create a bar graph that will attract the readers' attention. Be creative. Use imagination.

National Basketball Association
1992–1993 Regular Season Scoring Leaders

Player	Points per game
1. Michael Jordan (Chicago)	32.6
2. Dominique Wilkins (Atlanta)	29.9
3. Karl Malone (Utah)	27.0
4. Hakeem Olajuwon (Houston)	26.1
5. Charles Barkley (Phoenix)	25.6

Source: National Basketball Association

Short sleeves in winter? *An elaborate network of enclosed, elevated walkways makes it easy for shoppers and workers to travel in downtown Minneapolis, MN—especially on blustery days or on the cold, snowy days of winter.*

Some patterns involve many *pairs* of numbers. A bar for each pair would take up too much space. *Coordinate graphs* are used instead.

Example 1

Below is a table of temperatures for a cold January morning in Minneapolis. Put this information onto a coordinate graph.

Time of day	Temperature (°F)
1 A.M.	–2
2 A.M.	–2
3 A.M.	–2
4 A.M.	–1
5 A.M.	0
6 A.M.	3
7 A.M.	–1
8 A.M.	0
9 A.M.	3
10 A.M.	7
11 A.M.	11
12 noon	18

Solution

This coordinate graph is based on two number lines. The horizontal number line represents time of day. Its scale goes from 1 to 12 (from 1 A.M. to 12 noon). The vertical number line represents temperature. Its scale goes from -4°F to 18°F. Its interval is chosen as 2° so that all the temperatures will fit on the graph.

▶

Each row of the table corresponds to one point on the coordinate graph. So the graph contains 12 points.

At 1 A.M. the temperature was -2°. So we go over to 1 on the horizontal line and down to -2. The left point of the graph is the result. The other 11 points are found in the same way.

Time of Day	Temperature (°F)
1 A.M.	-2
2 A.M.	-2
3 A.M.	-2
4 A.M.	-1
5 A.M.	0
6 A.M.	3
7 A.M.	-1
8 A.M.	0
9 A.M.	3
10 A.M.	7
11 A.M.	11
12 noon	18

The coordinate graph has advantages over the table. It pictures the changes in temperature. As the temperature rises, so do the points on the graph. Also, you can insert points between the times on the coordinate graph without making it larger.

Naming Points as Ordered Pairs

When a pair of numbers is being graphed as a point, the numbers are put in parentheses with a comma between them. For instance, the point farthest left on the graph in Example 1 is (1, -2). This means at 1:00 A.M. the temperature was -2°F. The next three points are (2, -2), (3, -2), and (4, -1). The point farthest to the right is (12, 18). It means that at 12:00 noon the temperature was 18°F.

The symbol (a, b) is called an **ordered pair**. *The parentheses are necessary.* The **first coordinate** of the ordered pair is a; the **second coordinate** is b. *Order makes a difference.* In Example 1, the first coordinate is the time of day; the second coordinate is the temperature at that time. At 9 A.M. a temperature of 3°F is graphed as the point (9, 3). This is not the same as a temperature of 9°F at 3:00 A.M., which would be graphed as the point (3, 9).

Example 2

Name points *A, B,* and *C* in the graph below with ordered pairs.

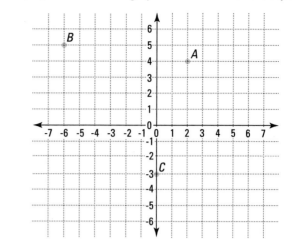

Solution

Find the first coordinate of *A* by looking at the horizontal number line. *A* is above the 2. The second coordinate is found by looking at the vertical number line. *A* is to the right of the 4. So A = (2, 4).

B is above the -6, so the first coordinate of *B* is -6. *B* is to the left of the 5, so the second coordinate of *B* is 5. B = (-6, 5).

C is below the 0 on the horizontal number line. *C* is at the -3 on the vertical number line. So C = (0, -3).

Example 3

Plot the points (-1, -3) and (7, 0) on a coordinate graph.

Solution

To plot (-1, -3), go left to -1 on the horizontal number line. Then go down to -3. That point is *V* on the graph below.

To plot (7, 0), go right to 7 on the horizontal number line. Then stay there! The 0 tells you to go neither up nor down. This is point *T* on the graph.

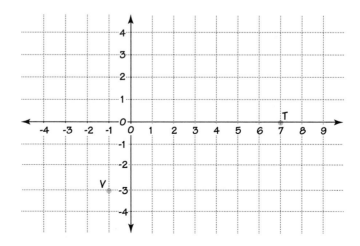

Names for Parts of a Coordinate Graph

When variables stand for coordinates, it is customary to let x stand for the first coordinate and y stand for the second coordinate. For this reason, the first coordinate of a point is called its **x-coordinate.** The second coordinate is called its **y-coordinate.** For example, the point (-20, 4.5) has x-coordinate -20 and y-coordinate 4.5. The horizontal number line is called the **x-axis.** The vertical number line is the **y-axis.** The x-axis and the y-axis intersect at a point called the *origin.* The **origin** has coordinates (0, 0).

The four areas of a graph determined by the axes are called **quadrants** I, II, III, and IV, as shown below.

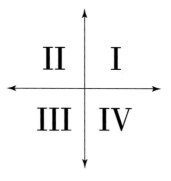

Activity

Graph the following 12 points. Then connect them *in order* by a smooth curve.
(13, 0), (12, 5), (5, 12), (0, 13), (-5, 12), (-12, 5), (-13, 0), (-12, -5), (-5, -12), (0, -13), (5, -12), (12, -5), (13, 0)

QUESTIONS

Covering the Reading

1. Trace the drawing below. Label the x-axis, y-axis, the origin, and the point (2, 4).

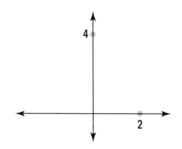

In 2–13, use the drawing below. The intervals on the graph are one unit. What letter names each point?

2. (2, 3)   **3.** (-3, 2)   **4.** (-2, 0)   **5.** (0, 3)

6. (0, -3)   **7.** (-2, -3)   **8.** (0, 0)   **9.** (-3, 0)

10. (3, -2)   **11.** (0, 2)   **12.** (3, 0)   **13.** (2, -3)

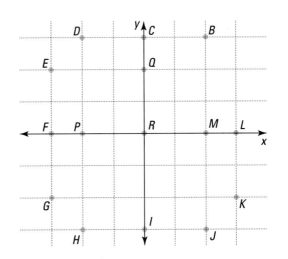

In 14 and 15, draw axes like those shown below. Plot the given points on your graph.

14. $A = (25, 10)$, $B = (25, 5)$, $C = (25, 0)$, $D = (25, -5)$

15. $F = (-5, -5)$, $G = (-10, -10)$, $H = (-15, -15)$, $I = (15, 15)$

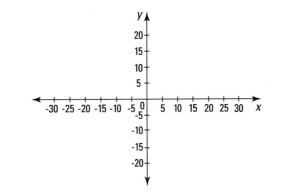

In 16 and 17, refer to Example 1.

16. a. From 3 A.M. to 5 A.M., did the temperature go up or down?
b. How is this reflected in the graph?

17. Estimate the temperature at 11:30 A.M.

18. The x-coordinate of (-3, -1) is __?__.

19. The point (0, 3) is on the __?__-axis.

20. The point (0, 0) is called the __?__.

21. The second coordinate of a point is also called its __?__-coordinate.

22. The x-axis contains the point (-4, __?__).

23. When a point is named as an ordered pair, its two coordinates are always placed inside __?__ with a __?__ between them.

In 24–27, tell in which quadrant the given point lies.

24. (-50, 400)

25. (78, -78)

26. (-.005, -4)

27. (14, 13)

28. Show what you found for the Activity in this lesson.

Applying the Mathematics

In 29–32, between which two quadrants does the point lie?

29. (0, -4000)

30. (1 millionth, 0)

31. (-30.43, 0)

32. (0, -989)

In 33–36, in which quadrant is the point (x, y) under the given circumstances?

33. x and y are both negative.

34. x is positive and y is negative.

35. x is negative and y is positive.

36. x and y are both positive.

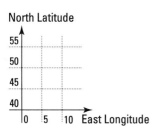

37. Given are the latitudes and longitudes of five European cities. Use these numbers as coordinates in graphing. Use axes like those at the left, but make your drawing bigger. (If correct, your drawing will approximate how these cities are located relative to each other.)

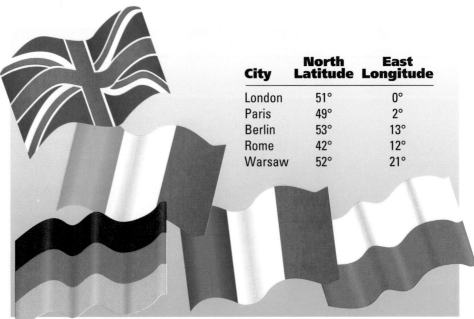

City	North Latitude	East Longitude
London	51°	0°
Paris	49°	2°
Berlin	53°	13°
Rome	42°	12°
Warsaw	52°	21°

38. The 30 tallest buildings in Atlanta, Georgia, as of 1993 were finished in the following years: 1988, 1990, 1973, 1981, 1989, 1980, 1988, 1968, 1985, 1967, 1975, 1991, 1990, 1986, 1986, 1989, 1961, 1975, 1979, 1974, 1987, 1978, 1974, 1975, 1975, 1985, 1987, 1968, 1971, 1972.

a. Put this information into a stem-and-leaf display.

b. Put this information into a bar graph in which the bars stand for the numbers of buildings built in the years 1960–69, 1970–79, 1980–89, 1990 or later.

c. Find the median of the data.

d. Find the range.

e. Which do you prefer, the stem-and-leaf display or the bar graph, and why? *(Lessons 8-1, 8-2)*

Standing tall. *The tallest building in Atlanta is the Nations Bank (1988) shown in the forefront.*

39. Draw a picture of two parallel lines and a transversal. Label one pair of alternate interior angles. *(Lesson 7-6)*

40. If one of two vertical angles has measure 20°, what is the measure of the other angle? *(Lesson 7-5)*

41. Evaluate $y - x$ when $y = -\frac{13}{16}$ and $x = -\frac{17}{32}$. *(Lessons 4-4, 7-2)*

42. If $a + b = 9$, find a when $b = 15$. *(Lesson 5-8)*

43. Convert this scale from miles to feet. *(Lesson 3-2)*

```
0      .5      1      1.5      2      2.5      3      3.5      4
|---+---+---+---+---+---+---+---+---+---+---+---+---+---+---+---+--- miles
```

Exploration

44. Most business sections of newspapers contain coordinate graphs. Cut out an example of a coordinate graph from a newspaper or a magazine.

a. What information is graphed?

b. Are the scales on the axes uniform?

c. If so, what are their intervals?

45. Find out the temperature for a recent 12-hour period in your area. Make a graph of this information similar to the graph in Example 1.

Lines of support. *A triangular arrangement of cables and support beams is widely used in bridges because of the rigidity and strength of triangles.*

Picturing a Relationship Between Two Variables

In a triangle, the sum of the measures of all three angles is 180°. In a right triangle, one angle has measure 90°. Suppose the other angle measures are x and y. Then x and y must add to 90°. Seven pairs of possible values of x and y are in the table below. These ordered pairs are graphed below the table.

x	y
10	80
20	70
35	55
52	38
45	45
80	10
62.8	27.2

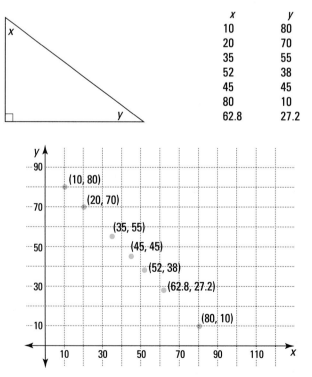

The *x*-coordinate and *y*-coordinate of each point satisfy the equation $x + y = 90$. The seven points lie on a line. This line is the *graph of all the solutions* to $x + y = 90$. The entire line is graphed below. The arrows on the line indicate that the line extends forever in both directions.

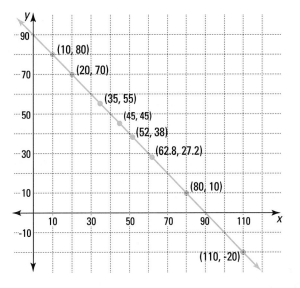

Graphing an Equation

In the equation $x + y = 90$, if you choose a value for one of the variables, you can determine the value of the other. Since there are an infinite number of choices for *x*, there will be an infinite number of solutions, *x* and *y*, to the equation. Some of them, like 110 and -20, involve negative numbers and would not be a solution to the right-triangle question. Only the part of the line in quadrant I contains possible angle measures for the acute angles in a right triangle.

Some graphs of sentences with two variables are beautiful curves. The simplest graphs are lines.

Example 1

Graph all solutions to the equation $x - y = 4$.

Solution

Find some pairs of numbers that work in the equation. You can start with some convenient values of *x*. Then for each *x* value, determine *y*. Keep track of the numbers you find by putting them in a table.

x	y
6	2
4	0
-1	-5

If $x = 6$, then $6 - y = 4$. So $y = 2$.
If $x = 4$, then $4 - y = 4$. So $y = 0$.
If $x = -1$, then $-1 - y = 4$. Solving this equation, $y = -5$.

Now graph the points (x, y) that you have found. For each equation in this lesson, the graph is a line unless you are told otherwise. You should use at least three points in your table. Two points enable the line to be drawn. The third point is a check.

▶

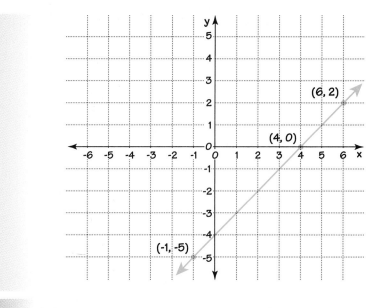

Activity 1

In Example 1, $x - y = 4$ is graphed. This equation is of the form $x - y = a$, with $a = 4$. Graph other lines of the form $x - y = a$. Can you make any conjectures about these lines?

You may have to solve an equation to find a pair of numbers that works in a sentence.

Example 2

Find a pair of numbers that satisfy the equation $y - x = \frac{1}{3}$.

Solution

Pick a value for x, say 6. Substitute 6 for x and solve for y.

When x = 6, y – 6 = $\frac{1}{3}$

$$y = 6\frac{1}{3}$$

Thus $(6, 6\frac{1}{3})$ is one point on the graph of y – x = $\frac{1}{3}$.

Substituting other numbers for x and solving for y will give the coordinates of other points that work.

Activity 2

Find three other pairs of numbers that satisfy $y - x = \frac{1}{3}$. Graph the line that contains them.

You are on the right track if your line in Activity 2 is parallel to the line in Example 1.

QUESTIONS

1. Name three pairs of numbers that satisfy $x + y = 90$.

2. When all pairs of numbers that satisfy $x + y = 90$ are graphed, they lie on a __?__.

3. Suppose x and y are measures of the two acute angles in a right triangle. How are x and y related?

4. Name three pairs of numbers that work in $x - y = 4$.

5. Describe the common feature of the graphs of lines with equations of the form $x - y = a$.

In 6–8, an equation and a value of x are given.
a. Find the corresponding value of y.
b. Tell what point these values determine on the line.

6. $x + y = 10$; $x = 2$ 7. $x - y = 5$; $x = 3$

8. $x = 6 - y$; $x = 40$

In 9 and 10, graph all solutions to each equation.

9. $x + y = 10$ 10. $x - y = 6$

11. *Multiple choice.*   The point (2, 3) is not on which line?
    (a) $x - y = -1$ (b) $-5 = -x - y$
    (c) $x - y = 1$ (d) $y - x = 1$

12. Give the coordinates of three points from Activity 2.

13. *Multiple choice.*   Which graph pictures the solutions to $y - x = 10$?

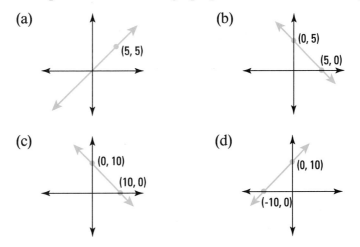

In 14 and 15, two variables and a situation are described.
a. Write an equation that relates x and y.
b. Graph all pairs of values of x and y that work in the situation.

14. Margie had x dollars at the beginning of the day. She spent y dollars. She wound up with $8.

15. There are 60 students, x of them in this classroom, y of them next door, none anywhere else.

16. If you need to find many points in a graph, it is a good idea first to solve for one of the variables.
a. Solve $-x - y = 0$ for y.
b. Use the solved equation to find three pairs of values of x and y that work in the equation.
c. Describe the graph of $-x - y = 0$.

17. A table like that on the first page in this lesson can be printed by a computer using the following program.

```
10 PRINT "X", "Y"
20 FOR X = 10 TO 80
30 Y = 90 - X
40 PRINT X, Y
50 NEXT X
60 END
```

a. Type and run this program, and describe the result.
b. Change line 20 as indicated here.

```
20 FOR X = 10 TO 80 STEP 10
```

How does this change what the computer does and what is printed?
c. Change the program so that it prints a different table of coordinates of points on the line. Record the changes you made and describe the table printed.

Review

18. Outlined at the right is a miniature golf hole. All the angles are right angles. Lengths of sides are given. Suppose this outline were graphed with A at $(0, 0)$ and B on the y-axis.
a. Give the coordinates of points B, C, D, E, and F.
b. Give the coordinates of the tee.
c. Estimate the coordinates of the hole. *(Lesson 8-3)*

19. In which quadrant is the point (-800, 403.28)? *(Lesson 8-3)*

20. Between which two quadrants is the point (-3, 0)? *(Lesson 8-3)*

21. A point is 5 units directly above (4, -3). What are its coordinates? *(Lesson 8-2)*

22. Consider the graph below. *(Lesson 8-2)*

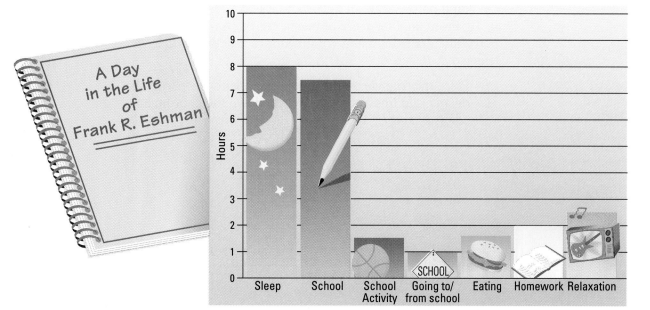

A Day in the Life of Frank R. Eshman

a. Describe what is being graphed.
b. What is the interval of the scale of the bar graph?
c. Are the bars on this graph horizontal or vertical?
d. Frank could spend more time on homework. Where could this time come from?
e. Graph a typical day for you, using bars like that in this graph. (Add other bars if you think they are needed.) *(Lesson 8-2)*

23. *Multiple choice.*   In this lesson, measures of angles are graphed whose sum is 90°. What are such angles called?
(a) complementary   (b) supplementary
(c) adjacent   (d) vertical *(Lesson 7-6)*

24. An integer is a **palindrome** if it reads the same forward and backward. For example, 252; 12,321; and 18,466,481 are palindromes. How many palindromes are there between 10 and 1000? *(Lesson 6-1)*

25. How many quarts are in a 2-liter bottle of mineral water? *(Lessons 3-4, 3-5)*

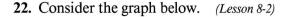

Exploration

26. Graph $y = |x|$. (Hint: The graph is not a line. Make a table with both positive and negative values for x.)

*Translations
(Slides)*

IN-CLASS
ACTIVITY

On graph paper, draw an *x*-axis and a *y*-axis and label each from -10 to 10. From another piece of paper, cut out a right triangle with perpendicular sides 5 and 7 units long. Label the vertices of this triangle *A, B,* and *C.*

1 Place the triangle on the graph paper with the perpendicular sides on any grid lines as shown. Write down the coordinates of the vertices *A, B,* and *C.*

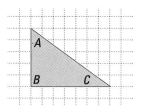

2 Slide the triangle 3 units to the left and 2 units up. Give the new coordinates of *A, B,* and *C.*

3 Compare the coordinates of the new positions of *A, B,* and *C* to the old coordinates.

Did the *x*-coordinate of each vertex change? If so, by how much?

Did the *y*-coordinate change? If so, by how much?

Royal patterns. *The* Omanhene *is the chief of an Ashanti village in Ghana. His robes show the patterns of the type often found in Kente cloth.*

Ashanti people of Ghana translate or slide a design when they paint or stamp patterns on fabric called Kente cloth. If a coordinate grid is placed on the fabric, the slide can be described by looking at the coordinates of points in the design. Any geometric figure can be placed on a coordinate grid. Adding the same number to the coordinates of the points in the figure yields a **translation image** or **slide image** of the original figure.

Horizontal Translations

Begin with triangle *MNO* shown below. Add 3 to each first coordinate. Then graph the new points. The result is a triangle 3 units to the right of *MNO*. We call this △*M′N′O′* (read "triangle *M* prime, *N* prime, *O* prime").

This procedure replaces (0, 0) by (3, 0), (2, 4) by (5, 4) and (-4, 6) by (-1, 6). Each **image** point is 3 units to the right of the **preimage** point.

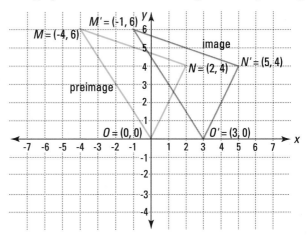

In general, if you add *h* to each first coordinate, you will get a slide image of the original figure that is *h* units to the right. (If *h* is negative, the image will move the opposite of right or left.) This is a two-dimensional version of the Slide Model for Addition that you studied in Chapter 5.

Translations that Are Not Horizontal

What happens if you add a particular number to the *second* coordinate? It's just what you might expect. The preimage slides up or down.

Below, a third triangle is now on the graph. It is the image of $\triangle M'N'O'$ when -5 is added to each second coordinate. We call the image $\triangle M^*N^*O^*$ (triangle *M* star, *N* star, *O* star).

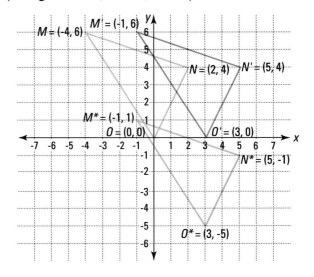

In general, if you add *k* to the second coordinate of all points in a figure, you will slide the figure *k* units up. If *k* is negative, as it is here, then the slide is "negative up," which is down.

Congruent figures are figures with the same size and shape. A translation image is always congruent to its preimage. The three triangles above are congruent.

QUESTIONS

In 1–4, what happens to the graph of a figure when the indicated action is taken?

1. 3 is added to the first coordinate of every point on it.

2. 10 is added to the second coordinate of every point on it.

3. -7 is added to the second coordinate of every point on it.

4. 6 is subtracted from the first coordinate of every point on it.

5. Another name for slide is ___?___.

6. When you change coordinates of points of a figure, the original figure is called the ___?___ and the resulting figure is called its ___?___.

In 7 and 8, graph triangle *ABC* shown below. Then, on the same axes, graph its image under the translation that is described.

7. Add 2 to the first coordinate of each point.

8. Add -3 to the first coordinate and 5 to the second coordinate of each point.

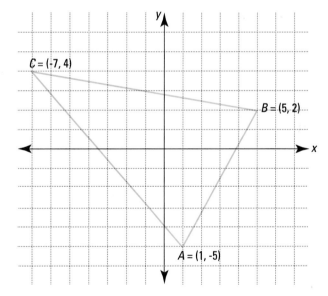

In 9 and 10, tell what happens to the graph of a figure when:

9. k is added to the second coordinate of each point and k is positive.

10. h is added to the first coordinate of each point and h is negative.

11. Congruent figures have the same ___?___ and ___?___.

12. *True or false.* A figure and its slide image are always congruent.

13.

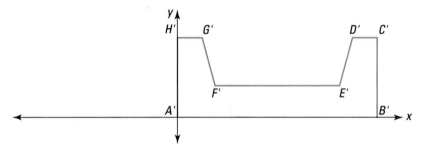

Polygon *ABCDEFGH* outlines a top view of a school building. The architect wishes to send this outline by a computer to a builder. To avoid negative numbers, the architect slides the graph so that the image of point *A* is at the origin.

What will be the coordinates of *B'*, *C'*, *D'*, *E'*, *F'*, *G'*, and *H'*?

14. a. Draw quadrilateral *PQRS* with *P* = (0, 0), *Q* = (5, 0), *R* = (5, 3), and *S* = (0, 4).
b. On the same axes, draw the image of *PQRS* when 3 is subtracted from each first coordinate and 2 is subtracted from each second coordinate.
c. The preimage and image are __?__.

15. Triangle *A'B'C'* is a slide image of triangle *ABC*. *A* = (0, 0), *B* = (3, 0), *C* = (0, -4), and *C'* = (-1, 5). What are the coordinates of *A'* and *B'*?

16. a. Give three instances of the following general pattern: Under a particular transformation, the image of (*x, y*) is (*x* + 4, *y* − 5).
b. Draw the three points and their images.

17. Graph the line with the equation $x + y = -8$. *(Lesson 8-4)*

18. Use the information below.

Year	Estimated World Population
1850	1.2 billion
1900	1.6 billion
1950	2.6 billion
2000	6.1 billion

    a. Display this data with a vertical bar graph.
    b. Graph this data on a coordinate graph.
    c. What advantage does the coordinate graph have over the vertical bar graph? *(Lessons 8-2, 8-3)*

19. Five people charge the following rates per hour of babysitting: $2.00, $1.50, $2.50, $1.75, $2.00.
    a. Give the median of the charges.
    b. Give the mode.
    c. Give the mean (average) charge.
    d. Give the range of the charges. *(Lesson 8-1)*

20. Jasmine scheduled 10 hours to work as a volunteer at the Evans City Nursing Home. She worked $1\frac{1}{2}$ hours on Monday, 2 hours 45 minutes on Tuesday, and 2 hours 15 minutes on Wednesday. Jasmine wants to get her hours in by Friday. How many hours does she need to work Thursday and Friday if she wants to work an equal amount of time on each of those days? *(Lessons 5-5, 7-4)*

These teens are visiting with a senior citizen in a nursing home.

21. The square below pictures a floor tile. A corner of the tile is shaded. Sixteen of these tiles are to be arranged to make a 4-by-4 square. Create at least two patterns.

22. An artist who used translations is M. C. Escher. Many libraries have books of his work. Find such a book and, from it, trace an example of a figure and its translation image.

Worth reflecting upon. *This extraordinary view is of Mt. Rainier and its reflection in Lake Tipsoo, in the state of Washington.*

Three Kinds of Transformations

Translating is not the only way to get an image of a figure. When you studied turns and their magnitudes in Chapter 5, you rotated points. Rotations, translations, and *reflections* are three ways that geometric figures can be changed or *transformed*. For this reason rotations, translations, and reflections are called **transformations.** You have seen computer-animated graphics. Animations are often done by beginning with a figure then transforming it. (In fact, most of the images in this chapter were drawn by a computer.)

The word "image" suggests a mirror. Mirrors are the idea behind reflections. $\triangle ABC$ and $\triangle A'B'C'$, its **reflection image over the line (or mirror) *m*,** are shown below.

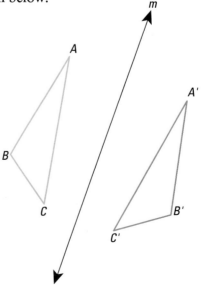

How to Find a Reflection Image

One way to draw the reflection image of a figure on paper is to fold the paper on the reflecting line. Put the figure on the outside of the fold. Hold the paper up to a light. Then trace the image.

You can also find a reflection image without folding. To do this, you need to find the images of some points. Examine the figure below. If you draw a segment from A to A', the segment is perpendicular to line m. Also, points A and A' are the same distance from line m. The same situation will exist for $\overline{BB'}$ and $\overline{CC'}$. The segments perpendicular to m are drawn below.

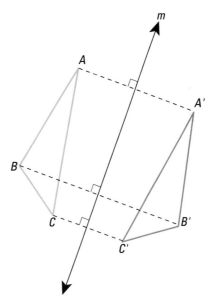

Reflecting a Point Over a Line

One way to locate the reflection image of a point over a line is shown below.

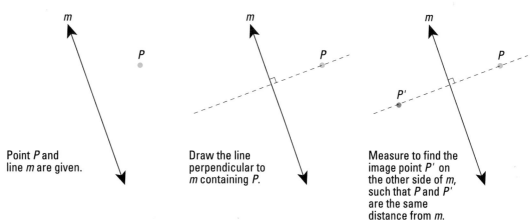

Point P and line m are given.

Draw the line perpendicular to m containing P.

Measure to find the image point P' on the other side of m, such that P and P' are the same distance from m.

What Happens to a Point on the Reflecting Line?

If a point is *on* the reflecting line, then it is its own image. We say that it **coincides** with (takes the same position as) its image. In the following figure, the reflecting line *m* is horizontal. The preimage consists of an oval and lines ℓ and *n*. Point *E* on line ℓ coincides with its image.

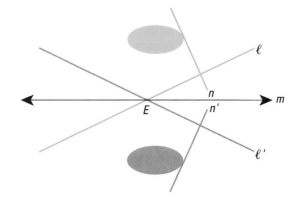

Line ℓ contains *E*, and so does its image, line ℓ′. The angles formed by ℓ and *n* have the same measures as the angles formed by ℓ′ and *n*′. In general, notice that figures and their reflection images have the same size and shape. They are congruent even though the reflection image may look reversed.

Mirrors can be used in humorous ways.

QUESTIONS

Covering the Reading

In 1–3, point *P*′ is the reflection image of point *P* over the reflecting line *m*.

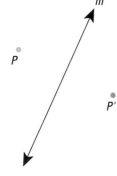

1. *m* is called the __?__.

2. $\overline{PP'}$ is __?__ to *m*.

3. *P* and *P*′ are the same __?__ from *m*.

4. If a point is on the reflecting line, then it __?__ with its image.

5. *True or false.* A figure to be reflected cannot intersect the reflecting line.

6. *True or false.* A figure and its reflection image are congruent.

7. Reflections, rotations, and translations are three types of __?__.

In 8 and 9, trace the drawing. Then draw the reflection image of each point over the given line.

8.

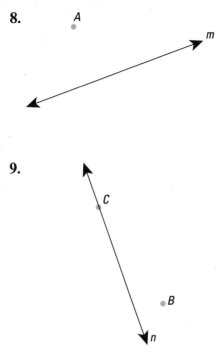

9.

In 10–11, trace the drawing. Then draw the reflection image of the given figure over the given line.

10.

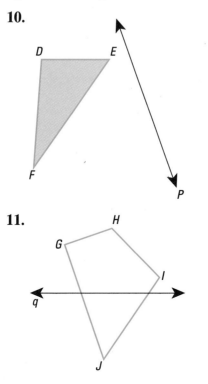

11.

12. The point (2, 4) is reflected over the *x*-axis. What are the coordinates of its image? (Hint: Draw a picture.)

13. a. Graph the quadrilateral with vertices (1, 2), (3, 4), (4, -2), and (5, 0).
   b. Change each first coordinate to its opposite and graph the quadrilateral with the new vertices.
   c. Over what line has the preimage been reflected?

14. a. Give four instances of the following general pattern: Under a particular transformation, the image of (*x, y*) is (*y, x*).
   b. Draw the four points from part **a** and their images.
   c. Describe the transformation.

In 15–18, trace the figure. What word results when the figure is reflected over the given line?

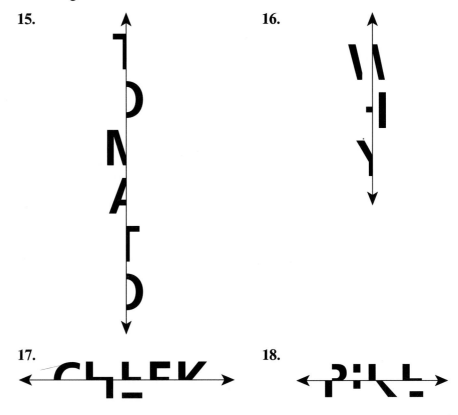

15.

16.

17.

18.

19. Wallpaper designs often involve reflections. Trace the figure below.

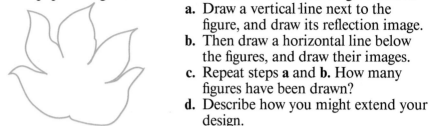

 a. Draw a vertical line next to the figure, and draw its reflection image.
 b. Then draw a horizontal line below the figures, and draw their images.
 c. Repeat steps **a** and **b**. How many figures have been drawn?
 d. Describe how you might extend your design.

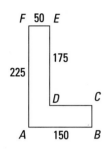

20. The picture of the letter L at the left is to be stored in a computer with point A having coordinates (0, 0) and point B on the x-axis. Give the coordinates of the other points.   *(Lesson 8-4)*

21. Graph the line with equation $x - y = 10$.   *(Lesson 8-4)*

In 22 and 23, use this information about the eight counties in the state of Connecticut.

County	Land area (sq mi)	Population (1990 census)
Fairfield	632	828,000
Hartford	739	852,000
Litchfield	921	174,000
Middlesex	373	143,000
New Haven	610	804,000
New London	669	255,000
Tolland	412	129,000
Windham	515	106,000

22. a. Graph the 8 ordered pairs (land area, population).
 b. Is it generally true that the larger counties in area have more people? Explain why or why not.   *(Lesson 8-3)*

23. Make a bar graph of the populations.   *(Lesson 8-2)*

24. Evaluate $(x - y)^2$ when $x = 6.9$ and $y = 3.4$.   *(Lessons 4-4, 4-5)*

25. Make up some hidden word puzzles like those in Questions 15–18.

Symmetry in textiles. *This Navajo blanket has two lines of symmetry.*

Nearly a thousand years ago the Diné people, ancestors of today's Navajo in the Southwest United States, learned to weave on looms from the neighboring Pueblo Indians. A Navajo blanket is shown in the photograph above. Notice that if you fold the blanket along a horizontal line through its center, the top and bottom halves match.

What Is Reflection Symmetry?

The same idea occurs with simpler figures. Consider triangle *ABC* shown below. If *ABC* is reflected over line *BD,* the image of *A* is *C.* The image of *C* is *A.* And the image of *B* is *B* itself. The points between these points have images between the points. So the entire triangle coincides with its reflection image. We say that the triangle is **symmetric with respect to line BD.** It has **reflection symmetry.**

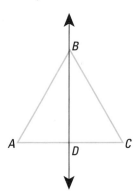

Symmetry lines do not have to be horizontal or vertical.

Check that rhombus *PQRS* has two symmetry lines, $\overleftrightarrow{PR}$ and $\overleftrightarrow{QS}$, by tracing *PQRS* and folding it over either of these lines.

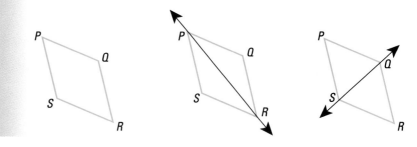

Testing for Reflection Symmetry

Many people think that the line *m* shown below at the left is another symmetry line for *PQRS*. But it is not. To test this, reflect *PQRS* over *m*. The image *P′Q′R′S′* is shown below at the right.

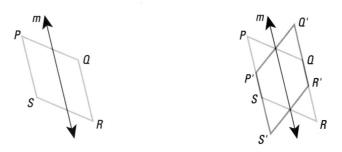

Since *P′Q′R′S′* does not coincide with the original rhombus, *m* is not a symmetry line.

Some Examples of Figures with Reflection Symmetry

Reflection symmetry is found in many places. Here are some common figures and their lines of symmetry.

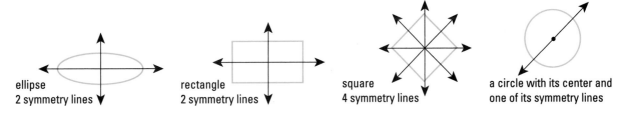

ellipse
2 symmetry lines

rectangle
2 symmetry lines

square
4 symmetry lines

a circle with its center and
one of its symmetry lines

Any line through the center of a circle is a line of symmetry. So a circle has infinitely many symmetry lines.

Trace these figures onto a piece of paper. Draw all the lines of symmetry.

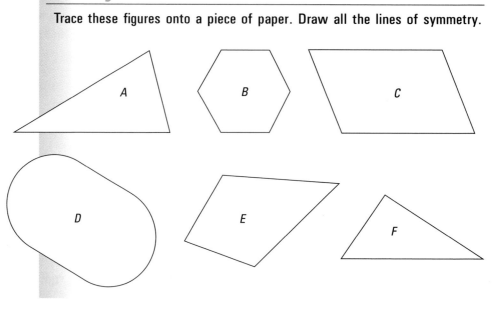

QUESTIONS

Covering the Reading

1. When does a figure have reflection symmetry?

2. How many lines of symmetry does a rhombus have?

3. Draw a figure with no lines of symmetry.

4. Draw a figure that has exactly one symmetry line.

5. Draw a figure that has exactly two lines of symmetry.

6. Draw a figure that has a symmetry line that is neither horizontal nor vertical.

7. How many symmetry lines does a circle have?

8. Trace and draw the lines of symmetry for each figure from Activity 2 in this lesson.

Applying the Mathematics

In 9–12, trace the figure, then draw all of its symmetry lines.

9.   **10.**

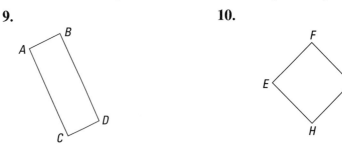

11.

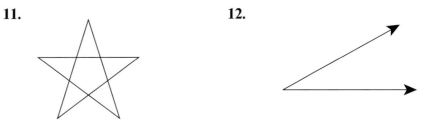

12.

In 13–15, examine these capital letters.

A B C D E F G H I J K L M N O P Q R S T U V W X Y Z

13. Which letters have a horizontal line of symmetry?

14. Which letters have a vertical line of symmetry?

15. Which letters have a symmetry line that is neither horizontal nor vertical?

In 16 and 17, a photograph of a living thing is given. In determining the symmetry of natural things like these, it is common to ignore small details.

How many lines of symmetry does the object have?

16.

17.

Passion flower. *There are about 400 species of the passion flower. Some are grown for their beauty and some, such as the papaya, for their fruit.*

Starfish *The starfish is not a fish but a marine invertebrate. Most are 8 to 12 in. across and live in the ocean. A starfish is capable of growing a new arm if it loses one.*

In 18 and 19, part of a symmetric figure is shown. All lines drawn are symmetry lines. Trace what is given, and then draw in the rest of the figure.

18.

19.

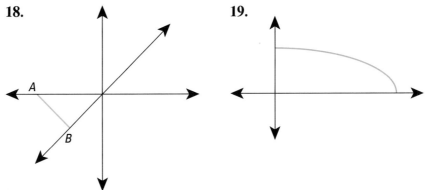

20. Draw a hexagon with no lines of symmetry.

21. Draw a hexagon with exactly two lines of symmetry.

In 22 and 23, explain why the given line is or is not a line of symmetry.

22.

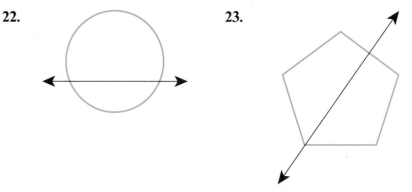

23.

24. Print your full name in capital letters. What percent of the letters in your name have a vertical line as a symmetry line? What percent have a horizontal line as a symmetry line?

Review

25. The point (-3, 8) is reflected over the *x*-axis. What is its image?
(Lesson 8-6)

26. a. Graph the quadrilateral with vertices (2, -3), (6, -1), (0, 0), and (-4, -2).
   b. Translate (slide) this quadrilateral up four units and give the vertices of the image quadrilateral. *(Lessons 8-3, 8-5)*

27. Graph the line with equation $x + y = -4$. *(Lesson 8-4)*

28. The triangle, quadrilateral, pentagon, and decagon pictured in the table on page 284 are **convex polygons**. The other polygons are not convex. Use this information to pick the correct description of convex polygon from this list. A convex polygon is:
   (a) a polygon in which two sides are parallel.
   (b) a polygon in which there is a right angle.
   (c) a polygon in which no diagonals lie outside the polygon.
   (d) a polygon with a number of sides that is not a prime number. *(Lesson 5-9)*

29. Here are average circulations for the last six months of 1992 of four of the best-selling magazines in the United States. Draw a bar graph with this information. *(Lesson 8-2)*

Magazine	Circulation
Better Homes and Gardens	8,002,585
National Geographic	9,708,254
Reader's Digest	16,258,478
TV Guide	14,498,341

30. Find the value of $a + 3(b + 4(c + 5))$ when $a = 1$, $b = 2$, and $c = 3$. *(Lesson 4-5)*

Exploration

31. A figure has **rotation symmetry** if it coincides with an image under a rotation or turn. For instance, all parallelograms have rotation symmetry. Under a 180° rotation (half turn) about point O, $ABCD$ coincides with its image.

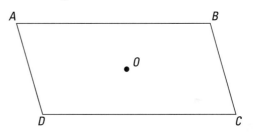

Tell whether the figure below has rotation symmetry. If so, give the magnitude of the smallest rotation under which the figure coincides with its image.

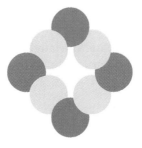

What Is a Tessellation?

Pictured here are some honeybees in their beehive. Each hexagon behind the bees is a cell in which the bees store honey. Bees are naturally talented at making such cells, so the pattern of hexagons is close to perfect. The pattern of hexagons is an example of a *tessellation.*

A **tessellation** is a filling up of a two-dimensional space by congruent copies of a figure that do not overlap. The figure that is copied is called the **fundamental region** or **fundamental shape** for the tessellation. Tessellations can be formed by combining translation, rotation, and reflection images of the fundamental region.

Shapes that Tessellate

In the beehive the fundamental region is a **regular hexagon. A regular polygon** is a convex polygon whose sides all have the same length and angles all have the same measure. A regular polygon with six sides is a regular hexagon.

Only two other regular polygons tessellate. They are the square and the **equilateral triangle.** Pictured here are parts of tessellations using them. A fundamental region is shaded in each drawing.

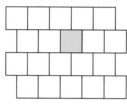

Variations of these regular polygons can also tessellate. There are many different ways to modify the sides of a regular fundamental region so that the resulting figure will tessellate.

Modify a rectangle to create a new tessellation.

Solution

1. Draw and cut any shape out of any side of an index card.

2. Slide the cut-out piece *straight across* to the opposite side. Do not flip it over. Tape it to the opposite side.

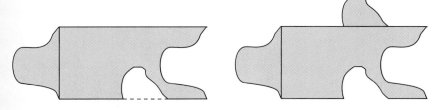

3. Pick an uncut side of the rectangle. Draw and cut another shape out of this side, slide it straight across to the opposite side, and tape it to the opposite side.

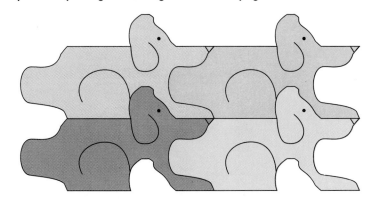

4. On a large sheet of paper, trace around the figure you made.
5. Slide the figure until part of it fits (like a jigsaw puzzle) into part of the tracing you drew. Trace around the figure again.
6. Repeat Step 5 again and again until the page is filled.

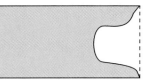

Trace the hexagon below onto a piece of cardboard. Modify it to create a new figure that will tessellate. Make a tessellation using the figure as a fundamental region. You may want to illustrate and/or color your design.

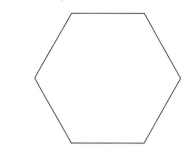

Save your modified figure.

Many other shapes can be fundamental regions for tessellations. The Dutch artist Maurits Escher became famous for the unusual shapes he used in tessellations. Here is part of one of his drawings.

M. C. Escher created this periodic design, Shells and Starfish, *in 1941. He based the design upon tesselations by congruent figures.*

Some shapes do not tessellate. For instance, there can be no tessellation using only congruent regular *pentagons.* (You may want to try, but you will not succeed.)

QUESTIONS

Covering the Reading

1. What is a tessellation?

2. What is the fundamental region for a tessellation?

3. In a beehive, what figure is the fundamental region?

4. What is a regular polygon?

5. Name three regular polygons that tessellate.

6. Can a polygon that is not a regular polygon tessellate?

7. If a figure is not a polygon, can it be a fundamental region for a tessellation?

8. Draw a tessellation with equilateral triangles.

Applying the Mathematics

9. Use your modified hexagon from the Activity in this lesson to make a tessellation.

In 10–13, trace the figure onto cardboard. Make a tessellation with at least 8 copies of the figure as a fundamental region. Figure 12 is from the drawings of Maurits Escher.

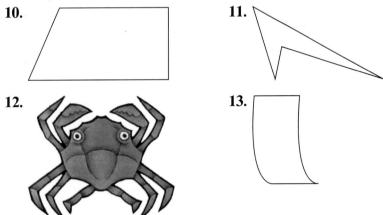

10.

11.

12.

13.

14. Pictured here is a postage stamp from the Polynesian kingdom of Tonga.
   a. Why are most postage stamps shaped like rectangles?
   b. Why do you think Tonga has stamps shaped like ellipses?

15. How many degrees are in each angle of an equilateral triangle? (Hint: What is the sum of the measures of the angles?)

16. a. Stop signs on highways are in the shape of what regular polygon?
   b. Can this shape be a fundamental shape for a tessellation?

17. Another name for *regular quadrilateral* is __?__.

18. Write a letter to a friend explaining how you modified the hexagon in the Activity in this lesson. Include a sample tessellation.

Review

In 19 and 20, draw all symmetry lines for the given figure. *(Lesson 8-7)*

19.

20.

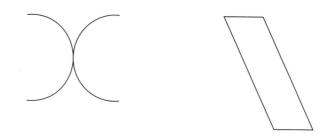

21. *Multiple choice.* Which is an expression for the measure of one angle of a regular *n*-gon? *(Lessons 5-9, 6-7, 8-7)*
   (a) $180n(n - 2)$
   (b) $\frac{360}{n}$
   (c) $\frac{180(n - 2)}{n}$
   (d) $\frac{180}{n}$

22. A quadrilateral is symmetric with respect to the *y*-axis. Two of the vertices of the quadrilateral are (4, 8) and (-2, 5). What are the other two vertices? *(Lessons 8-6, 8-7)*

23. Draw the reflection image of triangle *ABC* over line *m*.   *(Lesson 8-6)*

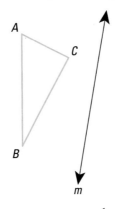

24. Tonya is 5 feet tall. If she grew $1\frac{1}{4}$ inches during the past year, and $\frac{7}{8}$ inch the year before, how tall was she two years ago?
(Lessons 3-2, 7-2)

25. Solve $a + b + c + d = 360$ for c.   *(Lesson 5-8)*

26. Write 0.00000 002 in scientific notation.   *(Lesson 2-9)*

Exploration

27. Tiles on floors or ceilings usually form tessellations. Find at least two examples of such tessellations. Sketch the design of each.

28. If you visit the Alhambra, you will see many examples of tessellations.
   a. Where is the Alhambra?   **b.** What is it?
   c. Who built it?   **d.** When was it built?

Shown here is a detail from the Alhambra.

PROJECTS

8

CHAPTER EIGHT

A project presents an opportunity for you to extend your knowledge of a topic related to the material of this chapter. You should allow more time for a project than you do for typical homework questions.

1 Tessellations

Make a tessellation in which the fundamental region is the shape of an animal or other thing. Design and color your tessellation.

2 Symmetry

Find or take pictures of objects that have different numbers of symmetry lines. For instance, a chandelier may have 10 or 12 symmetry lines; a coin from another country 8 symmetry lines. Find as many different kinds of objects with different symmetries as you can.

3 Automatic Graphers

Visual displays in mathematics are so helpful that some calculators and computer software have the capability of making coordinate graphs. We call these *automatic graphers,* but no grapher is completely automatic. Learn how to use an automatic grapher. Then use it to graph some of the equations from Lesson 8-4 or other equations of your own choosing. Give an oral or written presentation describing what you have learned.

4 Displays in Newspapers

Obtain a large-city Sunday newspaper or a copy of *USA Today*. Cut out every mathematical display in the newspaper. Organize the displays in some fashion, giving the purpose of each display.

5 Display of Temperature

Keep track of the high and low temperatures where you live each day for a week. Display the data in a coordinate graph. Summarize in prose what the temperature has been like for the week.

6 Displays of the Tchokwe

The design below comes from the Tchokwe people of northeast Angola, in the southwestern part of Africa. The Tchokwe draw in the sand and tell stories about what they draw. These drawings almost always begin with setting down a collection of equally spaced points. Then curves are added through and around the points. The drawing below represents a leopard with five cubs. Copy this design and decide where you think the leopard and five cubs are. Make two more designs of this type and write a short story for each.

7 Your Own Data

Ask a question that requires collecting some interesting data. Organize the data in several ways, and make at least three different graphs using the data. Report on how you found the data, and summarize what you found.

SUMMARY

Displays enable a lot of information to be presented in a small space. They can picture trends, are useful for exploring data, and can be used to sway opinions. In this chapter three kinds of displays are discussed: stem-and-leaf displays, bar graphs, and coordinate graphs.

Bar graphs compare quantities by using segments or bars on a specific scale. The scale uses the idea of a number line. By combining two number lines (usually one horizontal, one vertical), pairs of numbers can be pictured on a coordinate graph. Coordinate graphs can show trends and relationships between numbers. Some simple relationships, such as those pairs of numbers (x, y) satisfying equations of the form $x + y = k$ or $x - y = k,$ have graphs that are lines.

Relationships between geometric figures can also be displayed. A figure can be put on a coordinate graph. By changing the coordinates of points on the figure, an image of the figure can be drawn. By adding the same numbers to the coordinates, a slide or translation image results.

Any figure can be reflected over a line that acts like a mirror. The image is called a reflection image. If the figure coincides with its image, then it is said to be symmetric with respect to the reflecting line.

Reflections and translations are two types of transformations. So are the rotations you studied in an earlier chapter. Beautiful designs known as tessellations can be created using congruent images of figures under these transformations.

VOCABULARY

You should be able to give a general description and a specific example of each of the following ideas.

Lesson 8-1
stem-and-leaf display, stem, leaf
range
median
statistic
mode

Lesson 8-2
prose
bar graph
uniform scale

Lesson 8-3
coordinate graph
ordered pair
first coordinate
second coordinate
x-coordinate
y-coordinate
x-axis, y-axis
origin
quadrant

Lesson 8-4
graph of the solutions to an equation
palindrome

Lesson 8-5
translation image, slide image
preimage, image
congruent figures

Lesson 8-6
reflection image
reflecting line, mirror
coincide
transformation

Lesson 8-7
reflection symmetry, symmetry with respect to a line
rotation symmetry
convex polygon

Lesson 8-8
tessellation
fundamental region, fundamental shape
regular polygon
regular hexagon
equilateral triangle

PROGRESS SELF-TEST

Take this test as you would take a test in class. You will need graph paper and a ruler or straightedge. Then check your work with the solutions in the Selected Answers section in the back of the book.

In 1–4, use the graph below.

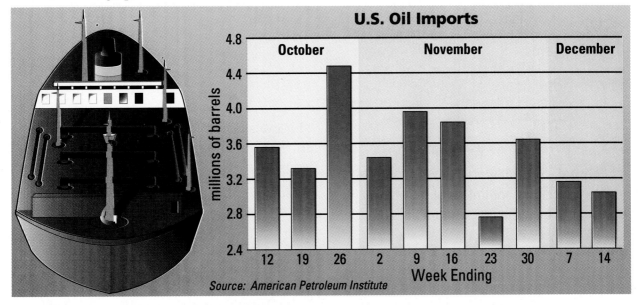

U.S. Oil Imports

millions of barrels

Source: American Petroleum Institute

Week Ending

1. In which week of November were the most barrels of oil imported into the U.S.?

2. About how many barrels of oil per day were imported into the U.S. during the week of December 14?

3. What is the interval of the scale on the vertical axis of the graph?

4. What could be misleading about this graph?

5. A Thanksgiving meal in 1991 cost the typical person about 57¢ for green beans, 81¢ for a beverage, 32¢ for cranberries, 22¢ for sweet potatoes, and $1.18 for turkey. Put this information into a bar graph.

6. Point Q has coordinates (2, 5). What are the coordinates of the point that is ten units directly above Q?

7. Graph the line with equation $x - y = 2$.

In 8–10, use the graph below. The interval of each scale is one unit.

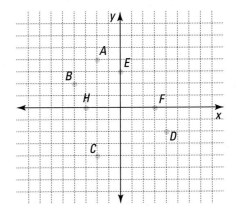

8. **a.** What letter names the point (0, 3)?
   b. What letter names the point (-2, 4)?

9. Which point or points are on the y-axis?

10. Which point or points are on the line $x + y = -2$?

PROGRESS SELF-TEST

In 11 and 12, use the graph of the miniature golf hole below. All angles are right angles. The intervals on the axes are uniform.

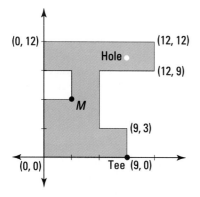

11. Give the coordinates of point *M*.

12. Estimate the coordinates of the hole.

13. Draw a quadrilateral with no lines of symmetry.

14. Draw a figure with exactly two lines of symmetry.

15. Draw a square and all its lines of symmetry.

16. Trace triangle *GHI* and line *m* below. Draw the reflection image of triangle *GHI* over line *m*.

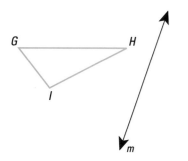

17. Draw part of a tessellation that uses triangle *GHI* as its fundamental region.

18. In 1993, it cost 29¢ to mail a letter weighing 1 oz, 52¢ for a 2-oz letter, 75¢ for a 3-oz letter, and 98¢ for a 4-oz letter. Graph the ordered pairs suggested by this information.

19. In the figure below, *V* is the reflection image of *W* over line ℓ. If m∠*UVW* = 72°, what is m∠*W*?

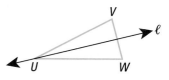

20. Give a reason for having graphs.

21. When are two figures congruent?

In 22–23, use these ages of the first 20 presidents of the United States at the time of inauguration.

57, 61, 57, 57, 57, 58, 57, 61, 54, 68, 51, 49, 64, 50, 48, 65, 52, 56, 46, 54

22. Make a stem-and-leaf display with these numbers.

23. a. What is the median of these numbers?
b. What is the range of these numbers?
c. What is the mode of these numbers?

CHAPTER REVIEW

Questions on SPUR Objectives

SPUR stands for **S**kills, **P**roperties, **U**ses, and **R**epresentations. The Chapter Review questions are grouped according to the SPUR Objectives for this chapter.

SKILLS DEAL WITH THE PROCEDURES USED TO GET ANSWERS.

Objective A: *Determine the median, range, and mode of a set of numbers.* *(Lesson 8-1)*

In 1–4, give the median, range, and mode of the given numbers.

1. 350, 568, 355, 504, 346, 350, 423, 351, 372 (the heights (in feet) of the 9 tallest buildings in Cincinnati, Ohio)

2. 55, 35, 30, 35, 40, 32, 35, 42, 28 (the numbers of stories in the 9 tallest buildings in Miami, Florida)

3. 7, 8, 7, 7, 6, 6, 7, 8, 5, 6, 4, 6, 7, 6, 5, 6, 6, 7 (the number of letters in the first names of 20th-century U.S. Presidents)

4. 8, 9, 4, 6, 7, 8, 6, 9, 6, 10, 7, 7, 5, 4, 6, 6, 4, 7 (the number of letters in the last names of 20th-century U.S. Presidents)

Objective B: *Draw the reflection image of a figure over a line.* *(Lesson 8-6)*

In 5–8, trace and use the figures below.

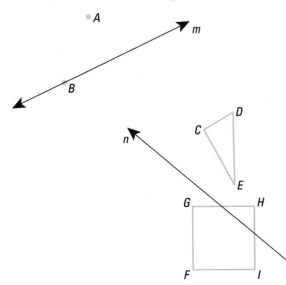

5. Draw the reflection image of point *A* over line *m*.

6. Draw the reflection image of point *B* over line *m*.

7. Draw the reflection image of triangle *CDE* over line *n*.

8. Draw the reflection image of square *FGHI* over line *n*.

Objective C: *Given a figure, identify its symmetry lines.* *(Lesson 8-7)*

In 9–12, draw all symmetry lines for the given figure.

9.

10.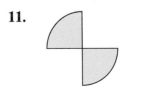

11.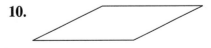

12.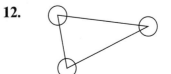

Objective D: *Make a tessellation using a given figure as a fundamental region.* *(Lesson 8-8)*

13. Make a tessellation using the figure of Question 10 as a fundamental region.

14. Make a tessellation using △*ABC* below as a fundamental region.

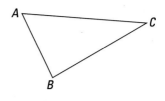

PROPERTIES DEAL WITH THE PRINCIPLES BEHIND THE MATHEMATICS.

Objective E: *Apply the relationships between figures and their reflection, rotation, and translation images.* *(Lessons 8-5, 8-6)*

In 15 and 16, use the figure below. *P′* is the reflection image of point *P* over line *t*. Answer with numbers.

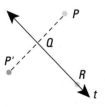

15. m∠*PQR* = __?__.

16. If *PQ* = 7, then *PP′* = __?__.

17. Triangle *ABC* is a translation image of triangle *DEF*. Must the two triangles be congruent? Explain your reasoning.

18. Can one of two congruent figures be the reflection image of the other? Why or why not?

USES DEAL WITH APPLICATIONS OF MATHEMATICS IN REAL SITUATIONS.

Objective F: *Interpret and display information in bar graphs.* *(Lesson 8-2)*

In 19–23, use the graph below. It shows the average heights of boys and girls for the ages 12–16.

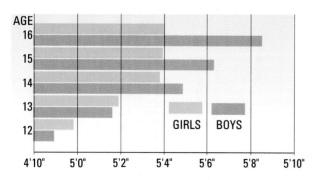

19. What is the interval of the horizontal scale of the graph?

20. At which age do girls and boys differ most in height?

21. What is the average height of 12-year-old boys?

22. Who grows more from age 12 to 13, girls or boys?

23. What about this graph could be misleading?

24. The calories in 1-cup servings of some vegetables are: green beans, 35; cauliflower, 32; lettuce, 5; peas, 110.

　　a. To display this information in a bar graph, what interval might you use on the scale?

　　b. Display in a bar graph.

25. Use the population of the United States during the first four censuses: 1790, 3.9 million; 1800, 5.3 million; 1810, 7.2 million; 1820, 9.6 million.

　　a. To display this information in a bar graph, what interval might you use on the scale?

　　b. Display in a bar graph.

Objective G: *Interpret and display information in coordinate graphs.* *(Lesson 8-3)*

In 26–29, use the graph below.

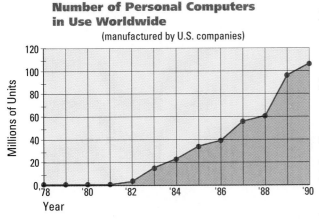

Number of Personal Computers in Use Worldwide
(manufactured by U.S. companies)

26. What major idea is shown in the graph?

27. Name the first year there were more than ten million personal computers in use.

28. In what year were about 30 million personal computers in use?

29. In what 2-year period did the number of personal computers in use increase the most?

30. Suppose it costs 23¢ a minute for a 1-minute call, 40¢ for a 2-minute call, and 57¢ for a 3-minute call. Graph three ordered pairs suggested by this information.

31. Graph the information of Question 25 on a coordinate graph.

Objective H: *Know reasons for having graphs.* *(Lesson 8-1)*

32. Give two reasons why the graph of Questions 19–23 might be useful.

33. Give two reasons why the graph of Questions 26–29 might be useful.

REPRESENTATIONS DEAL WITH PICTURES, GRAPHS, OR OBJECTS THAT ILLUSTRATE CONCEPTS.

Objective I: *Represent numerical data in a stem-and-leaf display.* *(Lesson 8-1)*

34. Represent the data of Question 2 in a stem-and-leaf display.

35. Here are the latitudes of some of the deepest trenches in the oceans north of the equator: 11°, 10°, 24°, 44°, 31°, 8°, 36°, 7°, 50°, 24°, 14°, 19°, 19°. Represent this information in a stem-and-leaf display.

Objective J: *Plot and name points on a coordinate graph.* *(Lessons 8-3, 8-4)*

In 36 and 37, use the graph below.

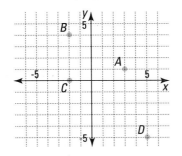

36. Give the coordinates of points *A* and *D*.

37. Give the coordinates of points *B* and *C*.

38. Graph the point (-4, -3).

39. Graph the point (2.5, 0).

Objective K: *Graph equations for lines of the form $x + y = k$ or $x - y = k$.* *(Lesson 8-4)*

40. Graph the line with equation $x + y = -5$.

41. Graph the line with equation $x - y = 3$.

42. Graph the line with equation $y = x + 2$.

43. Which of the three lines of Questions 40–42 contains the point (-2, -5)?

Objective L: *Interpret reflections and translations on a coordinate graph.*
(Lessons 8-5, 8-6)

44. What are the coordinates of the point 5 units directly above (4, 0)?

45. If (2, -3) is reflected over the *y*-axis, what are the coordinates of its image?

46. If (5, 6) is reflected over the *x*-axis, what are the coordinates of its image?

47. A triangle with vertices (2, 5), (0, 9), and (3, 7), is slid 2 units down and 7 units to the right. What are the coordinates of the vertices of the image triangle?

CHAPTER

9

PATTERNS LEADING TO MULTIPLICATION

Multiplication has a wide range of applications. Five quite different situations are described below, each leading to the multiplication $\frac{2}{3}$ times $\frac{4}{5}$. Can you see what the numbers $\frac{2}{3}$ and $\frac{4}{5}$ mean in each situation? Can you answer the questions? In this chapter, you will study several important models for multiplication. They will help you to answer these and other similar questions.

(1) *Area.* A farm is in the shape of a rectangle $\frac{2}{3}$ of a mile long and $\frac{4}{5}$ of a mile wide. What is its area?

(2) *Independent events.* Suppose one of your teachers gives homework 2 out of 3 days and another gives homework 4 of every 5 days on the average. Suppose also that the homework days are random, and the teachers do not discuss with each other when they give assignments. If a day is picked at random, what is the probability that you get homework from both?

(3) *Independent events and area.* About $\frac{2}{3}$ of the surface area of the earth is water. If the probability that an artificial satellite burns up today is $\frac{4}{5}$, what is the probability that it burns up today over water?

(4) *Rate factor.* A bug is traveling at a rate of $\frac{2}{3}$ meter per minute. At this rate, how far will it travel in 48 seconds?

(5) *Size change.* $\frac{4}{5}$ of the students in this class are here today. $\frac{2}{3}$ of those here took the bus to school. What part of the students in the class took the bus to school today?

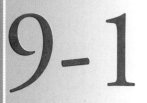

Paper by the square foot. *Allowing for trim, a typical double roll of wallpaper with 55 sq ft will cover a 6-ft section of a wall that is 8 ft high.*

Recall that the area of a plane figure tells you how much space is inside it. The area of any figure is measured in square units. The square shown below is actual size. Its area is 1 square centimeter. It is a *unit square.*

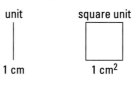

unit   square unit

1 cm   1 cm^2

How to Find the Area of a Rectangle

At the left below, the unit squares fit nicely inside the rectangle. So it is easy to find its area. You can count to get 12 square centimeters. However, at the right the unit squares do not fit so nicely. The area cannot be found just by counting. But in both rectangles, the area can be found by multiplying the length of the rectangle by its width.

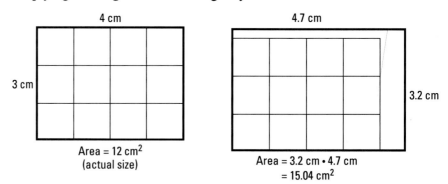

4 cm

3 cm

Area = 12 cm^2
(actual size)

4.7 cm

3.2 cm

Area = 3.2 cm • 4.7 cm
    = 15.04 cm^2

These examples are instances of a general pattern. The lengths of the sides of a rectangle are called its **dimensions.** Its area is the product of its dimensions. This pattern is a basic application of multiplication, the *Area Model for Multiplication.*

> **Area Model for Multiplication**
> The area of a rectangle with length ℓ units and width w units is $\ell \cdot w$ (or ℓw) square units.

Arrays

In the picture below, dots have been placed in the unit squares in a 3×4 rectangle. The result looks like a box of cans seen from the top. The dots form a **rectangular array** with 3 rows and 4 columns. The numbers 3 and 4 are the *dimensions of the array.* It is called a 3-by-4 array. The total number of dots is 12, the product of the dimensions of the array.

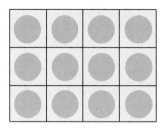

Dimensions and Area

In the physical world, a rectangle with dimensions ℓ and w is called an **ℓ-by-w or $\ell \times w$ rectangle.** A room that is $9'3'' \times 15'$, read "9 feet 3 inches by 15 feet", has area $9\frac{3}{12} \times 15$, or 138.75 square feet. The room has the same area regardless of where it is located. Below are two possible pictures.

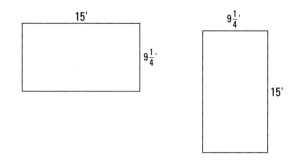

The two rectangles pictured above have the same dimensions. You could say the length and width have been switched. You could also think that one of the rectangles has been rotated 90°. Since the two rectangles have the same area,

$$15 \cdot 9\frac{1}{4} = 9\frac{1}{4} \cdot 15.$$

The area will be the same regardless of the order of the dimensions: $\ell w = w\ell$. This pictures a fundamental property of multiplication.

> **Commutative Property of Multiplication**
> For any numbers a and b, $ab = ba$.

What Is the Area of a Right Triangle?

It is easy to see that the area of every *right* triangle is half the area of a rectangle.

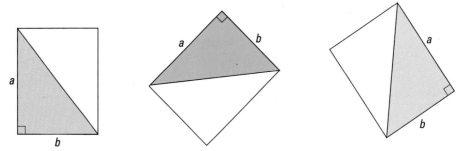

The two sides of a right triangle that are part of the right angle are called the **legs** of the right triangle. The legs in each picture above have lengths a and b. So the area of each rectangle is ab. The area of each right triangle is half of that.

> **Area Formula for a Right Triangle**
> Let A be the area of a right triangle with legs of lengths a and b. Then $A = \frac{1}{2} ab$.

The longest side of a right triangle is called its **hypotenuse.** In each right triangle above, the hypotenuse is a diagonal of the rectangle. You should ignore the hypotenuse when calculating area.

Example 1

Find the area of a right triangle with sides of length 3 cm, 4 cm, and 5 cm.

Solution

The hypotenuse is the longest side. It has length 5 cm. So the legs have lengths 3 cm and 4 cm. Thus,

$$A = \frac{1}{2} ab$$
$$= \frac{1}{2} \cdot 3 \text{ cm} \cdot 4 \text{ cm}$$
$$= 6 \text{ cm}^2.$$

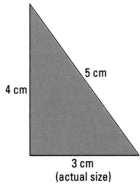

4 cm
5 cm
3 cm
(actual size)

By joining or cutting out right triangles, areas of more complicated figures can be found.

Example 2

A house is to be built on a lot with the following approximate dimensions. What is the area of the lot?

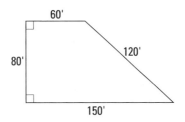

Solution 1

Think of the area as the sum of the areas of a rectangle and a right triangle.

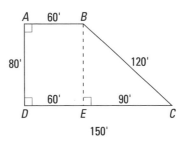

The opposite sides of the rectangle have the same length, so the legs of the right triangle have lengths 80′ and 90′.

$$\text{Area of } ABCD = \text{Area of } ABED + \text{Area of } \triangle BCE$$
$$= 60' \cdot 80' + \frac{1}{2} \cdot 80' \cdot 90'$$
$$= 4800 \text{ sq ft} + 3600 \text{ sq ft}$$
$$= 8400 \text{ sq ft}$$

Solution 2

Think of the area as the difference of the areas of a rectangle and a right triangle.

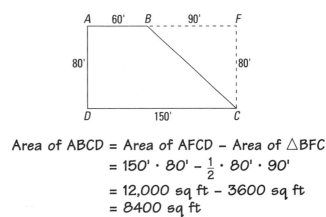

$$\text{Area of } ABCD = \text{Area of } AFCD - \text{Area of } \triangle BFC$$
$$= 150' \cdot 80' - \frac{1}{2} \cdot 80' \cdot 90'$$
$$= 12,000 \text{ sq ft} - 3600 \text{ sq ft}$$
$$= 8400 \text{ sq ft}$$

Covering the Reading

1. What is the area of a rectangle with dimensions 3 cm and 4 cm?

2. Make an accurate drawing of rectangles to show that the product of 3.2 and 4 is greater than the product of 3 and 4. (Use 1 cm as the unit.)

3. If the length and width of a rectangle are measured in centimeters, the area would probably be measured in what unit?

4. Draw a rectangle that is 4 cm by 5 cm, and give its area.

5. State the Area Model for Multiplication.

In 6 and 7, give the area of the rectangle drawn.

6.

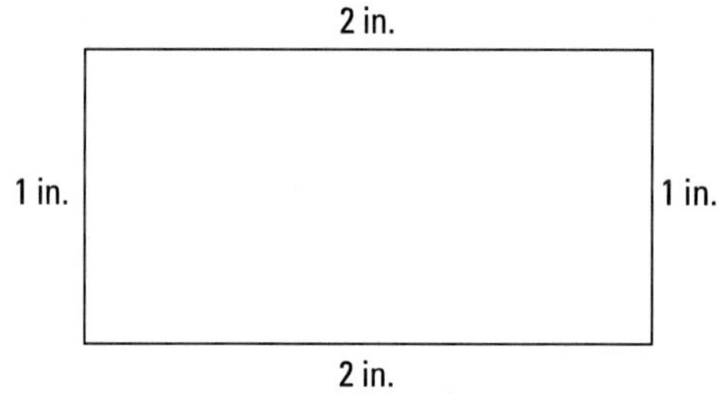

2 in.

1 in.   1 in.

2 in.

7.

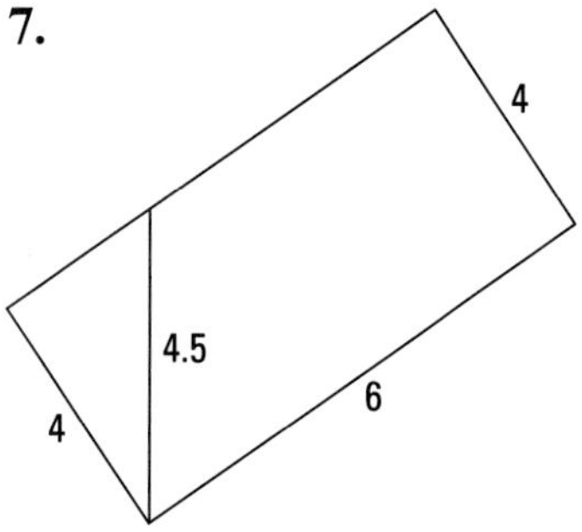

4

4.5

4   6

8. Felicia arranged some bottles in the array shown below.

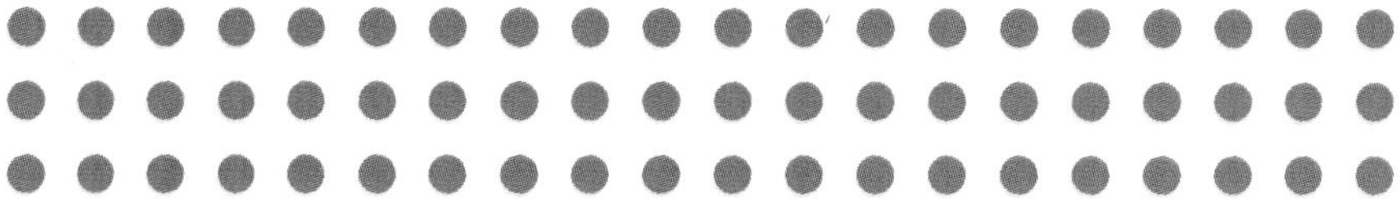

● ●
● ●
● ●

   a. How many rows are in this array?
   b. How many columns are in this array?
   c. How many bottles did she have in all?

9. a. Draw a rectangular array of dots with 7 rows and 5 columns.
   b. How many dots are in the array?

10. State the Commutative Property of Multiplication.

11. How is the Commutative Property of Multiplication related to area?

12. a. Draw a right triangle with sides of lengths 9 mm, 40 mm, and 41 mm.
    b. What are the lengths of the legs of this triangle?
    c. What is the length of its hypotenuse?
    d. What is its area?
    e. What is its perimeter?

13. One wall of E.J.'s bedroom has dimensions as shown below. What is the area of the wall?

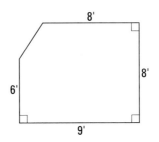

14. Arrange 30 dots in two rectangular arrays with different dimensions.

15. Give the dimensions of two noncongruent rectangles each of whose area is 15 square inches.

16. A rectangular array of dots has c columns, d dots, and r rows. Write an equation relating c, d, and r.

17. *Multiple choice.* Two stores are shaped like rectangles. Better Bargains is 60 feet wide and 120 feet long. Sales Central is 90 feet wide and 90 feet long. Both stores are all on one floor. Which is true?
    (a) Better Bargains has the greater floor space.
    (b) Sales Central has the greater floor space.
    (c) The two stores have equal floor space.
    (d) Not enough information is given to decide which store has more floor space.

18. Find the area of the shaded region between rectangles *MNOP* and *RSTQ*.

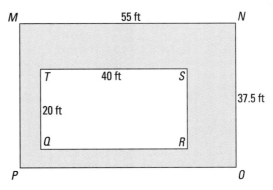

19. A tree is in the middle of a lawn. Around the tree there is a square region where no grass will be seeded. The dimensions of the lawn are shown in the drawing below. How much area will be seeded with grass?

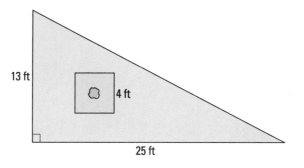

20. *RECT* is a rectangle with length 6 and height *h*.

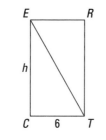

    a. What is the area of triangle *ECT*?
    b. What is the area of triangle *ERT*?

21. The design below is made up entirely of squares. The little black square has side of length 1 unit. What is the area of rectangle *ABCD*?

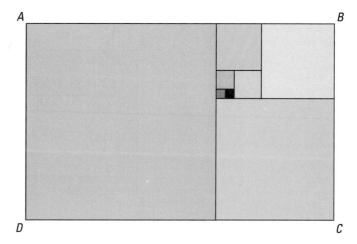

Northwest Territory, 1785

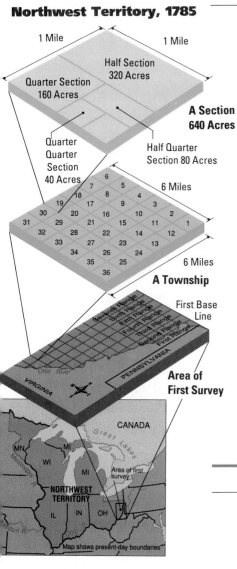

Prime real estate.
The Northwest Ordinance of 1787 set the terms under which the sections of the Northwest Territory could become states. Eventually the territory became 5 states: Ohio, Indiana, Illinois, Wisconsin, and Michigan.

22. a. Use a protractor to draw a right triangle with a 40° angle.
   b. What is the measure of the third angle of this triangle? *(Lessons 3-6, 7-9)*

23. The Northwest Ordinance of 1787 divided many areas of the central United States into square-shaped townships 6 miles on a side.
   a. What is the perimeter of such a township?
   b. What is the area?   *(Lessons 3-8, 5-10)*

24. Solve the equation $x + \frac{22}{7} = -\frac{3}{2}$.   *(Lesson 5-8)*

25. Evaluate $14 - 3(x + 3y)$ when $x = -2$ and $y = 5$.   *(Lesson 4-5)*

26. How much greater is the volume of a cube with edge 3.01 cm than the volume of a cube with edge 3 cm? *(Lesson 3-9)*

27. One kilogram is approximately equal to how many pounds? *(Lesson 3-5)*

28. a. Rewrite 48.49 as a simple fraction.   *(Lesson 2-6)*
   b. Round 48.49 to the nearest integer.   *(Lesson 1-4)*

29. There were about 2,400,000 high school graduates in the United States in 1990. Of these, about 60% went to a 2-year or 4-year college the next year. How many college students is this?   *(Lesson 2-5)*

Exploration

30. a. How many entries are needed to complete this multiplication table?

×	3.0	3.2	3.4	3.6	3.8	4.0
7.0						
7.2						
7.4						
7.6						
7.8						
8.0						

   b. Fill in the table.
   c. Find at least two patterns in the numbers you put into the table.

Abode sweet adobe. *Many people of Pueblo and Hopi Tribes live in traditional adobe structures. These homes, with their flat roofs, are shaped like rectangular solids.*

How to Describe Boxes

A **box** is a 3-dimensional figure with six **faces** that are all rectangles. A solid box is called a **rectangular solid.** When a box is situated like the one here, we can name the faces *front, back, right, left, top,* and *bottom.* The bottom is sometimes called the **base.** The sides of the faces are called the **edges** of the box. A box has 12 edges. A point at a corner of the box is called a **vertex** of the box. A box has 8 vertices.

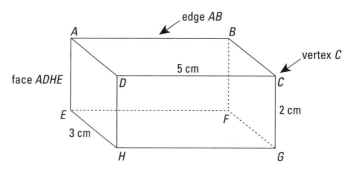

Three edges intersect at each vertex. Their lengths are the **dimensions** of the box. The above box has dimensions 2 cm, 3 cm, and 5 cm. The dimensions are called its **length, width,** and **height.** Sometimes one dimension (either front to back, or top to bottom) is called its **depth.** Notice that the length of every edge of a box equals one of the dimensions. When all the edges have the same length, the box is a cube.

How Is the Volume of a Box Found?

Recall that *volume* measures the space inside a 3-dimensional figure, or the space occupied by a solid figure. In Lesson 3-9, you saw that the volume of a cube with edge *s* was s^3. Now we seek to find the volume of a box whose measurements are known.

Example 1

Find the volume of the box pictured below.

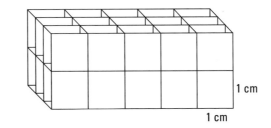

Solution

Since the dimensions are whole numbers, we can split the box into unit cube compartments. The unit cubes fit evenly into the box. You can count the cubes. There are 15 cubes in each layer. The volume is therefore 30 cubic centimeters, or 30 cm³.

Notice that in Example 1, the volume is the product of the three dimensions.

$$\text{Volume} = 5 \text{ cm} \cdot 3 \text{ cm} \cdot 2 \text{ cm}$$
$$= 30 \text{ cm}^3$$

The volume of a box or a rectangular solid is always the product of its three dimensions, even when the dimensions are not whole numbers.

Let *V* be the volume of a box or a rectangular solid with dimensions *a*, *b*, and *c*.

Then $V = abc$.
Or $V = Bh$,

where *B* is the area of the base and *h* is the height.

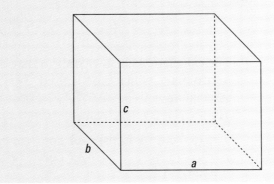

Example 2

Find the volume V of a rectangular solid 5.3 cm long, 2.7 cm high, and 3.1 cm deep.

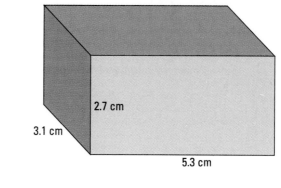

Solution

$V = 5.3 \text{ cm} \cdot 3.1 \text{ cm} \cdot 2.7 \text{ cm} = 44.361 \text{ cm}^3$

Finding the Volume of a Box Using Its Base

Notice in Example 2 that the volume of the box is the area of the base ($5.3 \cdot 3.1$) times its height (2.7). Below, the same box is shown with two different faces as its base. The volume of the figure on the left is the area of the base ($2.7 \cdot 5.3$) times the height (3.1). The volume of the figure on the right is the area of the base ($3.1 \cdot 2.7$) times the height (5.3).

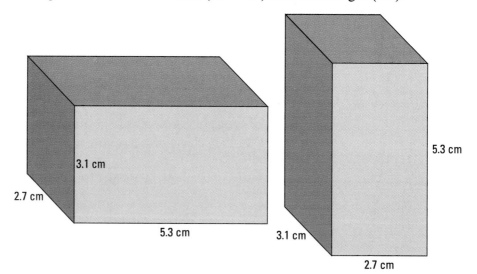

The calculations below confirm that the volume of the box is the same regardless of which base is used. This verifies that multiplication is associative.

$$
\begin{aligned}
V &= \quad (a \quad \cdot \quad b) \quad \cdot \quad c \\
&= (5.3 \text{ cm} \cdot 3.1 \text{ cm}) \cdot 2.7 \text{ cm} \\
&= 16.43 \text{ cm}^2 \qquad \cdot 2.7 \text{ cm} \\
&= 44.361 \text{ cm}^3
\end{aligned}
$$

$$
\begin{aligned}
V &= \quad a \quad \cdot \quad (b \quad \cdot \quad c) \\
&= 5.3 \text{ cm} \cdot (3.1 \text{ cm} \cdot 2.7 \text{ cm}) \\
&= 5.3 \text{ cm} \cdot 8.37 \text{ cm}^2 \\
&= 44.361 \text{ cm}^3
\end{aligned}
$$

> **Associative Property of Multiplication**
> For any numbers *a*, *b*, and *c*, (*ab*)*c* = *a*(*bc*) = *abc*.

Because multiplication is both commutative and associative, numbers can be multiplied in any order without affecting the product. These properties can shorten multiplications.

Example 3

Multiply in your head: 25 · 35 · 4 · 4 · 25.

Solution

Notice that 4 · 25 = 100. Multiply these first.

$$\begin{aligned}
\text{Think} \quad & 25 \cdot 35 \cdot 4 \cdot 4 \cdot 25 \\
= \ & 35 \cdot (25 \cdot 4) \cdot (4 \cdot 25) \\
= \ & 35 \cdot 100 \cdot 100 \\
= \ & 350{,}000
\end{aligned}$$

QUESTIONS

Covering the Reading

In 1 and 2, refer to the box below. A box of its size and shape was used to pack a laser printer.

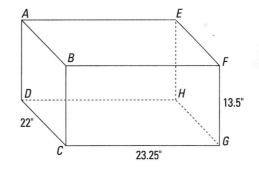

1. **a.** Name its faces.
   b. Name its edges.
   c. Name its vertices.

2. **a.** What are its dimensions?
   b. What is its volume?

In 3 and 4, give the volume of a rectangular solid with the given dimensions.

3. height 6 cm, width 3 cm, and depth 4 cm

4. height 10 meters, width *w* meters, and depth *d* meters

In 5–7, give the volume of a rectangular solid in which:

5. the area of the base is 40 square centimeters and the height is 4 centimeters.

6. the base is 10 inches by 2 feet, and the height is 1.5 inches.

7. the base has area A and the height is h.

8. State the Associative Property of Multiplication.

9. Give an instance of the Associative Property of Multiplication.

In 10–12, use the properties of multiplication to do these problems in your head. Do no calculator or pencil-and-paper calculations.

10. $5 \cdot 437 \cdot 2$

11. $6 \cdot 7 \cdot 8 \cdot 9 - 9 \cdot 8 \cdot 7 \cdot 6$

12. 50% of 67 times 2

Applying the Mathematics

13. The floor of a rectangular-shaped room is 9 feet by 12 feet. The ceiling is 8 feet high. How much space is in the room?

14. Give two possible sets of dimensions for a rectangular solid whose volume is 144 cubic units.

15. a. Draw a rectangular solid whose volume is 8 cm^3 and which has all dimensions the same length.
    b. What is this rectangular solid called?
    c. Draw a second rectangular solid whose volume is 8 cm^3 but whose dimensions are not equal.

16. Give the volume of this box.

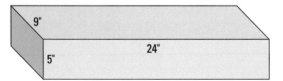

17. *Multiple choice.* A portable cassette player is nearly a rectangular solid with dimensions 12 cm × 9 cm × 4 cm. One "boom box" is also nearly rectangular, with dimensions 18 cm × 55 cm × 19 cm. About how many times as much volume does the boom box occupy as the portable cassette player?
    (a) 4   (b) 6   (c) 12   (d) 24   (e) 40

Play on. *Since a boom box is portable and has excellent sound quality, it is often used in exercise classes.*

18. **a.** *True or false.* If $x = 7.6092$, then $3(x + 4) = (x + 4) \cdot 3$.
    b. *True or false.* $\frac{2}{3} \cdot \frac{4}{5} = \frac{4}{5} \cdot \frac{2}{3}$
    c. What general principle is exemplified by parts **a** and **b**? *(Lesson 9-1)*

19. A computer has a display screen that can show up to 364 rows and 720 columns of dots. How many dots can be on that screen? *(Lesson 9-1)*

20. Find the area of the rectangle with vertices (2, 4), (5, 4), (5, 10), and (2, 10). *(Lessons 8-3, 9-1)*

21. **a.** In which quadrant is the point (4, -1)?
    b. Give an equation for a line that goes through (4, -1). *(Lessons 6-1, 8-3, 8-4)*

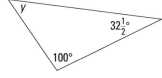

22. Refer to the triangle at the left.
    a. What is the value of y?
    b. How many angles are acute?
    c. How many angles are obtuse? *(Lessons 3-7, 7-9)*

23. Solve the equation $83 - x = 110$. *(Lesson 7-4)*

24. To divide a decimal by .001, how many places should you move the decimal point, and in what direction? *(Lessons 2-4, 2-6, 6-7)*

25. If $n = 10$, what is the value of $(3n + 5)(2n - 3)$? *(Lesson 4-5)*

26. 2.6 kilometers is how many meters? *(Lesson 3-4)*

In 27–29, put one of the signs <, =, or > in the blank. *(Lesson 1-9)*

27. -4 − 4 _?_ -4 + 4

28. -7.352 _?_ -7.351

29. 2/3 _?_ .66666666

30. What digit is in the thousandths place of 54,321.09876? *(Lesson 1-2)*

31. Choose a room where you live that is shaped like a rectangular solid. Find its dimensions and calculate its volume.

32. Suppose the dimensions of the box given in Question 16 were estimates, the result of rounding the lengths *to the nearest inch*. For instance, the shortest edge might have had a length as small as 4.5 in. or as large as 5.5 in. To the nearest cubic inch, find the smallest and largest volumes the rectangular solid could have.

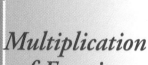

Can you cut it? *Think about how you would cut this cake if you wanted a large number of pieces.*

Multiplication of Unit Fractions

A large cake was bought for a class party. First the cake was cut into 20 pieces as shown. These pieces were too large, so each was cut into 2 smaller pieces.

Example 1

At the party, Kris ate one of the smaller pieces. What part of the cake did she eat?

Solution 1

Since each of 20 pieces was cut into 2, there were 40 smaller pieces. If all smaller pieces were the same size, Kris ate $\frac{1}{40}$ of the cake.

Solution 2

The original pieces were each $\frac{1}{20}$ of the cake. Kris ate $\frac{1}{2}$ of $\frac{1}{20}$. To find $\frac{1}{2}$ of $\frac{1}{20}$, multiply the fractions: $\frac{1}{2} \cdot \frac{1}{20} = \frac{1}{40}$. Again, Kris ate $\frac{1}{40}$ of the cake.

The fractions $\frac{1}{2}, \frac{1}{3}, \frac{1}{4}$, and so on, are unit fractions. A unit fraction is a fraction with 1 in its numerator and a positive integer in its denominator. The rule for multiplying two unit fractions is quite simple to state.

> **Multiplication of Unit Fractions Property**
> For all nonzero numbers a and b, $\frac{1}{a} \cdot \frac{1}{b} = \frac{1}{ab}$.

Fractions and Division

The ancient Egyptians used unit fractions extensively. In fact, they used no fractions other than unit fractions and the fraction $\frac{2}{3}$. Through some very ingenious methods, they wrote all other fractions as sums of unit fractions.

Today we look at fractions differently. We look at $\frac{2}{3}$ as being 2 times $\frac{1}{3}$, and in general we look at any fraction $\frac{a}{b}$ as being a times $\frac{1}{b}$. Because the fraction $\frac{a}{b}$ is equal to the quotient $a \div b$, this pattern gives a fundamental relationship between multiplication and division. It is sometimes called the *Algebraic Definition of Division*.

> **Algebraic Definition of Division**
> For all numbers a and nonzero numbers b, $a \cdot \frac{1}{b} = \frac{a}{b} = a \div b$.

Example 2

The 23 students in Kris's class each ate one piece of the cake. How much of the cake did these students eat altogether?

Solution

They ate 23 times $\frac{1}{40}$ of the cake. $23 \cdot \frac{1}{40} = \frac{23}{40}$.

When 40 of the pieces were eaten, then $40 \cdot \frac{1}{40}$ of the cake was eaten. This is 100% of the cake. Thus $40 \cdot \frac{1}{40} = 100\%$. More simply, $40 \cdot \frac{1}{40} = 1$.

What Are Reciprocals?

Two numbers whose product is 1 are called **reciprocals** or **multiplicative inverses** of each other. Above we explained why $\frac{1}{40}$ and 40 are reciprocals of each other. This is one instance of a general pattern involving reciprocals.

Notice that *the reciprocal of a number is 1 divided by that number.* All nonzero numbers have reciprocals, no matter how they are written.

Example 3

Write the reciprocal of 12.5 as a decimal.

Solution

The reciprocal of 12.5 is $\frac{1}{12.5}$. A calculator shows that $1 \div 12.5$ is equal to 0.08. The reciprocal of 12.5 is 0.08.

Check

Does $12.5 \cdot 0.08 = 1$? Yes.

Most scientific calculators have a reciprocal key [1/x]. Enter 12.5 and then press this key. The reciprocal of 12.5 should be displayed.

Multiplication of Any Fractions

The most general pattern involving multiplication of fractions is that the product of the fractions $\frac{a}{b}$ and $\frac{c}{d}$ equals $\frac{ac}{bd}$. That is, the product can be written as a fraction whose numerator is the product of the numerators of the given fractions, and whose denominator is the product of the denominators of the given fractions. The pattern is much easier to state with variables than in words. We call it the *Multiplication of Fractions Property*.

> **Multiplication of Fractions Property**
> For all numbers a and c, and nonzero numbers b and d, $\frac{a}{b} \cdot \frac{c}{d} = \frac{ac}{bd}$.

Corn country. *The main crop grown on this farm is corn. The U.S. is the leading exporter of corn, producing $\frac{2}{3}$ of the world's supply.*

Example 4

A farm is in the shape of a rectangle $\frac{2}{3}$ of a mile long and $\frac{4}{5}$ of a mile wide. What is its area?

Solution

The area (in square miles) is the product of $\frac{2}{3}$ and $\frac{4}{5}$. Using the Multiplication of Fractions Property, $\frac{2}{3} \cdot \frac{4}{5} = \frac{2 \cdot 4}{3 \cdot 5} = \frac{8}{15}$.

The area is $\frac{8}{15}$ square mile, or about .53 square mile.

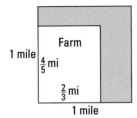

Figure 1

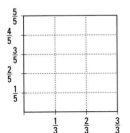

Figure 2

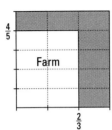

Figure 3

Picturing Multiplication of Fractions

There is probably no algorithm in arithmetic simpler than the Multiplication of Fractions Property. But why does it work? One way to explain why it works uses the Area Model for Multiplication. Look back at the situation of Example 4.

Think of the farm as being inside a square mile (Figure 1 at the left). Divide that square mile into rectangles that are $\frac{1}{5}$ mile by $\frac{1}{3}$ mile (Figure 2). You can see that 15 of these rectangles fit into the square mile. So the area of each small rectangle is $\frac{1}{15}$ square mile.

The farm, because it is $2 \cdot \frac{1}{3}$ miles wide, and $4 \cdot \frac{1}{5}$ miles long, includes $2 \cdot 4$, or 8 of these rectangles (Figure 3). So the farm has area $8 \cdot \frac{1}{15}$ square mile, or $\frac{8}{15}$ square mile.

QUESTIONS

Covering the Reading

1. Consider the case of Example 1. Suppose each of the 20 pieces had been cut into 3 equal pieces. Then each smaller piece is what part of the original cake?

2. What is meant by a *unit fraction?*

In 3–8, give the product as a fraction in lowest terms or as a whole number.

3. $\frac{1}{3} \cdot \frac{1}{4}$

4. $\frac{1}{40} \cdot \frac{1}{x}$

5. $\frac{1}{a} \cdot \frac{1}{b}$

6. $2 \cdot \frac{1}{3}$

7. $\frac{1}{8} \cdot 16$

8. $\frac{1}{x} \cdot 5$

9. State the Algebraic Definition of Division.

In 10–15, simplify.

10. $\frac{6}{5} \cdot \frac{2}{9}$

11. $\frac{3}{5} \cdot \frac{5}{3}$

12. $\frac{1}{9} \cdot \frac{9}{2}$

13. $(7 \cdot \frac{1}{12}) \cdot (3 \cdot \frac{1}{4})$

14. $12 \cdot \frac{3}{4}$

15. $\frac{a}{b} \cdot \frac{c}{d}$

16. A farm is rectangular in shape, $\frac{1}{2}$ mile by $\frac{2}{3}$ mile. Use a drawing to explain why the area of the farm is $\frac{1}{3}$ square mile.

In 17 and 18, give the reciprocal of the number.

17. 10

18. 4.5

19. Write the reciprocal of 40 as a decimal.

20. Kyle presses 16 $\boxed{1/x}$.
    a. What will the calculator do?
    b. How can he check his answer?

21. State the Property of Reciprocals.

22. Stephen has to blow up $\frac{1}{4}$ of all balloons needed for a celebration. He has a friend who will equally share Stephen's part of the job.
    a. How much of the total job will the friend do?
    b. If the total job pays $100, how much should Stephen's friend receive?

Applying the Mathematics

23. Verify that $\frac{5}{8}$ is the reciprocal of $\frac{8}{5}$ by converting both fractions to decimals and multiplying.

24. Write the reciprocal of $1\frac{1}{2}$ as a fraction.

25. Explain why 0 has no reciprocal.

26. Each edge of a cubical box is $\frac{1}{2}$ meter long. What is the volume of the box?

27. a. Multiply $\frac{1}{2} \cdot \frac{1}{3} \cdot \frac{1}{4}$.
    b. How do you know that the answer to part **a** must be a number less than 1, without doing any multiplication?

28. a. Multiply $\frac{3}{2} \cdot \frac{5}{4} \cdot \frac{7}{6}$.
    b. How do you know that the answer to part **a** must be a number greater than 1, without doing any multiplication?

29. a. Multiply $3\frac{1}{2} \cdot 2\frac{3}{4}$. Write your answer as a mixed number.
    b. How can you check your answer to part **a**?

Review

30. Give the volume of a box with dimensions 9 cm, 15 cm, and 20 cm. *(Lesson 9-2)*

31. Find the area of right triangle ABC. *(Lessons 6-2, 9-1)*

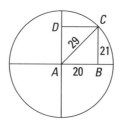

486

In 32 and 33, $\overline{PQ} \parallel \overline{ST}$ and angle measures are as shown on the figure.

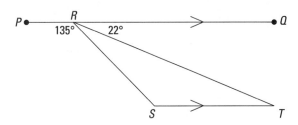

32. Trace the figure and draw the reflection image of $\triangle RST$ over the line containing $\overline{PQ}$. *(Lesson 8-6)*

33. Find the measures of the three angles of triangle RST. *(Lessons 7-6, 7-7, 7-9)*

34. Two ordinary dice are tossed.
A = the sum of the numbers showing on the dice is 5.
B = one of the dice shows a 5.
Are A and B mutually exclusive? Why or why not? *(Lesson 5-6)*

35. If it snowed three days last week in Barrow, Alaska, what percent of the week did it not snow? Answer to the nearest percent.
(Lesson 4-8)

36. a. Order from smallest to largest: $\frac{1}{4}$ $\frac{4}{7}$ $\frac{7}{40}$.
   b. Order from smallest to largest: $-\frac{1}{4}$ $-\frac{4}{7}$ $-\frac{7}{40}$.
   (Lessons 1-6, 1-8)

Way up north. *These children in Barrow, Alaska, are playing on a trampoline made from animal skins. Barrow, on the coast of the Arctic Ocean, is the northernmost community in the United States.*

Exploration

---

37. Examine this pattern involving unit fractions.
$$\frac{1}{2} = \frac{1}{3} + \frac{1}{6}$$
$$\frac{1}{3} = \frac{1}{4} + \frac{1}{12}$$
$$\frac{1}{4} = \frac{1}{5} + \underline{\quad ? \quad}$$

   a. Fill in the blank and then write the next two equations in the pattern.
   b. Use the pattern to write $\frac{1}{2}$ as the sum of ten different unit fractions.

A probable favorite? *Field hockey is a fast-paced team sport that became popular with women in the U.S. in the early 1900s. Women's field hockey became an Olympic sport at the 1980 Summer Games.*

In Lesson 5-6, you learned that when events *A* and *B* are mutually exclusive, the probability of the event *A or B* is the sum of the probabilities of *A* and *B*. For instance, in tossing two dice:

$$\text{probability of } A \text{ or } B = \text{probability of } A + \text{probability of } B$$

$$\text{probability of a sum of 7 or 11} = \text{probability of a sum of 7} + \text{probability of a sum of 11}$$

This is possible only because the sums of 7 and 11 cannot occur at the same time.

The Probability of *A and B*

What happens if *A* and *B can* occur at the same time? That is, what is the **probability of the event *A and B*** if the probability of *A* and the probability of *B* are known? It may surprise you that the probability can sometimes be found by multiplication. In this lesson, we examine when that can be done and when it cannot. We use the abbreviation *Prob(A)* for the probability of the event *A*.

Example 1

Two school teams are competing today. It is thought that the baseball team, with the amazing Kendall Hepitch on the mound, has an 85% chance of winning. The field hockey team, however, is thought to have only a 1 in 4 chance of winning. What is the probability that both teams will win?

▶

Solution

First name the individual events.

Let B = the baseball team wins. From the information, *Prob(B)* = 85%.

Let H = the field hockey team wins. From the information, *Prob(H)* = $\frac{1}{4}$.

B and H is the event that both teams win. The question asks for *Prob(B and H)*.

In this situation, we can assume that if one team wins, it does not affect the other team's result. So $\frac{1}{4}$ of the time the baseball team wins, the field hockey team will also win. This tells us how to find the answer:

$$Prob(B \text{ and } H) = 85\% \cdot \frac{1}{4} = 21.25\% \approx 21\%.$$

The probability that both teams will win is about 21%.

When Are Events Independent?

In Example 1, we say that winning the baseball game and winning the field hockey game are *independent events*. The idea of independent events is that the occurrence of one does not depend on the occurrence of the other. In general, two events *A* and *B* are **independent events** when and only when *Prob(A and B)* = *Prob(A)* · *Prob(B)*.

Example 2

On a TV game show there are three doors. The grand prize for the day is behind one of them. What is the probability that contestants will guess the correct door on the next two shows?

Solution 1

Let A be the event of guessing correctly on the next show. Let B be the event of guessing correctly on the second show. It is realistic to think that the guesses of contestants are random and independent.

Because of randomness, *Prob(A)* = $\frac{1}{3}$ and *Prob(B)* = $\frac{1}{3}$. Because of independence,

$$Prob(A \text{ and } B) = Prob(A) \cdot Prob(B) = \frac{1}{3} \cdot \frac{1}{3} = \frac{1}{9}.$$

Solution 2

Name the doors 1, 2, and 3. Then the possible guesses of doors for the two days are (1, 1), (1, 2), (1, 3), (2, 1), (2, 2), (2, 3), (3, 1), (3, 2), and (3, 3). Only one of these nine pairs will be correct, so The probability is $\frac{1}{9}$.

Notice how easy it is to calculate the probability of *A and B* when the events *A* and *B* are independent. The difficulty is deciding whether *A* and *B* are independent. For instance, consider the probability of rain. Storm systems tend to be in an area for more than one day. So rain today increases the probability of rain tomorrow. Rain today and rain tomorrow are not independent events.

Suppose the probability of rain today is $\frac{1}{3}$ and the probability of rain tomorrow is $\frac{1}{3}$. Is the probability of rain both days then $\frac{1}{9}$, as in Example 2? No. The events are not independent and there is no simple way to calculate the probability that they will both happen.

An Example with More than Two Independent Events

Arithmetic errors are usually not independent. If you make an error in one place, you are more likely to make an error someplace else. But for the next example, we assume that errors are independent. Notice that it is possible for three or more events to be independent.

Example 3

Suppose you are 98% accurate on basic arithmetic facts. A problem requires 10 basic facts. What is the probability that you will get all 10 facts correct?

Solution

Let E_1 = getting the first fact correct. Then $Prob(E_1)$ = 98% = .98. Similarly, let E_2 = getting the second fact correct, and so on, with E_{10} = getting the tenth fact correct. Then the event of getting all 10 facts correct is (E_1 and E_2 and E_3 and . . . and E_{10}). Assume the facts are independent. Then

$$Prob(E_1 \text{ and } E_2 \text{ and } E_3 \text{ and } \ldots \text{ and } E_{10})$$
$$= Prob(E_1) \cdot Prob(E_2) \cdot Prob(E_3) \cdot \ldots \cdot Prob(E_{10})$$
$$= \underbrace{.98 \cdot .98 \cdot .98 \cdot \ldots \cdot .98}_{10 \text{ factors}}$$
$$= .98^{10}$$
$$\approx .817$$

So there is about an 82% chance of getting all 10 facts correct.

There are many occasions when you might need to get 10 or more basic facts correct to solve a problem. But 82% accuracy in computation is not very high. This is why teachers want you to have even greater accuracy than 98% on basic facts, and why calculators are used today for complicated computations.

QUESTIONS

Covering the Reading

1. Translate into English words: *Prob(A)*.

2. Let Y = you wear a brown coat today.
   Let F = your best friend wears a brown coat today.
   a. Describe the event *Y or F*.   **b.** Describe the event *Y and F*.

3. Give a definition of *independent events*.

4. Suppose on a TV game show there are four doors. The grand prize for the day is behind one of them. One contestant each day is able to guess which door has the prize behind it.
  a. What is the probability that a contestant will guess the correct door on a particular day?
  b. What is the probability that the contestants will guess the correct door two days in a row?
  c. What is the probability that the contestants will guess the correct door three days in a row?
  d. Are guessing the correct doors today and tomorrow independent events?

5. Let A = it snows today. Let B = it snows tomorrow. Explain why A and B are not independent events.

6. Suppose you are 98% accurate when doing arithmetic computation. If a situation requires 5 independent arithmetic computations, what is the probability that you will get all 5 correct?

7. Explain why 98% accuracy in basic facts is not very high.

Applying the Mathematics

8. Consider the situation of Example 1.
  a. What is the probability that the baseball team will lose?
  b. What is the probability that the field hockey team will lose?
  c. What is the probability that both teams will lose?

9. Answer Question (2) on page 467.

10. Suppose a basketball player makes 74% of his free throws. If free throws are independent events, what is the probability that the player makes two free throws in a row?

11. About $\frac{9}{10}$ of people are right-handed, and $\frac{1}{10}$ are left-handed. People can also be "right-eyed" or "left-eyed." According to the *Encyclopedia Britannica,* "about three fourths of right-handed and one third of left-handed persons are right-eyed in sighting." Let A = a person is right-handed, and let B = a person is right-eyed. Are A and B independent events? Explain why or why not.

12. One summer camp has 136 applicants for 50 openings. A second camp has 200 applicants for 65 openings. Suppose each camp fills its openings randomly from the applicants. If you were to apply to both camps, what is the probability that both would accept you?

13. Suppose you are 98% accurate at arithmetic computation without a calculator and 99% accurate with a calculator. How much larger is the probability of getting 100 independent arithmetic computations correct with the calculator?

14. Two dice are tossed once. Explain why tossing a sum of 7 and tossing a sum of 11 are not independent events.

Review

In 15–20, calculate. *(Lessons 5-5, 9-3)*

15. $\frac{3}{4} \cdot 4$

16. $\frac{1}{2} \cdot \frac{3}{4} \cdot \frac{2}{3}$

17. $\frac{1}{2} + \frac{1}{2} \cdot \frac{1}{5}$

18. $2.6 + \frac{7}{25}$

19. $x \cdot \frac{1}{x}$

20. $\frac{1}{a} \cdot \frac{1}{b}$

21. If the dimensions of a rectangle are multiplied by 5, what happens to its area? Give two examples and the general pattern. *(Lessons 4-2, 9-1)*

22. Here are the seating capacities of American League baseball stadiums:

Team	Stadium (year built)	Seating Capacity
Baltimore Orioles	Oriole Park at Camden Yards (1992)	48,000
Boston Red Sox	Fenway Park (1912)	33,925
California Angels	Anaheim Stadium (1966)	64,593
Chicago White Sox	Comiskey Park (1991)	44,321
Cleveland Indians	Cleveland Stadium (1932)	74,483
Detroit Tigers	Tiger Stadium (1912)	52,416
Kansas City Royals	Kauffman Stadium (1973)	40,625
Milwaukee Brewers	Milwaukee County Stadium (1953)	53,192
Minnesota Twins	Hubert H. Humphrey Metrodome (1982)	55,883
New York Yankees	Yankee Stadium (1923)	57,545
Oakland A's	Oakland-Alameda County Coliseum (1968)	47,313
Seattle Mariners	Kingdome (1977)	58,823
Texas Rangers	Arlington Stadium (1965)	43,521
Toronto Blue Jays	SkyDome (1989)	50,516

More than just a stadium. *The SkyDome is home to Toronto's professional football team, the Argonauts, and baseball team, the Blue Jays. The SkyDome covers 8 acres; features a retractable roof that can be closed in a matter of minutes; and includes a hotel and entertainment mall.*

a. In your head, round the capacities down to the nearest thousand. Then put the rounded values into a stem-and-leaf display.

b. What is the median capacity? *(Lessons 1-3, 8-1)*

Exploration

23. a. Give an example different from any in this lesson of two events that are not independent.

b. Give an example different from any in this lesson of two events that are independent.

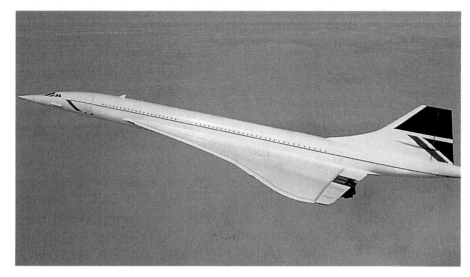

Faster than sound. *The BAC/Aérospatiale Concorde, shown here, is the only supersonic jet used for passenger service. It can cruise at speeds up to Mach 2.2— which is 2.2 times as fast as the speed of sound, or about 1450 mph at 36,000 ft.*

What Is a Rate?

A quantity is a rate when its unit contains the word "per" or "for each" or some synonym. Here are some examples of *rates*.

> 55 *miles per hour* (speed limit)
> 25.3 *students per class* (average class size)
> 107 *centimeters* of snow *per year* (average snowfall in Boston)
> $2\frac{1}{2}$ *pieces for each student* (result of splitting up a pie)

Every rate has a **rate unit.** Above, the rate units are in italics.

When used in an expression, the slash / or horizontal bar—means "per." Notice how the above rate units are written in fraction notation.

Using the slash		Using the bar
55 mi/hr	$=$	$55\,\frac{\text{mi}}{\text{hr}}$
25.3 students/class	$=$	$25.3\,\frac{\text{students}}{\text{class}}$
107 cm/yr	$=$	$107\,\frac{\text{cm}}{\text{yr}}$
$2\frac{1}{2}$ pieces/student	$=$	$2\frac{1}{2}\,\frac{\text{pieces}}{\text{student}}$

Multiplying by a Rate

Suppose a woman gains 2 pounds per month during pregnancy. Her rate of weight gain is then

$$2\,\frac{\text{pounds}}{\text{month}}.$$

Suppose she keeps this rate up for 5 months. Multiplication gives the total she gains.

$$5 \text{ months} \cdot 2 \frac{\text{pounds}}{\text{month}} = 10 \text{ pounds}$$

Look at the units in the multiplication above. They work as if they were whole numbers and fractions. The unit "months" at left cancels the unit "month" in the denominator. The unit that remains is pounds. You can see where the 10 comes from.

Here is a similar example of multiplying by a rate.

Example 1

A person buys 7 cans of tuna at $1.39 per can. What is the total cost?

Solution

Here is the way it looks with all units included.

$$7 \text{ cans} \cdot 1.39 \frac{\text{dollars}}{\text{can}} = 9.73 \text{ dollars}$$

Here is how it looks using the usual dollar signs and a slash.

$$7 \text{ cans} \cdot \$1.39/\text{can} = \$9.73$$

The Rate Factor Model for Multiplication

The quantity $1.39 \frac{\text{dollars}}{\text{can}}$ in Example 1 is a rate that is multiplied. So it is a **rate factor.** The general idea behind Example 1 is the Rate Factor Model for Multiplication.

Rate Factor Model for Multiplication
The product of (a $unit_1$) and $\left(b \frac{unit_2}{unit_1}\right)$ is (ab $unit_2$), signifying the total amount of $unit_2$ in the situation.

Some multiplications from earlier lessons in this chapter can be thought of as rate factor multiplications.

Example 2

(Array) An auditorium contains 12 rows with 19 seats per row. How many seats are in the auditorium?

Solution

Here $unit_1$ is rows, and $\frac{unit_2}{unit_1}$ is $\frac{\text{seats}}{\text{row}}$. $Unit_2$, the unit of the product, is seats.

$$12 \text{ rows} \cdot 19 \frac{\text{seats}}{\text{row}} = 228 \text{ seats}$$

A run for your money. *In recent years, the New York City Marathon has begun with about 26,000 entrants crossing the 4260-ft-long Verrazano-Narrows Bridge. Usually over 95% of the entrants complete the marathon.*

Converting Units Using Rates

A *conversion factor* is a special kind of rate factor. Start with a conversion formula. Here is an example.

$$1 \text{ mile} = 5280 \text{ feet}$$

Since the quantities are equal, dividing one by the other gives the number 1.

$$\frac{1 \text{ mile}}{5280 \text{ feet}} = 1 \quad \text{and} \quad \frac{5280 \text{ feet}}{1 \text{ mile}} = 1$$

You could say there is 1 mile for every 5280 feet or there are 5280 feet for every mile. The quantities $\frac{1 \text{ mile}}{5280 \text{ feet}}$ and $\frac{5280 \text{ feet}}{1 \text{ mile}}$ are conversion factors. A conversion factor is a rate factor that equals the number 1.

Example 3

(Conversion) 40,000 feet equals how many miles?

Solution

Multiply 40,000 feet by the conversion factor with miles in its numerator and feet in its denominator. The mile unit is needed in the product, so feet needs to be in the denominator to cancel.

$$40,000 \text{ feet} \cdot \frac{1 \text{ mile}}{5280 \text{ feet}} = \frac{40,000}{5280} \text{ miles}$$
$$= 7.5\overline{7} \text{ miles}$$

Example 4

(Conversion) 26.2 miles, the approximate length of a marathon, is how many feet?

Solution

Use the conversion factor with miles in its denominator.

$$26.2 \text{ miles} \cdot \frac{5280 \text{ feet}}{1 \text{ mile}} = 138,336 \text{ feet}$$

Uta Pippig of Germany, on the left wearing F3, won the women's division of the 1993 New York City Marathon with a time of 2:26:24. Andres Espinosa of Mexico won the men's division with a time of 2:10:04.

Multiplying Two or More Rates

In some situations, two rates are to be multiplied. Again the units are multiplied as you would multiply fractions.

Example 5

An administrative assistant has 8 letters to type. Each letter is 4 pages long. Each page takes 6 minutes to type. How long will it take to type all the letters?

Solution

Think rates. There are 4 pages per letter. There are 6 minutes per page.

$$8 \text{ letters} \cdot 4 \frac{\text{pages}}{\text{letter}} \cdot 6 \frac{\text{minutes}}{\text{page}}$$

$$= 32 \text{ pages} \cdot 6 \frac{\text{minutes}}{\text{page}}$$

$$= 192 \text{ minutes}$$

QUESTIONS

Covering the Reading

In 1 and 2, a sentence is given.
a. Copy the sentence and underline the rate.
b. Write the rate with its unit in fraction notation using the slash.
c. Write the rate with its unit in fraction notation using the fraction bar.

1. Only 4 tickets per student are available for the basketball game.

2. The speed limit there is 45 miles per hour.

3. Consider the multiplication $3 \text{ hr} \cdot 50 \frac{\text{km}}{\text{hr}} = 150 \text{ km}$.
   a. Identify the rate factor.
   b. Identify the unit of the rate factor.
   c. Make up a situation that could lead to this multiplication.

In 4–7, do the multiplication.
4. $5 \text{ classes} \cdot 25 \frac{\text{students}}{\text{class}} \cdot 30 \frac{\text{questions}}{\text{student}}$

5. $8.2 \text{ pounds} \cdot \$2.29/\text{pound}$

6. $\frac{1.06 \text{ quarts}}{\text{liter}} \cdot 6 \text{ liters}$

7. $2 \frac{\text{games}}{\text{week}} \cdot 8 \frac{\text{weeks}}{\text{season}}$

8. a. Do this multiplication. $30{,}000 \text{ ft} \cdot \frac{1 \text{ mile}}{5280 \text{ ft}}$
   b. Make up a question that leads to this multiplication.

9. If an animal gains 1.5 kg a month for 7 months, how many kg will it have gained in the seven months?

10. A typist can type 40 words a minute. At this rate, how many words can be typed in 20 minutes?

11. Name the two conversion factors for converting between feet and miles.

12. An airplane flying 35,000 feet up is how many miles up?

13. A section of a football stadium has 50 rows with 20 seats in each row. How many people can be seated in this section?

Applying the Mathematics

14. Recall that 1 inch = 2.54 cm.
    a. What two conversion factors does this formula suggest?
    b. Use one of the factors to convert 10 cm to inches.

15. If a heart beats 70 times a minute, how many times will it beat in a day?

16. Solve for *x:* 4 hours · *x* = 120 miles.

17. When the twins went to France in August of 1993, they found that 5.7 francs were worth 1 dollar. In Spain, they found that 1 franc was worth 22.8 pesetas. How many pesetas could they get for $10?

18. Suppose a particular car has a 14-gallon tank and can get about 25 miles per gallon. What do you get when you multiply these quantities?

Review

19. Sarah feels she has a 2 out of 7 chance of getting a B in English and a 4 out of 5 chance of getting an A in mathematics. If these events are independent, what is the probability that both will happen? *(Lesson 9-4)*

20. Suppose the probability that you win a game of bingo is $\frac{1}{15}$.
    a. What is the probability that you will win the first two games you play?
    b. What is the probability that you will win the first three games you play? *(Lesson 9-4)*

21. Three shirts are in a drawer and four skirts are on hangers. In the dark, you pick out a shirt and skirt. What are your chances of choosing the particular shirt and skirt you want? *(Lesson 9-4)*

22. Rebecca had $86.48 to spend during her vacation. She spent half of her money the first day and one fourth of what was left on the second day. How much money did she have left after the second day? *(Lesson 9-3)*

BINGO!

23. Central Park in New York City is shaped like a rectangle, about $2\frac{1}{2}$ miles long and $\frac{1}{2}$ mile wide. What is its area? *(Lessons 9-1, 9-3)*

A downtown haven.
Central Park, in the middle of Manhattan, is the largest of New York City's parks. It includes the Mall, where outdoor concerts are held, and the Ramble, an enclosed bird sanctuary.

24. *Multiple choice.* $\frac{2}{3} - \left(-\frac{2}{3}\right) =$

(a) $-\frac{4}{3}$ (b) 0 (c) $\frac{4}{9}$ (d) $\frac{4}{3}$ (e) none of these

(Lessons 5-5, 7-2)

25. Seven teams are to play each other in a tournament. Each team will play every other team once. How many games will be played? *(Lesson 6-3)*

26. A diving team went 267.8 feet below sea level and then rose 67.9 feet.
a. What addition tells where the team would be?
b. What was their final position? *(Lessons 5-1, 5-3)*

In 27–29, give an abbreviation for the unit. *(Lessons 3-3, 3-4)*

27. pound **28.** kilogram **29.** milliliter

Exploration

30. Depending on the number of words it contains, a work of prose is often classified as a short story, a novelette, or a novel. A short story is usually less than 10,000 words. To find the number of words in a work of prose, editors rarely count them all, but use rate factors.
a. Describe how you might estimate the number of words in a work of prose.
b. Estimate the number of words in this chapter.

Multiplication with Negative Numbers and Zero

A negative rate. *Sometime lakes are "drawn down" or drained in winter to prevent shoreline damage from the freezing water. These park rangers are controlling the rate at which water flows out of Pine Barrens Lake in New Jersey.*

A Situation Involving Multiplication by a Negative Number

Any loss can lead to a negative rate. Suppose a person loses 2 pounds per month on a diet. The rate of change in weight is then

$$-2 \frac{\text{pounds}}{\text{month}}.$$

If the person keeps losing weight at this rate for 5 months, multiplying gives the total loss.

$$5 \text{ months} \cdot -2 \frac{\text{pounds}}{\text{month}} = -10 \text{ pounds}$$

Ignoring the units, a positive and a negative number are being multiplied. The product is negative.

$$5 \cdot -2 = -10$$

The number of months does not have to be a whole number. The rate does not have to be an integer.

Example 1

On a medically supervised diet lasting $3\frac{1}{2}$ months, a person lost 2.4 kg per month. What was the net change in weight?

Solution

Use the Rate Factor Model for Multiplication. The rate is $-2.4 \frac{\text{kg}}{\text{month}}$.

$$-2.4 \frac{\text{kg}}{\text{month}} \cdot 3\frac{1}{2} \text{ months}$$
$$= -2.4 \frac{\text{kg}}{\text{month}} \cdot 3.5 \text{ months}$$
$$= -8.4 \text{ kg}$$

The net change is a loss of 8.4 kg.

Example 1 demonstrates that when a positive number is multiplied by a negative number, the product is negative.

Example 2

What is 5 · -4.8?

Solution

Think of this as a loss of 4.8 kg per month for 5 months. The result is a loss, so the answer is negative. Since 5 · 4.8 = 24,

$$5 \cdot -4.8 = -24.$$

Example 3

What is -7 · 4?

Solution

-7 · 4 = 4 · -7. Think of this as a loss of 7 kg each month for 4 months.

$$-7 \cdot 4 = -28$$

Multiplication with Two Negative Numbers

What happens if both numbers are negative? Again think of a weight-loss situation. Suppose a person loses 2 pounds a month. At this rate, 5 months *ago* the person weighed 10 pounds more. Going back is the negative direction for time. So here is another multiplication for this situation.

$$5 \text{ months ago} \cdot \text{loss of } 2 \frac{\text{pounds}}{\text{month}} = 10 \text{ pounds more}$$
$$-5 \text{ months} \cdot -2 \frac{\text{pounds}}{\text{month}} = 10 \text{ pounds}$$
Ignoring the units: $\quad -5 \cdot -2 \qquad\qquad = 10$

In general, the product of two negative numbers is a positive number.

There is another way to think of multiplication with positive and negative numbers.

When a number is multiplied by a positive number, the sign of the product (positive or negative) is the same as the sign of the number.

When a number is multiplied by a negative number, the sign of the product (positive or negative) is the opposite of the sign of the number.

Example 4

Is $-2 \cdot -5 \cdot 3 \cdot -4 \cdot 6 \cdot -2 \cdot -1 \cdot -3$ positive or negative?

Solution

Each multiplication by a negative changes the product's sign. Each two changes keeps it positive. Since there are 6 negative numbers, the sign is positive.

Check

Use your calculator to verify that the product is 4320.

Caution: The kind of thinking in the Solution to Example 4 works because the numbers are all multiplied. It does not work for addition or subtraction.

Multiplication by -1

Multiplication by -1 follows the above pattern, but it is even more special.

$$-1 \cdot 53 = -53$$
$$-1 \cdot -4 = 4$$
$$-1 \cdot \frac{2}{3} = -\frac{2}{3}$$

Multiplication by -1 changes a number to its opposite. Here is a description with variables.

> **Multiplication Property of -1**
> For any number x, $-1 \cdot x = -x$.

You have now seen that the product of a positive number and a negative number is negative. You have also seen that the product of two negative numbers is positive. Of course you already know that the product of two positive numbers is positive. These facts can be summarized as follows:

> The product of two numbers with the same sign is positive. The product of two numbers having opposite signs is negative.

Multiplication by 0

What about multiplying numbers by zero, a number which is neither positive nor negative? Notice what happens when the number 5 is multiplied by positive numbers that get smaller and smaller.

$$5 \cdot 2 = 10$$
$$5 \cdot 0.43 = 2.15$$
$$5 \cdot 0.07 = 0.35$$
$$5 \cdot 0.000148 = 0.00074$$

The smaller the positive number, the closer the product is to zero. You know what happens when 5 is multiplied by 0. The product is 0.

$$5 \cdot 0 = 0$$

There is a similar pattern if you begin with a negative number. Multiply -7 by the same numbers as above.

$$-7 \cdot 2 = -14$$
$$-7 \cdot 0.43 = -3.01$$
$$-7 \cdot 0.07 = -0.49$$
$$-7 \cdot 0.000148 = -0.001036$$

The products are all negative, but again they get closer and closer to zero. You can check the following multiplication on your calculator.

$$-7 \cdot 0 = 0$$

These examples are instances of an important property of zero.

Multiplication Property of Zero
For any number x, $x \cdot 0 = 0$.

QUESTIONS

Covering the Reading

1. What is the negative rate in this sentence? A person loses 3.8 pounds per month for 3 months.

In 2–4, a situation is given.
a. What multiplication problem involving negative numbers is suggested by the situation?
b. What is the product and what does it mean?

2. A person loses 3 pounds a month for 2 months. How much will the person lose in all?

3. A person loses 5 pounds a month. How will the person's weight 4 months from now compare with the weight now?

4. A person has been losing 6 pounds a month. How does the person's weight 2 months ago compare with the weight now?

Controlling weight.
Bicycle riding is an excellent way to exercise and maintain good health.

In 5 and 6, fill in the blank with one of these choices.
(a) is always positive   (b) is always negative
(c) is sometimes positive, sometimes negative

5. The product of two negative numbers __?__.

6. The product of a positive and a negative number __?__.

In 7–12, simplify.

7. $-4 \cdot 8$ **8.** $73 \cdot -45$ **9.** $-6 \cdot -3$

10. $-2 \cdot 5$ **11.** $1.8 \cdot -3.6$ **12.** $-4.1 \cdot -0.3$

In 13–16, tell whether xy is positive or negative.

13. x is positive and y is positive.   **14.** x is positive and y is negative.

15. x is negative and y is positive.   **16.** x is negative and y is negative.

17. State the Multiplication Property of 0.

In 18–21, simplify.

18. $8 \cdot -1$ **19.** $-1 \cdot -1 \cdot -1$

20. $0 \cdot -6$ **21.** $-3 \cdot -2 \cdot -1 \cdot 0$

22. Multiplication of a number by __?__ results in the opposite of the number.

23. Tell whether $-5 \cdot -4 \cdot -3 \cdot -2 \cdot -1 \cdot 1 \cdot 2 \cdot 3 \cdot 4 \cdot 5$ is positive, negative, or zero. Then explain why.

24. Calculate.
    a. $3 + -3$ **b.** $3 - (-3)$ **c.** $3 \cdot -3$

Applying the Mathematics

25. Find the value of $-5x$ for the given value of x.
    a. 2   **b.** 1   **c.** 0   **d.** -1   **e.** -2

26. Evaluate $3 + -7a + 2b$ when $a = -4$ and $b = -10$.

27. In Europe, the number zero is sometimes called the *annihilator*. What is the reason for this name?

In 28–31, simplify.

28. $-5 \cdot -5 \cdot -5 \cdot -5$ **29.** $(-4)^3$

30. $-\frac{1}{2} \cdot \frac{7}{3}$ **31.** $\left(\frac{2}{5}\right)\left(-\frac{5}{7}\right)\left(-\frac{7}{2}\right)$

32. *Multiple choice.* Morry is now $2500 in debt. He has been incurring additional debt at the rate of $200/week. Which expression tells how he was doing 5 weeks ago?
    (a) $2500 - 5 \cdot 200$ (b) $-2500 + -200 \cdot -5$
    (c) $-2500 + 5 \cdot -200$ (d) $2500 + -5 \cdot -200$

In 33 and 34, write the rate using fraction notation and abbreviations.   *(Lesson 9-5)*

33. kilometers per second

34. grams per cubic centimeter

35. A computer prints at the rate of 12 pages/min. How long will it take to print 2400 documents with 3 pages per document?   *(Lesson 9-5)*

36. Joann runs 5.85 miles each hour. At the same rate, how far will she run in 3 hours?   *(Lesson 9-5)*

37. If your probability of spelling a hard word correctly is 90%, what is the probability that you will correctly spell all 10 hard words on a test?   *(Lesson 9-4)*

38. **a.** Picture a rectangular park that is $\frac{2}{3}$ mile by $\frac{3}{5}$ mile.
b. Calculate the area of the park.   *(Lesson 9-3)*

39. If the edges of a cube are tripled in length, what happens to its volume?   *(Lessons 9-2, 3-9)*

40. What are supplementary angles?   *(Lesson 7-6)*

41. Give an example of a positive number and a negative number whose sum is positive.   *(Lessons 5-1, 5-3)*

42. Graph all solutions to $x > 5$ on a number line.   *(Lesson 4-10)*

43. Find an example of a rate in a newspaper or magazine.

An expanded view. *This projection of a computer image onto the large screen is an example of an expansion. The two images are similar.*

Size Changes of Lengths

The segment below has length *L*.

L ────

Place 2 such segments end to end to form a new segment. The total length of the new segment is *L* + *L* or 2*L*. Multiplying by 2 lengthens the segment. The new segment has twice the length of the original.

L + *L* ●━━━━●━━━━●
2*L* ━━━━━━━

Place another copy. The length is 2*L* + *L* or 3*L*. Multiplying by 3 expands the segment even more.

3*L* ━━━━━━━━━━

Activity 1

Consider the segment above with length *L*. Draw a segment with length 2.5*L*.

These examples show that multiplying by a number can be pictured as changing the size of things. The number is called the **size change factor.** Above, the size change factors are 2, 3, and 2.5.

Size Changes with Other Quantities

The idea of size change can involve quantities other than lengths.

Example 1

John weighed 6.3 pounds at birth. By the time he was a year old, his weight had tripled. How much did he weigh at age 1?

Solution

"Tripled" means a *size change of magnitude 3,* so multiply John's birth weight by 3. He weighed 3 · 6.3, or 18.9 pounds.

Example 2

Maureen works in a store for $7 an hour. She gets time and a half for overtime. What does she make per hour when she works overtime?

Solution

"Time and a half" means to multiply by $1\frac{1}{2}$ or 1.5. Working overtime, she makes 1.5 · $7.00, or $10.50 per hour.

In each of the above examples, there is a beginning quantity and a size change factor. They are multiplied together to obtain a final quantity. This is the idea behind the Size Change Model for Multiplication.

> **Size Change Model for Multiplication**
> Let k be a nonzero number without a unit. Then ka is the result of applying a **size change of magnitude k** to the quantity a.

Picturing Size Changes

A size change can be nicely pictured using coordinates. We begin with a quadrilateral. Then we multiply both coordinates of all its vertices by 2. The image has sides 2 times as long as the preimage. Also, the image sides are parallel to the corresponding preimage sides. The image quadrilateral has the same shape as the preimage.

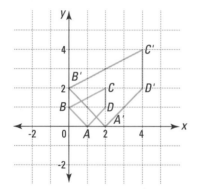

Preimage point	Image point
$A = (1, 0)$	$A' = (2, 0)$
$B = (0, 1)$	$B' = (0, 2)$
$C = (2, 2)$	$C' = (4, 4)$
$D = (2, 1)$	$D' = (4, 2)$

This transformation is called a *size change of magnitude 2*. The image of any point (x, y) is $(2x, 2y)$. Because the image figure is bigger, this size change is also called an **expansion** of magnitude 2. The number 2 is the size change factor.

Activity 2

Draw quadrilateral *ABCD* as shown on page 506 on graph paper. Multiply the coordinates of its vertices by 3. Call the image points *A**, *B**, *C**, and *D**. Draw *A*B*C*D**.

The transformation in Activity 2 is a size change of magnitude 3. In general, under a size change of magnitude 3, the image of (x, y) is $(3x, 3y)$.

Two figures that have the same shape are called **similar figures.** Under a size change, the preimage and the image are always similar. In Activity 2, you should find that $A^*B^*C^*D^*$ is bigger than $A'B'C'D'$. But it is still similar to *ABCD*.

Activity 3

On a new graph, multiply the coordinates of *ABCD* by 2.25. Draw the preimage and image. How do they compare?

In Activity 3, you have performed a size change of magnitude 2.25. Under this transformation, the image of (x, y) is $(2.25x, 2.25y)$. Activities 2 and 3 are instances of a two-dimensional size change.

Size Change Model for Multiplication (two-dimensional version)
Under a size change of magnitude k, the image of (x, y) is (kx, ky).

Size Changes with Magnitude 1

If the magnitude of a size change is 1, then the image of (x, y) is $(1 \cdot x, 1 \cdot y)$, or (x, y) itself. That is, each point is its own image. It keeps its identity. That is how you can remember the name of the *Multiplicative Identity Property of One.*

Multiplicative Identity Property of One
For any number x, $1 \cdot x = x$.

Now you see it, now you don't. *Some people take up magic as a hobby and earn money performing at parties.*

QUESTIONS

Covering the Reading

1. Suppose this segment has length x.

   ——————

   a. Copy this segment and then draw a segment of length $2x$.
   b. Draw a segment of length $3x$.
   c. Draw a segment of length $3.5x$.

2. Show the results of Activity 1.

3. Draw a line segment that is 1.5 times as long as the segment at the right.

4. David saved $25.50. After doing three magic shows, his savings had tripled.
   a. What size change factor is meant by the word *tripled?*
   b. How much money does he have now?

5. Megan earns $2.50 an hour for baby-sitting. After 10 P.M., she gets time and a half.
   a. What does she make per hour after 10 P.M.?
   b. What will Megan earn for baby-sitting from 7 P.M. to 11 P.M.?

6. Show the results of Activity 2.

7. Show the results of Activity 3.

8. Under a size change, a figure and its image have the same shape. They are called __?__ figures.

9. A triangle has vertices (1, 4), (2, 3), and (1, 2). Graph the triangle and its image under a size change with magnitude 3.

10. Under a size change with magnitude k, the image of (x, y) is __?__.

11. a. Graph the quadrilateral with vertices (1.5, 4), (4, 0), (2, 1), and (1, 2).
    b. Find the image of this quadrilateral under an expansion of magnitude 2.
    c. How do lengths of the sides of the image compare with lengths of the sides of the original quadrilateral?
    d. How do the measures of the angles in the preimage and image compare?

12. State the Multiplicative Identity Property of One.

13. A microscope lens magnifies 150 times. Some human hair is about 0.1 mm thick.
    a. How thick would the hair appear under this microscope?
    b. What is the size change factor in this situation?

14. The bookcase pictured is $\frac{1}{40}$ of the actual size. This means that the actual bookcase is 40 times as wide, 40 times as high, and 40 times as deep. Will the actual bookcase fit a space 5 feet wide?

15. An object is 4 cm long. A picture of the object is 12 cm long. The image of the object is __?__ times actual size.

16. Each word suggests an expansion. What is the magnitude of the expansion?
    a. doubled
    b. quintupled
    c. octupled
    d. quadrupled

In 17–20, perform the indicated operations. *(Lessons 9-6, 9-3, 5-5)*

17. $-1 \cdot -8 + -11 \cdot -8$

18. $-\frac{1}{2} \cdot -\frac{2}{3}$

19. $\left(\frac{1}{2} + -\frac{1}{6}\right) \cdot (-30)$

20. $6 \cdot -\frac{1}{6} + 5 \cdot 0 \cdot -\frac{1}{5}$

In 21–26, let $x = 10$ and $y = -1$. Give the value of each expression. *(Lessons 9-6, 4-4)*

21. xy

22. $-x \cdot -y$

23. x^y

24. y^x

25. $x^2 + y^2$

26. $x^3 + y^3$

27. About 51% of the babies born in the United States are boys. Suppose a family has two children. Assume the births of babies are independent events. How much more likely is a family to have two boys than two girls? *(Lesson 9-4)*

28. Here are the films that have generated the highest video rental incomes (as of January, 1993) of all those released in the given year. (Source: *Variety*, January, 1993)

Year Released	Movie Title	Total Video Rentals (in millions of dollars)
1980	The Empire Strikes Back	141.7
1981	Raiders of the Lost Ark	115.6
1982	E.T., the Extra-Terrestrial	228.6
1983	Return of the Jedi	169.2
1984	Ghostbusters	132.7
1985	Back to the Future	105.5
1986	Top Gun	79.4
1987	Three Men and a Baby	81.4
1988	Rainman	86.8
1989	Batman	150.5
1990	Home Alone	140.1
1991	Terminator 2	112.5
1992	Home Alone 2	102.0

A shot from *Ghostbusters*. *Shown here is a scene with the loveable marshmallow man.*

a. Give the mean of the total rental amounts.
b. Give the median of the total rental amounts.   *(Lessons 8-1, 4-6)*

29. Solve $x - 64.2 = -18.9$.   *(Lesson 7-3)*

30. A horse race was $1\frac{1}{8}$ miles long. For the first third of a mile, Lover's Kiss led the race. For the next $\frac{3}{4}$ mile, Kisser's Hug was in the lead. For the rest of the race, Hugger's Love was in front. For how long a distance did Hugger's Love lead the race?   *(Lessons 7-1, 5-5)*

31. Give an example of a negative number whose absolute value is between 0 and 1.   *(Lesson 5-3)*

32. Marlene needs to drive 500 km to visit her grandfather in Canada. Can she do this in a day without speeding? Explain why or why not. *(Lessons 9-5, 3-5)*

Exploration

33. What is the magnitude of a size change you might find in an electron microscope?

Poly-contractions. *The colorful bird in the contractions is a macaw, the largest member of the parrot family. Macaws may be found from Mexico to Bolivia. Many of the 315 parrot species are endangered because habitats have been destroyed by humans.*

What Are Contractions?

In Lesson 9-7, all of the size change factors were numbers larger than 1. So the products are larger than the original values. In this lesson, we multiply by numbers between 0 and 1. This results in products with smaller values. For instance, if 8 is multiplied by 0.35, the product is 2.8. The number 2.8 is smaller than 8.

This can be pictured. Consider the segments below.

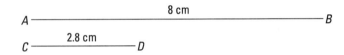

$AB = 8$ cm. $CD = 0.35 \cdot 8$ cm $= 2.8$ cm. CD is shorter.

Think of $\overline{AB}$ as having been reduced in size to 2.8 cm by a shrinking, or a *contraction*. A **contraction** is a size change with a magnitude whose absolute value is between 0 and 1. In the situation above, the *magnitude of the contraction* is 0.35.

Picturing Contractions

Below a two-dimensional contraction is pictured. The preimage is the large pentagon *PENTA*. We have multiplied the coordinates of all its vertices by $\frac{1}{2}$. Since $\frac{1}{2}$ is between 0 and 1, the resulting image $P'E'N'T'A'$ is smaller than the original. Again the preimage and image are similar figures. In this case, sides of the image have $\frac{1}{2}$ the length of sides of the preimage. Corresponding angles on the preimage and image have the same measure.

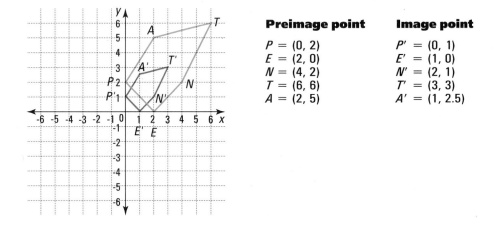

Preimage point	Image point
$P = (0, 2)$	$P' = (0, 1)$
$E = (2, 0)$	$E' = (1, 0)$
$N = (4, 2)$	$N' = (2, 1)$
$T = (6, 6)$	$T' = (3, 3)$
$A = (2, 5)$	$A' = (1, 2.5)$

So the image of (x, y) is $\left(\frac{1}{2}x, \frac{1}{2}y\right)$.

Contractions with Other Quantities

Contractions also occur with quantities that are not lengths.

Example 1

Suppose an airline has a plan in which a spouse can travel for $\frac{2}{3}$ the regular price and a child under 12 can travel for $\frac{1}{2}$ the regular price. Under this plan, what will it cost a married couple and their 11-year-old child to fly if the regular price of a ticket is $150?

Solution

It will cost one adult $150.
It will cost the spouse 2/3 of $150. This can be calculated by multiplying fractions: $\frac{2}{3} \cdot \$150 = \frac{\$300}{3} = \$100$.
It will cost the child $\frac{1}{2}$ of $150. That is $\frac{1}{2} \cdot \$150$, or $75.
The total cost will be $150 + $100 + $75, or $325.

In Example 1, the size change factors are $\frac{2}{3}$ and $\frac{1}{2}$. Size change factors for contractions are often written as fractions or percents. You have seen the type of question in Example 2 before. Now you can think of it as an example of size change multiplication.

Example 2

About 1.3% of the population of the United States are 85 or older. If the U.S. population is about 250,000,000, how many people are 85 or older?

Solution

The size change factor is 1.3%, or 0.013.
0.013 · 250,000,000 people = 3,250,000 people.
About 3.25 million people in the United States are over 85.

QUESTIONS

Covering the Reading

Twins Millie Reiger and Addie Moran were 100 years old on April 8, 1993.

1. **a.** By measuring, determine t in the drawing at the right to the nearest millimeter.
   b. Draw a segment with length $0.3t$.
   c. What is the magnitude of the contraction?

2. What is a *contraction?*

In 3–5, give the image of the point (6, 4) under the contraction with the given magnitude.

3. 0.5   4. 0.75   5. $\frac{1}{4}$

6. **a.** Graph the pentagon with vertices (10, 10), (10, 5), (0, 0), (0, 5), and (5, 10).
   b. Graph its image under a contraction with magnitude 0.8.
   c. How do the lengths of the corresponding sides of the preimage and image compare?
   d. How do the measures of corresponding angles in the preimage and image compare?

7. Repeat Example 1 if the regular cost of a ticket for an adult is $210.

8. About 9.0% of the U.S. population in 1990 were classified as being of Hispanic origin. If the population then was about 248,700,000, about how many people in the U.S. in 1990 were of Hispanic origin?

9. In Question 8, what is the size change factor?

Applying the Mathematics

10. The actual damselfly is similar but half the length of its picture at the left. How long is the actual damselfly?

11. In doing Question 10, some students measure lengths of this insect in centimeters. Others measure in inches. If they measure accurately, will they get equal values for the length of the damselfly?

12. What is the image of (12, 5) under a contraction with magnitude k?

13. *Multiple choice.* A car was priced at $7000. The salesperson offers a discount of $350. What size change factor, applied to the original price, gives the discount?
(a) 0.02   (b) 0.05   (c) 0.20   (d) 0.245

14. *Multiple choice.* A car was priced at $7000. The salesperson offers a discount of $350. What size change factor, applied to the original price, gives the offered price?
(a) 0.05   (b) 0.20   (c) 0.80   (d) 0.95

15. A size change of magnitude $\frac{1}{2}$ was applied to a pentagon in this lesson. Suppose the preimage and image were switched.
a. A size change of what magnitude would now be pictured?
b. Generalize the idea of this question.

Review

16. Jay asks for $5 an hour to mow a lawn. On holidays, he asks for time and a half. How much will he earn for a three-hour job on Memorial Day? *(Lesson 9-7)*

17. If the length of a 12.5 cm segment is quintupled, what is its new length? *(Lesson 9-7)*

18. Give the value of $-4(x - 1)^2$ when x has the indicated value. *(Lessons 9-6, 5-3, 4-5)*
a. 2   **b.** 1   **c.** 0   **d.** -1   **e.** -2

19. According to the Bureau of the Census, in 1990 about 13.5% of people in the United States were classified as "poor." About 80.3% of the population were classified as "White." *(Lesson 9-4)*
a. If being poor is independent of being white, what percent of the population would you expect to be classified as "poor and white"?
b. Actually, about 9.0% of the population were classified as "poor and white." Explain why this number is lower than the answer to part **a.**

20. How many boxes 50 cm long, 30 cm wide, and 20 cm high can be packed into a railroad car 3 m wide, 3 m high, and 15 m long? *(Lesson 9-2)*

21. Negative five is added to the second coordinate of every point on the graph of a figure. What happens to the graph of that figure? *(Lesson 8-5)*

22. Graph the line $x - y = 6$. *(Lesson 8-4)*

23. List three reasons for using graphs. *(Lesson 8-1)*

In 24–26, use the figure below. *(Lesson 7-7)*

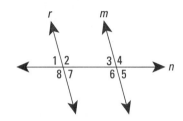

24. Angle 3 and __?__ are corresponding angles.

25. Name the pairs of alternate interior angles.

26. If lines r and m are parallel and m$\angle 3 = 70°$, what are the measures of the other angles?

27. Joe is saving to buy a present for his father's birthday. He has $5 now and adds $2 a week. *(Lesson 6-5)*
    a. How much will he have in 6 weeks?
    b. How much will he have in w weeks?

28. List all the divisors of 48. *(Lesson 6-2)*

29. Helen walked $3\frac{1}{4}$ miles on Monday, $3\frac{1}{8}$ miles on Tuesday, $2\frac{1}{3}$ miles on Wednesday, and x miles on Thursday. She walked a total of $12\frac{2}{3}$ miles. How far did she walk on Thursday? *(Lesson 5-8)*

Exploration

30. A map (like a road map or a map of the United States) is a drawing of a contraction of the world. The scale of the map is the size change factor. If 1″ on the map represents 1 mile on Earth, then the scale is,

$$\frac{1''}{1 \text{ mile}} = \frac{1''}{5280 \text{ feet}} = \frac{1''}{5280 \cdot 12 \text{ in.}} = \frac{1''}{63,360''} = \frac{1}{63,360}.$$

    This is sometimes written as 1:63,360. Find a map and determine its scale.

31. Below is a drawing. Make a drawing that is similar and 2.5 times the size.

9-9

Picturing Multiplication with Negative Numbers

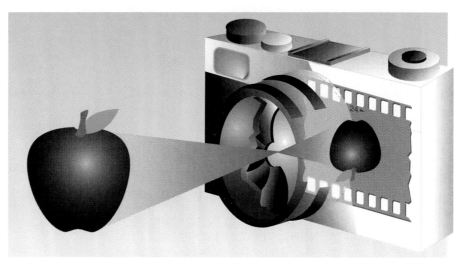

Negatives are negatives. *All cameras use the same basic principle: light is focused by the lenses onto the film to form an image in which up is down, and left is right. This is just like the image formed when multiplying by negative numbers.*

Remember that multiplication can be pictured by expansions or contractions. In Lessons 9-7 and 9-8, the coordinates of preimage points were positive. But coordinates can be any number. So the preimage can be anywhere. Here the preimage is a quadrilateral with a shaded region near one vertex.

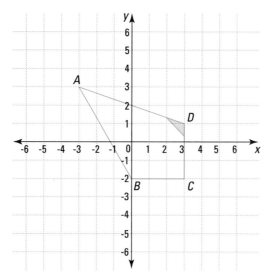

Preimage vertices

$A = (-3, 3)$
$B = (0, -2)$
$C = (3, -2)$
$D = (3, 1)$

An Expansion of Magnitude 2

First we multiply all coordinates by 2. You know what to expect.

1. The image is similar to the preimage.
2. The sides of the image are 2 times as long as the corresponding sides of the preimage.
3. The sides of the image are parallel to corresponding sides of the preimage.

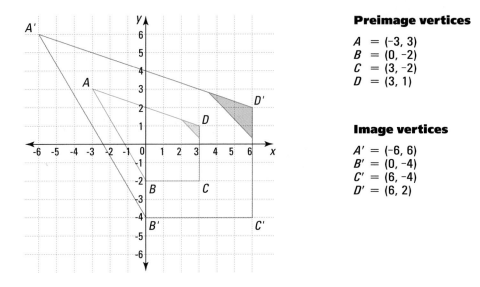

Preimage vertices

$A = (-3, 3)$
$B = (0, -2)$
$C = (3, -2)$
$D = (3, 1)$

Image vertices

$A' = (-6, 6)$
$B' = (0, -4)$
$C' = (6, -4)$
$D' = (6, 2)$

Notice also:

4. The line containing a preimage point and its image also contains (0, 0).
5. Each image point is 2 times as far from (0, 0) as its preimage.
6. The preimage and the image quadrilateral have the same tilt.

An Expansion of Magnitude -2

Now we find the image of the same quadrilateral under an expansion with the negative magnitude, -2. This is done by multiplying all coordinates by -2. We call this image $A''B''C''D''$. (Read this "A double prime, B double prime, and so on.")

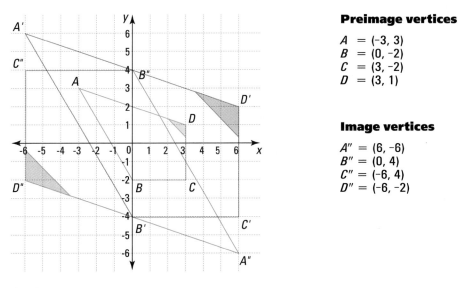

Preimage vertices

$A = (-3, 3)$
$B = (0, -2)$
$C = (3, -2)$
$D = (3, 1)$

Image vertices

$A'' = (6, -6)$
$B'' = (0, 4)$
$C'' = (-6, 4)$
$D'' = (-6, -2)$

Again:

1. The preimage and image are similar.
2. The sides of $A''B''C''D''$ are 2 times as long as the corresponding sides of $ABCD$.

3. Corresponding preimage and image sides are parallel.
4. The line containing a preimage point and its image contains (0, 0).
5. Each image point is 2 times as far from (0, 0) as its preimage.

But one new thing has happened.

6. The figure has been rotated 180°. It has been turned upside down. What was right is now left. What was up is now down.

> **Multiplication by a negative number reverses directions.**

Multiplying the coordinates of all points on a figure by the number k performs the transformation known as a *size change of magnitude k*. The number k may be positive or negative. But in all cases:

1. The resulting figure (the image) is similar to its preimage.
2. Lengths of sides of the resulting figure are $|k|$ times lengths of corresponding sides of the preimage.
3. Corresponding sides are parallel.
4. Corresponding preimage and image points lie on a line that goes through the origin.
5. Image points are $|k|$ times as far from the origin as corresponding preimage points.
6. If k is negative, the figure is rotated 180°.

The examples have pictured what happens when $k = 2$ and when $k = -2$.

Activity

Draw quadrilateral *ABCD* from the beginning of this lesson. Apply a size change of magnitude $-\frac{1}{2}$ to the quadrilateral. Name the image *A*B*C*D**. Verify the properties numbered 1–6 mentioned above for *ABCD* and *A*B*C*D**.

QUESTIONS

Covering the Reading

1. In a size change of magnitude -2, the image of (4, 5) is (__?__).

2. In a size change of magnitude -5, the image of (-10, 7) is (__?__).

3. In any size change, a figure and its image are __?__.

4. In a size change of magnitude -3, lengths on the image will be __?__ times the corresponding lengths on the preimage.

5. In any size change, a segment and its image are __?__.

6. What is the major difference between a size change of magnitude -3 and one of magnitude 3?

7. Let $A = (-2, -2)$, $B = (2, -2)$, and $C = (-2, 0)$. Graph $\triangle ABC$ and its image under an expansion of magnitude -3.

8. Multiplication by a negative number __?__ directions.

9. a. Show what you obtained as a result of the Activity of this lesson.
b. Is the result an expansion, a contraction, or neither?

10. A size change of magnitude -2 is like one of magnitude 2 followed by a rotation of what magnitude?

Applying the Mathematics

11. Let $P = (3.45, -82)$. Suppose that P' is the image of P under a size change of magnitude -10. *True or false?* $\overleftrightarrow{PP'}$ contains $(0, 0)$.

12. a. Are the quadrilaterals $A'B'C'D'$ and $A''B''C''D''$ in this lesson congruent?
b. If so, why? If not, why not?

13. Recall that the four quadrants of the coordinate plane are numbered from 1 to 4, as shown at the left. Consider a size change with magnitude -2.3. If a preimage is in quadrant 4, in which quadrant is its image?

Review

14. Suppose $\frac{4}{5}$ of the students in Mr. Mboya's class are here today, and $\frac{2}{3}$ of them took the bus to school. What fraction of the students in that class took the bus to school today? *(Lesson 9-8)*

15. Simplify: $0 \cdot a + 1 \cdot b + -1 \cdot c + -1 \cdot -d$ *(Lesson 9-6)*

16. Translate each phrase into a numerical expression or equation.
a. spending $50 a day
b. 4 days from now
c. 4 days ago
d. If you spend $50 a day for 4 days, you will have $200 less than you have now.
e. If you have been spending $50 a day, 4 days ago you had $200 more than you have now. *(Lesson 9-5)*

17. About 71% of the earth's surface is covered by water. (Some people think we should call our planet Water, not Earth.) If a large meteor were to hit the earth, what are the chances it would hit land on a Sunday? *(Lessons 9-4, 9-3)*

18. Draw part of a tessellation that uses quadrilateral $WXYZ$. *(Lesson 8-8)*

Big blue marble. *This shot of Earth was taken by an astronaut aboard Apollo 16. In the middle of the picture is southern California.*

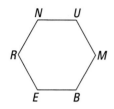

19. Trace hexagon *NUMBER* at the left and then draw all of its lines of symmetry. *(Lesson 8-7)*

20. Trace the figure below. Draw the reflection image of triangle *XYZ* over line *p*. *(Lesson 8-6)*

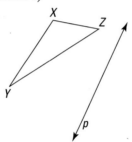

21. List all composite numbers *n* where $15 < n < 25$. *(Lesson 6-2)*

22. Is 41 prime or composite? Explain your answer. *(Lesson 6-2)*

In 23–25, name the kind of polygon. *(Lesson 5-9)*

23.   **24.**   **25.**

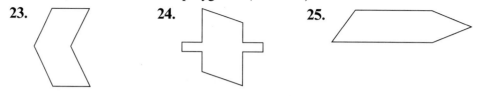

26. Find the value of $-5x + 6$ when $x = -9$. *(Lesson 4-4)*

27. Find the value of $3a + 2a$ when $a = -7$. *(Lesson 4-4)*

28. Three instances of a pattern are given. Describe the pattern using one variable. *(Lesson 4-2)*

$$-3 \cdot 5 + 4 \cdot 5 = 5$$
$$-3 \cdot 7.8 + 4 \cdot 7.8 = 7.8$$
$$-3 \cdot -1 + 4 \cdot -1 = -1$$

29. One thousand two hundred sixty students were asked whether they thought Hardnox High would win its next football game. Seventy-five percent thought it would win.
  a. How many students is this?
  b. Is the answer to part **a** greater than, equal to, or less than $\frac{3}{4}$ of 1260? *(Lesson 2-5)*

30. Write 5^{10} as a decimal. *(Lesson 2-2)*

31. Which is largest: $-\frac{2}{3}$, -0.6, or -0.66? *(Lesson 1-9)*

Exploration

32. Draw a figure on a coordinate grid. Then draw the images, in different colors, of the figure under size changes of magnitudes 3, $\frac{1}{3}$, $-\frac{1}{3}$, and -3. Describe how the four images and the preimage compare.

PROJECTS

9

CHAPTER NINE

A project presents an opportunity for you to extend your knowledge of a topic related to the material of this chapter. You should allow more time for a project than you do for typical homework questions.

1 Area of Your Residence

The floor area of an apartment or house is the area enclosed by the outer walls of the residence. Sketch the shape of the place where you live and, through measuring, multiplication, and addition, calculate this area for your residence.

2 Copies of All Sizes

Some copy machines enlarge and/or reduce. Start with an interesting original picture. Make copies of various sizes. Then make copies of copies to obtain still other sizes. Put all the copies together in a display explaining what you have done and giving the magnitudes of the size changes from the original picture.

3 Effects of Different Rates

Suppose you have to type up an essay that is 1000 words long. Then if you can type 25 words per minute, it will take you 40 minutes. Fill in at least fifteen rows of the following table. (One row is already filled in.)

typing speed	time to type
$25 \frac{\text{words}}{\text{min}}$	40 min

Graph the pairs of numbers you find. (For instance, (25, 40) should be on your graph.) Use the graph to describe how an increase in typing speed affects the amount of time it takes to type up the essay.

4 Streaks

Toss a coin at least 250 times, recording heads or tails for every toss. Answer the following questions to help you decide whether or not the results of your tosses are independent.

a. What is the relative frequency of heads for your tosses?

b. After each head, what is the relative frequency that the next toss is heads?

c. After each pair of heads, what is the relative frequency that the third toss is heads?

d. After each three heads in a row, what is the relative frequency that the fourth toss is heads?

e. Judging from what you find in parts **a** through **d,** do you think that tossing heads on one toss affects what happens on the next toss? Explain why or why not.

5 Half-size You

Make a two-dimensional contraction of yourself by measuring the length and width of your legs, toes, arms, hands, fingers, neck, face, nose, mouth, eyes, and so on. Make a paper model of yourself that is half your real size.

6 Multiplying Fractions Versus Multiplying Decimals or Percents

Instead of multiplying 2.5 by 3.75, you could multiply $\frac{5}{2}$ by $\frac{15}{4}$. Instead of taking 20% of some quantity, you could take $\frac{1}{5}$ of that quantity. When is it easier to multiply fractions? When is it easier to multiply with decimals or percents? Write an essay giving examples and your opinions.

SUMMARY

Like addition, multiplication is commutative and associative. There is an identity (the number 1) and every number x but zero has a multiplicative inverse (its reciprocal $\frac{1}{x}$).

Multiplying any number by 0 gives a product of 0. Multiplying by -1 changes a number to its opposite.

Given the dimensions, you can multiply to find the area of a rectangle, the area of a right triangle, the number of elements in a rectangular array, and the volume of a rectangular solid. Areas of parts of rectangles and rectangular solids can picture multiplication of fractions: the product of $\frac{a}{b}$ and $\frac{c}{d}$ is $\frac{ac}{bd}$.

Multiplication is also used to determine probabilities. If two or more events are independent, the product of their probabilities is the likelihood that they will both occur.

Another important use of multiplication involves rate factors. Examples of rate factors are the quantities 600 miles/hour, $2.50 per bottle, 27.3 students per class, and -3 kg/month. When rate factors are multiplied by other quantities, the units are multiplied like fractions. Rate factors like 5280 feet/mile can be used to convert units and to explain how to multiply with negative numbers.

Size changes can picture multiplication by a positive number or by a negative number. They show direction as well as size. Multiplying by a number greater than 1 can be pictured as an expansion. Multiplying by a number between 0 and 1 can be pictured as a contraction. Multiplying by a negative number rotates the picture 180°. So negative · positive = negative, and negative · negative = positive. In all cases, the figure and its size-change image are similar.

VOCABULARY

You should be able to give a general description and a specific example of each of the following ideas.

Lesson 9-1
Area Model for Multiplication
dimensions
rectangular array
Commutative Property of
    Multiplication
right triangle, legs, hypotenuse
Area Formula for a Right
    Triangle

Lesson 9-2
box, rectangular solid
edge, vertex, face
dimensions, length, width,
    height, depth
volume, base
Associative Property of
    Multiplication

Lesson 9-3
unit fraction
Multiplication of Unit
    Fractions Property
Algebraic Definition of
    Division
multiplicative inverse,
    reciprocal
Property of Reciprocals
Multiplication of Fractions
    Property

Lesson 9-4
independent events
Probability of the event *A and B*

Lesson 9-5
rate, rate unit, rate factor
Rate Factor Model for
    Multiplication
conversion factor

Lesson 9-6
Multiplication Property of -1
Multiplication Property of
    Zero

Lesson 9-7
size change factor
size change of magnitude k
expansion
similar figures
Size Change Model for
    Multiplication
Multiplicative Identity
    Property of One

Lesson 9-8
contraction

PROGRESS SELF-TEST

Take this test as you would take a test in class. Then check your work with the solutions in the Selected Answers section in the back of the book.

In 1–5, simplify.

1. $4 \cdot \frac{1}{3} \cdot \frac{5}{4}$ **2.** $\frac{1}{x} \cdot \frac{7}{16}$

3. $5 \cdot -9$ **4.** $-3 \cdot 3 + -2 \cdot -2$

5. $a + 1 \cdot a + b + 0 \cdot b + -21 \cdot c + c$

In 6 and 7, name the general property.

6. $(12 \cdot 43) \cdot 225 = 12 \cdot (43 \cdot 225)$

7. $m \cdot \frac{1}{m} = 1$

8. An auditorium has r rows with s seats in each row. Five seats are broken. How many seats are there to sit in?

9. Give the area of the right triangle pictured here.

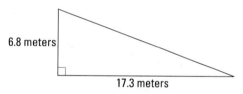

6.8 meters

17.3 meters

10. What multiplication with negative numbers is suggested by the following situation? For four hours, the flood waters receded three centimeters an hour.

11. What is the volume of a rectangular solid with dimensions 3 feet, 4 feet, and 5 feet?

12. Could a jewelry box with a volume of 3375 cm^3 fit in a dresser drawer with dimensions 40 cm, 12 cm, and 45 cm? Explain your answer.

13. Give dimensions for two noncongruent rectangles, each with an area of 16.

14. Round the reciprocal of 38 to the nearest thousandth.

15. What is the image of (8, 2) under an expansion of magnitude 4?

16. Alta earns $5.80 per hour. She gets time and a half for overtime. How much does she make per hour of overtime?

17. a. What multiplication of fractions is pictured below?

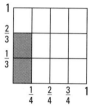

b. What is the product?

18. A person makes $8.50 an hour and works 37.5 hours a week. **a.** Multiply the two rates in this situation. **b.** Explain the meaning of the result.

In 19 and 20, a triangle has vertices (2, 0), (-4, -4), and (2, -4).

19. Graph this triangle and its image under an expansion of magnitude -2.5.

20. Name one thing that is the same about the preimage and image and one thing that is different.

21. $\frac{3}{25}$ of Clemente High students are in the band. There are 850 students at Clemente High. How many students are in the band?

22. Larry earned a scholarship which will pay for $\frac{2}{3}$ of his $8400 college tuition. How much is his scholarship?

23. You have a one in ten chance of guessing a correct digit. What are your chances of guessing three correct digits in a row?

24. You feel you have a 2 in 3 chance of getting an A on a history test and a 2 in 5 chance of getting an A on your upcoming math test. If these are independent events, what are your chances of getting an A on both tests?

25. *Multiple choice.* Which number is the reciprocal of 1.25?
(a) 0.125   (b) -1.25   (c) 0.8   (d) 0.75

26. The highest mountain in Africa, Mount Kilimanjaro, is 19,340 ft high. How many miles is this?

CHAPTER REVIEW

Questions on SPUR Objectives

SPUR stands for **S**kills, **P**roperties, **U**ses, and **R**epresentations. The Chapter Review questions are grouped according to the SPUR Objectives for this chapter.

SKILLS DEAL WITH THE PROCEDURES USED TO GET ANSWERS.

Objective A: *Find the area of a rectangle or a right triangle, given its dimensions.* *(Lesson 9-1)*

1. What is the area of a rectangle with length 7 cm and width 3.5 cm?

2. What is the area of a rectangle with length 5 and width w?

3. What is the area of a right triangle with legs 6 and 9?

4. What is the area of the right triangle pictured here?

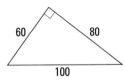

Objective B: *Find the volume of a rectangular solid, given its dimensions.* *(Lesson 9-2)*

5. What is the volume of the rectangular solid pictured below?

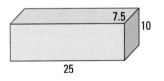

6. What is the volume of a rectangular solid with dimensions x, y, and z?

Objective C: *Multiply fractions.* *(Lesson 9-3)*

In 7–15, perform the multiplication and write the answer in lowest terms.

7. $3 \cdot \frac{1}{6}$ 8. $\frac{4}{5} \cdot \frac{4}{3}$ 9. $100 \cdot \frac{3}{8}$

10. $\frac{2}{5} \cdot \frac{2}{5} \cdot \frac{2}{5}$ 11. $10\frac{1}{8} \cdot 10$ 12. $2\frac{1}{2} \cdot 3\frac{2}{3}$

13. $\frac{6}{7} \cdot \frac{7}{8} \cdot \frac{8}{9}$ 14. $\frac{3}{10} \cdot 3.25$ 15. $1\frac{1}{3} \cdot 1\frac{1}{5}$

Objective D: *Multiply positive and negative numbers.* *(Lesson 9-6)*

In 16–21, simplify.

16. $-8 \cdot -2$ 17. $3 + -10 \cdot 9$

18. $5 \cdot -5 \cdot 4 \cdot -4$ 19. $6x - 5$ when $x = -1$

20. $(-8)^2$ 21. $5 \cdot (-4)^3$

22. If $x = -3$ and $y = -16$, what is the value of $5x - 2y$?

PROPERTIES DEAL WITH THE PRINCIPLES BEHIND THE MATHEMATICS.

Objective E: *Identify properties of multiplication.* *(Lessons 9-1, 9-2, 9-3, 9-6)*

23. State the Multiplication Property of Zero.

24. What property justifies that $(78 \cdot 4) \cdot 25 = 78 \cdot (4 \cdot 25)$?

25. What property justifies that $a \cdot b = b \cdot a$?

26. A number not equal to zero multiplied by its reciprocal equals what value?

27. *True or false.* The product of any two negative numbers is a negative number.

28. What is the reciprocal of $\frac{2}{3}$?

Objective F: *Use properties of multiplication to simplify expressions and check calculations.*
(Lessons 9-1, 9-2, 9-3, 9-6)

In 29–33, simplify without using a calculator.

29. $ab - ba$

30. $\frac{1}{2} \cdot 2 \cdot \frac{1}{3} \cdot 3$

31. $\frac{7}{8} \cdot \frac{8}{7}$

32. $-1 \cdot -1 \cdot -1 \cdot -1 \cdot 7$

33. $\frac{1}{2} \cdot \frac{1}{5} \cdot 5 \cdot x$

34. Use the Commutative Property of Multiplication to check whether $2.48 \cdot 6.54 = 16.2192$.

35. a. Find the product of $\frac{1}{8}$ and $\frac{1}{5}$.

b. Check by converting the fractions to decimals.

36. a. What is the reciprocal of $\frac{9}{10}$?

b. Check your answer by converting the fractions to decimals and finding their product.

USES DEAL WITH APPLICATIONS OF MATHEMATICS IN REAL SITUATIONS.

Objective G: *Find areas of right triangles and rectangles, and the number of elements in rectangular arrays in applied situations.*
(Lesson 9-1)

37. The flag of the United States has as many stars as states. How many states did the U.S. have when the flag below was in use?

38. The smaller rectangle below is the space for a kitchen in a small restaurant. The larger rectangle is the space for the restaurant. The shaded region is space for seating. What is the area available for seating?

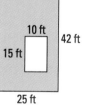

10 ft

42 ft

15 ft

25 ft

39. An 8″ by 10.5″ rectangular sheet of paper is cut in half along its diagonal. What is the area of each half?

Objective H: *Find the volume of a rectangular solid in real contexts.* *(Lesson 9-2)*

40. A stick of margarine is approximately 11.5 cm long, 3 cm wide, and 3 cm high. What is the volume of the stick?

41. A computer was packed in a box 51 cm by 48 cm by 35 cm. What is the volume of this box?

42. A plastic container has a base with area 72 square centimeters and height 14 cm. Can it hold a liter of soup? Explain why or why not.

43. A typical yardstick is 3 feet long (of course), $1\frac{1}{8}″$ wide, and $\frac{1}{8}″$ thick. To the nearest cubic inch, what is its volume?

Objective I: *Calculate probabilities of independent events.* *(Lesson 9-4)*

44. In 1993, Mark Price missed only about 5% of the free throws he attempted. Suppose his free throws are independent events. What is the probability that he would miss two free throws in a row at the end of a ball game?

45. At the 1992 U.S. National Indoor Rifle and Pistol Championships, Air Rifle champion Debra Sinclair hit the bull's eye 595 out of 600 times. Assume her shots are independent events. If you took a picture of two random shots in a row, what is the probability that both shots hit the bull's eye?

46. Suppose that, for a particular plant, the probability that a seed will not sprout is $\frac{1}{20}$. What is the probability that two seeds will both not sprout?

47. Five of seven days are weekdays. Resttown gets hit by a snowstorm about once every 5 years. What are the chances of Resttown being hit by a snowstorm next year on a weekday?

Objective J: *Apply the Rate Factor Model for Multiplication.*   *(Lesson 9-5)*

In 48 and 49, a multiplication problem is given. **a.** Do the multiplication. **b.** Make up a question that leads to this multiplication.

48. $2 \frac{\text{cookies}}{\text{day}} \cdot 365 \frac{\text{days}}{\text{year}}$

49. $5 \text{ hours} \cdot 25 \frac{\text{miles}}{\text{hour}}$

50. A person makes $10.50 an hour and works 37.5 hours a week. How much does the person earn per year?

51. Fire laws say there is a maximum of 60 people allowed per small conference room. There are 6 small conference rooms. Altogether, how many people can meet in them?

52. If Lois has been losing weight at the rate of 2.3 kg per month, how did her weight 4 months ago compare with her weight now?

Objective K: *Use conversion factors to convert from one unit to another.*   *(Lesson 9-5)*

53. Name the two conversion factors from the conversion equation 1 foot = 30.48 cm.

54. 500 cm equals about how many ft?

55. 150 hours equals how many days?

56. Name the two conversion factors from the conversion equation 1 mile = 1760 yards.

Objective L: *Apply the Size Change Model for Multiplication in real situations.*   *(Lessons 9-7, 9-8)*

57. Mrs. Kennedy expects to save $\frac{1}{8}$ of her weekly grocery bill of $150 a week by using coupons from a newspaper. How much money does she expect to save?

58. On the average, the cost of medical care in the U.S. quintupled from 1970 to 1991. If an item cost x dollars in 1970, what did it cost in 1991?

59. If you make $4.75 an hour, what will you make per hour of overtime, if you are paid time and a half?

60. Mr. Jones tithes. That is, he gives one-tenth of what he makes to charity. If he makes $500 a *week*, how much does he give every *year* to charity?

REPRESENTATIONS DEAL WITH PICTURES, GRAPHS, OR OBJECTS THAT ILLUSTRATE CONCEPTS.

Objective M: *Picture multiplication using arrays or area.*   *(Lessons 9-1, 9-3)*

61. Show $5 \cdot 4 = 20$ using a rectangular array.

62. Show that $6.5 \cdot 3$ is larger than $6 \cdot 3$ using rectangles with accurate length and width. (Use centimeters as the unit.)

63. Picture a rectangular park $\frac{1}{2}$ km by $\frac{2}{3}$ km and find its area.

64. Explain why $75 \cdot 23 = 23 \cdot 75$ using ideas of area.

Objective N: *Perform expansions or contractions on a coordinate graph.*
(Lessons 9-7, 9-8, 9-9)

65. Graph the triangle with vertices (0, 5), (6, 2), and (4, 4) and its image under an expansion of magnitude 2.5.

66. What is the image of (x, y) under an expansion of magnitude 1000?

67. Is a size change of magnitude $\frac{3}{7}$ an expansion or a contraction?

68. Graph the segment with endpoints (4, 9), and (2, 3). Graph its image under a size change of magnitude $\frac{1}{3}$.

69. What is the image of (40, -80) under a size change of magnitude -0.2?

70. Under a size change of magnitude -5, how will a quadrilateral and its image be the same and how will they be different?

71. Let $A = (-4, 5)$, $B = (2, 0)$, and $C = (0, -3)$. Graph $\triangle ABC$ and its image under a size change of magnitude -2.

72. A size change of magnitude -12 is like a size change of magnitude 12 followed by a rotation of what magnitude?

CHAPTER

10

MULTIPLICATION AND OTHER OPERATIONS

Chapter 9 covered uses of multiplication without reference to other operations. But multiplication is related to all the other basic operations of arithmetic. Each of these relationships can be pictured using area.

Multiplication and Addition

Multiplication provides a shortcut for adding many instances of the same number.

$$3.5 + 3.5 + 3.5 + 3.5 + 3.5 + 3.5 + 3.5 + 3.5 = 8 \cdot 3.5$$

Multiplication and Subtraction

Sometimes subtraction can help with a multiplication. Here is a way to multiply by 99 in your head.

$$62 \cdot 99 = 62 \cdot 100 - 62 \cdot 1$$
$$= 6200 - 62$$
$$= 6138$$

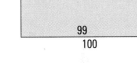

Multiplication and Division

Every multiplication fact gives rise to division facts.

$7 \cdot 9 = 63$ means $\frac{63}{9} = 7$

and $\frac{63}{7} = 9$.

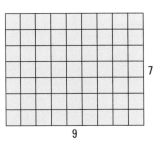

In this chapter, you will study the general algebraic patterns that relate multiplication to addition and subtraction. These relationships have applications to the areas of two-dimensional figures and the surface areas of three-dimensional figures.

10-1

*Multiplication
as Shortcut
Addition*

Food for thought. *The FDA now recommends a minimum of five daily servings of fruits and vegetables to maintain good health.*

Shortcut Addition with One Variable

Suppose you buy 5 different items at a grocery store.

The clerk will add the costs to arrive at your total cost.

$$a + b + c + d + e$$

But if you buy 5 items that are the same, the checkout clerk has a choice.

The prices can be added:

$$x + x + x + x + x.$$

Or, the price of the one item can be multiplied by 5. The total cost to you is $5x$.

Multiplication can be a shortcut for addition because the addends are equal. This property of multiplication is called the *Repeated Addition Property*.

> **Repeated Addition Property of Multiplication**
> If n is a positive integer, then
> $$nx = \underbrace{x + x + \ldots + x.}_{n \text{ addends}}$$

Specifically:
$$1x = x$$
$$2x = x + x$$
$$3x = x + x + x$$
$$4x = x + x + x + x$$

and so on.

Example 1

What is the perimeter of a square with side length s?

Solution
The perimeter is the sum of the lengths of the sides: $s + s + s + s$.
By repeated addition, The perimeter is s + s + s + s = 4s.

Shortcut Addition with Two Variables

Think again about the grocery situation. Suppose you buy amounts of two different items.

What is the total cost? By adding, the total cost is
$$(e + e + e + e + e) + (x + x + x).$$

This, due to the Repeated Addition Property of Multiplication, equals
$$5e + 3x.$$

The clerk can multiply the first cost by 5 and the second cost by 3, and then add them together. No other simplification can be made.

This idea can be used to obtain a formula for the perimeter of a rectangle.

Example 2

What is the perimeter of a rectangle with length ℓ and width w?

Solution

$$\text{perimeter} = \ell + w + \ell + w$$

Commutative and Associative Properties of Addition	$= \ell + \ell + w + w$
Repeated Addition Property of Multiplication	$= 2\ell + 2w$

From Examples 1 and 2 come the formulas $p = 4s$ for the perimeter of a square and $p = 2\ell + 2w$ for the perimeter of a rectangle.

Example 3

Simplify $8x + 4y + x + 3 + 2y + 3x + 6$.

Solution 1

Write the multiplications as repeated addition, then rearrange.
$8x + 4y + x + 3 + 2y + 3x + 6$
$= x + x + x + x + x + x + x + x + y + y + y + y + x + 3 + y + y + x + x + x + 6$
$= x + x + x + x + x + x + x + x + x + x + x + x + y + y + y + y + y + y + 3 + 6$
$= 12x + 6y + 9$

Solution 2

Think repeated addition. Group the like terms together.
$8x + 4y + x + 3 + 2y + 3x + 6$
$= (8x + x + 3x) + (4y + 2y) + 3 + 6$
$= 12x + 6y + 9$

As Solution 2 indicates, $8x$, x, and $3x$ are called *like terms* because they are multiples of the same variable. $12x$ and $6y$ are *unlike terms*. Like terms can be combined into a single term. Unlike terms cannot be simplified into a single term.

QUESTIONS

Covering the Reading

1. Suppose you buy 3 shirts at $10.99 each and 2 more shirts at $4.99 each.
   a. Write out the cost as an addition.
   b. Write out the cost using multiplication for the repeated additions.

In 2–4, simplify.

2. $x + x + x$ 3. $\ell + w + \ell + w$

4. $25 + 20 + 20 + 25 + 20 + 20 + 25 + 20 + 25$

5. The ℓ and w in Question 3 might represent what real-world quantities?

6. Give the perimeter of this outline of a pencil.

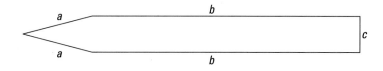

In 7 and 8, change the multiplications to additions.

7. $6y$ **8.** $2x + 4z$

In 9 and 10, simplify.

9. $m + 1 + n + 2m + 4n + 6m$ **10.** $4 + A + 2A + A + 6 + B$

Applying the Mathematics

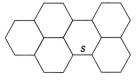

11. Part of a beehive is drawn at the left. Each cell is a regular hexagon. What is the sum of the lengths of the pictured segments?

12. Suppose frozen orange juice costs $1.39 per can in a grocery store.
  a. A shopper buys 6 cans. Name two ways the shopper can figure out the total cost.
  b. A shopper buys c cans. What is the total cost?

In 13 and 14, simplify.

13. $4e - e$ (Hint: Change all operations to additions.)

14. $11x - 2x + 3x$

15. a. Write $8 \cdot -\frac{1}{2}$ as repeated addition.
  b. Calculate $8 \cdot -\frac{1}{2}$.

16. Give a formula for the perimeter p of an equilateral triangle if one side has length s.

Review

17. Solve $5 = 80 + u$.   *(Lesson 5-8)*

18. Solve $-3 + x = 1140$.   *(Lesson 5-8)*

19. Which pairs of numbers are reciprocals?   *(Lesson 9-3)*
  (a) 100 and 0.01   (b) 2 and $\frac{1}{2}$
  (c) 2.5 and $\frac{2}{5}$ (d) $\frac{7}{4}$ and $\frac{4}{7}$
  (e) 3.5 and $\frac{3}{5}$ (f) 16 and -16
  (g) 1.5 and $0.\overline{6}$ (h) 3 and .3

In 20–23, name the general property. *(Lessons 3-2, 5-7, 7-2, 9-2)*

20. $a + (b + c) = (a + b) + c$ **21.** $\frac{1}{5}(5w) = \left(\frac{1}{5} \cdot 5\right)w$

22. $35 + {}^-78 = 35 - 78$ **23.** Since 1 in. = 2.54 cm,
3 in. = 3 · 2.54 cm.

24. A box has dimensions 2′, 3′, and 4′. Consider the largest face of the box. *(Lessons 5-10, 9-1)*
  a. What are its dimensions?
  b. What is the area of this face?
  c. What is the perimeter of this face?

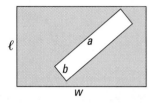

25. A rectangle with dimensions a and b is tilted inside a rectangle with dimensions ℓ and w.
  a. Is this enough information to give a formula for the area of the shaded region?
  b. If so, what is the formula? If not, why not? *(Lessons 7-1, 9-1)*

26. Find the measures of the numbered angles, given that $m \parallel n$.
(Lesson 7-7)

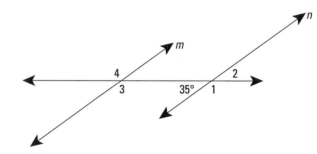

27. What is the expected temperature? *(Lesson 7-2)*
  a. The temperature is 7°F and is expected to drop 8°.
  b. The temperature is w°F and is expected to drop z°.

28. *True or false.* $1.005 \cdot 2.34567893456 > 2.34567893456$ *(Lesson 4-10)*

Exploration

29. Enter 5 on your calculator. Press the reciprocal key $\boxed{1/n}$ twice.
  a. What number is displayed?
  b. What has happened?
  c. What will happen if you press the reciprocal key 75 times?
  d. Make a generalization.

30. Some calculators have a constant key, often named $\boxed{K}$.
  a. If your calculator has such a key, press 5 $\boxed{+}$ $\boxed{K}$ $\boxed{=}$ $\boxed{=}$ $\boxed{=}$ . . .
  What happens as you continue to press $\boxed{=}$?
  b. Press 5 $\boxed{+}$ $\boxed{K}$ 8 $\boxed{=}$ $\boxed{=}$ $\boxed{=}$ . . . What happens now?
  c. Explore what happens in parts **a** and **b** if the $\boxed{+}$ key is replaced by keys for other operations.

Laws of equality. *This statue of the* Scales of Justice, *which is atop New York City Hall, symbolizes the belief that justice should be applied equally to all. In this lesson, you will also see scales used to show equality.*

Picturing the Solving of $ax = b$

Here is an equation you should be able to solve in your head. (Think of a solution before reading on.)

$$3w = 6$$

This equation can be represented with a balance scale.

Three boxes with unknown weight w per box are on the left side of the scale. They balance the 6 one-kilogram weights on the right side. This situation pictures $3w = 6$.

Since the weight is shared equally among the three boxes, one box weighs 2 kilograms.

Sharing equally gives the same result as multiplying both sides of the equation $3w = 6$ by $\frac{1}{3}$. Example 1 shows the same steps without the balance scale.

Example 1

Solve $3w = 6$.

Solution

$$3w = 6$$
$$\frac{1}{3} \cdot 3w = \frac{1}{3} \cdot 6 \qquad \text{Multiply both sides by } \tfrac{1}{3}.$$
$$w = 2 \qquad \text{Arithmetic}$$

Check

Substitute 2 for w in the original equation. Does $3 \cdot 2 = 6$? Yes.

Notice why both sides of $3w = 6$ are multiplied by $\frac{1}{3}$. The product of 3 and $\frac{1}{3}$ is 1. So, $\frac{1}{3} \cdot 3w = \left(\frac{1}{3} \cdot 3\right)w = 1w = w$. Thus, multiplying both sides of the equation in Example 1 by $\frac{1}{3}$ leads to an equivalent equation of the form $x = \underline{\quad}$. This solves the equation.

Solving $ax = b$ Using the Multiplication Property of Equality

In general, multiplying both sides of an equation by any nonzero number will not affect its solutions. This situation is an instance of the *Multiplication Property of Equality* which you studied in Lesson 3-2.

Multiplication Property of Equality
If $x = y$, then $ax = ay$.

The Multiplication Property of Equality can be used to solve any equation of the form $ax = b$, if $a \neq 0$. To solve for x, just multiply both sides by $\frac{1}{a}$, the reciprocal of a. This method is particularly effective in solving equations which don't have obvious solutions.

Example 2

Solve $\frac{3}{5}m = \frac{1}{4}$ and check the solution.

Solution
First write the equation, leaving room for work below it.

$$\frac{3}{5}m = \frac{1}{4}$$

This is an equation of the form $ax = b$. Here $a = \frac{3}{5}$, $x = m$, and $b = \frac{1}{4}$. We want to multiply both sides by a number that will leave m by itself on the left side. That number is $\frac{5}{3}$, the reciprocal of $\frac{3}{5}$. ▶

$$\frac{5}{3} \cdot \left(\frac{3}{5} \cdot m\right) = \frac{5}{3} \cdot \frac{1}{4}$$

Because all numbers on the left-hand side are multiplied, the Associative Property of Multiplication lets us drop the parentheses.

$$\frac{5}{3} \cdot \frac{3}{5} \cdot m = \frac{5}{3} \cdot \frac{1}{4}$$

The hard part is done. Now you just simplify.

$$1 \cdot m = \frac{5}{3} \cdot \frac{1}{4}$$

Using $1 \cdot m = m$, and multiplying the fractions on the right side,

$$m = \frac{5}{12}.$$

Check

Substitute $\frac{5}{12}$ for m in the original equation.

Does $\frac{3}{5} \cdot \frac{5}{12} = \frac{1}{4}$?

The product of the fractions on the left side is $\frac{1}{4}$ in lowest terms. **Yes, it does. So** $\frac{5}{12}$ **is the solution.**

The next example may seem hard, but it can be solved using an equation.

Example 3

Six percent of a number is 30. What is the number?

Solution

First translate into an equation. **Let the number be n.**

Six percent   of   a number   is   30.

$$.06 \quad \cdot \quad n \quad = \quad 30$$

Here we solve using the Multiplication Property of Equality. Multiply both sides by the reciprocal of .06.

$$\frac{1}{.06} \cdot .06n = \frac{1}{.06} \cdot 30$$

Now simplify.
$$n = \frac{30}{.06}$$
$$n = 500$$

Check
6% of 500 = .06 × 500 = 30

To use the Multiplication Property of Equality, you need to know how to find reciprocals, how to multiply fractions, and how to find equal fractions. Now you can see why these ideas were discussed in previous chapters! But you still might wonder when these equations are solved outside of mathematics classes. That is the subject of the next lesson.

QUESTIONS

Covering the Reading

1. a. Draw a balance scale diagram representing the equation $4w = 12$.
  b. Solve the equation.

2. a. To solve $3x = 0.12$ using the Multiplication Property of Equality, by what number would both sides be multiplied?
  b. Solve $3x = 0.12$ using the Multiplication Property of Equality.

3. a. To solve $5y = 80$ using the Multiplication Property of Equality, by what number should you multiply both sides?
  b. Solve $5y = 80$.

4. If $x = y$, then $6x = \underline{\ ?\ }$.

5. Question 4 is an instance of what property?

6. Delilah said the solution to $\frac{2}{3}x = 5$ is $\frac{15}{2}$. Is she correct?

7. a. To solve $\frac{6}{25} \cdot A = \frac{2}{9}$, it is most convenient to multiply both sides by what number?
  b. Solve this equation.

In 8–10, solve and check.

8. $\frac{2}{3}x = 8$ **9.** $\frac{t}{9} = 40$ **10.** $\frac{4}{7}y = \frac{1}{8}$

11. Seven times a number is 413. What is the number?

12. Seven percent of a number is 84. What is the number?

13. 40% of what number is 25?

Applying the Mathematics

14. Ten-thirds of a number is 30. What is the number?

In 15–17, solve and check.

15. $16.56 = 7.2y$ **16.** $\frac{2}{3}k = 62\%$ **17.** $x + x + x = 5.736$

18. Give the property telling why each step follows from the previous one.

$$\frac{7}{4} \cdot A = \frac{1}{5}$$

Step 1 $\frac{4}{7} \cdot \left(\frac{7}{4} \cdot A\right) = \frac{4}{7} \cdot \frac{1}{5}$

Step 2 $\left(\frac{4}{7} \cdot \frac{7}{4}\right) \cdot A = \frac{4}{7} \cdot \frac{1}{5}$

Step 3 $1 \cdot A = \frac{4}{7} \cdot \frac{1}{5}$

Step 4 $A = \frac{4}{35}$

19. If $\frac{4}{3}$ of a number is 1200, what is $\frac{3}{4}$ of that number?

In 20 and 21, simplify. *(Lesson 10-1)*

20. $7 + 2x + 5x + 3$

21. $y + z + 2z$

22. Draw a figure that has exactly 3 symmetry lines. *(Lesson 8-7)*

In 23 and 24, consider the following information about the U.S. Senate.

Years	Democrats	Republicans
1983–85	46	54
1985–87	47	53
1987–89	54	46
1989–91	57	43
1991–93	57	43
1993–95	58	42

Changing Congressional faces. *In 1948, Margaret Chase Smith became the first woman elected to a full six-year Senate term. When the 103rd Congress convened in January 1993, there were 6 female senators and 47 female representatives.*

23. What is the mean number of Democrats in the U.S. Senate from 1983–95? *(Lesson 4-6)*

24. The ordered pairs (Democrats, Republicans) for these six two-year periods all lie on a line. What is an equation for that line? *(Lesson 8-4)*

25. Since there was no year that we number 0, how many years were between the given years?
   a. 2 A.D. and 2 B.C.
   b. 4 A.D. and 4 B.C.
   c. 10 A.D. and 10 B.C.
   d. n A.D. and n B.C. *(Lesson 7-2)*

26. a. To solve $\frac{31}{6} + m = -\frac{5}{8}$, what number can be added to both sides?
   b. Solve $\frac{31}{6} + m = -\frac{5}{8}$. *(Lesson 5-8)*

27. Use a protractor to measure $\angle ABC$ to the nearest degree. *(Lesson 3-6)*

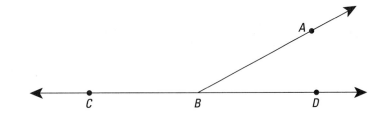

28. Write one millionth in the requested form.
   a. as a decimal
   b. as a power of 10 *(Lessons 1-2, 2-8)*

29. 144 seats can be arranged in a rectangular array with 12 rows and 12 columns. How many other rectangular arrays are possible?

LESSON

10-3

Using $ax = b$

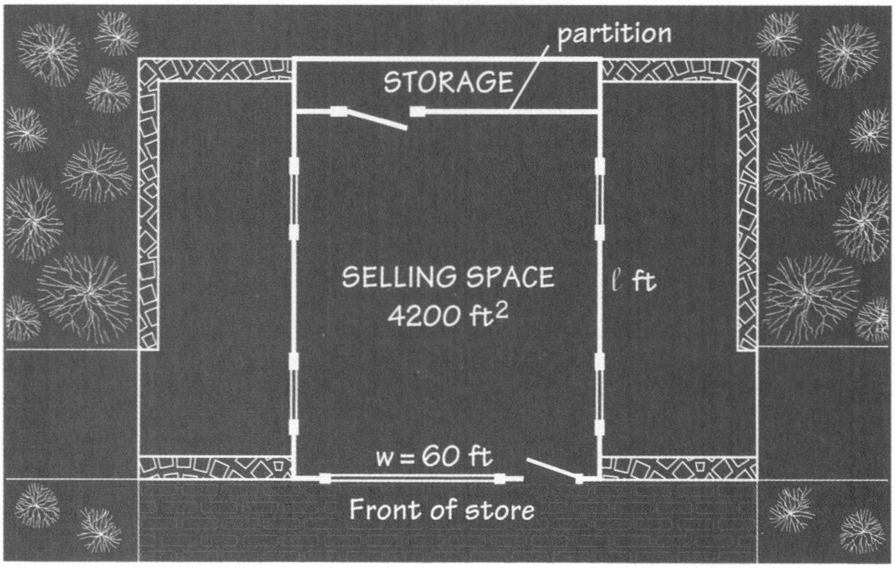

You have learned a variety of applications of multiplication:

areas of rectangles
arrays
rate factors
expansions (times as many)
contractions (part of)
percents of

Any of these applications can lead to an equation of the form $ax = b$. Here is an example involving areas of rectangles.

An Example Involving Area

Example 1

Claus's Clothes occupies a rectangular floor space. The store is 60 feet wide as shown in the diagram above. The store must be split into selling space and storage space by a partition. Mrs. Claus wants 4200 square feet for selling space. Where should the partition be located?

Solution

The selling space is a rectangle. One dimension of the rectangle is known to be 60 feet. The area is to be 4200 square feet. Now apply the formula for the area of a rectangle.

$$A = \ell w$$

Substitute 4200 for A and 60 for w in the formula.

$$4200 = \ell \cdot 60$$

▶

▶ Solve this equation in your head or by using the Multiplication Property of Equality.

$$\ell = \frac{4200}{60} \text{ feet}$$
$$\ell = 70 \text{ feet}$$

The partition should be 70 ft from the front of the store.

Check

You should always check answers with the original question. If the depth (length) of the selling space is 70 feet, will there be 4200 square feet of selling space? Yes, because $60 \cdot 70 = 4200$.

An Example Involving Size Changes

The next example involves a size change. Before you begin, read the question carefully. Without solving, you should know whether the answer will be greater than 12.60 or less than 12.60.

Example 2

A worker receives time and a half for overtime. If a worker gets $12.60 an hour for overtime, what is the worker's normal hourly wage?

Solution

Recall that time and a half means the worker's normal wage is multiplied by $1\frac{1}{2}$ to calculate the overtime wage. Let W be the normal hourly wage.

Then $1\frac{1}{2} \cdot W = \$12.60$.

Now the job is to solve this equation. First change the fraction to a decimal.

$$1.5 \cdot W = 12.60$$
$$W = \frac{12.60}{1.5}$$
$$W = \$8.40$$

The worker's normal hourly wage is $8.40.

Check

Ask: If a worker makes $8.40 an hour, will that worker get $12.60 for overtime? Yes; half of $8.40 is $4.20, and $4.20 plus $8.40 is $12.60.

Job requirements.
To compete for permanent future jobs, young people will need good communication, interpersonal, and analytic skills.

General Advice

In answering these kinds of questions, many people find it useful to think of the following steps.

(1) Read carefully. Determine what is to be found and what is given.

(2) Let a variable equal the unknown quantity.

(3) Write an equation.

(4) Solve the equation.

(5) Check your answer back in the original question.

An Example Involving Percents

Here is a size-change example using percents.

Example 3

In a small-town election, it was reported that Phineas Foghorn got about 38% of the votes. It was also reported that he received 405 votes. How many people voted?

Solution

Let v be the number of people who voted. Given that 38% of v equals 405, translate this into an equation.

$$.38v = 405$$
$$\frac{.38v}{.38} = \frac{405}{.38} \qquad \text{Multiply both sides by } \frac{1}{.38}$$
$$v \approx 1065.8$$

There were approximately 1066 voters.

Check

Suppose 1066 people voted and Phineas received 405 votes. Is this 38%?

$$\frac{405}{1066} = 0.37992\ldots$$

When rounded, this number equals 38%. So the answer checks.

QUESTIONS

Covering the Reading

1. What is the first step you should do in solving a problem like those in this lesson?

2. You solve a problem using an equation. Why should you first check the equation's solution in the original problem, not in the equation?

3. In Example 1, suppose the width of Claus's Clothes was increased to 80 feet. Where should the partition now be located?

4. Stacey gets paid time and a half for each hour she baby-sits after midnight. She made $3.60 per hour after midnight on New Year's Eve. How much did she make for each hour before midnight?

5. In an election, it was reported that Belinda Bellows received 1,912 votes, 54% of the total. How many votes were cast?

Applying the Mathematics

In 6–9, write an equation, solve it, and check your work.

6. A pair of in-line skates is on sale at 70% of the original price. The sale price is $189.95. What was the original price?

7. At an average speed of 550 mph, how long will it take an airplane to fly from Los Angeles to Manila, a distance of 7300 miles?

8. Doll house models are often $\frac{1}{12}$ actual size. If a doll house window is 188 mm wide, how wide is the real window?

9. A Macintosh Classic screen has 175,104 pixels in a rectangular array. There are 512 columns in this array. How many rows are there?

10. From 1932 until 1992, the largest percent of people eligible to vote who actually voted in a presidential election was 62.8%. This occurred in the 1960 election between John Kennedy and Richard Nixon. About 68,839,000 votes were cast. How many people were old enough to vote then but didn't?

11. In the Chinese game of mah-jongg, the player "going mah-jongg" (or winning) receives the value w of the winning score from two of the other players and double the winning score from the player designated East. If East wins, the other three players give that person double the winning score. Let T be the total value received by the winner.
    a. Write an equation relating w and T if East does not win.
    b. If the winning score in part **a** is 900, find T.
    c. If East wins, write an equation relating w and T.
    d. If the winning score in part **c** is 900, find T.

Going mah-jongg. Shown is one example of a winning mah-jongg hand of tiles. In Chinese, mah-jongg means "sparrow." Therefore, a sparrow or mythical "bird of 100 intelligences" appears on the 1s tiles.

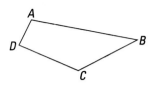

Review

In 12–14, solve. *(Lessons 10-1, 10-2)*

12. $14n = 4$
13. $a + a + a + a = 30$
14. $2170 = 2\frac{1}{3}x$

15. Simplify $1 - 5 - 4$. *(Lessons 4-1, 7-2)*

16. As of January 1993, the world record for the 5000-meter run was 12 minutes, 58.39 seconds for men (set by Said Aouita of Morocco in 1987) and 14 minutes 37.33 seconds for women (set by Ingrid Kristiansen of Norway in 1986). Who ran faster, and by how much? *(Lesson 7-1)*

17. Make a tessellation using copies of quadrilateral $ABCD$. *(Lesson 8-8)*

In 18–20, evaluate if $x = 3.2$ and $y = 7.8$. *(Lesson 4-4)*
18. $100x + 200$
19. $100x + 200x$
20. $100x + 200y$

21. 2 cm is how many meters? *(Lesson 3-4)*

Exploration

22. Suppose the area of a rectangularly shaped farm is 240 acres, or $\frac{3}{8}$ of a square mile. What might be the dimensions of the farm, in feet?

10-4

Solving $ax + b = c$

Algebra in action. *Suppose each kitten weighs the same. If the cat weighs 9 pounds, and the cat and kittens together weigh 19 pounds, then the equation $4w + 9 = 19$ could be used to find the weight of each kitten.*

Picturing the Solving of $ax + b = c$

Here is a balance-scale picture of the equation $3w + 5 = 14$. On the scale, three boxes of equal unknown weight and 5 one-kilogram weights balance with 14 one-kilogram weights.

You can find the weight w of one box in two steps. Each step keeps the scale balanced.

Step 1   Remove 5 one-kilogram weights from each side.

Step 2 Distribute the remaining weight equally among the boxes.

Therefore, one box weighs 3 kilograms. Example 1 shows the same steps without the balance scale.

Example 1

Solve $3w + 5 = 14$.

Solution

$$3w + 5 + -5 = 14 + -5 \qquad \text{Addition Property of Equality}$$
(Add -5 to each side.)
$$3w = 9 \qquad \text{Simplify.}$$
$$\tfrac{1}{3} \cdot 3w = \tfrac{1}{3} \cdot 9 \qquad \text{Multiplication Property of Equality}$$
(Multiply both sides by $\tfrac{1}{3}$.)
$$w = 3 \qquad \text{Simplify.}$$

Check

Substitute 3 for w in the original equation.

$$\text{Does } 3w + 5 = 14?$$
$$3 \cdot 3 + 5 = 14?$$
$$9 + 5 = 14?$$
$$14 = 14? \text{ Yes.}$$

Solving $ax + b = c$ Using Properties

In the equation $3w + 5 = 14$, the unknown w is multiplied by a number. Then a second number is added. We call this an **equation of the form** $ax + b = c$. In this case, $a = 3$, $b = 5$, $c = 14$, and $x = w$. All equations in this form can be solved for x with two major steps. First add $-b$ to both sides. This step gets the term with the variable alone on one side. Then multiply both sides by $\frac{1}{a}$.

Example 2 illustrates this strategy. The original sentence may look complicated, but it can be simplified to a sentence of the form $ax + b = c$.

Example 2

Solve $10 + 6h - 14 = 32$.

Solution

First simplify the left-hand side. You may not need to write all the steps that are shown here.

$$10 + 6h - 14 = 32$$
$$10 + 6h + \text{-}14 = 32 \qquad \text{Algebraic Definition of Subtraction}$$
$$10 + \text{-}14 + 6h = 32 \qquad \text{Associative and Commutative Properties of Addition}$$
$$\text{-}4 + 6h = 32 \qquad \text{Arithmetic}$$

This is an equation like the one solved in Example 1. Add 4 to both sides.

$$4 + \text{-}4 + 6h = 4 + 32 \qquad \text{Addition Property of Equality}$$
$$6h = 36 \qquad \text{Arithmetic}$$

You can solve this equation in your head. Otherwise, either multiply both sides by $\frac{1}{6}$ or divide both sides by 6.

$$h = 6$$

Check

Substitute: Does $10 + 6 \cdot 6 - 14 = 32$?
$$10 + 36 - 14 = 32?$$
$$46 - 14 = 32? \text{ Yes.}$$

Here is a typical problem that leads to an equation of the form $ax + b = c$.

Example 3

In a recent year, taxicabs in New York City began each trip by setting the meter at $1.50. The meter added 25¢ for each $\frac{1}{5}$ mile traveled. If a trip cost $17.75, how far did the cab travel?

Solution

Let n be the number of $\frac{1}{5}$ miles traveled. Then
$$\text{total cost} = 1.50 + .25n$$
$$17.75 = 1.50 + .25n$$

Subtract 1.50 from each side.
$$16.25 = .25n$$

Multiply each side by $\frac{1}{.25}$. This is the same as dividing by .25.
$$65 = n$$

The taxi traveled $65 \cdot \frac{1}{5}$ miles, or 13 miles altogether.

Fair fares. *Wilhelm Bruhn invented the taximeter in 1891, enabling commercial drivers to determine accurately a fee for each ride. The modernization of taxicabs (which were named after the taximeter) paralleled that of the automobile.*

If you cannot write an equation right away, a good strategy is to make a table. For Example 3, you might make a table like this one.

$\frac{1}{5}$ miles traveled	Cost
0	1.50
1	1.50 + .25
2	1.50 + .25 · 2
3	1.50 + .25 · 3
.	.
.	.
.	.

Continue writing until you see the pattern.

n	1.50 + .25n

Then, since the cost was $17.75, the equation to be solved is
$17.75 = 1.50 + .25n$.

QUESTIONS

Covering the Reading

1. Illustrate how to solve the equation $5w + 2 = 12$ using a balance scale.

In 2–5, the equation is of the form $ax + b = c$. Identify a, b, and c.

2. $3x + 8 = 12$

3. $17 = 11 + \frac{1}{2}x$

4. $-4 + 2y = 9$

5. $2x - 8 = 12$

6. a. To solve the equation $4x + 25 = 85$, first add ___?___ to both sides.
   b. Solve $4x + 25 = 85$. Show your work.

7. a. What should be the first step in solving $3 + 4x + 5 = 6$?
   b. Solve this equation. Show your work.

In 8–11, solve the equations of Questions 2–5.

In 12–17, solve and check. Show your work.

12. $2y + 7 = 41$

13. $300 + 120t = 1500$

14. $8y - 2 = 0$

15. $17 + 60m + 3 = 100$

16. $\frac{5}{8}x + 20 = 85$

17. $-20 + 4x = -12$

18. Use the situation in Example 3.
   a. How much would it cost to take a cab 2 miles?
   b. How far did the cab travel if a trip cost $9.50?

19. To take a cab in a certain city, it costs 75¢ plus 15¢ for every $\frac{1}{5}$ mile traveled. If a cab fare is $10.05, how long was the trip?

20. French fries have about 11 calories apiece. So, if you eat F French fries, you take in about $11F$ calories. A plain 4-oz hamburger with a bun has about 500 calories.
   a. Together, how many calories do the hamburger and F French fries have?
   b. How many calories are in a plain 4-oz hamburger on a bun and 20 French fries?
   c. How many French fries can you eat with a plain 4-oz hamburger on a bun for 800 total calories?

21. Paolo estimates that a trip to Brazil to see relatives will cost $1500 for air fare and $90 a day for living expenses.
   a. What will it cost to stay n days?
   b. How long can Paolo stay for $2500?

22. Use the formula $F = \frac{9}{5}C + 32$ to find the Celsius equivalent of 68°F.

23. Meticulous Matilda likes to put in every step in solving equations. Here is her solution to $5m + 7 = 17$. Give a reason for each step.
   a. $(5m + 7) + \text{-}7 = 17 + \text{-}7$
   b. $5m + (7 + \text{-}7) = 17 + \text{-}7$
   c. $\qquad 5m + 0 = 17 + \text{-}7$
   d. $\qquad\qquad 5m = 10$
   e. $\qquad \frac{1}{5} \cdot (5m) = \frac{1}{5} \cdot 10$
   f. $\qquad \left(\frac{1}{5} \cdot 5\right)m = \frac{1}{5} \cdot 10$
   g. $\qquad\qquad 1 \cdot m = 2$
   h. $\qquad\qquad\quad m = 2$

24. Solve $2x + 3 + 4x + 5 = 6$.

25. a. To solve $mx + n = p$ for x, what could you add to both sides?
   b. Solve this equation.

Relaxing in Rio. *The city Rio de Janeiro lies along part of Brazil's 9700 km (over 6000 mi) of coastline. One of the main attractions of Rio is its beautiful white-sand beaches, such as the famous Copacabana Beach shown here.*

Review

26. If the area of a rectangular plot of land is $\frac{3}{8}$ sq mi and one dimension of the plot is $\frac{1}{2}$ mi, what is the other dimension?   *(Lesson 10-3)*

27. If $11x = 1331$, what is x?   *(Lesson 10-2)*

28. a. Give the perimeter of this polygon in simplified form.   *(Lesson 10-1)*

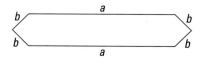

   b. What kind of polygon is this figure?   *(Lesson 5-9)*

29. A size change has magnitude $\frac{5}{6}$. Is it a contraction or an expansion? *(Lessons 9-7, 9-8)*

30. a. Graph the quadrilateral with vertices (1, 2), (-2, 3), (0, -1), and (1, -1). *(Lesson 8-3)*
   b. Graph the image of this quadrilateral under a size change of magnitude 3. *(Lesson 9-7)*

31. One hat contains 4 slips of paper labeled 0 and 6 slips of paper labeled 1. A second hat contains 26 slips, each labeled with a different letter of the alphabet. You draw a 1 from the first hat and an *R* from the second hat. A person is allowed one guess. What are the chances that the person will guess what you drew? *(Lesson 9-4)*

32. Trace the figure below and draw its lines of symmetry. *(Lesson 8-7)*

33. On January 1 about 2500 deer were in the White Forest. During the year, *D* deer died, *B* were born, and *S* were relocated. At the end of the year about 2300 deer were in the forest. Give an equation relating all of these numbers and variables. *(Lessons 5-1, 7-1)*

Exploration

34. a. Find rates for taxi rides near where you live.
   b. How far can you travel for $10.00?

10-5

Solving $ax + b = c$ When a Is Negative

Food distribution. *Shown here are volunteers from a relief organization as they distribute food to the needy. Food pantries in large cities may give out 1000 lb of food per day. See Question 22.*

Solving $ax = b$ When a Is Negative

The equation $-4x = 3$ is an equation of the form $ax = b$, with $a = -4$ and $b = 3$. This equation can be solved using the Multiplication Property of Equality. But first you must be able to find the reciprocals of negative numbers.

The reciprocal of n is $\frac{1}{n}$, because $n \cdot \frac{1}{n} = 1$. The reciprocal of $-n$ is $-\frac{1}{n}$, because $-n \cdot -\frac{1}{n} = 1$.

For example, the reciprocal of -4 is $-\frac{1}{4}$, or -0.25. The reciprocal of $-\frac{2}{3}$ is $-\frac{3}{2}$, or -1.5.

Activity

a. Enter -5 into a calculator. Press the reciprocal key to see what the calculator shows.

b. Repeat part **a** for $-\frac{2}{7}$.

Example 1

Solve $-4x = 30$.

Solution

Work as you did before, with equations of the form $ax = b$. Multiply both sides by the reciprocal of -4. Because of the Multiplication Property of Equality, this does not change the solution.

$$-\frac{1}{4} \cdot -4x = -\frac{1}{4} \cdot 30$$

▶

Because $-\frac{1}{4} \cdot -4 = 1$, the left-hand side simplifies to $1 \cdot x$, which equals *x*. The numbers on the right-hand side can be multiplied as usual.

$$x = -\frac{30}{4}$$
$$= -7.5$$

Check

Substitute -7.5 for *x* in the original equation.

Does -4 · -7.5 = 30?

Yes, so the solution checks.

When solving equations with negative numbers, you should first decide whether the solution is positive or negative. Is it obvious in Example 1 that *x* is negative? In Example 2, you should realize before solving that *m* is positive.

Example 2

Solve $-4.5m = -27$.

Solution

$$-4.5m = -27$$

Multiply both sides by $-\frac{1}{4.5}$, the reciprocal of -4.5.

$$-\frac{1}{4.5} \cdot -4.5m = -\frac{1}{4.5} \cdot -27$$

All that is left is to simplify both sides.

$$1 \cdot m = -\frac{-27}{4.5}$$
$$m = 6$$

Check

Does -4.5 · 6 = -27? Yes, it does.

Solving $ax + b = c$ When a Is Negative

In the next example, there is a starting amount from which a certain amount is subtracted at a constant rate. The question leads to a sentence of the form $ax + b = c$ where *a* is negative.

Example 3

Water in a swimming pool is currently 2.8 meters deep. After opening a valve, the depth of the water will decrease 2 centimeters, or .02 meter, per minute. In how many minutes will the water be 1.5 meters deep?

Solution

Let *t* be the time (in minutes) the valve is open. The height (in meters) of the water after *t* minutes is

$$2.8 - .02t.$$

▶

So we need to solve

$$2.8 - .02t = 1.5.$$

First convert the subtraction to addition using the Algebraic Definition of Subtraction.

$$2.8 + -.02t = 1.5$$

Add -2.8 to both sides.

$$-.02t = -2.8 + 1.5$$
$$-.02t = -1.3$$

Now solve the equation as you did Example 2. Multiply both sides by the reciprocal of -.02.

$$-\frac{1}{.02} \cdot -.02t = -\frac{1}{.02} \cdot -1.3$$
$$t = \frac{1.3}{.02}$$
$$t = 65$$

In 65 minutes the water will be 1.5 meters deep.

QUESTIONS

Covering the Reading

In 1–6, write the reciprocal of each number.

1. -8 **2.** $-\frac{1}{3}$ **3.** $\frac{9}{5}$

4. -1.2 **5.** $-x$ **6.** $-\frac{1}{x}$

7. What results did you get for the Activity in this lesson?

8. To solve $-ax = b$, by what number should you multiply both sides?

In 9–12, solve and check.

9. $-4x = 8$ **10.** $-4.5y = -81$

11. $1.2 = -1.2B$ **12.** $-\frac{2}{15} = -\frac{5}{3}t$

13. a. To solve the equation $8 - 2v = -50$, first __?__ to both sides.
 b. Solve $8 - 2v = -50$.

In 14 and 15, solve and check.

14. $100 = 160 - 20h$ **15.** $-2c + 5 = 17$

16. What property enables both sides of an equation to be multiplied by the same number without the solutions being affected?

17. Consider the situation of Example 3.
 a. Where does the number -.02 come from?
 b. In how many minutes will the water be 1 meter deep?
 c. In how many minutes will the pool be empty?

In 18–21, solve and check.

18. $-\frac{2}{3}m + \frac{1}{3} = 11$

19. $6 - 30n - 18 = 69$

20. $-4 = -4 - 4a$

21. $2b + 2 + 3b + 3 - 4b - 4 = 5$

22. A relief agency began the year with 20 tons of food. Suppose 0.3 ton of food was used each day.
   a. After d days, how much food was left?
   b. After how many days are only 1.5 tons of food left?

23. Suppose the original length of a pencil is $7\frac{1}{2}$ inches. Marna thinks she can use the pencil until it is 3 inches long. If she uses $\frac{1}{5}$ inch a day, how many days will it last?

24. Is the solution to $5.83146 = -24.987t$ positive or negative?

25. Suppose $xy = z$ and both x and z are negative. What can be said about y?

26. Find the integer between -1111 and 1111 that is the solution to the equation $-11x - 111 = -11111$.

27. Solve $9u - 47 = 214$.   *(Lesson 10-4)*

28. a. At Elm Grove High School, 600 students are on at least one of the school teams. This is 30% of all students. How many students are at Elm Grove High?
   b. At Maple Middle School, 30% of the 600 students are enrolled in a computer class. How many students is this?   *(Lessons 2-5, 10-3)*

29. Under a size change of magnitude -4, how will a quadrilateral and its image be the same, and how will they be different?   *(Lesson 9-9)*

30. Solve $0x = 7$.   *(Lesson 9-6)*

31. Suppose a person makes $9.25 an hour and works 28 hours a week. How much does this person earn per year?   *(Lesson 9-5)*

In 32 and 33, use the drawing below.

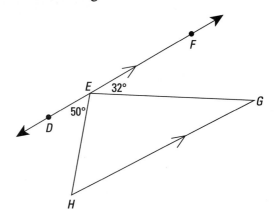

32. Trace the drawing and draw the reflection image of △*EGH* over $\overleftrightarrow{DF}$.
(Lesson 8-6)

33. Find the measures of all angles of △*EGH*.   *(Lessons 7-7, 7-9)*

34. Find m∠*DBC* in terms of *x*.   *(Lesson 7-6)*

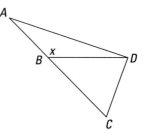

35. The probability that a sum *S* will appear when two ordinary dice are thrown is $(6 - |S - 7|)/36$. (*S* will always be an integer from 2 to 12.) Use this formula to calculate the probability that a sum of 3 will appear.   *(Lessons 4-6, 5-3)*

<hr>

Exploration

36. a. Solve for *x*: $ax + b = c$. (Assume $a \neq 0$.)
   b. Without solving, how can you predict when the solution will be positive?
   c. Without solving, how can you predict when the solution will be negative?
   d. Without solving, how can you predict when the solution will be zero?

LESSON

10-6

The Distributive Property

Mathematics in art. *The Dutch painter Piet Mondrian specialized in abstract forms using the simple harmonies of straight lines, right angles, and the primary colors. How could the Distributive Property be used to relate the areas of some of the rectangles in his painting above?*

Suppose you need to find the total area of these rectangles.

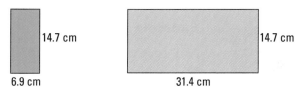

14.7 cm

6.9 cm

14.7 cm

31.4 cm

One obvious way is to find $6.9 \cdot 14.7$ and $31.4 \cdot 14.7$. Then add the two products.

Activity 1

Do these calculations to find the total area.

Another way to find the area involves a simpler addition and only one multiplication. Join the rectangles.

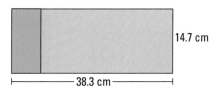

14.7 cm

38.3 cm

One dimension is still 14.7 cm. The other dimension is 6.9 cm + 31.4 cm, or 38.3 cm. Multiplying the two dimensions, 14.7 cm and 38.3 cm, gives the area.

Find 38.3 × 14.7. Compare your answer to what you found in Activity 1.

Distributing Multiplication over Addition

The above activities indicate that

$$6.9 \cdot 14.7 + 31.4 \cdot 14.7 = (6.9 + 31.4) \cdot 14.7.$$

Since multiplication is commutative and associative, the 14.7 could be at the left of each multiplication. Thus the rectangles also picture

$$14.7 \cdot 6.9 + 14.7 \cdot 31.4 = 14.7(6.9 + 31.4).$$

The general patterns are known as the *Distributive Property of Multiplication over Addition.*

The Distributive Property of Multiplication over Addition
For any numbers *a*, *b*, and *x*,

$$ax + bx = (a + b)x$$
$$\text{and } x(a + b) = xa + xb.$$

The Distributive Property gets its name because it says that multiplications "distributed" over several quantities can be combined into one multiplication. In Example 1 below, the 4 is distributed over 1.75 and 0.29 as a common factor. The two multiplications can be written as one.

Example 1

April bought 4 greeting cards at $1.75 each and 4 stamps at $0.29 each. What was the total cost?

Solution
By the Distributive Property, $a \cdot x + b \cdot x = (a + b)x$.

$$\$1.75 \cdot 4 + \$0.29 \cdot 4 = (\$1.75 + \$0.29)4$$
$$= (\$2.04) \cdot 4$$
$$= \$8.16$$

The total cost was $8.16.

Using the Distributive Property to Calculate in Your Head

In Example 2, the distributive property is used to multiply 18 and 10.50. It enables you to do the calculation in your head.

Example 2

A store owner bought 18 blouses for $10.50 each. Mentally calculate the total cost to the store owner.

Solution

Think of 10.50 as 10 and 0.50. Here are the mental steps.

$$18 \cdot 10.50 = 18(10 + 0.50)$$
$$= 18 \cdot 10 + 18 \cdot 0.50$$
$$= 180 + 9$$
$$= 189$$

The total cost to the store owner was $189.

Distributing Multiplication over Subtraction

There is also a Distributive Property of Multiplication over Subtraction. You may have already used it in doing mental arithmetic.

> **The Distributive Property of Multiplication over Subtraction**
> For any numbers *a*, *b*, and *x*,
> $$ax - bx = (a - b)x.$$
> $$\text{and } x(a - b) = xa - xb.$$

Example 3

What will 7 pairs of socks at $3.98 a pair cost? Do this in your head.

Solution

Think of 3.98 as $4 - 0.02$.
Apply the Distributive Property of Multiplication over Subtraction.
These multiplications can be done mentally.
You are taking 14¢ from 28 dollars.

$$7 \cdot 3.98 = 7(4 - 0.02)$$
$$= 7 \cdot 4 - 7 \cdot 0.02$$
$$= 28 - 0.14$$
$$= 27.86$$

Seven pairs of socks will cost $27.86.

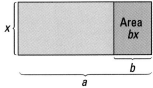

At the left is an area picture of the Distributive Property of Multiplication over Subtraction. What is the area of the blue rectangle?

Answer 1: The dimensions of the blue rectangle are *x* and $a - b$. So its area is $(a - b)x$.

Answer 2: The area of the whole figure is *ax*. The area of the orange rectangle is *bx*. So the area of the blue rectangle is $ax - bx$.

Since the answers must be equal, $(a - b)x = ax - bx$.

Distributing Multiplication over Addition and Subtraction

The distributive properties over addition and subtraction can be combined.

Example 4

Multiply $3(5x + y - 8)$.

Solution

Use the distributive properties.
$$3(5x + y - 8) = 3 \cdot (5x) + 3 \cdot y - 3 \cdot 8$$
$$= 15x + 3y - 24$$

Check

Substitute a different number for each variable. We use $x = 2$ and $y = 4$. The values for the given expression and the answer should be the same.

Given expression: $3(5x + y - 8) = 3(5 \cdot 2 + 4 - 8) =$
$3(10 + 4 - 8) = 3(6) = 18$
Answer: $15x + 3y - 24 = 15 \cdot 2 + 3 \cdot 4 - 24 =$
$30 + 12 - 24 = 18$

Since the values are equal, the answer checks.

QUESTIONS

Covering the Reading

1. **a.** What is the total area of these rectangles?

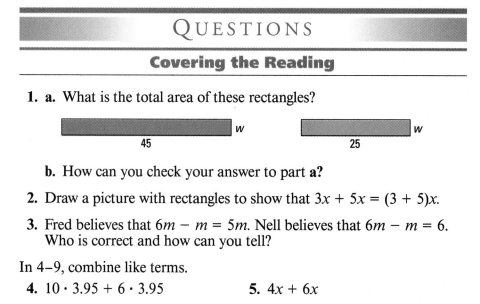

    b. How can you check your answer to part **a?**

2. Draw a picture with rectangles to show that $3x + 5x = (3 + 5)x$.

3. Fred believes that $6m - m = 5m$. Nell believes that $6m - m = 6$. Who is correct and how can you tell?

In 4–9, combine like terms.

4. $10 \cdot 3.95 + 6 \cdot 3.95$ 5. $4x + 6x$

6. $4y + 6y + 8$ 7. $2.4v + 3.5v + v$

8. $8x + 8y$ 9. $11A + 11B$

In 10–13, multiply and check.

10. $50(x + 4)$ 11. $2(6 + 7n)$

12. $a(b + 2c)$ 13. $5(100 + t + u)$

14. Give an instance of the Distributive Property of Multiplication over Subtraction.

15. Draw a picture with rectangles to show that $10y - 3y = 7y$.

In 16 and 17, show how you can apply the Distributive Property of Multiplication over Subtraction to find the total cost in your head.

16. You buy three sweaters at $29.95 each.

17. You purchase 8 tapes at $8.99 each.

In 18–21, use the Distributive Property to rewrite the expression.

18. $5x - 2x$

19. $12 \cdot \$3.75 - 2 \cdot \$3.75$

20. $a(b - c)$

21. $4(a - 5b + 3c)$

Applying the Mathematics

22. Use a calculator to verify that

$$3.29(853 + 268) = 3.29 \cdot 853 + 3.29 \cdot 268.$$

As you evaluate each side, write down the key sequence you use and the final display.

23. A hamburger costs x cents. Dave bought 3 hamburgers. Sue bought 2. How much did they spend altogether?

24. Suppose an acre of land will yield B bushels of wheat. The McAllisters planted 400 acres. Mr. Padilla planted 120 acres. Mrs. Smith planted 240 acres. How many bushels of wheat can the three expect to harvest altogether?

In 25 and 26, apply the Distributive Properties.

25. $a(a + a^2)$

26. $3x(3y - 3z)$

27. Multiply 732 by 999,999,999,999,999.

28. Phil, Gil, and Will each bought a CD player costing t dollars, an amplifier costing a dollars, and speakers costing s dollars. How much did they spend altogether?

In 29 and 30, apply the Distributive Properties. Then simplify if possible.

29. $-2(3 + x) + 5(0.4x)$

30. $\frac{1}{3}(3m + n) + \frac{1}{3}(3m - n)$

Combin*ing* the work.
Combines (com'bines) are machines used to cut and thresh wheat. Teams of people with combines travel from Texas to Canada to harvest the wheat as it ripens.

Review

In 31–33, solve.   *(Lessons 10-2, 10-5)*

31. $14x = 42$

32. $5y - 3 = 20$

33. $6 - 4y = 20$

34. A swimming pool has a depth of $2\frac{1}{2}$ ft now. It is being filled at a rate of $\frac{1}{20}$ ft per minute.
a. What will its depth be in m minutes?   *(Lesson 6-5)*
b. When will it be 9 feet deep?   *(Lesson 10-4)*

35. a. What is the reciprocal of $\frac{6}{7}$?   *(Lesson 9-3)*
   b. Solve $\frac{6}{7}x = 84$.   *(Lesson 10-2)*

36. Consider triangle *ABC*.

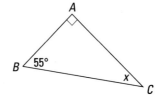

   a. What is the value of *x*?
   b. How many angles of $\triangle ABC$ are acute?
   c. How many angles of $\triangle ABC$ are obtuse?   *(Lessons 3-7, 7-9)*

Exploration

37. Here is a worked-out multiplication of 372×681.

$$
\begin{array}{r}
681 \\
\times\ 372 \\
\hline
1362 \\
4767 \\
2043 \\
\hline
253332
\end{array}
$$

Explain how this process is an application of the Distributive Property of Multiplication over Addition.

38. One way to express the area of the entire figure is $(a + b)(c + d)$. What is another expression for the area?

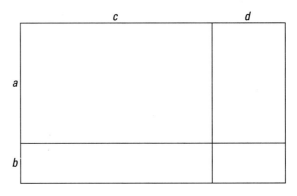

LESSON

10-7

*The
Surface
Area of a
Box*

From the grape vine. *These claymation raisins are used by the California Raisin Advisory Board to promote consumption of raisins. California, the only state producing raisins commercially, is the world's leading raisin-grape producer.*

Each face of the raisin box at the right is a rectangle. (In a perspective drawing, you view some faces at an angle. So some of the faces may not seem to be rectangles.)

The surface of this raisin box can be constructed by cutting a pattern out of a flat piece of cardboard. The pattern is called a **net** for the 3-dimensional surface. The construction is pictured below. The dashed segments show where folds are made. The folds become the edges of the box. (For a real box, some overlap is needed. But the same idea is used.)

the net   after one fold

A rectangular solid consists of a box and all points inside it. The **surface area** of a rectangular solid is the sum of the areas of its faces. The raisin box has six faces: top, bottom, right, left, front, and back. The surface area tells you how much cardboard is needed to make the box.

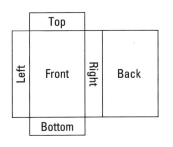

Activity

Find the surface area of a miniature raisin box

Step 1   Take the raisins out and eat them (if allowed), or put them away.

Step 2   Flatten the box by cutting along edges.

Step 3   Copy and complete the chart below. Use your ruler to measure edges to the nearest millimeter.

Face	Length (mm)	Width (mm)	Area (mm²)
Left			
Right			
Front			
Back			
Top			
Bottom			

Step 4   Adding the six areas gives the total surface area. Do this.

Example

Find the surface area of a rectangular solid 10 in. high, 4 in. wide, and 3 in. deep.

Solution

First, draw a picture. You may want to separate the faces as shown here.

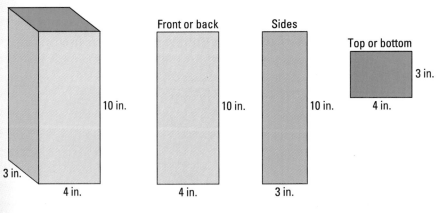

Now just add the areas of the faces. The surface area is

$$10 \cdot 4 + 10 \cdot 4 + 10 \cdot 3 + 10 \cdot 3 + 3 \cdot 4 + 3 \cdot 4$$
$$= 40 + 40 + 30 + 30 + 12 + 12$$
$$= 164 \text{ square inches}$$

The surface area is 164 square inches.

QUESTIONS

1. A raisin box and all points inside it is an example of a figure called a ___?___.

2. What is the surface area of a solid?

3. How many faces does a rectangular solid have?

4. Give the surface area, to the nearest square millimeter, that you found for the miniature raisin box in the Activity.

5. What is a net for a 3-dimensional figure?

In 6 and 7, dimensions for a box are given.
a. Make a sketch of the box.
b. Draw a net for the box.
c. Find the surface area of the box.

6. dimensions 9 cm, 12 cm, and 7 cm

7. dimensions 3″, 3″, and 3″

Applying the Mathematics

8. A box is pictured.

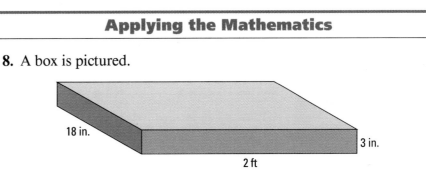

a. Find its surface area in square inches.
b. Find its volume in cubic inches.

9. A rectangular solid has length ℓ, width w, and height h.
a. What is its volume?
b. What is its surface area?

10. A shirt to be given as a gift is put into a box. The dimensions of the box are 14″, 9″, and 2.5″. What is the least amount of wrapping paper needed to wrap the gift? (You are allowed to use a lot of tape.)

11. Refer to the Example in this lesson. Suppose that the surface of the box is made from a rectangular sheet of cardboard.
a. What are the smallest dimensions the cardboard could have?
b. How much material will be wasted?

12. Use a sheet of notebook paper, a ruler, tape, and scissors to construct a box. You may use any dimensions you wish.

13. Patricia Tern wanted to make a rectangular box out of cardboard. She drew the following net. It is not correct. Draw a corrected net by moving one rectangle.

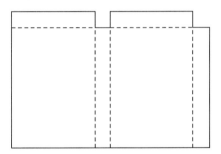

14. A net for a cube is shown below. All sides of the squares have length 5. What is the surface area of the cube?

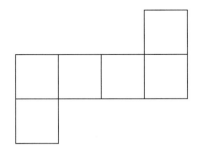

15. The bottom of a box has dimensions 24 cm and 20 cm. The box is 10 cm high and has no top. What is the surface area of the box?

Review

16. Bill bought a dozen roses for Mother's Day at $2.50 each and a dozen carnations at $.50 each. Explain how the total amount he spent can be calculated with one multiplication. *(Lesson 10-6)*

In 17–19, simplify. *(Lessons 10-1, 10-6)*

17. $3x + 8x$

18. $2y + y - 5y$

19. $8m + 2n - 20m - n$

20. Mel weighs 180 lb. If he goes on a diet and loses 2 lb every thirty days, how long will it take him to reach 150 lb? *(Lesson 10-5)*

21. Solve $3x - x = 1$. *(Lesson 10-5)*

22. Solve for y: $x - y = 34$. *(Lesson 7-4)*

23. Simplify $\frac{1}{6} - \frac{3}{18} - \frac{2}{3} + \frac{5}{9}$. *(Lesson 7-2)*

24. What is the perimeter of a rectangle with length 40 and width 60? *(Lesson 5-10)*

25. What property guarantees that $x + 45$ equals $45 + x$? *(Lesson 5-7)*

26. Let *n* be a number. Translate into mathematics.
   a. 2 less than a number
   b. 2 is less than a number.
   c. 2 less a number   *(Lesson 4-3)*

27. What is the volume of the cube illustrated in Question 14?
   (Lesson 3-9)

Exploration

28. In Question 14 a net was given for a cube. Here is a different net. (Nets are different if they are not congruent.)

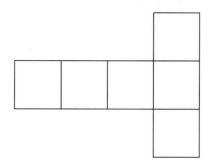

Draw all possible different nets for a cube.

Fishing room galore. *With an area of 8320 km² (about 3200 mi²), Lake Titicaca is the second largest lake in South America. It is also the world's highest navigable lake for large vessels, located 3810 m (about 12,500 ft) above sea level.*

Measuring the Size of a One-Dimensional Figure

The size of a line segment is given by its length.

Measuring the Size of Two-Dimensional Figures

Rectangles and other polygons are two-dimensional. There are two common ways of measuring these figures. Perimeter measures the boundary. Area measures the space inside the figure.

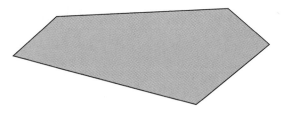

Perimeter and area are quite different. The perimeter of the rectangle pictured below is 6 ft + 100 ft + 6 ft + 100 ft, or 212 *feet*. That's how far you have to go to walk around the rectangle. Perimeter is measured in units of length. The area of the rectangle pictured below is 100 ft · 6 ft or 600 *square feet*. That's how much space it occupies. Area is measured in square units.

100 ft

6 ft

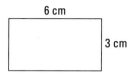
6 cm

3 cm

The perimeter of the rectangle at the left is 6 cm + 3 cm + 6 cm + 3 cm, or 18 cm. Its area is 6 cm · 3 cm, or 18 cm^2. Perimeter and area can never actually be equal because the units are different. In a situation like this one, the perimeter and area are called *numerically equal.*

In lakes, perimeter measures shoreline. Area measures fishing room. Below are maps of Lake Volta in the African country of Ghana, and Lake Titicaca, which lies in the South American countries of Peru and Bolivia. Lake Volta has only a little more fishing room despite having much more shoreline. It has a small area for its perimeter.

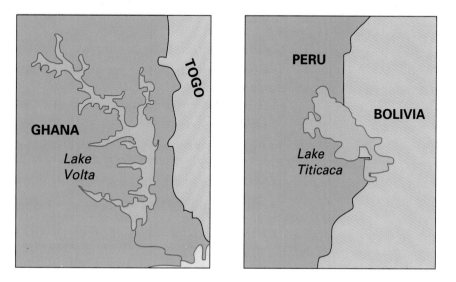

Measuring the Size of Three-Dimensional Figures

There are three different common ways to measure the size of a three-dimensional figure. Consider the shoe box below. We can measure the total length of its edges. Like perimeter, this is one-dimensional. We can measure its surface area, which is two-dimensional. Or we can measure its volume, which is three-dimensional.

The shoebox has edges of lengths 14 inches, 5 inches, and 7 inches. There are four edges of each length. (Some are hidden from view.) The total length of all edges is 4 · 5″ + 4 · 7″ + 4 · 14″, or 104 *inches.*

5 in.

7 in.

14 in.

The surface area of this shoe box is found by adding the areas of the six rectangular faces. You should check that it is 406 *square inches.*

The volume of this box measures the space occupied by the box. It is 5 in. · 7 in. · 14 in., or 490 *cubic inches.*

Surface area and volume have different units. They measure different things, as the example below shows.

Example

The spaghetti box and tea bag box each have a volume of 750 cubic centimeters. Which box requires more cardboard?

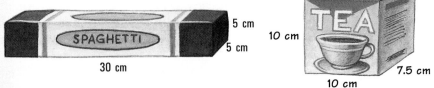

Solution

The surface area of the spaghetti box is the sum of the areas of its faces.

front + back + top + bottom + right + left =
$30 \cdot 5 + 30 \cdot 5 + 30 \cdot 5 + 30 \cdot 5 + 5 \cdot 5 + 5 \cdot 5 = 650$ cm^2.

The surface area of the box of tea bags is
$10 \cdot 10 + 10 \cdot 10 + 10 \cdot 7.5 + 10 \cdot 7.5 + 10 \cdot 7.5 + 10 \cdot 7.5$, or 500 cm^2.

So the spaghetti box requires more cardboard.

QUESTIONS

Covering the Reading

In 1–4, suppose that length is measured in centimeters. In what unit would you expect each to be measured?

1. perimeter

2. volume

3. area

4. surface area

5. __?__ measures the space inside a 2-dimensional figure.

6. __?__ measures the space inside a 3-dimensional figure.

7. __?__ measures the distance around a 2-dimensional figure.

8. __?__ measures the surface of a 3-dimensional figure.

9. In this lesson, the area of Lake Titicaca is measured in square kilometers.
a. In what unit would you expect its shoreline to be measured?
b. In what unit would you expect its fishing room to be measured?

10. Draw a lake that has a large perimeter but very little area.

11. Suppose a side of a square has length 4.
a. *True or false.* The perimeter and area of the square are equal.
b. Explain your answer.

12. Find the area of each face of the shoe box in this lesson.

13. Give an example of two rectangular solids with the same volume but with different surface areas.

14. A small box has dimensions 11 mm, 12 mm, and 13 mm.
    a. Find the total length of its edges.
    b. Find its surface area.
    c. Find its volume.

Applying the Mathematics

15. A box has dimensions ℓ, w, and h. What is the total length of its edges?

16. A rectangle has a perimeter of more than 100 inches. But its area is less than 1 square inch.
    a. Is this possible?
    b. If so, give the dimensions of such a rectangle. If not, tell why it is not possible.

In 17–19, tell whether the idea concerns surface area or volume.

17. the amount of wrapping paper needed for a gift

18. how many paper clips a small box can hold

19. the weight of a rock

20. 640 acres = 1 square mile. A farm is measured in acres.
    Multiple choice. What is being measured about this farm?
    (a) its land area   (b) its perimeter   (c) its volume

21. Recall that a liter is a metric unit equal to 1000 cm^3.
    Multiple choice. What does a liter measure?
    (a) length   (b) area   (c) volume

Review

22. This program calculates and prints the surface area and volume of a box. Two lines are incomplete.

```
10  PRINT "SIDES OF BOX"
15  INPUT A, B, C
20  PRINT "VOLUME", "SURFACE AREA"
30  VOL = ____
40  SA = ____
50  PRINT VOL,SA
60  END
```

    a. Complete lines 30 and 40 to make this program give correct values for *VOL* and *SA*.
    b. What will the computer print if $A = 2$, $B = 3$, and $C = 4$?
    c. Run the completed program with values of your own choosing to test it. *(Lessons 9-2, 10-7)*

23. *Multiple choice.* Which expression is equal to $\frac{2}{3}x - x$?
(Lesson 10-6)

(a) $\frac{2}{3}$ (b) $-\frac{1}{3}$ (c) $-\frac{1}{3}x$ (d) none of these

24. Simplify $100x - 5y + 8x - 23 + 9y - 1 + x + 80y$.   *(Lesson 10-6)*

25. Solve $3c + 5c + 12 = 240$.   *(Lessons 10-4, 10-6)*

26. Find the area of the triangle with vertices (4, -8), (-3, -8), and (-3, 11).   *(Lessons 8-3, 9-1)*

27. Trace the drawing. Then reflect triangle *ABC* over the line *m*.
(Lesson 8-6)

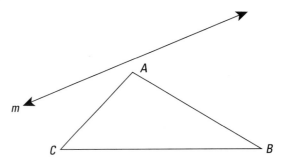

28. The point (8, -4) is translated 10 units down. What is its image?
(Lesson 8-5)

29. Graph the line with equation $y = x$.   *(Lesson 8-4)*

30. In this figure, $\angle 2$ is a right angle and $m\angle 1 = 140°$. What is $m\angle 3$?
(Lessons 7-6, 7-7)

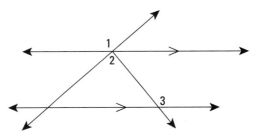

31. Graph all solutions to the inequality $x > 4$.   *(Lesson 4-10)*

32. a. Rewrite 3 trillion in scientific notation.
b. Rewrite 3 trillionths in scientific notation.   *(Lessons 2-3, 2-8)*

33. Rewrite each number as a percent. **a.** $\frac{1}{3}$ **b.** $\frac{2}{3}$ **c.** $\frac{4}{5}$ *(Lessons 1-6, 2-4)*

Exploration

34. The "D" in 3-D movies is short for "dimensional."
   a. Are these movies actually three dimensional?
   b. When you watch a 3-D movie in a theater, you must wear special glasses. How do these glasses work?

Shades of the 50s. *The first movie requiring 3-D glasses was produced in 1952 by Arch Obeler.*

Triangular support. *Buckminster Fuller invented the geodesic dome. Geodesic spheres, like this one at EPCOT Center, consist of lightweight triangular faces.*

Finding the Area of a Triangle from the Area of a Related Rectangle

You saw in Lesson 9-1 that the area of any right triangle is half the area of a rectangle. The area of rectangle *DFEG* below is *bh*. So the area of triangle *DEG* is $\frac{1}{2}$ *bh*.

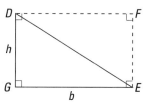

Every triangle has half the area of a related rectangle.

Activity 1

Trace two copies of triangle *ABC* onto a sheet of paper. Trace rectangle *ABDE* onto another sheet of paper.

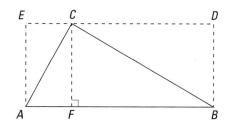

Cut out both triangles. Then cut one of the triangles at $\overline{CF}$. Show that the three triangles you now have fill the entire rectangle.

The Activity verifies that the area of △ABC is $\frac{1}{2}$ the area of rectangle ABDE. One dimension of the rectangle is AB. The other dimension equals CF. $\overline{CF}$ is the segment from point C perpendicular to the opposite side $\overline{AB}$. It is called the **height,** or **altitude,** of △ABC.

A height (or altitude) of a triangle does not have to be the length of a vertical line segment. Every triangle has three altitudes, one from each vertex perpendicular to the opposite side. The side to which the altitude is drawn is called the **base** for that altitude. Below are three copies of △PQR. The three altitudes are drawn with a blue line. Also drawn are three rectangles. The area of the triangle is always one-half the area of the rectangle. That is, the area of a triangle is half the product of a side (the base) and the altitude drawn to that side.

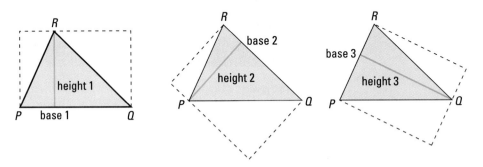

Area Formula for Any Triangle
Let *b* be the length of a side of a triangle with area *A*. Let *h* be the length of the altitude drawn to that side. Then

$$A = \frac{1}{2}bh.$$

Using an Area Formula to Find the Area of a Triangle

Example 1

$\overline{VY}$ is perpendicular to $\overline{XZ}$. Find the area of △XYZ.

Solution
$\overline{VY}$ is the only altitude shown. Its base is $\overline{XZ}$. So use $\overline{XZ}$ as the base in the formula.

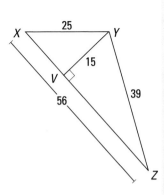

$$\text{Area of } \triangle XYZ = \frac{1}{2} \cdot bh$$
$$= \frac{1}{2} \cdot XZ \cdot VY$$
$$= \frac{1}{2} \cdot 56 \cdot 15$$
$$= 420$$

The area of △XYZ is 420 square units.

Example 2

Find the area of △*ABC*.

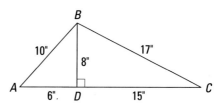

Solution 1

Separate △*ABC* into the two right triangles *ABD* and *BDC*.

Area of △ABD = $\frac{1}{2}$ · 6 · 8 square inches

Area of △BDC = $\frac{1}{2}$ · 8 · 15 square inches

Add the areas to get the area of △ABC.

Area of △ABC = $\frac{1}{2}$ · 6 · 8 + $\frac{1}{2}$ · 8 · 15

   =   24   +   60

   = 84 square inches

Solution 2

Use the formula for the area of a triangle. Since $\overline{BD} \perp \overline{AC}$, *BD* is the height for the base $\overline{AC}$.

Area = $\frac{1}{2}$ bh = $\frac{1}{2}$ · (6 + 15) · 8 = $\frac{1}{2}$ · 21 · 8 = 84 square inches

Measuring to Find the Area of a Triangle

In Example 2, more information was given than was needed. The 10″ and 17″ lengths were not used. In the next Activity, no lengths are given. But by measuring, you can estimate the area.

Activity 2

Trace this triangle. By drawing an altitude and measuring lengths, estimate the area of this triangle to the nearest square inch.

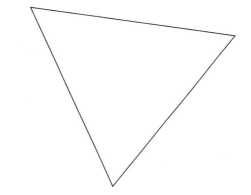

If your estimate is about 2 square inches, you probably did Activity 2 correctly.

Covering the Reading

1. Find the area of $\triangle DEF$.

2. Find the area of $\triangle ABC$.

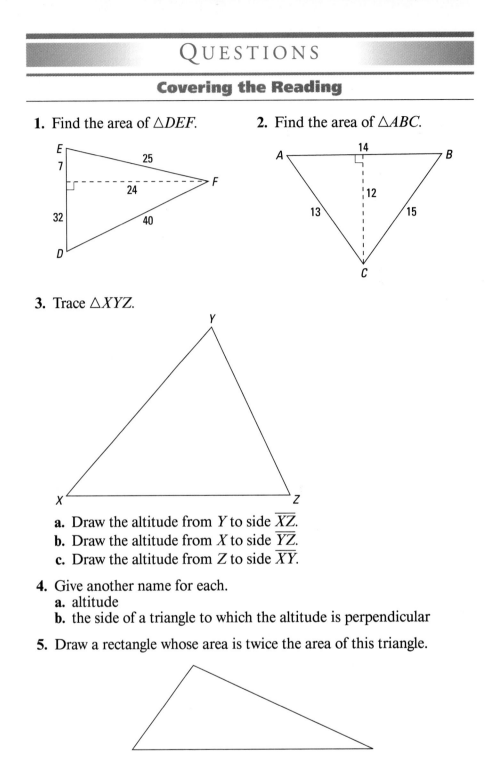

3. Trace $\triangle XYZ$.

a. Draw the altitude from Y to side $\overline{XZ}$.
b. Draw the altitude from X to side $\overline{YZ}$.
c. Draw the altitude from Z to side $\overline{XY}$.

4. Give another name for each.
a. altitude
b. the side of a triangle to which the altitude is perpendicular

5. Draw a rectangle whose area is twice the area of this triangle.

6. Give a formula for the area of a triangle.

7. In Activity 2, what are the dimensions of the side you measured and of the altitude to that side?

8. Each square in the grid below is 1 unit on a side. Find the area of △*MNP*. Explain the method you used.

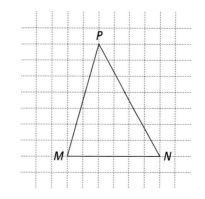

9. The triangle below has a base with length 40 cm and an area of 300 cm². What is the length of the altitude to that base?

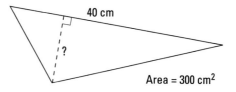

10. Below is a photograph of three of the great pyramids of Giza in Egypt. Each face of the largest pyramid is a triangle whose base is about 230 meters long. The height of each face is about 92 meters. What is the area of a face?

The pyramids of ancient Egypt, built as royal tombs, are the largest stone structures in the world.

11. Find the area of △*EAR* below.

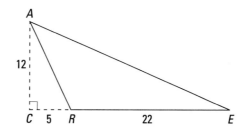

12. A box has dimensions 30 cm, 26 cm, and 12 cm. *(Lessons 3-9, 10-7)*
   a. What is the area of the largest face of the box?
   b. What is the total surface area of the box?
   c. What is the volume of the box?

13. Multiply $a(a + 2)$. *(Lesson 10-6)*

14. Solve $1.7x - 0.4 = 3.0$. *(Lesson 10-4)*

15. Evaluate $3x + 14x + 2x + x$ when $x = 5.671$. *(Lesson 10-1)*

16. What is the value of $-4m$ when $m = -4$? *(Lessons 4-4, 9-6)*

17. How many two-digit numbers satisfy both of the following conditions?
   (1) The first digit is 1, 3, 5, or 7.
   (2) The second digit is 2, 4, or 6. *(Lessons 6-1, 9-4)*

In 18–22, refer to these polygons. *(Lesson 7-8)*

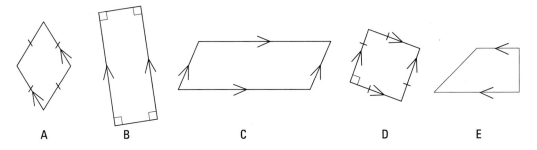

A   B   C   D   E

18. The polygons are all __?__.

19. Which of the polygons are parallelograms?

20. Which of the polygons are rectangles?

21. Which of the polygons are rhombuses?

22. Which of the polygons are squares?

23. Add: $\frac{2}{5} + \frac{1}{3} + \frac{1}{4}$. *(Lesson 5-5)*

24. What simple fraction in lowest terms equals $\frac{0.46}{9.2}$? *(Lesson 1-10)*

25. Translate into English: 345.29. *(Lesson 1-2)*

26. Identify an object where you live that has the shape of a triangle. (For example, you might use part of a roof or a side of a piece of furniture.) Draw a copy of the object, give or estimate the lengths of its sides and at least one altitude, and find its area.

Areas of Trapezoids

Trapezoids in architecture. *The design of this hotel in Cancun, Mexico, illustrates the use of trapezoidal shapes in depicting a thunderbird. In North American Indian mythology, thunderbirds are powerful spirits.*

Pentagon *ABCDE* is split into three triangles.

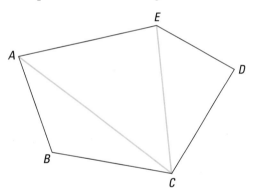

To find the area of *ABCDE,* you could add the areas of the triangles. This is one reason you learned how to find the area of a triangle. *Any* polygon can be split into triangles.

Finding the Area of a Trapezoid by Splitting It into Triangles

When a figure is split into two or more triangles that have the same altitude, their areas can be added using the Distributive Property. A simple formula can be the result. This can always be done if the figure is a *trapezoid.* A **trapezoid** is a quadrilateral that has at least one pair of parallel sides.

Example

Find the area of trapezoid *TRAP*.

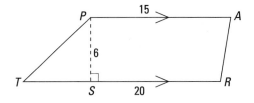

Solution

Draw the diagonal $\overline{PR}$. This splits the trapezoid into two triangles, *PAR* and *PRT*. Each triangle has altitude 6.

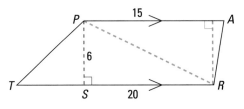

$$
\begin{aligned}
\text{Area of trapezoid TRAP} &= \text{Area of } \triangle\text{PAR} + \text{Area of } \triangle\text{PRT} \\
&= \tfrac{1}{2} \cdot 6 \cdot 15 + \tfrac{1}{2} \cdot 6 \cdot 20 \\
&= 3 \cdot 15 + 3 \cdot 20 \\
&= 45 + 60 \\
&= 105
\end{aligned}
$$

The area of TRAP is 105 square units.

A Formula for the Area of a Trapezoid

The parallel sides of a trapezoid are called its **bases.** The distance between the bases is the **height,** or **altitude,** of the trapezoid. The example shows that you need only the lengths of the bases and the height to find the area of a trapezoid.

To find a general formula, replace the specific lengths with variables. Let the bases have lengths b_1 and b_2. b_1 means the "first base," b_2 means the "second base." Let the height be h.

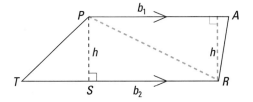

$$
\begin{aligned}
\text{Area of trapezoid } TRAP &= \text{Area of } \triangle PAR + \text{Area } \triangle PRT \\
&= \tfrac{1}{2} \cdot h \cdot b_1 + \tfrac{1}{2} \cdot h \cdot b_2
\end{aligned}
$$

Now use the Distributive Property. The following formula is the result.

Area Formula for a Trapezoid
Let A be the area of a trapezoid with bases b_1 and b_2 and height h.
Then $A = \frac{1}{2} \cdot h(b_1 + b_2)$.

You can check that this formula gives the answer found in the Example.
There $h = 6$, $b_1 = 15$, and $b_2 = 20$.

$$A = \tfrac{1}{2}h(b_1 + b_2)$$
$$= \tfrac{1}{2} \cdot 6(15 + 20)$$
$$= 105$$

Some people prefer the formula in words.

The area of a trapezoid is one half the product of its height and the
sum of the lengths of its bases.

For What Figures Does This Formula Work?

Trapezoids come in many different sizes and shapes. Here are some
trapezoids with bases drawn in blue. Heights of the trapezoids below are
dashed. In the trapezoid at the far right, one base has to be extended to
meet the height.

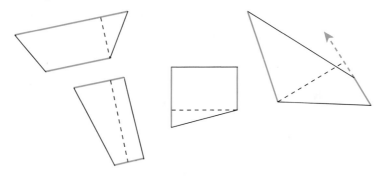

Rectangles, squares, rhombuses, and parallelograms have parallel sides,
so they are all trapezoids. Therefore, whatever is true for all trapezoids is
true for all rectangles, squares, rhombuses, and parallelograms. The area
formula for a trapezoid works for all these figures.

However, there is no general formula that works for the area of
all polygons.

Covering the Reading

1. Why is the area formula for a triangle so important?

2. How can you tell if a figure is a trapezoid?

In 3–7, *true or false*.

3. All quadrilaterals are trapezoids.

4. All squares are trapezoids.

5. All trapezoids are rectangles.

6. The formula for the area of a trapezoid can be used to find the area of a parallelogram.

7. There is a formula for the area of any polygon.

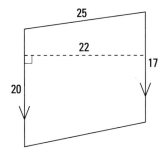

8. Use the trapezoid at the left.
   a. Give the lengths of its bases.
   b. Give its height.
   c. Give its area.

9. a. Draw a trapezoid with bases of length 4 cm and 5 cm and height of 2 cm.
   b. Find the area of this trapezoid.

10. A trapezoid has bases b_1 and b_2 and height h. What is its area?

11. In words, state the formula for the area of a trapezoid.

Applying the Mathematics

12. Find the area of the parallelogram drawn below.

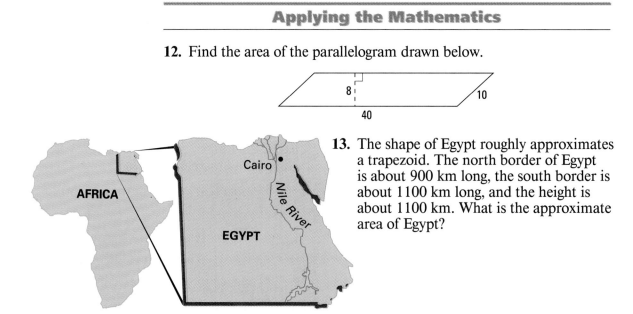

13. The shape of Egypt roughly approximates a trapezoid. The north border of Egypt is about 900 km long, the south border is about 1100 km long, and the height is about 1100 km. What is the approximate area of Egypt?

14. Draw a trapezoid with exactly two sides of equal length.

15. A trapezoid has an area of 60 square meters. Its height is 10 meters. One of the bases is 5 meters long. What is the length of the other base?

16. Order these seven terms from most general to most specific.
polygon   square   trapezoid   figure
rectangle   quadrilateral   parallelogram

17. a. Trace trapezoid *ABCD* and draw an altitude.
    b. Measure the bases and the altitude to the nearest millimeter. Use these measurements to estimate the area of *ABCD*.

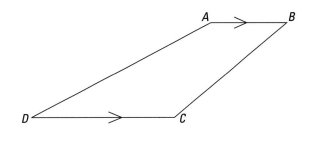

Review

18. Two-thirds of a piece of cloth was used. The starting amount was *A*. How much is left? *(Lesson 10-6)*

19. a. Solve $8x - 10 = 70$.
    b. Solve $-8y - 10 = 70$.
    c. Solve $-8(z + 3) - 10 = 70$.   *(Lessons 10-4, 10-5, 10-6)*

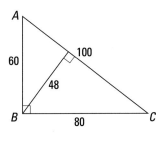

20. Use the figure shown at the left. Find the area of $\triangle ABC$ in two different ways. *(Lesson 9-1)*

21. Find the area of a right triangle with sides of lengths 12 ft, 16 ft, and 20 ft. *(Lesson 9-1)*

22. *True or false.* If the diagonal of a rectangle is drawn, the perimeter of each right triangle is one-half the perimeter of the rectangle. *(Lesson 5-10)*

23. To the nearest integer, what is the sum of $2\frac{1}{3}$, $3\frac{1}{4}$, and $4\frac{1}{5}$? *(Lesson 5-5)*

Exploration

24. Words often have many meanings. Sometimes those meanings do not agree with each other. One of the meanings of trapezoid is "trapezium."
    a. Look in a dictionary to find another meaning of "trapezium."
    b. Draw a trapezium.

A project presents an opportunity for you to extend your knowledge of a topic related to the material of this chapter. You should allow more time for a project than you do for typical homework questions.

1 Units for Land Area

In the metric system, land is often measured using a unit called the *hectare*. In the U.S. system, a commonly used unit of land area is the *acre*. Write an essay telling how these units are related to other units within their system, and determine how hectare and acre are related to each other.

2 Areas of Special Types of Trapezoids

Parallelograms, squares, rectangles, and rhombuses are all special types of trapezoids. Each of these types of figures has its own area formula. Find a formula for the area of each type. Then find the area of a specific example of each type, using the formula you have found and then using the formula for the area of a trapezoid given in Lesson 10-10.

3 How Often Is a Solution to an Equation an Integer?

Even if the numbers you deal with are small positive integers, solutions to many equations are not integers. To verify this, try this experiment. Find three six-sided dice and call them *a, b,* and *c.* Toss the dice and substitute the numbers into the equation $ax + b = c$. For instance, if the die *a* shows 2, *b* shows 5, and *c* shows 1, then the equation will be $2x + 5 = 1$. Write down 50 equations using this procedure. Determine the relative frequencies of the following events:

a. The solution is positive.

b. The solution is an integer.

c. The solution is a positive integer. Write down what you think are the probabilities of these events, and explain how you have come to your conclusions.

4 Surface Areas and Volumes of Boxes

Find at least six boxes of various sizes and calculate their surface areas and volumes. Explain why a box that has greater volume might not have greater surface area.

6 Different Methods of Multiplication

A variety of ways of multiplying two whole numbers have been developed over the centuries. Find out about one of the following methods of multiplication:

(a) using an abacus,
(b) Napier's bones (shown below),
(c) the Russian Peasant algorithm.

Show how to multiply 364×27 using the method you have chosen.

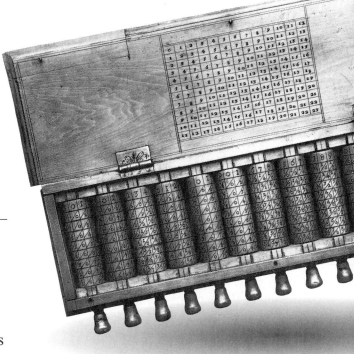

5 Multiplication of Numbers in Scientific Notation

Suppose two numbers are given in scientific notation: $a \cdot 10^m$ and $b \cdot 10^n$. What is the product of those numbers in scientific notation? By experimenting with your calculator or by hand, come up with rules for multiplying numbers in this form.

SUMMARY

This chapter has three major themes: solving equations, the Distributive Property, and areas of common figures.

The many uses of multiplication—areas, arrays, and volumes; size changes; and rate factors—lead to situations in which one factor and a product are known, but the other factor is not. These situations translate into equations of the form $ax = b$. If you cannot solve such an equation in your head, you can use the Multiplication Property of Equality:

If $a \neq 0$, the solution to $ax = b$ is $x = \frac{b}{a}$. Other situations combine multiplication and addition and lead to equations of the form $ax + b = c$. These equations can be solved by first adding $-b$ to both sides.

The Distributive Property is the basic property connecting multiplication to addition. Its many forms are all consequences of the basic form

$$ax + bx = (a + b)x$$

that arises from adding like terms. Using the Commutative Property of Multiplication, a second form arises:

$$xa + xb = x(a + b).$$

Two other forms involve subtraction.

$$ax - bx = (a - b)x$$
$$xa - xb = x(a - b)$$

The surface area of a rectangular solid can be found by adding the areas of its six faces. Because it is an area, surface area is measured in square units. (The Distributive Property can make it easier to calculate this area.) The area of any triangle is $\frac{1}{2} bh$, where b is the base and h the height to that base. Adding the areas of two triangles and using the Distributive Property leads to the formula $A = \frac{1}{2}h(b_1 + b_2)$ for the area of a trapezoid.

VOCABULARY

You should be able to give a general description and specific example for each of the following ideas.

Lesson 10-1
Repeated Addition Property of Multiplication
like terms
unlike terms

Lesson 10-2
Multiplication Property of Equality

Lesson 10-4
equation of the form $ax + b = c$

Lesson 10-6
Distributive Property of Multiplication over Addition
Distributive Property of Multiplication over Subtraction

Lesson 10-7
net
surface area

Lesson 10-8
numerically equal
acre

Lesson 10-9
height, altitude, base of a triangle
Area Formula for Any Triangle

Lesson 10-10
trapezoid
Area Formula for a Trapezoid
trapezoid—height, altitude, base

PROGRESS SELF-TEST

Take this test as you would take a test in class. Then check your work with the solutions in the Selected Answers section in the back of the book.

In 1–4, simplify.

1. $-y + y + y + 3 + y + y$

2. $m - 3m$

3. $7k - 2j - 4k + j$

4. $x + 2x + 3x + 4x$

5. Write the perimeter of the hexagon below in simplest form.

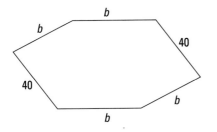

6. Explain how the Distributive Property can be used to calculate $49 \cdot 7$ mentally.

7. The Distributive Property is used here in going from what line to what line?
   Line 1 $5x + 2 - 2x - 2$
   Line 2 $= 5x + 2 + {-2x} + {-2}$
   Line 3 $= 5x + {-2x} + 2 + {-2}$
   Line 4 $= 5x + {-2x} + 0$
   Line 5 $= 5x + {-2x}$
   Line 6 $= (5 + {-2})x$
   Line 7 $= 3x$

8. Suppose the shoreline of a lake is measured in kilometers. In what unit would you expect to measure the amount of room for fishing?

9. *Multiple choice.* Which is *not* necessarily a trapezoid?
   (a) quadrilateral   (b) rectangle
   (c) square   (d) parallelogram

10. The area of a triangle is 400 square inches, and the height to one base is 25 inches. What must be that length of that base?

11. Explain how the Distributive Property can be used to find the total combined area of all these rectangles.

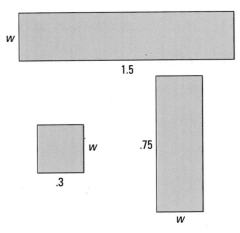

12. How much cardboard is needed to make the closed carton pictured below?

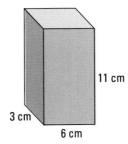

In 13–16, solve.

13. $35.1 = -9t$

14. $\frac{2}{5}m = -\frac{3}{4}$

15. $2 + 3a = 17$

16. $12 - 4h = 10$

17. A person made $2000 more this year than last. This is 8% of last year's income. What was last year's income?

PROGRESS SELF-TEST

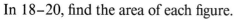

In 18–20, find the area of each figure.

18.

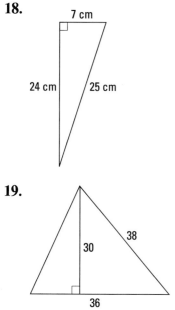

7 cm

24 cm   25 cm

19.

38

30

36

20.

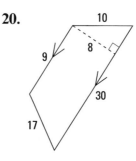

10

9   8

30

17

21. An ant is 4 meters up a tree and is crawling down at .02 meters per second. Assume this rate continues.

   a. How high will the ant be after 5 seconds?

   b. When will the ant be 2.5 meters high?

22. The balance scale diagram below represents what equation?

w   w   w   kg

kg kg kg kg kg
kg kg kg kg kg

CHAPTER REVIEW

Questions on SPUR Objectives

SPUR stands for **S**kills, **P**roperties, **U**ses, and **R**epresentations. The Chapter Review questions are grouped according to the SPUR Objectives for this chapter.

SKILLS DEAL WITH THE PROCEDURES USED TO GET ANSWERS.

Objective A: *Solve and check equations of the form $ax = b$.*   *(Lessons 10-2, 10-5)*

In 1–8, solve and check.

1. $40t = 3000$ **2.** $-22 = 4A$

3. $0.02v = 0.8$ **4.** $\frac{2}{3}x = 18$

5. $-49 = -7y$ **6.** $2.4 + 3.6 = (5 - 0.2)n$

7. $\frac{4}{5}n = 12$ **8.** $700m = 14$

Objective B: *Solve and check equations of the form $ax + b = c$.*   *(Lessons 10-4, 10-5)*

In 9–16, solve and check.

9. $8m + 2 = 18$ **10.** $-2.5 + .5y = 4.2$
11. $11 - 6u = -7$ **12.** $23 + 4x - 10 = -39$
13. $2x + 3x + 5 = 17$ **14.** $44 = 10 - 2z$
15. $1.3 = 0.8 + 2x$ **16.** $160 = 9a - a + 16$

Objective C: *Apply properties of multiplication to simplify expressions.*   *(Lessons 10-1, 10-6)*

In 17–22, simplify.

17. $x + x + x + y + y + z + z$ **18.** $2v + 8v$
19. $5x - x - 2x$ **20.** $13a + 4b + 7a$
21. $-9 + 5m + 2 - 8m + m$ **22.** $m(1 + n) - m$
23. Multiply $6(a - b + 2c)$.
24. Multiply $2(3x + 5 - 2y)$.

Objective D: *Find the area of a triangle.*
(Lesson 10-9)

25. Find the area of $\triangle CAT$.

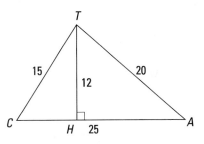

26. *SNOW* is a rectangle with dimensions in meters. Find the area of $\triangle SEW$.

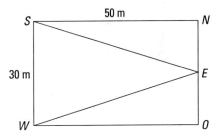

27. Use the triangles below.
   a. Find the area of $\triangle BUI$.

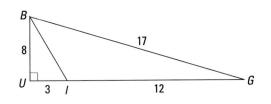

   b. Find the area of $\triangle BIG$.

28. Find the area of $\triangle ABC$.

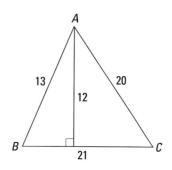

Objective E: *Find the area of a trapezoid.*
(Lesson 10-10)

In 29–32, *WEYZ* is a parallelogram.

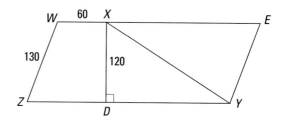

29. If $ZY = 350$, find the area of *WXYZ*.
30. If $ZY = 350$, find the area of *WEYZ*.
31. If $XE = 180$, find the area of *WEYZ*.
32. If $ZD = 110$, find the area of *WXDZ*.

PROPERTIES DEAL WITH THE PRINCIPLES BEHIND THE MATHEMATICS.

Objective F: *Recognize and use the Distributive Property, the Repeated Addition Property of Multiplication, and the Multiplication Property of Equality.*
(Lessons 10-1, 10-2, 10-6)

33. What property of multiplication can be used to solve $\frac{11}{3}y = \frac{2}{9}$?

34. The Repeated Addition Property of Multiplication is a special case of what property of multiplication?

35. The Distributive Property is used here in going from what line to what other line?
Line 1 $3x - x$
Line 2 $= 3x - 1 \cdot x$
Line 3 $= (3 - 1)x$
Line 4 $= 2x$

36. Explain how the Distributive Property can be applied to calculate $\$19.95 \cdot 4$ in your head.

USES DEAL WITH APPLICATIONS OF MATHEMATICS IN REAL SITUATIONS.

Objective G: *Find unknowns in real situations involving multiplication.*   *(Lessons 10-3, 10-4, 10-5)*

37. The floor of a store is in the shape of a rectangle. Find the length of the floor if its area is 3500 sq ft and the width is 40 ft.

38. A movie theater has 500 seats and 20 rows. There are the same number of seats in each row. How many seats are there per row?

39. Find the height of this box if its volume is 2400 cubic centimeters.

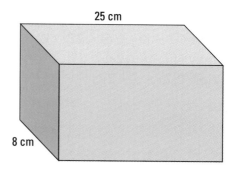

40. Elvis is currently 63 inches tall and is growing at the rate of $\frac{1}{4}$ inch every month.

  a. How tall will Elvis be in 6 months?

  b. At this rate of growth, how long will it take Elvis to reach a height of 74 inches?

41. Melissa's car can hold 45 liters of gas. When she is driving on the highway, she uses 125 mL per minute.

  a. How much gas will Melissa have left after 1 hour?

  b. How long can she drive before she has less than 10 liters left?

Objective H: *Find the surface area of a rectangular solid in real contexts.* *(Lesson 10-7)*

42. A stick of margarine is approximately 11.5 cm long, 3 cm wide, and 3 cm high. To wrap the margarine in aluminum foil, what is the least area of foil needed?

43. The mattress of a water bed is 7' long, 6' wide, and 3/4' high. Explain why a quilt with an area of 48 square feet cannot cover the top and all sides of the mattress.

44. How much gift wrap is needed to cover a box that is 12" by 8" by 3"?

Objective I: *Pick appropriate units in measurement situations.* *(Lesson 10-8)*

45. The perimeter of a vegetable garden is measured in feet. In what unit would you expect to measure the amount of space you have for planting?

46. If the dimensions of a rectangular solid are measured in inches, in what unit would the surface area most probably be measured?

47. Name two units of volume.

48. *Multiple choice.* The acre is a unit of:
  (a) length.   (b) land area.
  (c) volume.   (d) weight.

REPRESENTATIONS DEAL WITH PICTURES, GRAPHS, OR OBJECTS THAT ILLUSTRATE CONCEPTS.

Objective J: *Represent equations of the form $ax = b$ and $ax + b = c$ with a balance scale diagram.* *(Lessons 10-2, 10-4)*

49. a. Draw a balance scale diagram representing the equation $3w = 9$.

  b. Solve the equation.

50. a. The balance scale diagram below represents what equation?

  b. Solve the equation.

Objective K: *Represent the Distributive Property by areas of rectangles.* *(Lesson 10-6)*

51. What instance of the Distributive Property is pictured here?

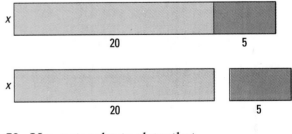

52. Use rectangles to show that $8.2 \cdot 13.6 + 9 \cdot 13.6 = (8.2 + 9)13.6$.

CHAPTER

11

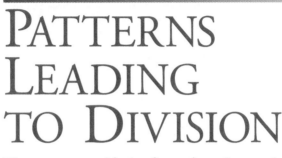

PATTERNS LEADING TO DIVISION

There are many kinds of questions that can be answered by division. Here are seven questions that can be answered by dividing 20 by 7.

A. In Brazil, land is currently being offered to people for farming. Suppose a subdivision of 20 lots is split equally among 7 people. What will be the area of each person's share?

B. The distance from El Paso, Texas, to Boston is about 2000 miles, and from El Paso to Los Angeles is about 700 miles. How many times as far is it from El Paso to Boston as it is from El Paso to Los Angeles?

C. You expect 20 people at dinner. Seven people can be seated around each table you have. How many tables are needed?

D. Mr. and Mrs. Torrence have 3 daughters, 4 sons, and 20 grandchildren. On the average, how many children does each of their children have?

E. If $WX = 7$ cm, how long is $\overline{WZ}$?

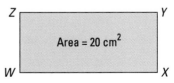

F. If you run at a constant rate of 7 mi/hr, how long will it take you to run 20 miles?

G. If you wished to expand the smaller figure below into the larger figure, what is the magnitude of the size change you would need?

This chapter is concerned with these and other situations that lead to division.

Division in a theme park. *Integer division can be used to determine the number of roller coaster cars needed to seat all members of a certain party. Shown here is the Scream Machine from Great America theme park in New Jersey.*

What Is Real-Number Division?

To answer the questions on page 591, you need to divide 20 by 7. But what are the answers? So far in this book, we have emphasized that

20 divided by 7 equals the quotient $\frac{20}{7}$, which equals $2\frac{6}{7}$, or 2.857142. . . .

This is the answer to questions **A, B, D, E, F,** and **G.** We call this *real-number division.* In **real-number division,** the result of dividing a by b is the single number $\frac{a}{b}$. The variable a can be any number; b can be any number except zero.

What Is Integer Division?

For some situations, however, a correct answer to 20 divided by 7 is

a quotient of 2 with remainder 6.

We call this *integer division.* Notice that the integer part of the quotient is the same regardless of the type of division. This is an appropriate division to use for situation **C** on page 591. If 20 people are to be seated at tables that seat 7 each, then 3 tables are needed. Of these, 2 tables will be filled, and there will be 6 people left over.

In **integer division** of a by b, the number a must be an integer and b must be a whole number. The result of dividing a by b is given by two numbers: an integer quotient and an integer remainder that is less than b. Although integer division may seem more complicated than real-number division, it is probably how you first learned to divide.

Here are other examples of the two types of division.

Division problem	Real-number division answer	Integer division answer
184 divided by 5	quotient 36.8	quotient 36, remainder 4
12 divided by 4	quotient 3	quotient 3, remainder 0
7 divided by 12	quotient $\frac{7}{12}$ = 0.58$\overline{3}$	quotient 0, remainder 7

Finding Answers in Integer Division

Most calculators give answers to real-number divisions, but not to integer divisions. Because some situations call for integer division, it is useful to be able to convert real-number division answers to integer division answers (sometimes called **quotient-remainder form**). The first example shows how to do this using arithmetic.

Example 1

Some school buses seat 44 people. If 600 people ride the buses to a game, how many buses could be filled? How many people would then be in an unfilled bus?

Solution

Divide 600 by 44. Our calculator shows $\boxed{13.636364}$. This indicates that 13 buses could be filled. Multiply 44 by 13 to determine how many students could be in those buses. $44 \cdot 13 = 572$. Now subtract 572 from 600 to indicate how many students will be left over. $600 - 572 = 28$. Thus, the answer to the integer division is: quotient 13, remainder 28. 13 buses could be filled, and 28 people would then be in the 14th bus.

The Quotient-Remainder Formula

Some people prefer using formulas. Notice in Example 1 how the divisor, dividend, quotient, and remainder are related.

$$d = \text{divisor} = 44$$
$$n = \text{dividend} = 600$$
$$q = \text{integer quotient} = 13$$
$$r = \text{remainder} = 28$$

In the solution, 600 was the sum of 572, the number of students on the full buses, and the remainder 28. Since $572 = 44 \cdot 13$, this tells us that

$$600 = 44 \cdot 13 + 28.$$

The more general formula is known as the **Quotient-Remainder Formula**. It relates the dividend n, divisor d, integer quotient q, and remainder r:

$$n = d \cdot q + r.$$

In the integer division problem 184 divided by 5, identify *n, d, q,* and *r.*
Show that these four numbers satisfy the Quotient-Remainder Formula.

The power of the Quotient-Remainder Formula is that given any three of
the divisor, dividend, quotient, and remainder, the fourth can be found
by solving an equation.

Example 2

The deepest spot in the Atlantic Ocean is the Puerto Rico Trench, 30,246
feet below sea level. How many miles and feet is this?

Solution

We need to find the integer quotient and remainder when
30,246 is divided by 5,280, the number of feet in a mile.
Here n = 30,246 and d = 5,280.

A calculator displays the real-number quotient as 5.7284091. This means
that The integer quotient q = 5. That is, the trench is between 5 and 6
miles below sea level. (You could have done this part in your head.)

Now substitute into the Quotient-Remainder Formula.

$$n = d \cdot q + r$$
$$30{,}246 = 5{,}280 \cdot 5 + r$$

Solve the equation.

$$30{,}246 = 26{,}400 + r$$
$$3{,}846 = r$$

The Puerto Rico Trench is 5 miles, 3,846 feet below sea level.

*An oceanic trench, such as
the Puerto Rico Trench, is
a long, narrow, steep-sided
depression in the ocean
bottom. The Puerto Rico
Trench lies about 75 mi
north of Puerto Rico.
Shown here is San Juan
Bay on the northern
coastline of Puerto Rico.*

QUESTIONS

Covering the Reading

1. Give the answer to 45 divided by 8:
   a. as a real-number division.
   b. as integer division.

In 2–4, *multiple choice.*   Use these choices.
(a) integer division only
(b) real-number division only
(c) both integer and real-number division
(d) neither integer nor real-number division

2. has a quotient and a remainder

3. has an answer that is a single number

4. can be used with fractions

5. A class of 172 students is having a picnic. The school buses seat 44 people, including the 2 teachers who must ride on each bus.
   a. How many buses could be filled?
   b. How many people would then be on an unfilled bus?

6. Write your answer for the Activity in this Lesson.

7. A number w is divided by x, giving an integer quotient of y and a remainder of z. From the Quotient-Remainder Formula, how are these four numbers related?

8. The deepest spot in the Indian Ocean is the Java Trench, 23,376 feet below sea level. How many miles and feet is this?

Applying the Mathematics

9. When 365 is divided by 7, the quotient is 52 and the remainder is 1.
   a. Show how these numbers are related using the Quotient-Remainder Formula.
   b. What is the everyday significance of these numbers?

10. **a.** How many years and days old is a person who is 10,000 days old?
    b. Is it possible for a person to live to be 50,000 days old? Why or why not?

11. A teacher has 1000 sheets of paper and decides to distribute them equally to the 23 students in the class. How many sheets will each student get? How many sheets will be left over?

12. What is the quotient and remainder when 31 is divided by 67?

13. 450 is divided by an unknown number x, leaving a quotient of 32 and a remainder of 2. Is this possible? If so, what is x? If not, why not?

14. **a.** Complete the following table.

dividend	divisor	real-number quotient	integer quotient	integer remainder
40	9			
41	9			
42	9			
43	9			
44	9			
45	9			
46	9			
47	9			
48	9			
49	9			
50	9			

   b. Describe patterns you see in the table.

15. Perform a size change of magnitude $\frac{3}{2}$ on *ABCD*. *(Lesson 9-7)*

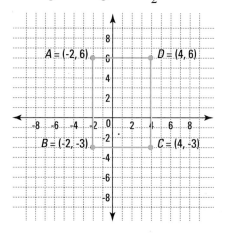

$A = (-2, 6)$
$D = (4, 6)$
$B = (-2, -3)$
$C = (4, -3)$

16. Suppose you have applied a size change of magnitude $\frac{3}{2}$ to a rectangle. *(Lessons 9-1, 9-7)*
  a. How do the perimeter of the rectangle and its image compare?
  b. How do the area of the rectangle and its image compare?

17. In the United States in 1991, about one in 4.5 births was by cesarean section. About one in 87 births was of twins. *(Lessons 4-8, 9-4)*
  a. Which of these, a cesarean section or twins, was more likely?
  b. If these events are independent, about one in how many births would be a cesarean section birth of twins?

In 18 and 19, do the divisions in parts **a–d.**

18. Perform the divisions without a calculator if you can. *(Lesson 2-4)*
  a. $\frac{20}{5}$ **b.** $\frac{20}{0.5}$ **c.** $\frac{20}{0.05}$ **d.** $\frac{20}{0.00005}$

19. Use a calculator only on part **a,** if needed. *(Lesson 2-1)*
  a. $\frac{3}{16}$ **b.** $\frac{0.3}{16}$ **c.** $\frac{0.03}{16}$ **d.** $\frac{0.003}{16}$

20. To divide a decimal D by 10^x, you should move the decimal point in D __?__ spaces to the __?__. *(Lessons 6-5, 6-7)*

21. Find the smallest positive integer n with *all* of the following properties.

The remainder is 1 when n is divided by 2.
The remainder is 2 when n is divided by 3.
The remainder is 3 when n is divided by 4.
The remainder is 4 when n is divided by 5.
The remainder is 5 when n is divided by 6.
The remainder is 6 when n is divided by 7.

11-2

The Rate Model for Division

Lower rates. *In 1960, the average family size was 3.67 people. Due to lower birth rates, the average family size had dropped to 3.17 by 1990.*

Here are some examples of **rates.**

55 miles per hour   1.9 children per family   \$1.95 for each 2-liter bottle

The **rate units** can be written as fractions.

$$\frac{miles}{hour} \qquad \frac{children}{family} \qquad \frac{dollars}{liter}$$

Example 1

A car is driven 250 miles in 5 hours. What is the average rate?

Solution

Since the unit of the answer will be $\frac{miles}{hour}$, divide miles by hours.

$$\frac{250 \text{ miles}}{5 \text{ hours}} = 50 \frac{miles}{hour}$$

People do Example 1 so easily that sometimes they do not realize they divided to get the answer. Example 2 is of the same type. But the numbers are not so simple.

Example 2

A car goes 283.4 miles on 15.2 gallons of gas. How many miles per gallon is the car getting?

Solution

The key to the division is "miles per gallon." This indicates to divide miles by gallons.

$$\frac{283.4 \text{ miles}}{15.2 \text{ gallons}} \approx 18.6 \frac{miles}{gallon}$$

The car is getting about 18.6 miles per gallon.

Do not let the numbers in rate problems scare you. If you get confused, try simpler numbers and examine how you got the answer.

Examples 1 and 2 are instances of the Rate Model for Division.

> **Rate Model for Division**
> If *a* and *b* are quantities with different units, then $\frac{a}{b}$ is the amount of quantity *a* per quantity *b*.

One of the most common examples of rate is *unit cost*.

Example 3

A 6-oz can of peaches sells for 89¢. An 8-oz can sells for $1.17. Which is the better buy?

Solution

Calculate the cost per ounce. That is, divide the total cost by the number of ounces.

Cost per ounce for 6-oz can: $\frac{89¢}{6 \text{ oz}}$ = 14.83 . . . cents per ounce.

Cost per ounce for 8-oz can: $\frac{117¢}{8 \text{ oz}}$ = 14.625 cents per ounce.

The cost per ounce is called the **unit cost**. The 8-oz can has a slightly lower unit cost. So the 8-oz can is the better buy.

Another common rate is items per person.

Example 4

Eleven people in a scout troop must deliver flyers to 325 households. If the job is split equally, how many flyers will each scout deliver? How many will be left over?

Solution

Think: flyers per scout. This means to divide the number of flyers by the number of scouts.

$$\frac{325 \text{ flyers}}{11 \text{ scouts}} = 29.54 \ldots \text{ flyers/scout}$$

Each scout will have to deliver 29 flyers. To find the remainder, use integer division.

$$n = dq + r$$

Here n = 325, d = 11, and q = 29.
$$325 = 11 \cdot 29 + r$$
$$325 = 319 + r$$

So r = 6, meaning that 6 flyers would be left over. The job could be finished if six of the scouts delivered 30 flyers.

Notice that the unit in Example 4 is *flyers per scout*. The number of flyers and the number of scouts are whole numbers. But the number of flyers per scout is a rate. Rates do not have to be whole numbers, they may also involve fractions or negative numbers.

Good citizens. *Juliette Gordon Low formed the first girl scout troop in the U.S. in 1912. Then, as now, girls promised to follow a certain code of behavior, to participate in community service projects, and to develop skills by earning proficiency badges.*

Of course, rates can involve variables.

Example 5

Hal typed *W* words in *M* minutes. What is his typing speed?

Solution

The usual unit of typing speed is words per minute. So divide the total number of words by the total number of minutes.

His typing speed is $\frac{W}{M}$ words per minute.

QUESTIONS

Covering the Reading

In 1–7, calculate a rate suggested by each situation.

1. A family drove 400 miles in 8 hours.

2. A family drove 400 miles in 8.5 hours.

3. A family drove 600 kilometers in 9 hours.

4. An animal traveled *d* meters in *m* minutes.

5. There were 28 boys and 14 girls at the party.

6. 150 students signed up for 7 geometry classes.

7. Six people live on 120 acres.

8. State the Rate Model for Division.

9. The Smith family went to the grocery store. An 18-oz box of corn flakes costs $2.89, and a 24-oz box costs $3.99.
   a. Give the unit cost of the 18-oz box to the nearest tenth of a cent.
   b. Give the unit cost of the 24-oz box to the nearest tenth of a cent.
   c. Based on unit cost, which box is the better buy?

10. a. Answer the questions of Example 4 if one of the troop members is sick and unable to deliver any flyers.
    b. How would you allocate the remaining flyers?

11. If in *h* hours you travel *m* miles, what is your rate in miles per hour?

Applying the Mathematics

12. Nine nannies need to nail nine hundred nineteen nails into a nook.
    a. If the job is split evenly, how many nails will each nanny nail?
    b. How many nails will be left over?

©Walt Disney Studios

Supercalifragilistic-expialidocious. *Mary Poppins is probably the most famous fictional nanny of all time. She is shown here with her friend Bert and her two charges in a scene from the movie* Mary Poppins.

13. A person earns $222 for working 34 hours. How much is the person earning per hour? (Answer to the nearest penny.)

14. In Bangladesh, a laborer may earn $20 for working 80 hours. How much is this person earning per hour?

15. According to the *Statistical Abstract of the United States: 1992,* Hawaii had 2,491 active non-federal physicians and a population of 1,108,000 in 1990. California had 70,062 doctors for a population of 29,760,000. Which state had more doctors *per capita?* (The phrase per capita means "per person." It comes from the Latin words meaning "per head.")

16. The school nurse used 140 bandages in 3 weeks. On the average, how many bandages were used per school day?

17. Susannah calculated that she used an average of $8\frac{2}{5}$ sheets of paper per day. How many sheets of paper could she have used in how many days to get this rate?

Jute capital.
Bangladesh is a country about the size of Wisconsin, situated between Burma and India. It is the world's largest exporter of jute, a plant whose fiber is used to make carpets.

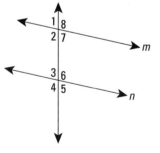

Review
---

18. Consider the division of 100 by 15.   *(Lesson 11-1)*
   a. Give the real-number division answer.
   b. Give the integer division answer.

19. a. How many feet are in a yard?
   b. How many square feet are in a square yard?
   c. How many full square yards are in 250 square feet? How many square feet are left over?   *(Lessons 3-2, 11-1)*

20. Solve $-5 - 5m = 5$.   *(Lesson 10-5)*

In 21–24, lines *m* and *n* are parallel.   *(Lesson 7-7)*

21. Name all pairs of alternate interior angles.

22. Name all pairs of corresponding angles.

23. If m∠5 = 75°, what is m∠1?   **24.** If m∠4 = *x*, what is m∠7?

25. If $a = -2$, $b = 5$, and $c = 1$, what is $|a - b + c| - |a|$?   *(Lessons 5-3, 7-2)*

26. An octagon has __?__ sides, __?__ angles, and __?__ diagonals.
(Lessons 5-9, 6-3)

Exploration
---

27. Look in an almanac or other book to find at least one of these rates.
   a. the minimum number of grams of protein per day recommended for someone your age
   b. the number of people per square mile in the United States
   c. the number of miles per gallon you should expect to get from a certain car (you pick the car)

Dividing the tab. *A knowledge of fractions is useful when dividing a restaurant bill among the people who ate.*

The Algebraic Definition of Division

May went with her friends June, Julius, and Augie to a restaurant. The bill totalled $27.30. Each person owed $\frac{1}{4}$ of the bill. Everyone knew that to find $\frac{1}{4}$ of $27.30 they could divide $27.30 by 4.

That is, $\frac{\$27.30}{4} = 27.30 \cdot \frac{1}{4}$.

In general, instead of dividing by b you can multiply by $\frac{1}{b}$, the reciprocal of b. We call this the *Algebraic Definition of Division.*

> **Algebraic Definition of Division**
> For any numbers a and b, $b \neq 0$,
> a divided by $b = a$ times the reciprocal of b, or
> $$\frac{a}{b} = a \cdot \frac{1}{b}.$$

Why is this true? Just think of multiplying fractions. For any fraction $\frac{a}{b}$,

$$\frac{a}{b} = \frac{a}{1} \cdot \frac{1}{b}$$
$$= a \cdot \frac{1}{b}.$$

Using the Algebraic Definition of Division to Divide Fractions

Since the property holds for *any* numbers, it holds when a and b are themselves fractions. Then, it is critical to remember that the reciprocal of the fraction $\frac{x}{y}$ is the fraction $\frac{y}{x}$.

Example 1

What is $\frac{6}{5}$ divided by $\frac{3}{4}$?

Solution

By the Algebraic Definition of Division,
$\frac{6}{5}$ divided by $\frac{3}{4}$ = $\frac{6}{5}$ times the reciprocal of $\frac{3}{4}$

You can think of the division in either of two ways.

$$\frac{\frac{6}{5}}{\frac{3}{4}} = \frac{6}{5} \cdot \frac{4}{3} \qquad\qquad \frac{6}{5} \div \frac{3}{4} = \frac{6}{5} \cdot \frac{4}{3}$$

$$= \frac{24}{15} \qquad\qquad\qquad\qquad = \frac{24}{15}$$

$$= \frac{8}{5} \qquad\qquad\qquad\qquad\quad = \frac{8}{5}$$

Check

To check that $\frac{8}{5}$ is the answer, you can change the original fractions to decimals. $\frac{6}{5} = 1.2$ and $\frac{3}{4} = 0.75$, so

$$\frac{\frac{6}{5}}{\frac{3}{4}} = \frac{1.2}{0.75} = 1.6.$$

Since $\frac{8}{5} = 1.6$, the answer checks.

Example 2

Simplify $\frac{\frac{2}{3}}{15}$.

Solution

Think: Instead of dividing by 15, multiply by $\frac{1}{15}$.

$$\frac{\frac{2}{3}}{15} = \frac{2}{3} \div 15 = \frac{2}{3} \cdot \frac{1}{15} = \frac{2}{45}$$

Rate situations can lead to division of fractions.

Example 3

Suppose you make $8 for babysitting 1 hour and 40 minutes. How much are you making per hour?

Solution

Earnings per hour is a rate, so divide the total earned by the number of hours. First change the hours and minutes to hours.

40 minutes $= \frac{40}{60}$ hour $= \frac{2}{3}$ hour, so 1 hour 40 minutes $= 1\frac{2}{3}$ hours $= \frac{5}{3}$ hours.

$$\text{Salary per hour} = \frac{8 \text{ dollars}}{1 \text{ hour 40 min}} = \frac{8 \text{ dollars}}{\frac{5}{3} \text{ hours}}$$

To divide 8 by $\frac{5}{3}$, use the Algebraic Definition of Division.

$$\frac{8 \text{ dollars}}{\frac{5}{3} \text{ hours}} = 8 \cdot \frac{3}{5} \frac{\text{dollars}}{\text{hour}}$$

$$= \frac{24}{5} \frac{\text{dollars}}{\text{hour}}$$

$$= 4.8 \frac{\text{dollars}}{\text{hour}}$$

You earn $4.80 per hour.

The next example shows how powerful division of fractions can be.

Example 4

Two-thirds of the way through the 1993 baseball season, the Atlanta Braves had won 64 games. At this rate, how many games would they win in the entire season?

Solution

Games won per entire season $= \dfrac{64 \text{ games}}{\frac{2}{3} \text{ season}}$

$$= 64 \cdot \frac{3}{2} \frac{\text{games}}{\text{season}}$$

$$= \frac{192}{2} \frac{\text{games}}{\text{season}}$$

$$= 96 \frac{\text{games}}{\text{season}}$$

At this rate, the Braves would win 96 games in the entire season.

Even better. *Actually, the Atlanta Braves won 104 games during the 1993 regular season and were the National League Western Division Champions for the third year in a row.*

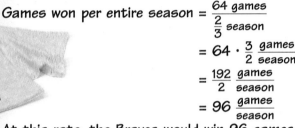

QUESTIONS

Covering the Reading

1. Instead of multiplying a number by $\frac{1}{4}$, you can divide the number by __?__.

2. State the Algebraic Definition of Division.

3. Suppose a person earns $17 for working two and a half hours. How much did the person earn per hour?

In 4–7, simplify.

4. $\dfrac{\frac{8}{9}}{\frac{4}{3}}$ 5. $\dfrac{\frac{2}{5}}{7}$ 6. $\dfrac{\frac{17}{6}}{\frac{2}{3}}$ 7. $\dfrac{\frac{10}{3}}{\frac{6}{5}}$

8. About three-fourths of the way into the basketball season, the Hornets have won 44 games. At this rate how many games will they win in the entire season?

9. With one-fifth of the hockey season gone, the Red Wings have lost 7 games. At this rate, how many games will they lose in the season?

10. a. Divide $4\frac{1}{4}$ by $3\frac{2}{5}$.

   b. The answer you get to part **a** should be bigger than 1. How could you tell this before you did any division?

   c. Check your answer to part **a** by changing each fraction to a decimal.

11. a. Divide $1\frac{2}{3}$ by $5\frac{4}{7}$.

   b. The answer you get to part **a** should be less than 1. How could you tell this before you did any division?

   c. Check your answer to part **a** by changing each fraction to a decimal.

   d. In this question, what advantage do the fractions have over the decimals?

12. Jay walked $\frac{1}{3}$ of a mile in $\frac{1}{6}$ of an hour.

   a. At this rate, how far would Jay walk in an hour?

   b. Explain how you obtained your answer to part **a**.

In 13 and 14, simplify.

13. $\dfrac{\frac{2}{x}}{\frac{1}{y}}$

14. $\dfrac{\frac{a}{b}}{\frac{c}{d}}$

15. Seven and one-half times a number is 375. What is the number?

Review

---

16. What is your hourly wage if you earn $21.12 in 3.3 hours? *(Lesson 11-2)*

In 17–18, use the cereal box pictured below.

30.8 cm

8.4 cm

20.4 cm

17. Find its total surface area. *(Lesson 10-7)*

18. a. Find its volume. *(Lesson 9-2)*

   b. Find its volume after rounding each dimension to the nearest cm.

In 19–21, solve. *(Lessons 7-4, 7-5, 10-2)*

19. $-4G = -3$

20. $-4 - G = -3$

21. $-4 + G = -3$

22. The probability that an amateur golfer will make a hole-in-one on a par 3 golf hole is estimated to be 1 in 12,600. What then is the probability of making two holes-in-one in a row? *(Lesson 9-4)*

23. Nyasha planted 20 tomato plant seeds in her garden. Of these, 17 developed into tomato plants. Write the relative frequency that a seed developed into a plant:
a. as a fraction.
b. as a decimal.
c. as a percent. *(Lessons 1-6, 2-6, 4-8)*

In 24–26, let $x = \frac{5}{4}$ and $y = \frac{1}{2}$. Write as a simple fraction.
(Lessons 5-5, 7-2, 9-3)

24. $8xy$ **25.** $x - y$ **26.** $x + y$

Exploration

27. Here is a famous puzzle. If a hen and a half can lay an egg and a half in a day and a half, at this rate how many eggs can be laid by two dozen hens in two dozen days?

Question 28? *Which came first—the chicken or the egg?*

11-4

Division with Negative Numbers

A negative cash flow. *Although a person may spend only 50 cents to play a video game, Americans spend $5 billion a year playing arcade video games.*

Dividing a Negative Number by a Positive Number

A person spends 10 dollars in a video arcade in 2 hours. What is the rate? The answer is given by division.

$$\frac{\text{spend 10 dollars}}{2 \text{ hours}} = \text{spend 5 dollars per hour}$$

You can translate the dollars spent into a negative number.

$$\frac{-10 \text{ dollars}}{2 \text{ hours}} = -5 \frac{\text{dollars}}{\text{hour}}$$

This situation is an instance of the division $\frac{-10}{2} = -5$. Another way to do the division is to think as follows: Dividing by 2 is the same as multiplying by its reciprocal, $\frac{1}{2}$.

$$\frac{-10}{2} = -10 \cdot \frac{1}{2} = -5$$

In general, if a negative number is divided by a positive number, the quotient is negative.

Example 1

On five consecutive days, the low temperatures in a city were 3°C, -4°C, -6°C, -2°C, and 0°C. What was the mean low temperature for the five days?

Solution

Recall that the mean (or average) temperature is found by adding up the numbers and dividing by 5.

$$\frac{3 + -4 + -6 + -2 + 0}{5} = \frac{-9}{5} = -1.8$$

The mean low temperature was -1.8°C, or about -2°C. This temperature is a little below freezing.

Dividing a Positive Number by a Negative Number

In Example 2, a positive number is divided by a negative number. The quotient is again negative.

Example 2

What is 10 divided by -2?

Solution

Dividing by -2 is the same as multiplying by $-\frac{1}{2}$, the reciprocal of -2.

$$\frac{10}{-2} = 10 \cdot -\frac{1}{2} = -5$$

Check

Think of the video-arcade player at the beginning of this lesson, but go back in time for the 2 hours. The spender had $10 more 2 hours ago. What is the loss rate?

$$\frac{10 \text{ dollars more}}{2 \text{ hours ago}} = \frac{10 \text{ dollars}}{-2 \text{ hours}} = -5 \frac{\text{dollars}}{\text{hour}}$$

Heavy weights. *The typical weight of full grown African elephants is 3600 kg (8000 lb) for cows and 5400 kg (12,000 lb) for bulls. A baby elephant usually weights about 90 kg (200 lb).*

Dividing a Negative Number by a Negative Number

Here is a division question in which both numbers are negative. Is it obvious whether the answer will be positive or negative?

Example 3

What is -150 divided by -7?

Solution

$$\frac{-150}{-7} = -150 \cdot -\frac{1}{7} \text{ (Now we know the answer will be positive.)}$$
$$= \frac{150}{7}$$
$$\approx 21.43$$

Division with two negative numbers can also be thought of using the rate model. Suppose an elephant gains 14 kg in weight in 3.5 months. It has gained weight at the rate of 4 kg per month. These are all positive quantities.

$$\frac{14 \text{ kg more}}{3.5 \text{ months later}} = \frac{4 \text{ kg}}{\text{month}} \text{ gain}$$

Another way of looking at the situation is that 3.5 months ago the elephant weighed 14 kg less. These are negative quantities.

$$\frac{14 \text{ kg less}}{3.5 \text{ months ago}} = \frac{4 \text{ kg}}{\text{month}} \text{ gain}$$

Notice the equal rates. $\frac{14 \text{ kg more}}{3.5 \text{ months later}} = \frac{14 \text{ kg less}}{3.5 \text{ months ago}}$

Ignoring the units but using negative numbers when appropriate:

$$\frac{14}{3.5} = \frac{-14}{-3.5}.$$

That is, $\dfrac{-14}{-3.5} = 4$.

Because $\dfrac{-1}{-1} = 1$, we can multiply a fraction by $\dfrac{-1}{-1}$ and the value of the fraction is not changed. For instance, begin with $\dfrac{-10}{2}$. Multiplying both numerator and denominator by -1 yields the equal fraction $\dfrac{10}{-2}$.

Altogether, the rules for dividing with negative numbers are just like those for multiplying. If both divisor and dividend are negative, the quotient is positive. If one is positive and the other is negative, the quotient is negative. These properties can be stated with variables.

For all numbers a and b, and $b \neq 0$,

$$\frac{a}{b} = \frac{-a}{-b}, \text{ and } \frac{-a}{b} = \frac{a}{-b} = -\frac{a}{b}.$$

If you forget how to do operations with negative numbers, there are two things you can do. (1) Change subtractions to additions; change divisions to multiplications. (2) Think of a real situation using the negative numbers. Use the situation to help you find the answer.

QUESTIONS

Covering the Reading

In 1–6, simplify.

1. $\dfrac{-14}{7}$

2. $\dfrac{-100}{300}$

3. $\dfrac{-56}{-8}$

4. $\dfrac{60}{-2}$

5. $-\dfrac{144}{12}$

6. $\dfrac{-80}{-100}$

In 7–9, find the mean of each set of numbers.

7. 40, 60, 80, 100

8. -40, -60, -80, -100

9. -11, 14, -17, -20, 6, -30

In 10–12, tell whether the number is positive or negative.

10. $\dfrac{-2.5}{6}$

11. $-\dfrac{-100}{300}$

12. $-\dfrac{-54}{-81}$

13. $\dfrac{53}{-2} \cdot \dfrac{55}{-2}$

14. Separate the numbers below into two collections of equal numbers.

$$-\frac{1}{2} \qquad \frac{-1}{2} \qquad \frac{-1}{-2} \qquad \frac{1}{-2} \qquad -\frac{-1}{-2} \qquad \frac{1}{2} \qquad -\frac{-1}{2}$$

15. Four inches of snow on the mountain melted after two sunny days.
    a. Calculate a rate from this information.
    b. What division problem with negative numbers gives this rate?
    c. Copy and complete: __?__ days ago, the mountain had __?__ inches of snow __?__ than it does now.
    d. What division problem is suggested by part **c?**

16. What two things can you do if you forget how to calculate with negative numbers?

Applying the Mathematics

17. In the twenty years from 1970 to 1990, Arizona's population rose by about 1,900,000.
    a. Calculate a rate from this information.
    b. Copy and complete: __?__ years before 1990, Arizona's population was 1.9 million __?__ than it was in 1990.
    c. What division problem is suggested by part **b?** What is the quotient?

In 18–21, calculate $x + y$, $x - y$, xy, and $\frac{x}{y}$ for the given values.

18. $x = -6$ and $y = -9$

19. $x = -12$ and $y = -12$

20. $x = -\frac{15}{8}$ and $y = -\frac{1}{8}$

21. $x = -\frac{1}{3}$ and $y = \frac{1}{2}$

22. Use $C = 5(F - 32)/9$ to convert $-40°$ Fahrenheit to Celsius.

23. Round $\frac{350}{-6}$ to the nearest integer.

24. The *center of gravity* of a set of given points is the point whose first coordinate is the mean of the first coordinates of the given points, and whose second coordinate is the mean of the second coordinates of the given points.
    a. Find the coordinates of the center of gravity of the four points graphed below.
    b. Copy the drawing and plot the center of gravity on your copy.

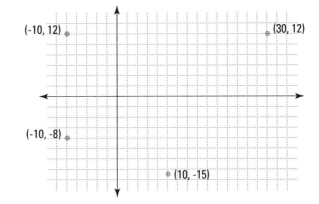

25. In $3\frac{1}{2}$ days, a boat sailed the 500 miles from Kingston, Jamaica, to Santo Domingo in the Dominican Republic. How many miles did the boat average per day? *(Lessons 11-2, 11-3)*

26. For the first 70 days of the year, the high temperature was above normal on 21 days. If this rate continues, for how many days will the temperature be above normal during the entire year? *(Lesson 11-2)*

27. a. 40% of a number is 500. What is the number? *(Lessons 2-6, 10-3)*
b. What is 40% of 500? *(Lesson 2-6)*

28. Begin at the point (14, 6). Move 20 units to the left and 3 units down. What are the coordinates of the image point? *(Lesson 8-5)*

29. Draw a hexagon *ABCDEF* in which $\overline{AB} \perp \overline{BC}$ and $\overline{AB} \parallel \overline{CD}$.
(Lessons 5-9, 7-7)

30. a. Write 44% as a decimal.
b. Write 44% as a simple fraction in lowest terms. *(Lesson 2-4)*

These Jamaican entertainers are performing a traditional dance for tourists.

31. a. Press 5 [+/−] [÷] 0 on your calculator. What does the display show?
b. Press 5 [+/−] [÷] 0.001 on your calculator. What is displayed?
c. Press 5 [+/−] [÷] 0.001 [+/−] on your calculator. What is displayed?
d. Divide -5 by other numbers near zero. Record your results.
e. Explain what happens when the divisor gets nearer to zero and is positive.
f. Explain what happens when the divisor gets nearer to zero and is negative.

The Ratio Comparison Model for Division

Open the gates! *Located at the junction of the Missouri and Mississippi Rivers, the St. Louis area was the site of record flood levels in 1993. Shown here is the Gateway Arch on the historic riverfront of St. Louis.*

In St. Louis, Missouri, it rained 11 of the 30 days of June, 1993. This rain saturated the ground and contributed to floods in July. Dividing 11 by 30 gives the relative frequency of rain.

$$\frac{11}{30} \approx .37$$

Relative frequency is often expressed as a percent. It rained about 37% of the days in June. You have calculated relative frequencies like this before.

What Are Ratios?

Notice what happens when the units are put into the numerator and denominator.

$$\frac{11 \text{ days}}{30 \text{ days}}$$

The units are the same. They cancel in the division, so the answer has no unit. Therefore, this is not an example of a rate. The answer is an example of a **ratio.** Ratios have no units. Because the 11 days is being compared to the 30 days, this use of division is called **ratio comparison.**

> **Ratio Comparison Model for Division**
> If a and b are quantities with the same units, then $\frac{a}{b}$ compares a to b.

All percents can be considered as ratio comparisons.

Example 1

Suppose the tax is $0.42 on a $7.00 purchase. What is the tax rate? (It's called a tax rate even though it is technically a ratio.)

Solution

Divide $0.42 by $7.00 to compare them. The units are the same, so they cancel.

$$\frac{\$0.42}{\$7.00} = \frac{0.42}{7.00} = .06$$

.06 = 6%, so the tax rate is 6%.

Check

A tax of 6% means you pay 6¢ tax for every dollar spent. So for $7, the tax should be 7 · 6¢, or 42¢.

Some comparisons can be done in either order.

Example 2

Compare the 1990 estimated populations of the United States (250,000,000 people) and Canada (27,000,000 people).

Solution

The units (people) are the same. Divide one of the numbers by the other to compare them.

$\frac{250,000,000}{27,000,000} \approx 9.3$, so the population of the U.S. was about 9.3 times that of Canada.

Dividing in the other order gives the reciprocal.

$\frac{27,000,000}{250,000,000} \approx .11$, so Canada's population was about 11% the population of the U.S.

Either answer is correct.

Percents as Ratios

Example 3

5 is what percent of 40?

Solution

This problem asks you to compare 5 to 40. So divide, and then convert the answer to a percent.

$$\frac{5}{40} = 0.125 = 12.5\%$$

Therefore, 5 is 12.5% of 40.

Check

Calculate 12.5% of 40.
12.5% of 40 = 0.125 · 40 = 5.

No horsing around. *A mounted policeman is shown here on Parliament Hill in Ottawa, the capital of Canada.*

QUESTIONS

1. In a city, it rained 70 of the 365 days in a year. What percent of the time is this?

2. A supermarket charges $0.64 tax on a grocery bill of $31.98. What is the tax rate?

3. Suppose there is a sales tax of 35¢ on a purchase of $5.83.
   a. What is the sales tax rate?
   b. 35 is what percent of 583?

4. a. According to the Ratio Comparison Model for Division, what can you do to compare the numbers 6 and 25?
   b. 6 is what percent of 25?
   c. 6 is what part of 25?

5. a. What percent is 6 of 12?
   b. What percent is 6 of 6?
   c. 6 is what percent of 3?

In 6 and 7, answer to the nearest whole-number percent.

6. 41 is what percent of 300?

7. 250 is what percent of 300?

8. The 1990 population of New York City was about 7.3 million. The population of York, England (from which New York got its name) was about 100,000.
   a. Then York, England had __?__ percent the number of people of New York City.
   b. New York City had about __?__ times as many people as York, England.

9. What is the difference between a rate and a ratio?

10. 14 of the 25 students in the class are boys.
    a. What percent are boys?
    b. What percent are girls?

11. What number is 12 percent of 90?

12. a. Banner High School won 36 of its last 40 games. What percent is this?
    b. What is 36 percent of 40?
    c. 40 is 36 percent of what number?

13. If a $60 jacket is reduced $13.50, what is the percent of discount?

14. If a population of 350,000 increases by 7000, what is the percent of increase?

Walking is not only a popular form of exercise; it is a sport. Walking races have been a part of the Olympics since 1906.

Review

15. If $x = \frac{2}{3}$ and $y = \frac{3}{4}$, give the value of each expression.
(Lessons 7-2, 9-3, 9-6, 11-3, 11-4)
   a. $3x + 4y$ **b.** $3x - 4y$ **c.** $(-x)(-y)$
   d. $\frac{xy}{5}$ **e.** $\frac{x}{-y}$ **f.** $\frac{-y}{-x}$

16. Let $a = 10$, $b = 20$, $c = 30$, and $d = 40$. What is the value of $\frac{a-c}{b-d}$?
(Lessons 4-6, 7-2, 11-4)

17. Suppose a person earns time and three quarters for overtime. If the overtime wage is $17.15 per hour, what is the person's normal hourly wage? *(Lessons 9-5, 10-3)*

18. Solve $10y = 43.8$. *(Lesson 10-2)*

19. A person walked for 2.5 hours at 2.5 miles per hour. How far did the person travel? *(Lesson 9-5)*

20. Determine the angle measures of $\triangle ABC$ below. *(Lessons 7-1, 10-2)*

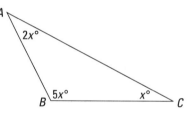

21. There are 12 members on the school math team. Sending them to the state tournament will cost C dollars per team member. If $200 has been collected, how much more needs to be collected?
(Lessons 4-3, 7-1)

22. Write 5 simple fractions equal to $\frac{6}{21}$. *(Lesson 1-10)*

Exploration

23. What percent of English words have more consonants than vowels? Carry out an experiment using words from at least two different sources.

614

Take a survey. *In this classroom, $\frac{8}{18}$ of the students are male, $\frac{1}{18}$ is wearing a green shirt, and $\frac{1}{12}$ of the girls are wearing glasses. Do these equal the ratios in your classroom?*

Equal Ratios and Equal Rates

Remember that the fraction $\frac{a}{b}$ is the result of dividing a by b. This means that every use of division gives rise to fractions. For instance, when 9 out of 12 students in a class ride a bus to school, then $\frac{9}{12}$ of the class ride the bus.

You know that $\frac{9}{12}$ is equal to $\frac{3}{4}$. In this case you could say that 3 of 4 students ride a bus to school. Equal fractions mean equal ratios.

Now consider a rate. A car goes 9 miles in 12 minutes. The fraction

$$\frac{9 \text{ miles}}{12 \text{ minutes}}$$

is the car's average rate. Simplify the fraction. Going 9 miles in 12 minutes is the same rate as going 3 miles in 4 minutes. Equal fractions mean equal rates.

$$\frac{9 \text{ miles}}{12 \text{ minutes}} = \frac{3 \text{ miles}}{4 \text{ minutes}}$$

What Is a Proportion?

A **proportion** is a statement that two fractions are equal. Here are three more examples.

$$\frac{730}{365} = \frac{2}{1} \qquad \frac{12ab}{3b} = \frac{4ab}{b} \qquad \frac{3 \text{ km}}{4 \text{ liters}} = \frac{d \text{ km}}{10 \text{ liters}}$$

Some equations with fractions are not proportions. Examine these two equations.

$$\frac{x+5}{8} = \frac{2}{1}$$
a proportion

$$\frac{x}{8} + \frac{5}{8} = \frac{x}{2}$$
not a proportion

The equation on the left is a proportion because on each side there is one fraction. The equation on the right is equivalent to the left equation. But it is not a proportion because its left-hand side is not a single fraction.

Solving Proportions

Like other equations, proportions can be true or false.

$$\frac{30}{100} = \frac{1}{3}$$
False

$$\frac{320}{100} = \frac{16}{5}$$
True

When a proportion has variables in it, the question is often to **solve the proportion.** That means to find the value or values that make the proportion true. For example:

$$\frac{320}{100} = \frac{16}{x}$$ has the solution $x = 5$, because $\frac{320}{100} = \frac{16}{5}$.

Solving a proportion is just like solving any other equation. No new properties are needed. But it may take more steps. A nice feature is that there are many ways of checking proportions.

Example 1

Solve $\frac{32}{m} = \frac{24}{25}$.

Solution

Multiply both sides by 25m, the product of the denominators.

$$25m \cdot \frac{32}{m} = \frac{24}{25} \cdot 25m$$

We put in a step here that you may not need to write. It shows the products before the fractions are simplified.

$$\frac{25m \cdot 32}{m} = \frac{24 \cdot 25m}{25}$$

Now simplify the fractions.

$$25 \cdot 32 = 24m$$
$$800 = 24m$$
$$\frac{800}{24} = m$$

Rewrite the fraction in lowest terms.

$$\frac{100}{3} = m$$

Check 1

To do a rough check, change $\frac{100}{3}$ to a decimal.

$$\frac{100}{3} = 33.3\ldots$$

Then substitute this value for m in the original proportion.

Does it look right? Does $\frac{32}{33.3}$ seem equal to $\frac{24}{25}$? It seems about right, because in each fraction the numerator is just slightly less than the denominator.

Check 2

To do an exact check, substitute $\frac{100}{3}$ for m in the original proportion.

Does $\frac{32}{\frac{100}{3}} = \frac{24}{25}$?

Work out the division of fractions on the left-hand side.

$$\frac{32}{\frac{100}{3}} = 32 \cdot \frac{3}{100} = \frac{96}{100}$$

Since $\frac{96}{100}$ simplifies to $\frac{24}{25}$ (the right-hand side), the answer checks.

Many real situations lead to having to solve proportions.

Example 2

If you can stuff 100 envelopes in 8 minutes, how many could you stuff in 30 minutes? Assume that you can keep up the same rate.

Solution

Let N be the number of envelopes you can stuff in 30 minutes. Since the rates are equal,

100 envelopes in 8 minutes = N envelopes in 30 minutes.

$$\frac{100 \text{ envelopes}}{8 \text{ minutes}} = \frac{N \text{ envelopes}}{30 \text{ minutes}}$$

$$\frac{100}{8} = \frac{N}{30}$$

Multiply both sides by $30 \cdot 8$ to get rid of all fractions.

$$30 \cdot 8 \cdot \frac{100}{8} = \frac{N}{30} \cdot 30 \cdot 8$$
$$30 \cdot 100 = N \cdot 8$$
$$3000 = 8N$$
$$375 = N$$

At this rate, you can stuff 375 envelopes in 30 minutes.

Check

375 envelopes in 30 minutes is 12.5 envelopes per minute. 100 envelopes in 8 minutes is also 12.5 envelopes per minute.

Covering the Reading

1. Suppose 20 out of 60 students in a class are boys. In lowest terms, __?__ out of __?__ students are boys.

2. Suppose you type 300 words in 10 minutes. At this rate, you would type __?__ words in 20 minutes.

3. Define *proportion.*

4. *Multiple choice.* Which proportion is not true?
   (a) $\frac{100}{7} = \frac{50}{3.5}$
   (b) $\frac{1}{3} = \frac{33}{100}$
   (c) $\frac{24}{60} = \frac{14}{35}$

5. *Multiple choice.* Which equation is not a proportion?
   (a) $\frac{x}{5} = \frac{3}{4}$
   (b) $\frac{1}{9} = \frac{15}{23}$
   (c) $\frac{1}{2} + \frac{2}{2} = \frac{3}{2}$

6. Consider the equation $\frac{40}{t} = \frac{21}{15}$.
   a. By what can you multiply both sides to get rid of the fractions?
   b. What equation results after that multiplication?
   c. Solve this equation.
   d. Check your answer.

In 7–10, solve.

7. $\frac{8}{7} = \frac{112}{Q}$

8. $\frac{200}{8} = \frac{x}{22}$

9. $\frac{L}{24} = \frac{0.5}{4}$

10. $\frac{39}{a} = \frac{130}{7}$

11. Jennifer can assemble 3 cardboard boxes in 4 minutes. At this rate, how many boxes can she assemble in 24 minutes?

12. Victor can assemble 5 cardboard boxes in 6 minutes. At this rate, how many boxes can he assemble in 45 minutes?

Applying the Mathematics

13. In the book *Big Bucks for Kids,* the author says that a kid can earn $25 for singing 20 minutes at a wedding. At this rate, what should the fee be for singing 45 minutes?

14. If a car can travel 300 km on 40 liters of gas, can it travel 450 km on a tank of 50 liters?

15. A recipe says to use 2/3 of a teaspoon of salt for 6 people. In using this recipe for 25 people, how many teaspoons of salt should be used?

16. At the end of the 1992–1993 season, basketball player Michael Jordan had scored 21,541 points in 667 regular season games. He then announced his retirement. If he had continued at this rate, in what game would he have broken Kareem Abdul-Jabbar's record of 38,387 points?

17. In 1990, female armed forces personnel were distributed among the four services as shown below. *Source: The Universal Almanac, 1993*

83,681	Army
73,341	Air Force
56,970	Navy
9,305	Marine Corps

   a. What percent of all female armed forces personnel was in each service?

   b. Construct a circle graph showing this information.
(Lessons 2-6, 2-7, 11-5)

18. a. What percent of 15 is 30?

   b. What percent of 30 is 15?

   c. What is the relationship between the answers to parts **a** and **b**?
(Lessons 10-3, 11-5)

19. Felice and Felipe folded fancy napkins for a Friday feast. In $\frac{2}{3}$ of an hour Felice folded 32 napkins. In 25 minutes Felipe folded 20 napkins. Who was folding faster? *(Lessons 11-2, 11-3)*

20. Pieces of paper numbered from 1 to 80 are put into a hat. A piece of paper is taken out. Assume each piece is equally likely. Mike's favorite number is 7. What is the probability that the number chosen has a 7 as one of its digits? *(Lesson 4-8)*

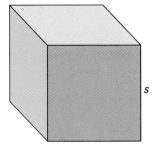

21. The amount of material needed to make a cubical box with edge of length s is $6s^2$. How much material is needed to make a cubical box with an edge of length 30 cm? *(Lesson 4-4)*

22. Suppose a 10″ pizza costs $5.

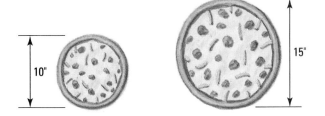

   a. Using a proportion based on the diameters of the pizzas, what should a 15″ pizza cost?

   b. Most pizza places charge more for a large pizza than the price calculated using proportions involving diameters. Why do they charge more?

   c. Find a menu from a place that sells pizza. Do the larger pizzas cost more per inch of diameter?

11-7

The Means–Extremes Property

Ageless calculations. *In Europe in the Middle Ages, pictured here by a detail from a painting by Pieter Brueghel the Elder, 1559, merchants used a shortcut called the* Rule of Three *to solve proportions.*

The Means and the Extremes in a Proportion

In this lesson, you will learn a property that is very useful in solving proportions. First, however, you need to learn some new terms.

Here is a true proportion.

$$\frac{4}{10} = \frac{6}{15}$$

In some places in the world, this proportion is written

$$4:10 = 6:15.$$

In the United States, the colon sign indicates a ratio. We say "the ratio of 4 to 10 equals the ratio of 6 to 15." Or, we say "4 is to 10 as 6 is to 15."

Look again at 4:10 = 6:15. Since the numbers 4 and 15 are on the outside they are called the **extremes** of the proportion. The numbers 10 and 6 are in the middle and are called the **means.** Notice that 4 times 15 equals 10 times 6. The product of the means equals the product of the extremes.

Activity

Write down another true proportion. Verify that the product of the means of your proportion equals the product of the extremes. For instance, if you write down

$$\frac{34.5}{23} = \frac{3}{2}$$

means   extremes

you would multiply to verify that

$$23 \cdot 3 = 34.5 \cdot 2.$$

The general pattern can be shown to be true using the Multiplication Property of Equality. Suppose

$$\frac{a}{b} = \frac{c}{d}.$$

Now multiply both sides of the proportion by *bd*.

$$bd \cdot \frac{a}{b} = bd \cdot \frac{c}{d}$$
$$da = bc$$

The right side of the equation $da = bc$ is the product of the means. The left side is the product of the extremes. This important property is stated below.

Means-Extremes Property

In any true proportion, the product of the means equals the product of the extremes. If $\frac{a}{b} = \frac{c}{d}$, then $b \cdot c = a \cdot d$.

Using the Means-Extremes Property to Solve Proportions

Example 1

Solve $\frac{2}{3} = \frac{12}{x}$.

Solution

You may be able to solve this equation in your head. But if you cannot, here is how to use the Means-Extremes Property.

The means are 3 and 12. The extremes are 2 and *x*. By the Means-Extremes Property,

$$3 \cdot 12 = 2 \cdot x.$$
$$36 = 2x$$

Solve this equation either mentally or by multiplying both sides by $\frac{1}{2}$.

$$18 = x$$

Check

Use substitution. Does $\frac{2}{3} = \frac{12}{18}$? Yes, because the fraction on the right simplifies to the fraction on the left.

Above is another detail of the Brueghel painting.

How Proportions Were Solved Before Algebra Was Invented

In the Middle Ages, there were no variables nor equations as we have them today. Here is a typical problem of that time. Can you solve it before turning the page?

Example 2

Suppose 6 bags of wheat cost 11 silver pieces. How much should 10 bags cost?

Solution

Set up the two equal rates.

$$\frac{11 \text{ pieces}}{6 \text{ bags}} = \frac{p \text{ pieces}}{10 \text{ bags}}$$

Now use the Means-Extremes Property.

$$6p = 110$$

Multiply both sides by $\frac{1}{6}$.

$$p = \frac{110}{6}$$
$$= 18\frac{2}{6}$$
$$= 18\frac{1}{3}$$

The 10 bags should cost a little more than 18 silver pieces.

Students who wanted to solve proportions in the Middle Ages used a shortcut called the Rule of Three. They were taught to memorize: Multiply 10 by 11, and then divide by 6. But how could they remember which two numbers to multiply? Students were usually confused. A poem by an anonymous writer in the late Middle Ages indicates the confusion.

> Multiplication is vexation,
> Division is as bad,
> The Rule of Three
> Does puzzle me
> And practice drives me mad.

Good problem solvers try to find shortcuts like the Means-Extremes Property, or even the Rule of Three. But they also try to understand why the shortcuts work. A shortcut in which you have to guess what to do is no good at all.

With proportions, as with all other equations, it is essential to check answers. Here is a check for Example 2. Does $\frac{11}{6} = \frac{18.\overline{3}}{10}$? Yes, each equals $1.8\overline{3}$. Here is another check. Since 6 bags cost 11 pieces, 12 bags will cost twice as much, or 22 pieces. Ten bags should cost closer to 22 than 11 pieces, and they do.

QUESTIONS

Covering the Reading

1. Consider the proportion $\frac{15}{t} = \frac{250}{400}$.
   a. Identify the means.
   b. Identify the extremes.
   c. According to the Means-Extremes Property, what equation will help you solve this proportion?
   d. Solve this proportion for t.
   e. Check your answer to part **d**.

2. Consider this situation. Eight small cans of grapefruit juice cost $1.79. You want to know how much ten small cans cost.
   a. Write a proportion that will help answer the question.
   b. Solve the proportion using the Means-Extremes Property.
   c. Check your answer.

3. Write down how the proportion 2:3 = 6:9 is read.

4. Why did the Rule of Three puzzle students in the Middle Ages?

5. Cyril went to market and found that 5 bags of salt cost 12 silver pieces. How much should he pay for 8 bags?

In 6–9, solve.

6. $\frac{n}{100} = \frac{72}{20}$ 7. $\frac{3}{10} = \frac{x}{25}$ 8. $\frac{200}{m} = \frac{21}{35}$ 9. $\frac{15}{8} = \frac{12}{v}$

Applying the Mathematics

10. Lannie tried to solve the proportion of Question 6 by multiplying both sides by 100. Will this work? Explain why or why not.

In 11–14, *true or false.*   Consider the proportion $\frac{a}{b} = \frac{c}{d}$.

11. $ad = bc$ 12. $ab = cd$ 13. $ac = bd$ 14. $da = cb$

15. If small cans of grapefruit juice are 5 for $1.69, how many cans can be bought for $10?

16. On the first two days of the week-long hunting season, 47 deer were taken.
    a. At this rate, how many deer will be taken during the week?
    b. Why might it be incorrect to assume the rate will stay the same all week?

This 17th-century engraving shows laborers harvesting rice in China.

17. Why won't the Means-Extremes Property work on the equation $x + \frac{2}{3} = \frac{4}{5}$?

18. During approximately the first century A.D. the most important Chinese mathematics text of its time, the *Chiu Chang,* was written. Here is a problem from that text. (The *picul* was the average weight a man could carry on his back, about 65 kg.) If two and one-half piculs of rice were purchased for $\frac{3}{7}$ of a tael of silver, about how many piculs of rice could be bought for 9 taels?

Giant of a man. *Robert Wadlow is shown with part of his family on his 21st birthday, February 22, 1939.*

19. A telephone survey was conducted in a small city. Of 240 households called on the phone, 119 were watching television at the time.
  a. To the nearest hundredth, what percent of households were watching television?   *(Lesson 11-5)*
  b. If 25,000 households are in this city, how many would you expect to have been watching TV?   *(Lesson 11-6)*
  c. What number might you choose as the probability that someone in a household was watching television?   *(Lesson 4-8)*

20. Simplify.   *(Lessons 1-10, 11-4)*
  a. $\frac{12}{18}$ **b.** $\frac{12}{-18}$ **c.** $\frac{-12}{-18}$ **d.** $\frac{-12}{18}$

In 21 and 22, simplify.   *(Lesson 11-3)*

21. $\frac{7}{10} \div \frac{2}{7}$

22. $1\frac{4}{5} \div \frac{1}{15}$

23. Robert Wadlow of Alton, Illinois, was $5\frac{1}{3}$ feet tall at the age of 5 and $8\frac{2}{3}$ feet tall at the age of 21. On the average, how fast did he grow during those years?   *(Lesson 11-2)*

24. Ru-Niteroi, in Guanabara Bay, Brazil, is the world's largest continuous box-and-plate girder bridge, with a length of 45,603 ft.
  a. What is the bridge's length to the nearest tenth of a mile?
  (Lesson 9-5)
  b. Find the length in miles and feet.   *(Lesson 11-1)*

25. a. Graph four ordered pairs that are the vertices of a quadrilateral that is not a trapezoid.   *(Lessons 8-3, 10-10)*
  b. Apply a size change of magnitude -3 to your quadrilateral.
  (Lesson 9-9)

26. Use the array of squares at the left. Assume each side of a little square has length $\frac{1}{8}$ inch.
  a. How many little squares are in the array?   *(Lesson 9-1)*
  b. What is the area of a little square?   *(Lesson 9-3)*
  c. What is the area of the entire figure?   *(Lesson 10-1)*
  d. What is the perimeter of a little square?   *(Lesson 10-8)*
  e. What is the total length of all the line segments in the figure?
  (Lesson 10-1)

Exploration

For 27 and 28, look again at the poem in this lesson.

27. What is the meaning of the word *vexation?*

28. Approximately when were the Middle Ages?

Proportions in Similar Figures

IN·CLASS
ACTIVITY

Materials: Ruler
Work with a partner.

In Chapter 9 you learned that the image of a figure under a size change of magnitude k is similar to the original figure. In this activity you will learn how to determine the size change factor when given two similar figures.

1 Polygons $ABCDE$ and $A'B'C'D'E'$ below are similar. Measure the lengths of corresponding sides. Then compute $\frac{A'B'}{AB}$, $\frac{B'C'}{BC}$, $\frac{C'D'}{CD}$, $\frac{D'E'}{DE}$, and $\frac{A'E'}{AE}$.

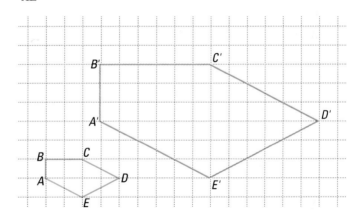

2 $A'B'C'D'E'$ is the size change image of $ABCDE$ under a size change of magnitude 3. How does this size change magnitude compare to the ratios of lengths of corresponding sides that you found?

3 Pick a size change magnitude to use to draw a picture of the house shown here that is similar to the original. Do not tell your partner the size change magnitude you chose. After you and your partner have both drawn the image of the house, trade pictures with your partner. Measure lengths of segments on your partner's drawing and compute appropriate ratios to try to find out what size change magnitude your partner used.

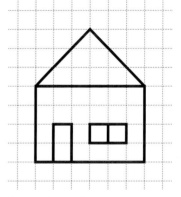

4 ***Draw conclusions.*** With your partner, determine what you can say about the ratios of lengths of corresponding sides of similar figures.

11-8

Proportions in Similar Figures

Scaled down. *A model airplane, such as the one being constructed in the photo, is called a scale model because its dimensions are scaled proportionally to those of the full-sized airplane it represents.*

In the In-class activity on page 625, you examined the lengths of corresponding sides of similar figures. We call such lengths *corresponding lengths.* In the activity, you should have found that the following property is true.

> In similar figures, ratios of corresponding lengths are equal.

Finding Lengths in Similar Figures

If you know which sides of similar figures correspond, you can find unknown lengths.

Example 1

Triangles *CAT* and *DOG* are similar. Name the corresponding sides.

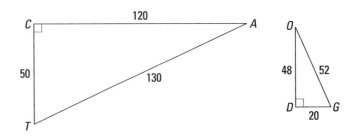

Solution

The sides will correspond in order of their lengths. The shortest sides, $\overline{CT}$ and $\overline{DG}$, correspond, as do the middle sides, $\overline{CA}$ and $\overline{DO}$, and the longest sides, $\overline{AT}$ and $\overline{OG}$. Notice that the ratios of these sides are equal: $\frac{50}{20} = \frac{120}{48} = \frac{130}{52}$.

You can find lengths of sides in similar figures by solving proportions.

Example 2

The figures below are similar with corresponding sides parallel. Find *EF*.

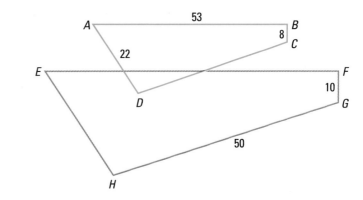

Solution

$\overline{EF}$ corresponds to $\overline{AB}$. The only corresponding sides with both lengths given are $\overline{FG}$ and $\overline{BC}$. Use them in the proportion.

$$\frac{EF}{AB} = \frac{FG}{BC}$$

Now substitute what is known. *FG* = 10, *BC* = 8, and *AB* = 53.

$$\frac{EF}{53} = \frac{10}{8}$$

Solve as you would any other proportion. Use the Means-Extremes Property.

$$53 \cdot 10 = 8 \cdot EF$$

Multiply both sides by $\frac{1}{8}$. $\frac{530}{8} = EF$

$$EF = 66.25$$

Check

For a rough check, look at the figure. Does a length of about 66 for $\overline{EF}$ seem correct? Yes, since *EF* should be longer than *AB*.

For an exact check, substitute the length of $\overline{EF}$ in the proportion.

$$\frac{66.25}{53} = \frac{10}{8}$$

Division shows each side to equal 1.25.

How Do Similar Figures Arise?

Begin with a real object. Take a photograph of it, or draw a picture of it. You may construct a scale model of a large object. You might magnify a small object. Any of these activities lead to similar figures in which you might want to find lengths of sides. In these ways, similar figures arise, and with them, situations involving proportions.

QUESTIONS

Covering the Reading

1. In the In-class activity on page 625, what value did you find for each ratio?
   a. $\frac{A'B'}{AB}$
   b. $\frac{B'C'}{BC}$
   c. $\frac{C'D'}{CD}$

2. In similar figures, what ratios are always equal?

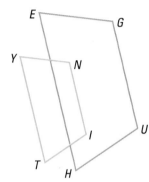

3. a. Polygon *HUGE* at the left is an expansion image of polygon *TINY*. Name the ratios equal to $\frac{IN}{UG}$.
   b. If $YN = 16$, $EG = 24$, and $TY = 26$, what is the value of *HE*?

In 4 and 5, refer to Example 2.

4. Find *EH*.

5. Find *CD*.

6. Draw an example of similar figures whose corresponding sides are not parallel.

7. The figures below are similar. What is the value of *x*?

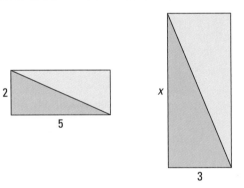

8. Name three places where similar figures can be found.

The pool shown here is in the Tuileries Gardens in Paris, France. Children often sail toy boats in this pool.

Applying the Mathematics

9. A right triangle has sides with lengths 3 cm, 4 cm, and 5 cm. The shortest side of a similar right triangle has length 9 cm.
   a. Draw an accurate picture of the original right triangle.
   b. Find the lengths of the other two sides of the similar right triangle.
   c. Draw an accurate picture of the similar right triangle.

10. A drawing is 20 cm long and 15 cm wide. It is put into a copy machine to be reduced to 60% of its original size.
    a. What will the dimensions of the copy be?
    b. Are the ratios of corresponding sides of the drawing and its copy equal?
    c. Are the drawing and its copy similar?

11. A boat is 220 feet long. A scale model of the boat is 20 inches long. If the scale model is 8 inches wide, how wide was the original boat?

0 10 20 30 miles

12. At the left is a map of the region around Santa Fe, New Mexico. The map is similar to the actual land. A distance of 14 mm on the map represents an actual distance of about 30 miles. The distance from Albuquerque to Santa Fe on the map is about 31 mm. How far is this in actual miles?

Review

In 13–15, solve. *(Lesson 11-6)*

13. $\frac{t}{40} = \frac{5}{16}$

14. $\frac{3.5}{11} = \frac{x}{132}$

15. $\frac{3}{4} = \frac{5}{A}$

16. A plane goes 300 miles in 36 minutes. At this rate, how far will it travel in an hour? *(Lesson 11-2)*

17. *Multiple choice.* If $x = 60$ and $y = -10$, which of the following are negative? *(Lessons 5-5, 7-2, 9-6, 11-4)*

(a) $x + y$ (b) $x - y$ (c) $y - x$ (d) xy (e) $\frac{x}{y}$ (f) $\frac{y}{x}$

18. Solve $10A - 50 = 87$. *(Lesson 10-4)*

19. Consistent Constance constantly tips servers 15% of the bill for dinner. You see Constance leave a tip of $2.25. Estimate the dinner bill. *(Lesson 10-3)*

In 20–23, let $u = \frac{8}{5}$ and $v = -\frac{7}{5}$. Evaluate. *(Lessons 5-5, 7-2, 9-3, 11-3)*

20. $u + v$

21. $u - v$

22. uv

23. $\frac{u}{v}$

24. Merv knows that he can multiply both the numerator and denominator of a fraction by the same nonzero number without changing its value. He wonders what else he can do without changing its value.
a. Can he add the same number to the numerator and denominator?
b. Can he subtract the same number from the numerator and denominator?
c. Can he divide both the numerator and denominator by the same number? *(Lessons 6-4, 6-7)*

25. Write 3,800,000,000 in scientific notation. *(Lesson 2-3)*

26. Write $17\frac{2}{11}$ as a simple fraction and as a decimal. *(Lessons 1-7, 1-10)*

27. Write $\frac{41}{8}$ as a decimal and as a mixed number. *(Lessons 1-6, 1-10)*

28. Between which two whole numbers is $\frac{41}{8}$? *(Lesson 1-2)*

Exploration

29. Look again at the drawing for Question 3. Some people think that the drawing might be a picture of a 3-dimensional object with missing lines. Trace the drawing, and add lines to picture an object.

How much forest is there? *Forests comprise a large portion of the land area of the United States. A forest near Corinth, Vermont, is shown above. See Example 2.*

Suppose 2 bags of peanuts cost 59¢. How much will 8 bags cost? You can answer this question by solving the following proportion.

$$\frac{2 \text{ bags}}{59¢} = \frac{8 \text{ bags}}{C}$$

Some people would notice that 8 is 4 times 2 and reason that the cost will be 4 times 59¢. These people have used *proportional thinking* to answer the question. **Proportional thinking** is the ability to get or estimate an answer to a proportion without solving an equation. Some people believe that proportional thinking is one of the most important kinds of thinking you can have in mathematics.

To improve your proportional thinking, try first to estimate answers to questions involving proportions. Then try to get the exact answer.

Example 1

If you travel at 45 mph, how much time will it take to go 80 miles?

Solution 1

Estimate. At 45 mph, you can go 45 miles in 1 hour. Therefore, you can go 90 miles in 2 hours. So it will take between 1 and 2 hours to go 80 miles. Since 80 is closer to 90 than to 45, it will take almost 2 hours.

Solution 2

Try simpler numbers. At 40 mph, it would take 2 hours. This answer is 80 divided by 40. This suggests that dividing 80 by 45 will get the answer.

$$\frac{80 \text{ miles}}{45 \frac{\text{miles}}{\text{hour}}} \approx 1.8 \text{ hours} = 1 \text{ hour } 48 \text{ minutes}$$

The answer seems right because it agrees with the estimate in Solution 1.

Solution 3

Set up a proportion.
$$\frac{45 \text{ miles}}{1 \text{ hour}} = \frac{80 \text{ miles}}{h \text{ hours}}$$

Use the Means-Extremes Property or multiply both sides by h.

$$45h = 80$$

Solve.
$$h = \frac{80}{45}$$
$$\approx 1.8 \text{ hours}$$

The important thing here is flexibility. You do not have to know every way there is of solving proportions. But you should have at least two ways to do these questions. Use one way to check the other.

Example 2 uses the hectare, a very common unit for measuring land area in the metric system.

Example 2

According to the *State of the World 1991,* about 296,000,000 hectares of the United States are forest. There are about 259 hectares in a square mile. About how many square miles of the U.S. are forest?

Solution 1

Estimate using proportional thinking. Since there are about 259 hectares in one square mile, there are 259 million hectares in 1 million square miles. 296,000,000 is more than 259 million, so more than 1 million square miles of the U.S. are forest.

Solution 2

The problem is to convert hectares to square miles. Use a method from Lesson 9-5. The abbreviation for hectare is ha.

$$296,000,000 \text{ ha} = 296,000,000 \text{ ha} \cdot \frac{1 \text{ square mile}}{259 \text{ ha}}$$
$$\approx 1,140,000 \text{ square miles}$$

Solution 3

Set up a proportion.

$$\frac{296,000,000 \text{ hectares}}{F \text{ square miles}} = \frac{259 \text{ hectares}}{1 \text{ square mile}}$$

Solve the proportion.

$$259F = 296,000,000$$
$$F = \frac{296,000,000}{259}$$
$$\approx 1,140,000 \text{ square miles}$$

The entire land area of the United States is about 3,536,000 square miles. Dividing 1,140,000 by 3,536,000, we get about 0.32. So about 32% (almost 1/3) of the land of the United States is forest.

QUESTIONS

1. What is proportional thinking?

2. Use proportional thinking to answer this question. If tuna is 2 cans for $2.50, how much will 6 cans cost?

3. What proportion will answer the question of Question 2?

4. About how many hectares equal one square mile?

5. *True or false.* In a thousand square miles, there are more than 500,000 hectares.

6. The area of the state of Illinois is about 56,000 square miles. About how many hectares is this?

7. Suppose you travel 45 kph (kilometers per hour). Then
   a. it takes __?__ hours to travel 90 km.
   b. it takes __?__ hours to travel 450 km.
   c. it takes between __?__ and __?__ to go 200 km.
   d. Exactly how long does it take to go 200 km?

In 8–10, **a.** answer the question using any method you wish.
b. Check your answer using a different method.

8. At 55 mph, how long will it take to drive 120 miles?

9. There are 12 eggs in a dozen. 100 eggs is how many dozen?

10. There are 640 acres in a square mile. Suppose a forest fire burned 12,000 acres of timberland. About how many square miles burned?

Applying the Mathematics

11. Speedy Spencer can type 30 words in 45 seconds. At this rate, how many words can he type in a minute?

12. Jack can mow his lawn in 40 minutes. Jill can mow the same lawn in 40 minutes. How long will it take Jack and Jill, working together to mow four lawns of the same size? Explain your reasoning.

13. Beatrice used 100 stamps in 2 weeks. At this rate, how many stamps will she use in 3 weeks?

14. If it costs $500/month to feed a family of 4, how much will it cost to feed a family of 5?

In 15 and 16, triangles *ABC* and *DEF* are similar with corresponding sides parallel. *(Lesson 11-8)*

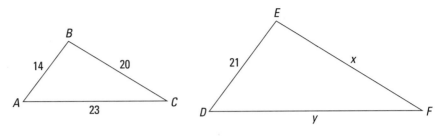

15. Find *x*.

16. Find *y*.

17. If $x = -4$ and $y = -2$, what is $-3x$ divided by $2y$? *(Lessons 4-4, 9-6, 11-4)*

18. If 30% of a number is 51, what is 40% of that same number? *(Lessons 2-5, 10-3)*

19. What is 14% of 200? *(Lesson 2-5)*

20. 200 is 14% of what number? *(Lesson 10-3)*

21. You often hear that the cost of living is going up. The government calculates a number called the *consumer price index* (CPI) to measure the cost of living. When the CPI doubles, it means that, on the average, prices have doubled. The CPI scale used in 1991 considered average prices in the years 1982–84 to be 100.
   a. In 1991, the CPI was 136.2. In 1970, the CPI was 38.8. Between 1970 and 1991, did prices more than triple or less than triple?
   b. If a new car cost $11,630 in 1970, what would you expect a new car of the same type to cost in 1991?

22. a. The following program prints the cost of peanuts if a bag costs $.59.

```
10 FOR N = 2 TO 20 STEP 2
20 PRINT N; "BAGS OF PEANUTS COST $";.59 * N
30 NEXT N
40 END
```

   Run the program and record its output.
   b. What could you do to print the costs for every whole number of bags from 1 to 20?
   c. Modify the program to print another similar table.

The farm shown here grows peanut plants. The peanut plant is unusual in that its pods develop underground.

A project presents an opportunity for you to extend your knowledge of a topic related to the material of this chapter. You should allow more time for a project than you do for typical homework questions.

1 Similar Objects

Find at least five pairs of similar objects around your house. (One pair might be different-sized photographs of the same scene.) Calculate the ratios of at least three corresponding lengths in each pair to verify that they are similar and to determine the ratio of similitude.

2 Best Buys

Go to a grocery store and find examples of at least five foods that are packaged in different sizes. Calculate the unit cost for each size to determine best buys. Write up what you find, and answer whether it is true that bigger packages are more economical.

3 Proportional Thinking

Invent ten real-world problems that can be solved with proportions and that are different from those given in this chapter. Include your solutions.

A B

4 Integer Arithmetic with Negative Divisors

Extend the idea of a quotient and remainder to cover division by -2, -3, and other negative integers. Assume that the remainder will always be positive, and calculate the quotient and remainder when 436 is divided by -7, and when -100 is divided by -6. Explain how you obtained your results and why you believe they are correct.

5 World-Record Rates

Look in an almanac or other reference book for all the world records in running or swimming, for men or for women. Calculate the average rate of speed of each of the record holders. Graph the ordered pairs (distance, average speed), and describe what you find.

6 A Different Way to Divide Fractions

One student had the following way of dividing fractions. First the student renamed the two fractions with a common denominator. Then the student ignored the denominators and just divided the numerators to obtain the answer. Does this method work? Try it on enough examples to form a conclusion about whether it *always, sometimes,* or *never* works.

SUMMARY

There are two types of division. For the integer division $x \div y$, x must be an integer, and y must be a positive integer. The answer is given as a quotient and remainder. For the real-number division $x \div y$, x and y can be any numbers with $y \neq 0$, and the answer is the single-number quotient $\frac{x}{y}$.

One basic use of division is to calculate rates. In a rate, the divisor and the dividend have different units. The unit of the rate is written as a fraction. For example, if x words are divided by y minutes, the quotient is $\frac{x}{y}$ words per minute, which can be written as $\frac{x}{y} \frac{\text{words}}{\text{minute}}$.

The numbers involved in rates can be fractions, or decimals, or negative numbers. So rates can help you learn how to divide these kinds of numbers. To divide fractions, you can use the Algebraic Definition of Division. Multiply by the reciprocal of the divisor. To divide with negative numbers, follow the same rules for signs that work for multiplication.

Another basic use of division is in calculating ratios. Ratios are numbers that compare two quantities with the same units. Since both divisor and dividend have the same units, the quotient has no unit. Ratios may be decimals, fractions, or integers. Percents are almost always ratios.

If two fractions, rates, or ratios are equal, a proportion results. The form of a proportion is $\frac{a}{b} = \frac{c}{d}$. Because it has many applications, proportional thinking is very important. You should be able to solve simple proportions mentally. If $\frac{a}{b} = \frac{c}{d}$, then $ad = bc$. This is known as the Means-Extremes Property. You could also solve the proportion by multiplying both sides of the equation by bd. Proportions occur in geometry and all other areas of mathematics. Proportions also occur in all sorts of daily experiences. So it helps to know more than one way to solve a proportion.

VOCABULARY

You should be able to give a general description and a specific example of each of the following ideas.

Lesson 11-1
real number division
integer division
Quotient-Remainder Form
Quotient-Remainder Formula

Lesson 11-2
rate, rate unit
Rate Model for Division
unit cost

Lesson 11-3
Algebraic Definition of
    Division

Lesson 11-5
ratio, ratio comparison
Ratio Comparison Model for
    Division

Lesson 11-6
proportion
solving a proportion

Lesson 11-7
extremes, means
Means-Extremes Property

Lesson 11-8
corresponding lengths

Lesson 11-9
proportional thinking

PROGRESS SELF-TEST

Take this test as you would take a test in class. Then check your work with the solutions in the back of the book.

1. What rate is suggested by this situation? A person types 300 words in 5 minutes.

2. If you travel k kilometers in h hours, what is your average speed in kilometers per hour?

3. Divide $\frac{4}{9}$ by $\frac{1}{3}$.

4. Simplify $\frac{-42}{-24}$.

5. Simplify $\frac{\frac{3}{5}}{\frac{6}{5}}$.

6. Give the value of $\frac{-2x}{8+y}$ when $x = -5$ and $y = -9$.

7. If a is negative and b is positive, which of the following are negative?
$$\frac{a}{b} \qquad \frac{-a}{b} \qquad \frac{-a}{-b}$$

8. Write a rate question about weight that results in division of -8 by 2.

9. Solve $\frac{40}{x} = \frac{8}{5}$.

10. Solve $\frac{5}{12} = \frac{p}{3}$.

11. To the nearest percent, 14 is what percent of 150?

12. If $0.30 is the tax on a $4 purchase, what percent tax rate is this?

13. If 60% of a number is 30, what is the number?

14. At 45 mph, how long does it take to travel 189 miles?

15. There are 640 acres in a square mile. How many acres are in 10,000 square miles?

16. State the Means-Extremes Property as it applies to the proportion $\frac{a}{b} = \frac{x}{y}$.

17. Name the means in the proportion of Question 16.

In 18 and 19, triangles ABC and DEF are similar with corresponding sides parallel.

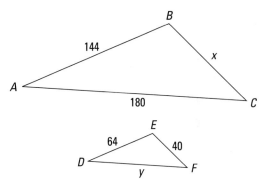

18. Find x.   19. Find y.

20. A recipe for 4 people says to use $\frac{1}{4}$ teaspoon of pepper. How much should be used for 10 people?

21. A book has 245 pages. Marissa feels she can read 15 pages a day. At this rate, how many days will it take her to read the book? How many pages will she need to read on the last day?

22. Explain how Question 21 relates to the Quotient-Remainder Formula.

CHAPTER REVIEW

Questions on SPUR Objectives

SPUR stands for **S**kills, **P**roperties, **U**ses, and **R**epresentations. The Chapter Review questions are grouped according to the SPUR Objectives for this chapter.

SKILLS DEAL WITH THE PROCEDURES USED TO GET ANSWERS.

Objective A: *Divide fractions with numbers or variables.* *(Lesson 11-3)*

1. What is 12 divided by $\frac{2}{3}$?   2. $\frac{3}{5} \div \frac{5}{3} = \underline{\ ?\ }$

In 3–5, simplify.

3. $\dfrac{\frac{2}{3}}{\frac{4}{3}}$ 4. $\dfrac{1\frac{1}{6}}{3\frac{2}{3}}$ 5. $\dfrac{\frac{x}{y}}{\frac{a}{b}}$

6. Divide $\frac{x}{3}$ by 2.   7. Divide $3\frac{3}{4}$ by $2\frac{1}{7}$.

Objective B: *Divide positive and negative numbers.* *(Lesson 11-4)*

8. Give the value of $\frac{2c}{d}$ when $c = 3$ and $d = -1$.

In 9–12, simplify.

9. $\frac{-10}{2}$ 10. $-\frac{-400}{-200}$ 11. $\frac{-40}{-30}$ 12. $\frac{6}{-9}$

Objective C: *Solve proportions.* *(Lessons 11-6, 11-7, 11-9)*

13. In solving $\frac{7}{t} = \frac{168}{192}$, Jenny got $t = 24$. Is she correct?

In 14–19, solve.

14. $\frac{40}{9} = \frac{12}{x}$ 15. $\frac{y}{2} = \frac{9}{7}$ 16. $\frac{700}{y} = \frac{20}{1}$

17. $\frac{7}{G} = \frac{1}{3}$ 18. $\frac{6}{20} = \frac{AB}{50}$ 19. $\dfrac{\frac{1}{2}}{CD} = \frac{5}{6}$

PROPERTIES DEAL WITH THE PRINCIPLES BEHIND THE MATHEMATICS.

Objective D: *Recognize the Means-Extremes Property and know why it works.* *(Lessons 11-6, 11-7)*

20. According to the Means-Extremes Property, if $\frac{a}{d} = \frac{b}{e}$, what else is true?

21. To eliminate fractions in the equation $\frac{3}{x} = \frac{5}{6}$, by what can you multiply both sides?

22. Why won't the Means-Extremes Property work on the equation $\frac{4}{.5} + x = \frac{3}{.4}$?

Objective E: *Know the general properties for dividing positive and negative numbers.* *(Lesson 11-4)*

23. *Multiple choice.* Which does *not* equal $\frac{-x}{y}$?

(a) $-\frac{x}{y}$ (b) $-x \cdot \frac{1}{y}$ (c) $\frac{x}{-y}$ (d) $\frac{-x}{-y}$

24. If $\frac{a}{b} = c$ and a is negative, and c is positive, then is b positive or negative?

USES DEAL WITH APPLICATIONS OF MATHEMATICS IN REAL SITUATIONS.

Objective F: *Use integer division in real situations.* *(Lesson 11-1)*

25. a. How many gallons are in 25 quarts?
    b. Relate part **a** to the Quotient-Remainder Formula.

26. a. How many feet and inches are in 218 inches?
    b. How does this relate to the Quotient-Remainder Formula?

27. Twelve *Transition Mathematics* books can fit in a box.
  a. If a school orders 70 books, how many boxes could be filled and how many would then be in the unfilled box?
  b. Relate this situation to the Quotient-Remainder Formula.

28. 700 students at a Model United Nations convention were split into groups of 16.
  a. How many full groups were formed?
  b. How many students were left for a smaller group?
  c. How does this relate to the Quotient-Remainder Formula?

Objective G: *Use the Rate Model for Division.*
(Lessons 11-2, 11-3, 11-4)

29. A car travels 200 km in 4 hours. What is its average speed?

30. Sixteen hamburgers are made from 5 pounds of ground beef. How much beef is this per hamburger?

31. On a diet, some people lose 10 pounds in 30 days. What rate is this?

32. In 1990, Boston had a population of about 578,000 people and an area of about 47 square miles. About how many people per square mile is this?

33. Two-thirds of the way through the summer, Jeremy had earned $120 mowing lawns. At this rate, how much will he earn mowing lawns during the entire summer?

34. Betty spent $200 on a 4-day vacation. Use this information to make up a question involving division and negative numbers. Answer your question.

35. Three days ago, the repaving crew was 6 km down the road. Use this information to make up a question involving division and negative numbers. Answer your question.

Objective H: *Use the Ratio Comparison Model for Division.* *(Lesson 11-5)*

36. Bill bought 5 CDs last year. This year he bought 8. This year's amount is how many times last year's amount?

37. If you get 32 right on a 40-question test, what percent have you missed?

38. In Seattle, Washington, it rains an average of 216 days a year. What percent of days in a year is this? (Answer to the nearest percent.)

39. In 1988, Ford Motor Company sold about 47,000 Lincoln Continentals. In 1989, the number of Continentals sold was about 9,000 more. In other words, in 1989 the company sold about __?__ times as many Continentals as in 1988. (Answer to the nearest tenth.)

Objective I: *Recognize and solve problems involving proportions in real situations.*
(Lessons 11-6, 11-7, 11-9)

40. You made 35 copies in 2.5 minutes using a duplicating machine. At this rate, how long will it take the machine to make 500 copies?

41. A recipe for 6 people calls for $1\frac{1}{3}$ cups of sugar. How much sugar is probably needed for a similar recipe for 10 people?

42. You want to buy 5 cans of tuna. You see that 8 cans cost $5. At this rate, what will the 5 cans cost?

REPRESENTATIONS DEAL WITH PICTURES, GRAPHS, OR OBJECTS THAT ILLUSTRATE CONCEPTS.

Objective J: *Find missing lengths in similar figures.* *(Lesson 11-8)*

In 43 and 44, a right triangle has sides with lengths 5, 12, and 13. The shortest side of a similar right triangle has length 6.5.

43. Find the lengths of the other two sides of the similar right triangle.

44. Draw accurate pictures of the two triangles using the same unit.

45. The figures below are similar. If $AB = 80$, $CD = 45$, and $EL = 56$, what is IK?

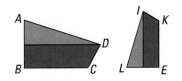

CHAPTER
12

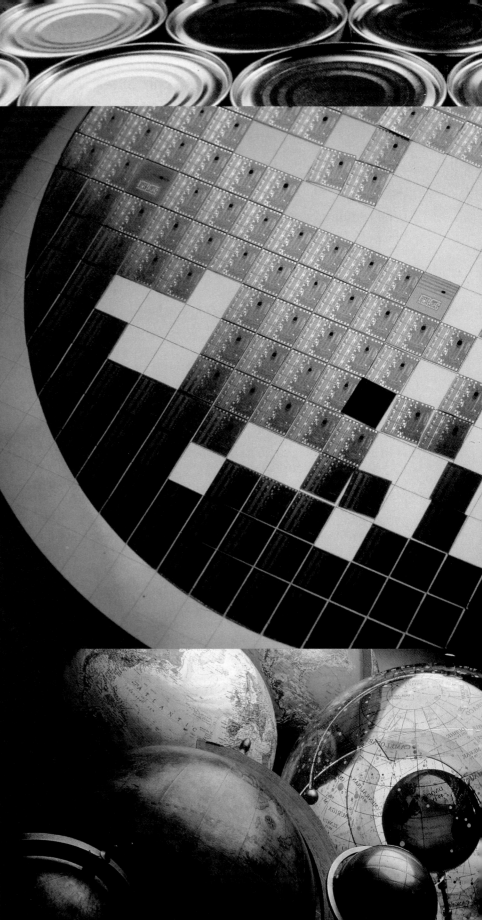

REAL NUMBERS, AREA, AND VOLUME

The numbers that can be represented by decimals are called real numbers. All of the numbers you have encountered in this book have been real numbers. Here are some examples.

Whole numbers, such as   6931

Simple fractions, such as $\frac{24}{25} = 0.96$

Negative numbers, such as $-7.3 \times 10^{-4} = -0.00073$

Infinite non-repeating decimals, such as $\pi = 3.141592\ldots$

Mixed numbers, such as $11\frac{4}{37} = 11.108108108108\ldots$

Some decimals are terminating; their decimals end. Others are infinite decimals. Some of the infinite decimals, such as the infinite decimal for $11\frac{4}{37}$ above, repeat. Others, like the decimal for π, have no repeating pattern. In this chapter, you will learn applications using some infinite non-repeating decimals. These numbers are important in finding the areas and volumes of common, everyday figures. For instance, the number π is involved in the volume formulas of both spheres and cylinders.

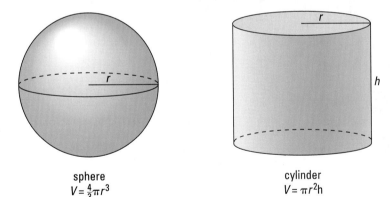

sphere
$V = \frac{4}{3}\pi r^3$

cylinder
$V = \pi r^2 h$

As a project, you can explore the *complex numbers*. These are numbers that are not real numbers. The complex numbers also have a variety of geometrical applications, but you will not usually study those uses until later mathematics courses.

PLATAFORMA	DAMAS	FINAL
PLATFORM	WOMEN	FINAL
1 WYLAND WENDY	USA	426.57
2 RIBOT VERONICA	ARG	404.01
3 CANSECO G.	MEX	385.20
4 NEYER MEGAN	USA	378.27
5 DE LA TTORE E.	MEX	373.80
6 FULLER DEBBIE	CAN	357.99
7 LOWRY REGINA	CUB	354.09
8 WITTMEIER ROBIN	CAN	351.66

Can you read these? *When you read a finite decimal aloud, you can write what you say as a fraction.*

Very early in this book, in Lesson 1-6, you studied how to convert a simple fraction $\frac{a}{b}$ into its decimal form. Just divide a by b. For instance, to find the decimal equal to $\frac{6}{11}$, divide 6 by 11. On a calculator, you will see something like $\boxed{0.5454545}$. Similarly, $\frac{873}{200} = 873 \div 200 = 4.365$. You should check these now on your calculator.

Fractions for Terminating Decimals

It is often useful to convert a decimal to its equivalent fraction. If the decimal is finite, you can do this by applying the Equal Fractions Property.

Example 1

Find a simple fraction equal to 4.365.

Solution

First write 4.365 as a fraction. $4.365 = \frac{4.365}{1}$

Multiply both numerator and denominator by a power of 10 large enough to obtain a whole-number numerator. In this case, the power is 10^3.

$$\frac{4.365}{1} = \frac{4.365 \cdot 10^3}{1 \cdot 10^3} = \frac{4365}{1000}$$

This simple fraction answers the question. If you want the fraction in lowest terms, look for common factors of the numerator and denominator. Here, 5 is obviously a common factor. Divide both the numerator and denominator by it.

So we find that $4.365 = \frac{4365}{1000} = \frac{4365 \div 5}{1000 \div 5} = \frac{873}{200}$.

Fractions for Repeating Decimals

However, suppose you wanted to convert an infinite repeating decimal such as 0.5148148148148 . . . to a fraction. The method of Example 1 will not work because there is no power of 10 that can be multiplied by the numerator to make a whole number. One way to find a fraction equal to an infinite repeating decimal uses the Distributive Property in a surprising way.

Example 2

Find a simple fraction that equals $0.5\overline{148}$.

Solution

The idea is to solve an equation whose solution is known to be $0.5\overline{148}$.

Step 1: Let $x = 0.5\overline{148}$. Write x with a few repetitions of the repetend 148.
$$x = 0.5148148148 \ldots$$

Step 2: The repetend has 3 digits, so multiply x by 1000. This moves the decimal point 3 places to the right.
$$1000x = 514.8148148148 \ldots$$

Step 3: Subtract the first equation from the second equation.
$$1000x - x = 514.8\overline{148} - 0.5\overline{148}$$

Step 4: Use the Distributive Property on the left side of the equation. Subtract on the right side.
$$(1000 - 1)x = 514.8\overline{148} - 0.5\overline{148}$$
$$999x = 514.3$$

Step 5: Solve the resulting equation.
$$x = \frac{514.3}{999} = \frac{5143}{9990}$$

This is a simple fraction, but it is not in lowest terms. Dividing numerator and denominator by 37, this fraction equals $\frac{139}{270}$. Therefore,
$$0.5\overline{148} = \frac{139}{270}.$$

Check

Convert both $\frac{5143}{9990}$ and $\frac{139}{270}$ to decimals using a calculator.

Example 3

Find a simple fraction equal to $5.0\overline{3}$.

Solution

Step 1: Write $x = 5.0333333333 \ldots$
Step 2: Multiply both sides by 10. $10x = 50.33333333 \ldots$
Step 3: Subtract. $10x - x = 50.3\overline{3} - 5.0\overline{3}$
Step 4: Simplify. $9x = 45.3$
Step 5: Solve. $x = \frac{45.3}{9} = \frac{453}{90} = \frac{151}{30}$

So, $5.0\overline{3} = \frac{151}{30}$

Example 4

Cleo had just done a division on her calculator when she was distracted. When she returned to her work she realized she had forgotten which numbers she had divided! The display on the calculator read 187.54545 What numbers might she have divided?

Solution

The display might be a rounding of the repeating decimal $187.\overline{54}$.

Step 1: Let $\qquad\qquad\qquad\qquad\qquad x = 187.\overline{54}.$
Step 2: Multiply both sides by 100. $\quad 100x = 18754.\overline{54}$
Step 3: Subtract. $\qquad\qquad 100x - x = 18754.\overline{54} - 187.\overline{54}$
Step 4: Simplify. $\qquad\qquad\qquad\qquad 99x = 18567$
Step 5: Solve. $\qquad\qquad\qquad\qquad\qquad x = \dfrac{18567}{99} = \dfrac{2063}{11}$

So Cleo may have divided 18,567 by 99. Or she may have divided 2063 by 11. Or, she divided any two other numbers whose quotient is $\dfrac{2063}{11}$.

Check

2063 ÷ 11 = 187.54545

Remember that all simple fractions equal either repeating or terminating decimals. So if a decimal is infinite and does not repeat, it cannot be converted to a simple fraction. For instance, no simple fraction equals π.

QUESTIONS

Covering the Reading

In 1–4, classify the decimal as repeating, terminating, or cannot tell.

1. 3.04444   **2.** $3.0\overline{4}$ **3.** 3.040404 . . .   **4.** $3.\overline{04}$

5. Which of the decimals in Questions 1–4 are finite, which infinite?

In 6–8, find the simple fraction in lowest terms equal to the decimal.

6. 11.6   **7.** 0.061   **8.** 0.24

9. a. If $x = 0.\overline{24}$, then $100x = \underline{\ ?\ }$.
   b. Find a simple fraction equal to $0.\overline{24}$.
   c. Find a simple fraction equal to $5.\overline{24}$.

10. Find a simple fraction equal to $0.\overline{810}$.

11. Convert $3.0\overline{405}$ to a fraction.

12. Name two numbers other than 2063 and 11 which could have given Cleo's calculator display in Example 4.

13. The digits of 0.12345678910111213 . . . are from the whole numbers. Explain why this decimal cannot be converted to a simple fraction.

Applying the Mathematics

14. A hat contains between 10 and 25 marbles. Some marbles are green, and the rest are yellow. Without looking you are to reach into the hat and pull out a marble. The probability of pulling out a green marble is $0.\overline{2}$.
   a. Rewrite the probability of pulling out a green marble as a simple fraction.
   b. Use your answer to part **a** to determine how many marbles are in the hat. Describe how you arrived at this answer.
   c. How many of the marbles are green?

In 15 and 16, refer to this calculator display: 58.833333 .

15. If the 3 repeats forever, name a division problem with this answer.

16. Name a division problem with this exact answer.

17. Use the method of this section to show that $1 = 0.\overline{9}$.

18. Write $99.\overline{4}\%$ as a simple fraction.

Review

In 19 and 20, use the drawing at the left.

19. If the measure of angle 1 is $48\frac{2}{5}^{\circ}$, what is the measure of angle 2? *(Lesson 7-6)*

20. If angle 1 has measure x and angle 2 has measure $2x$, what is the value of x? *(Lessons 7-6, 10-2, 10-6)*

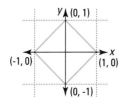

In 21 and 22, use the figure at the left.

21. How many lines of symmetry does the figure have? *(Lesson 8-7)*

22. Trace the figure and draw a tessellation using it as the fundamental region. *(Lesson 8-8)*

23. **a.** What is the area of a square with one side of length s?
   b. What is the area of the square with vertices $(1, 0)$, $(0, 1)$, $(-1, 0)$, and $(0, -1)$? *(Lessons 3-8, 8-3)*

Exploration

24. What fraction with 1 in its numerator and a two-digit integer in its denominator equals $0.\overline{012345679}$?

25. **a.** What decimal is the sum of $0.\overline{12}$ and $0.\overline{345}$?
   b. What simple fraction is the sum?

Mathmagic. *In this scene from the film* Donald in Mathmagic Land, *Donald chases a pencil and runs into many square roots.*

About 2500 years ago, Greek mathematicians began to organize the knowledge that today we call geometry. To the Greeks, numbers were lengths. They were probably the first to discover that some lengths could not be expressed as simple fractions. In this lesson you will learn about these lengths.

What Are Square Roots?

Here are two squares and their areas.

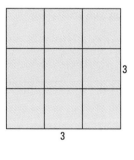

Area = 3 · 3 = 3^2 = 9 square units

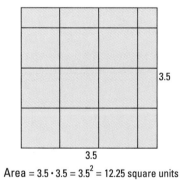

Area = 3.5 · 3.5 = 3.5^2 = 12.25 square units

Recall that 9 is called the *square* of 3. We say that 3 is a *square root* of 9. Similarly, 12.25 is the square of 3.5, and we say that 3.5 is a square root of 12.25.

> If $A = s^2$, then s is called a **square root** of A.

Since $-3 \cdot -3 = 9$, $(-3)^2 = 9$. So -3 and 3 are both square roots of 9. *Every positive number has two square roots.* The two square roots are opposites of each other. The symbol for the positive square root is $\sqrt{}$, called the **radical sign.** We write $\sqrt{9} = 3$. The negative square root of 9 is $-\sqrt{9}$, or -3. The symbol $\pm\sqrt{9}$, or ± 3, is sometimes used to refer to both square roots.

All scientific calculators have a square root key $\boxed{\sqrt{}}$. This key gives or estimates the positive square root of a number.

Activity 1

Find the positive square root of 12.25 using your calculator.

You should see 3.5 displayed. This tells you that the positive square root of 12.25 is 3.5. The two square roots of 12.25 are 3.5 and -3.5.

$$\sqrt{12.25} = 3.5 \text{ and } -\sqrt{12.25} = -3.5$$

Square Roots and Geometric Squares

area of square = 2
side of square = $\sqrt{2}$

The name square root comes from the geometry of squares. *If a square has area A, then its side has length $\sqrt{A}$.* In Question 23b of Lesson 12-1, you were asked to find the area of the square with vertices (1, 0), (0, 1), (-1, 0), (0, -1). One way to answer the question is to split the square into the four triangles shown at the left. Each triangle has area $\frac{1}{2}$ so the square has area 2. Therefore, each side of the square has length $\sqrt{2}$. The decimal for $\sqrt{2}$ is infinite and does not repeat. This means that although you can never write all the digits in the decimal for $\sqrt{2}$, you can draw a segment whose length is $\sqrt{2}$.

Estimating Square Roots

Activity 2

Determine what your calculator displays for $\sqrt{2}$.

Our calculator displays $\boxed{1.4142136}$, an estimate of $\sqrt{2}$. To check that 1.4142136 is a *good* estimate for $\sqrt{2}$, multiply this decimal by itself. Our calculator shows that

$$1.4142136 \cdot 1.4142136 \approx 2.0000001.$$

The negative square root of 2 is about -1.4142136.

Square roots of positive integers are either integers or infinite non-repeating decimals. If a square root is an integer, you should be able to find it without a calculator. If a square root is not an integer, then you should be able to estimate it without a calculator.

Square roots?

Example

Between what two whole numbers is $\sqrt{40}$?

Solution

Write down the squares of the whole numbers beginning with 1. Stop when the square is greater than 40.

$$1 \cdot 1 = 1$$
$$2 \cdot 2 = 4$$
$$3 \cdot 3 = 9$$
$$4 \cdot 4 = 16$$
$$5 \cdot 5 = 25$$
$$6 \cdot 6 = 36$$
$$7 \cdot 7 = 49$$

This indicates that $\sqrt{36} = 6$ and $\sqrt{49} = 7$. Because larger positive numbers have larger square roots, $\sqrt{40}$ must be between $\sqrt{36}$ and $\sqrt{49}$. So $\sqrt{40}$ is between 6 and 7.

The idea of square root was known to the ancient Egyptians thousands of years ago. The $\sqrt{}$ sign was invented in 1525 by the German mathematician Christoff Rudolff. The bar of the radical sign is a grouping symbol. Like parentheses you must work under the bar before doing anything else. For example, to simplify $\sqrt{36 + 49}$, you must add 36 and 49 first. Then find the square root.

$$\sqrt{36 + 49} = \sqrt{85} \approx 9.21954$$

QUESTIONS

Covering the Reading

1. Since $100 = 10 \cdot 10$, we call 10 a __?__ of 100.

2. When $A = s^2$, A is called the __?__ of s and s is called a __?__ of A.

3. $6.25 = 2.5 \cdot 2.5$. Which number is a square root of the other?

4. The two square roots of 9 are __?__ and __?__.

In 5–7, calculators are not allowed. Give the two square roots of each number.

5. 81   6. 4   7. 25

8. The $\sqrt{}$ sign is called the __?__ sign.

In 9–11, calculators are not allowed. Simplify.

9. $\sqrt{64}$ 10. $\pm\sqrt{1}$ 11. $-\sqrt{49}$

12. Find the positive square root of 54.76 using your calculator.

13. **a.** What does your calculator display for $\sqrt{2}$?
    b. Approximate $\sqrt{2}$ to the nearest thousandth.

14. *Multiple choice.* The decimal for $\sqrt{2}$ is
    (a) finite.
    (b) infinite and repeating.
    (c) infinite and nonrepeating.

15. **a.** Use your calculator to approximate $\sqrt{300}$.
    b. How can you check that your approximation is correct?
    c. Round the approximation to the nearest hundredth.

16. Suppose the area of a square is 400 square meters.
    a. Give the length of a side of the square.
    b. What is the positive square root of 400?
    c. Simplify: $-\sqrt{400}$.

Area
400 m²

17. A square has an area of 8 square units. To the nearest tenth, what is the length of a side of the square?

In 18–23, simplify.

18. $\sqrt{25} + \sqrt{16}$ 19. $\sqrt{25} - \sqrt{16}$ 20. $\sqrt{25 + 16}$

21. $\sqrt{25} \cdot \sqrt{25}$ 22. $\sqrt{25 - 16}$ 23. $\sqrt{5^2 + 4^2}$

24. $\sqrt{50}$ is between what two whole numbers?

Applying the Mathematics

25. A side of a square has length $\sqrt{10}$. What is the area of the square?

26. Which is larger, $\sqrt{2}$ or $\frac{239}{169}$?

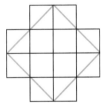

27. The length of each side of a little square at the left is 1 unit.
    a. What is the area of the large tilted square?
    b. What is the length of a side of this square?

In 28–31, use the table of numbers and squares at the left. Do not use a calculator. According to the table:

Number	Square
16.0	256.00
16.1	259.21
16.2	262.44
16.3	265.69
16.4	268.96
16.5	272.25
16.6	275.56
16.7	278.89
16.8	282.24
16.9	285.61
17.0	289.00

28. What is a square root of 268.96?

29. $\sqrt{285.6}$ is about __?__.

30. $\sqrt{270}$ is between __?__ and __?__.

31. $\sqrt{250}$ is less than __?__.

In 32 and 33, find a simple fraction equal to the given decimal.
(Lesson 12-1)

32. 4.26

33. 4.$\overline{26}$

34. Find the area of triangle *ABC*. *(Lesson 10-9)*

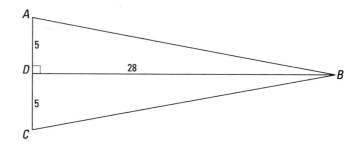

35. An article discussing some research in the learning of mathematics
had references to other research articles published in the following
years:
1982, 1987, 1987, 1989, 1982, 1990, 1992, 1989, 1974, 1987, 1988,
1981, 1990, 1969, 1989, 1992, 1988, 1987, 1989, 1991, 1989, 1989,
1981, 1989.
a. Plot the years in a stem-and-leaf display.
b. Give the median year for these references.
c. Give the mode year for these references.
d. In what year do you think the article was published?
(Lessons 6-1, 8-1)

36. *Multiple choice.* An item costs *C* dollars. The price is reduced by *R*
dollars. You could buy the item at the reduced price and pay *T* dollars
tax. How many dollars must you pay for the item? *(Lessons 5-1, 7-2)*
(a) $R - C + T$ (b) $C - R + T$
(c) $C - R - T$ (d) $R - C - T$

Exploration

37. a. Using your calculator, write down the square roots of the integers
from 1 to 15 to the nearest thousandth.
b. Verify that the product of $\sqrt{2}$ and $\sqrt{3}$ seems equal to $\sqrt{6}$.
c. The product of $\sqrt{3}$ and $\sqrt{5}$ seems equal to what square root?
d. Find three other instances like those in parts **b** and **c.**
e. Write the general pattern using variables.

*The
Pythagorean
Theorem*

IN-CLASS

ACTIVITY

Materials: Ruler, scissors, paste or glue

Work with a partner.

This activity shows a way of finding the length of the hypotenuse of a right triangle without measuring it.

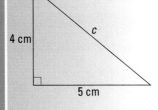

1 Cut out a right triangle with legs of length 4 cm and 5 cm. We want to find c, the length of the hypotenuse.

2 Make 3 copies of the triangle and paste the four triangles on a piece of paper as shown here. This forms a large square and a tilted square in the middle with side c.

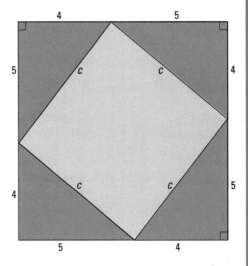

3 Do the following calculations:

a. What is the length of a side of the big square? What is its area?

b. What is the area of each corner right triangle? What is the total area of the four triangles?

c. What is the remaining area, the area of the tilted square?

d. What is the length of a side of the tilted square?

4 Measure a side of the tilted square (in centimeters). How close is your measure to what you calculated in Step 3?

Pythagorean action. *The carpenters are building triangular trusses, structural supports for a roof. They can use the Pythagorean Theorem to determine the lengths of the sides of the truss.*

In the In-class Activity, you began with a right triangle with legs 4 and 5. You should have found that the hypotenuse of the triangle has exact length $\sqrt{41}$. The numbers 4, 5, and $\sqrt{41}$ are related in a simple but not obvious way.

$$4^2 + 5^2 = (\sqrt{41})^2$$

That is, add the squares of the two legs. The sum is the square of the hypotenuse. *The Pythagorean Theorem* is the name commonly given to the general pattern. It is simple, surprising, and elegant.

Pythagorean Theorem
Let the legs of a right triangle have lengths a and b.
Let the hypotenuse have length c. Then
$$a^2 + b^2 = c^2.$$

A **theorem** is a statement that follows logically from other statements known or assumed to be true. The Pythagorean Theorem is the most famous theorem in all geometry. With the theorem, you do not have to make copies of the triangle to find the length of a third side. This theorem was known to some Chinese, Egyptians, and Babylonians before the Greeks. But it gets its name from the Greek mathematician Pythagoras who was born about 572 B.C. He and his followers may have been the first people in the Western world to prove that this theorem is true for any right triangle.

Example 1

What is the length of the hypotenuse of
the right triangle drawn at the right?

Solution

Use the formula of the Pythagorean Theorem.

So
$$11^2 + 15^2 = h^2$$
$$121 + 225 = h^2$$
$$346 = h^2$$
$$h = \sqrt{346} \text{ or } -\sqrt{346}.$$

But h must be positive, so
$$h = \sqrt{346}.$$

A calculator shows that $\sqrt{346} \approx 18.6$.

The length of the hypotenuse is exactly $\sqrt{346}$, or about 18.6.

Activity

Draw a right triangle with legs 11 and 15 units. Then measure the
hypotenuse. What is the difference between your measure and $\sqrt{346}$?

Applying the Pythagorean Theorem

Example 2 shows how to use the Pythagorean Theorem to find lengths
that are difficult or impossible to measure directly.

Example 2

The bottom of a 12-foot ladder is 3 feet from a wall. How high up does
the top of the ladder touch the wall?

Solution

First draw a picture. A possible picture is shown at the left. According to
the Pythagorean Theorem,

$$3^2 + x^2 = 12^2.$$

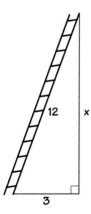

Now solve the equation.
First simplify.
$$9 + x^2 = 144$$
Add −9 to both sides.
$$x^2 = 135$$
Now, use the definition of square root.
$$x = \sqrt{135} \text{ or } -\sqrt{135}$$

Because x is a length, it cannot be negative. So
$$x = \sqrt{135}.$$

A calculator shows that $\sqrt{135} \approx 11.6$. This seems correct because the
ladder must be less than 12 feet up on the wall.

The top of the ladder touches the wall about 11.6 feet up.

Caution: The Pythagorean Theorem works only for right triangles. It
does not work for any other triangles.

Covering the Reading

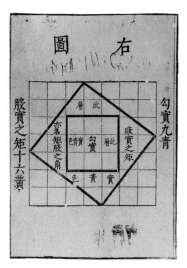

An arithmetic classic.
This Chinese block print shows a proof of the Pythagorean Theorem. The proof is from one of the oldest known sources of Chinese mathematics, the Chou Pei Suan Ching, *written about 500 B.C.*

1. What is a theorem?

2. *Multiple choice.*   Pythagoras was an ancient
   (a) Chinese.
   (b) Greek.
   (c) Egyptian.
   (d) Babylonian.

3. State the Pythagorean Theorem.

In 4 and 5, use the drawing below.

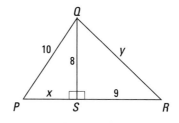

4. a. What relationship exists between 8, 9, and *y*?
   b. Find *y*.

5. a. What relationship exists between *x*, 8, and 10?
   b. Find *x*.

6. Lee had to find *LM* in triangle *LMN*, but she forgot the Pythagorean Theorem. Take Lee through these steps to find *LM*.

   a. Trace △*LMN*. Then draw three copies of the triangle to form a large square and a tilted square in the middle.
   b. Find the area of the large square.
   c. Find the area of △*LMN*.
   d. Find the area of the tilted square in the middle.
   e. Find *LM*.

7. Answer the question of the Activity in this lesson.

8. The bottom of a 3-meter ladder is 1 meter away from a wall. To the nearest cm, how high up on the wall will the ladder reach?

9. A right triangle has legs with lengths 6″ and 7″. Find the length of its hypotenuse, to the nearest tenth of an inch.

10. *Multiple choice.* For which of the following triangles can the Pythagorean Theorem be used?

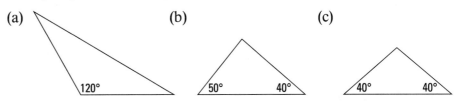

(a)   (b)   (c)

11. A rectangular farm is pictured below. It is 9 km long and 2 km wide. Danny wants to go from point *A* to point *B*.

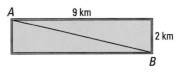

a. What is the shortest distance Danny can travel going along the roads that surround the farm?
b. What is the distance by tractor along the diagonal of the farm?
c. Which path is shorter? How much shorter is it?

12. The Chicago White Sox built a new stadium that opened in 1991. In the old ballpark, someone seated in the last row of the upper deck was about 150 feet away from home plate. Use the information about the new Comiskey Park in the diagram below. How much farther away from home plate is the last row of the new upper deck than it was in the old park?

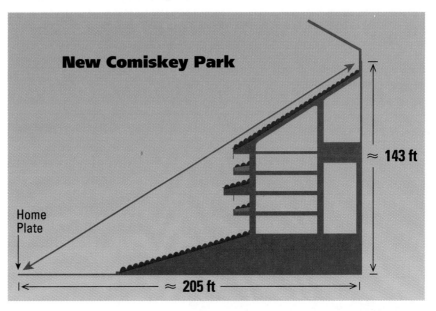

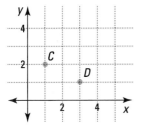

13. In the graph at the left, point *C* = (1, 2). *D* = (3, 1). What is the distance between these points? (Hint: Use a right triangle with hypotenuse $\overline{CD}$.)

14. Typing paper is often an 8.5″ by 11″ rectangle. Find the length of a diagonal of the rectangle using the given process.
a. measuring with a ruler
b. applying the Pythagorean Theorem

15. *Multiple choice.* There was no year 0. What year is 2500 years after the estimated year of Pythagoras's birth?
(a) 1928   (b) 1929   (c) 1972   (d) 2072

Review

16. Simplify $\sqrt{225 - 144}$. *(Lesson 12-2)*

17. Simplify $\sqrt{2^2 + 3^2}$. *(Lesson 12-2)*

18. *True or false.* $\sqrt{9} + \sqrt{16} = \sqrt{25}$. *(Lesson 12-2)*

19. a. The square of what number is 0?
b. Simplify $\sqrt{0}$. *(Lesson 12-2)*

20. In the first 40 games of the baseball season, Homer had 31 hits. At this rate, how many hits will he have for the entire 125-game season? *(Lesson 11-6)*

21. At 35 miles per hour, how long will it take to go 10 miles? *(Lessons 9-5, 10-2)*

22. Solve $5000 = 50 + x - 25$. *(Lessons 5-8, 7-2)*

23. *Multiple choice.* $0 + (x + 1) = 0 + (1 + x)$ is an instance of what property? *(Lessons 5-2, 5-7)*
(a) Commutative Property of Addition
(b) Associative Property of Addition
(c) Additive Identity Property of Zero
(d) Addition Property of Equality

Donald Duck meets the Pythagoreans during one of their jam sessions, in a scene from Donald in Mathmagic Land.

24. To the nearest integer, what is 2.5^7? *(Lesson 2-2)*

Exploration

25. The followers of Pythagoras, called the Pythagoreans, worshiped numbers and loved music. They discovered relationships between the lengths of strings and the musical tones they give. Look in an encyclopedia to find out one of the musical relationships discovered by the Pythagoreans.

*The
Circumference
of a Circle*

IN·CLASS
A C T I V I T Y

Materials: Circular objects, tape measure
Work with two or three other people, if possible,

Gather at least four objects in which you can see circles, such as cans,
paper towel tubes, clocks, and bicycle wheels. You will also need a tape
measure, (preferably cloth, so it bends).

1 Each person in the group should measure the distance across (the
diameter) and the distance around (the circumference) each circle,
as shown here.

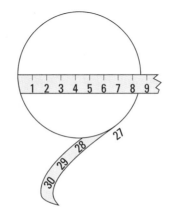

distance across ≈ 8.8 cm

distance around ≈ 27.6 cm

2 For each circle, each person in the group should calculate the ratio
$\frac{\text{distance around}}{\text{distance across}}$ rounded to two decimal places.

3 For each circle now find the average of the quotients found by the
members of your group. This will give you a ratio for each of
the circles.

4 As a group, decide which of the following is the correct choice.
As a circle gets larger, the ratio $\frac{\text{distance around}}{\text{distance across}}$
(a) becomes larger (b) stays about the same. (c) becomes smaller.

5 If the ratio becomes larger or smaller, how much larger or smaller
does it become? If the ratio stays the same, about what number
does it equal?

The Circumference of a Circle

Swing time. *Hoops like this one were first introduced in the 1950s for exercising or playing. What is the circumference of a hoop with a 4-ft diameter?*

The Idea Behind Circumference

Suppose you put a tape measure around your waist. Then you can measure the perimeter of your waist. You can do this even though your waist is curved.

The same idea works for circles. The perimeter of a circle can be measured. It is the distance around the circle, the distance you would travel if you walked around the circle.

The top of the aluminum can pictured below is a circle with diameter d. Think of the distance around this circle. (It is how far a can opener turns in opening the can.) This distance can be estimated by rolling the circle. We draw the circle with different shading so you can follow the circle as it rolls.

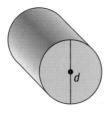

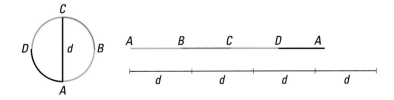

In the diagram above, the distance around the circle is compared to the diameter. The diagram shows that the distance around the circle is just a little more than 3 times the diameter. In fact, the unraveled length, called the **circumference** of the circle, is exactly 3.1415926535 . . . times the diameter. You have seen this number before. It is called π (pi), and its decimal never repeats or ends.

> **Formula for Circumference of a Circle**
> In a circle with diameter d and circumference C,
> $$C = \pi d.$$

The circumference of a circle is the circle's perimeter. Don't be fooled just because the word is long and different.

Example

To the nearest inch, what is the circumference of a can with a 4″ diameter?

4 in.

Solution

Let C be the circumference. Using the formula $C = \pi d$,

$C = \pi \cdot 4 = 4\pi$ inches exactly.

To approximate C to the nearest inch, use 3.14 for π.

$C \approx 3.14 \cdot 4$, or 12.56, inches, so C is about 13″.

Some Facts About π

When you divide both sides of the formula $C = \pi d$ by d, you get $\frac{C}{d} = \pi$. That is, the ratio of the circumference of a circle to its diameter is π. $\frac{C}{d}$ looks like a simple fraction. However, it was proved in 1767 that no whole numbers C and d can have a ratio equal to π. You should have found a number close to π for each ratio you calculated in the In-class Activity preceding this lesson.

The number π is so important that scientific calculators almost always have a key or keys you can press for π. Pressing that key will give you many decimal places of π. On one calculator, we press INV π and see 3.1415927. On another we press 2nd π ENTER and see 3.141592654. Without a calculator, people use various approximations. For rough estimates, you can use 3.14 or $\frac{22}{7}$. Use more decimal places if you want more accuracy.

Rational and Irrational Numbers

A number that *can* be written as a simple fraction is called a **rational number.** Finite decimals, repeating decimals, and mixed numbers all represent rational numbers. A number that *cannot* be written as a simple fraction is called an **irrational number.** Since π cannot be written as a simple fraction, it is irrational. Irrational numbers are those numbers whose decimals are infinite and do not repeat. When the square root of a positive integer is not an integer, then it is irrational. For instance, $\sqrt{2}$ and $\sqrt{3}$ are irrational. But $\sqrt{4}$ is rational, because $\sqrt{4} = 2$.

Irrational numbers are very important in many measurement formulas. Square roots often appear as lengths of sides of right triangles. As you will see in the next few lessons, many formulas involve π. As you study more mathematics, you will learn about other irrational numbers.

Every real number is either rational or irrational. The chart shows how some of the various kinds of numbers are related.

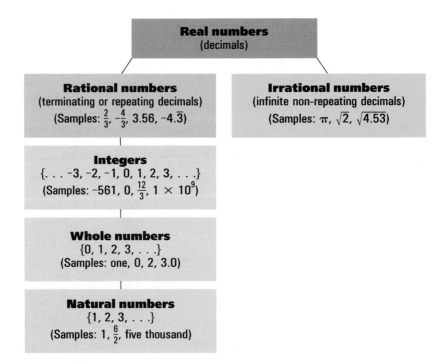

QUESTIONS

Covering the Reading

1. What is a *circle?*

2. The distance across a circle through its center is called the __?__ of the circle.

3. The perimeter of a circle is called the __?__ of the circle.

4. A circle has diameter 10 cm. To the nearest cm, what is its circumference?

5. The circumference of a circle with diameter s is __?__.

6. A simple fraction that is used as an approximation to π is __?__.

7. A calculator shows π to be 3.141592654.
   a. Is this an exact value or an estimate?
   b. Round this number to the nearest hundredth.
   c. Round this number to the nearest hundred thousandth.
   d. What does your calculator show for π?

8. The equator of the earth is approximately a circle whose diameter is about 7920 miles. What is the distance around the earth at its equator?

9. Solve the equation $C = \pi d$ for d.

10. What is a rational number?

11. What is an irrational number?

In 12–17, a real number is given.
a. Tell whether the number is rational or irrational.
b. If it is rational, give a fraction in lowest terms equal to it.

12. $\frac{2}{3}$

13. π

14. $6.\overline{87}$

15. $\sqrt{5}$

16. $5\frac{1}{2}$

17. 0.0004

18. What is a real number?

19. Which of the numbers of Questions 12–17 are real numbers?

20. Are there any numbers that are not real numbers? (The answer is given on page 641.)

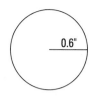

In 21–22, use the fact that the diameter of a circle is twice its radius.

21. The circle at the left has radius 0.6″. What is its circumference?

22. A circle has radius *r*. What is its exact circumference? (You need to remember this formula.)

23. The circle below is split into 8 congruent sectors. If the radius is 100, what is the length of the arc of one of the sectors, to the nearest tenth?

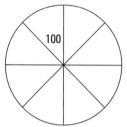

24. In the seventeenth century, the Dutch built a circular fort at the southernmost point of New Amsterdam, what is now Manhattan Island in New York City. The fort was also used to protect the city during the War of 1812. Today it is a national monument known as Castle Garden. Castle Garden is about 236 feet across at its widest. What is the length of its outside wall?

The Castle Garden national monument is the circular building on the left in this detail of the 1855 painting, Immigrants Debarking at the Battery, *by Samuel B. Waugh.*

25. A plastic tube 50″ long is bent to make a circular hoop. To the nearest inch, what is the diameter of the hoop?

26. Some bicycle wheels are 24″ in diameter. How far will the bike go in 10 revolutions of the wheels?

27. The number $\frac{355}{113}$ was used by the Chinese as an estimate for π.
   a. Write $\frac{355}{113}$ as a mixed number.
   b. Which is a better approximation to π, $\frac{355}{113}$ or $\frac{22}{7}$?

Inventor, too! *Although a great mathematician and physicist, Archimedes was probably best known during his lifetime as an inventor. His inventions included systems of levers and pulleys for moving great weights. This portrait of Archimedes is by a 16th-century artist, André Thevet.*

28. The Greek mathematician Archimedes knew that π was between $3\frac{1}{7}$ and $3\frac{10}{71}$. (Archimedes lived from about 287 B.C. to 212 B.C. and was one of the greatest mathematicians of all time.)
  a. Calculate the decimal equivalents of these mixed numbers.
  b. Why didn't Archimedes have a decimal approximation for π?

29. Which is larger, π or $\sqrt{10}$?

30. Approximate $\pi + 1$ to the nearest tenth.

Review

31. Simplify $\sqrt{841} - \sqrt{400}$.   *(Lesson 12-2)*

32. Evaluate $5 \cdot \sqrt{3}$ to the nearest integer.   *(Lesson 12-2)*

33. 6 is what percent of 30?   *(Lesson 11-5)*

34. 6% of what number is 30?   *(Lesson 10-2)*

35. Olivia biked m miles at 20 miles per hour. How many hours did it take her to do this?   *(Lessons 9-5, 10-2)*

36. Evaluate $|2 - n|$ when $n = 5$.   *(Lessons 5-3, 7-2)*

37. Evaluate $9x^3 - 7x^2 + 12$ when $x = 6$.   *(Lesson 4-4)*

Exploration

38. a. What does your calculator show for π?
  b. Subtract the whole-number part of the decimal. (The first time this is 3.) Then multiply the difference by 10. Your calculator may show another digit of π.
  c. Repeat part **b** several times if possible. Describe what happens.

39. a. Run this program, inputting a large number for the blank.

```
10 SUM = 0
20 FOR N = 1 TO ____
30 TERM = 1/(N*N)
40 SUM = SUM + TERM
50 NEARPI = SQR(6*SUM)
60 PRINT N, NEARPI
70 NEXT N
80 END
```

    Write down the last line the computer prints.
  b. What does the program do?
  c. Try a larger number in the blank. What happens?

LESSON

12-5

The Area of a Circle

The circular advantage. *Each rotating arm of length r units in this irrigation system waters πr² square units of land. If the circles do not overlap, then some regions get no water from the system.*

Finding the Area of a Circle

Recall that the formula for the area of a triangle, $A = \frac{1}{2} bh$, was found by showing that two copies of a triangular region exactly fill up a rectangle.

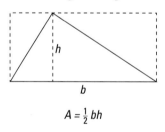

$$A = \tfrac{1}{2} bh$$

This idea does not work for finding the area of a circle. Sectors of a circle cannot be cut and pasted to fill up any rectangle exactly. However, the area of a circle can be estimated by splitting it up into small sectors, approximating each sector with a triangle, and adding up all the areas of the triangles.

Activity

The circle at the top of page 665 has radius 5 cm. It has been split into 16 congruent sectors. Measure the height *h* and base *b* of the triangle drawn. Use these measurements to estimate the area of one of the sectors. Then multiply by 16 to estimate the area inside the circle.

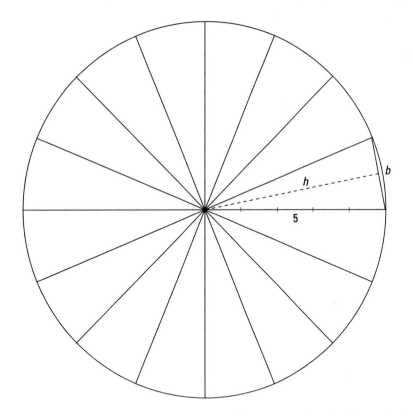

The activity does not indicate how π gets involved with the formula for the area of a circle. This is harder to do, but try to follow the argument.

We take the 16 sectors from the Activity and rearrange them.

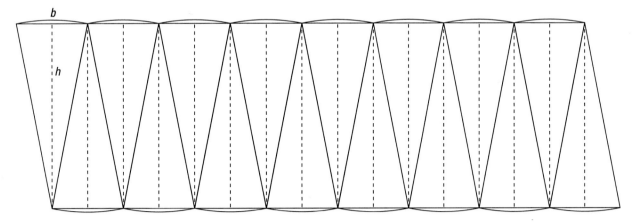

Now the figure resembles a parallelogram with height h. The top and bottom are not straight since they are made of arcs. The total length of the top and bottom arcs is the circumference of the circle, πd. So the top has length half that, or $\frac{1}{2}\pi d$, which equals πr.

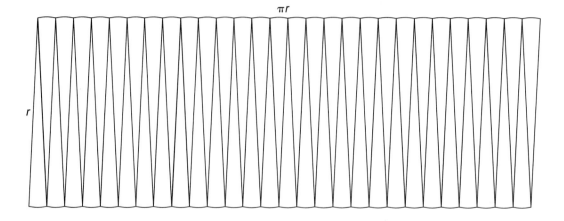

πr

r

If the circle is split into more and more triangles, the height of each triangle becomes closer and closer to the circle's radius r. The "parallelogram" becomes more and more like a rectangle with width πr. And the area of that rectangle is $\pi r \cdot r$, or πr^2. This is why the area of the circle is πr^2.

Area Formula for a Circle
Let A be the area of a circle with radius r. Then
$$A = \pi r^2.$$

Example 1

Find the exact area of a circle with a radius of 5 cm.

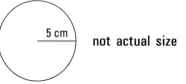

5 cm   not actual size

Solution
$$A = \pi \cdot (5 \text{ cm})^2 = 25\pi \text{ cm}^2$$

This is the exact answer. To estimate this to the nearest square centimeter, use an approximation to π.
$$\approx 25 \cdot 3.1416 \text{ cm}^2$$
$$\approx 78.54 \text{ cm}^2$$
$$\approx 79 \text{ cm}^2$$

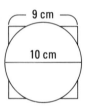

9 cm

10 cm

Check

Notice that a square with side 9 cm has about the same area as the circle. The area of the square is 81 cm^2. So an area of about 79 cm^2 for the circle seems about right.

Area of a Sector of a Circle

The area of a sector of a circle depends on the size of the central angle.

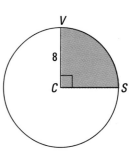

Example 2

In sector *VCS*, ∠*VCS* is a right angle, and *VC* = 8. Find the area of the sector.

Solution

It may help you to draw the entire circle, as shown at the right.

A right angle has measure 90°. This is $\frac{90}{360}$ or $\frac{1}{4}$ of a revolution.

So the sector has $\frac{1}{4}$ the circle area. $A = \frac{1}{4} \cdot \pi \cdot 8^2 = \frac{1}{4} \cdot 64 \cdot \pi = 16\pi$.

Check

Notice that $16\pi \approx 16 \cdot 3.14 \approx 50$. A square with sides $\overline{VC}$ and $\overline{CS}$ has area 64. This is larger than the area of the sector, as it should be.

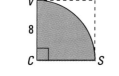

The area of any sector can be found if the radius and the measure of the central angle is known.

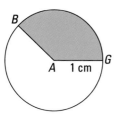

Example 3

Consider the figure at the left, with *AG* = 1 cm and m∠*BAG* = 145°. Find the area of sector *BAG*.

Solution

145 degrees is $\frac{145}{360}$ of a revolution. The area of circle A is $\pi \cdot 1^2 = \pi$ cm^2.

So the area of sector BAG is $\frac{145}{360} \cdot \pi$ cm^2, or approximately 1.27 cm^2.

The process of finding the area of a sector of a circle is generalized in the next formula.

Area Formula for a Sector
Let *A* be the area of a sector of a circle with radius *r*. Let *m* be the measure of the central angle of the sector in degrees. Then $A = \frac{m}{360} \cdot \pi r^2$.

QUESTIONS

Covering the Reading

1. Refer to the Activity on page 664. What value did you get for each quantity?
   a. h
   b. b
   c. the area of one of the sectors
   d. the area of the circle

2. The area of a circle with radius r is equal to the area of a rectangle with dimensions r and __?__.

3. What is the area of a circle with radius r?

4. To the nearest integer, what is the area of a circle with radius 4?

5. What is the area of one half of a circle with radius 6?

In 6 and 7, use the figure at the left. G is the center of the circle.

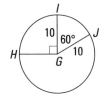

6. What is the area of region GHI?

7. What is the area of region GIJ?

Applying the Mathematics

8. A quarter has radius 12.15 mm. What is the area of one of its faces?

9. What calculator key sequence will give you an estimate to the area of a circle with radius 50?

10. A central angle of a sector has measure 50°. If the area of the circle is 180 square meters, what is the area of the sector?

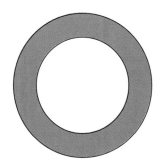

11. The smaller circle at the left has radius 0.5 inch, and the larger circle has radius 0.75 inch.
    a. What is the area of the ring?
    b. Which has the greater area, the smaller circle or the shaded ring?

12. Clyde decided to make a large spinner for a game. He wanted equal-sized sectors for each of three choices. Below is his plan.

    a. What should be the measure of the central angle of each sector?
    b. If the diameter of the circle is to be 11 in., what will be the area of each sector of the spinner?

13. Lisa read that there were about 10 calories in each square inch of a 12-inch diameter cheese pizza. If she eats $\frac{1}{8}$ of a pizza, about how many calories would she consume?

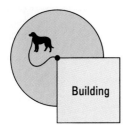

14. A dog is on a leash tied to the corner of a building. This enables the dog to roam anywhere in a sector of the circle, as shown at the left. If the leash is 9 meters long, what is the area of land on which the dog may roam?

15. Which gives you more for your money, one 14″ pizza, or three 7″ pizzas of the same kind?

Review

16. To the nearest tenth of an inch, find the circumferences of the two circles in Question 11. *(Lesson 12-4)*

17. When does a decimal represent an irrational number? *(Lesson 12-4)*

18. The hypotenuse of a right triangle has length 50. One leg has length 14. What is the length of the other leg? *(Lesson 12-3)*

19. Order from smallest to largest: $\sqrt{5}, 3\sqrt{12}, 2\sqrt{3}, \sqrt{6}$. *(Lesson 12-2)*

20. Find the simple fraction in lowest terms equal to $.0\overline{36}$. *(Lesson 12-1)*

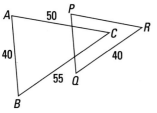

21. The triangles at the left are similar and tilted the same way. Find *PR*. *(Lesson 11-8)*

22. In $\triangle TON$ below, $NW = 80$, $TO = 60$, and $\overline{TO} \perp \overline{NT}$.
  a. Is enough information given to find the area of $\triangle NOW$?
  b. If so, find its area. If not, explain why not. *(Lesson 10-9)*

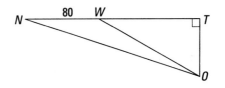

23. Multiply: $5(x - y + 12)$. *(Lesson 10-6)*

24. A certain variety of lichen grows about 0.01 inch in a year. At this rate, how many inches would it grow in a century? *(Lesson 9-5)*

Exploration

25. Take two pieces of string of equal length. Form the first string into a circle. Form the second string into a square.
  a. Which figure has the greater area? Explain why you think as you do.
  b. Does the answer to part **a** depend on the length of the string?

26. Copy the circle in the Activity on pages 664 and 665. Cut out the sectors and rearrange them to form a figure like that on page 665. Estimate the area of that figure in square centimeters. How close is your estimate to the area of the circle?

LESSON

12-6

*Surface
Areas of
Cylinders
and Prisms*

On a roll. *Large rotary printing presses use cylinders in the printing process. The greater the surface area of the cylinder, the larger each impression or printed sheet can be.*

Surface Areas of Cylinders

Many cans of food are shaped like *circular cylinders*. The top and bottom are the **bases** of the cylinder. Some cans have paper labels that can be peeled off. If you peel off the label, you will see that it is a rectangular piece of paper. The height of the rectangle is the height h of the can. If you ignore any overlap, the other dimension of the rectangle is the circumference of a base.

From this information, you can find the area of the side of the can, known as its **lateral area (*L.A.*).**

$$L.A. = hC,$$

where C is the circumference of a base. But since $C = \pi d$, another formula is

$$L.A. = \pi dh.$$

And since $d = 2r$, still another formula (and the most common) is

$$L.A. = 2\pi rh.$$

In these formulas d and r are the diameter and radius of a base, respectively.

If you add the areas of the top and bottom of the can, you have calculated the **total surface area (*S.A.*)** of the cylinder.

Example 1

A typical soft-drink can is almost a circular cylinder with height of about $4\frac{3}{4}$ inches and a diameter of about $2\frac{1}{2}$ inches. Find the total surface area of such a can.

Solution

First draw a picture, as at the left. For convenience we have changed the dimensions to decimals. The lateral area is the area of a rectangle with height 4.75 inches and width $\pi \cdot 2.5$ inches.

$$\begin{aligned} \text{L.A.} &= \pi dh \\ &\approx \pi(2.5)(4.75) \\ &\approx 37.3 \text{ square inches} \end{aligned}$$

The area of the circular top is $\pi(1.25)^2 \approx 4.9$ square inches. The bottom has the same area. So

$$\begin{aligned} \text{S.A.} &\approx 37.3 + 2 \cdot 4.9 \\ &= 47.1 \text{ square inches.} \end{aligned}$$

The total surface area of this typical soft-drink can is about 47.1 square inches.

To make a cylinder, you can reverse the process of unwrapping the paper off the side. If the bases are not included, then you will get something resembling the shape of a straw. Below we show a net for a circular cylinder including its bases. Notice that the length of the rectangle has to be the same as the circumference of the circular bases.

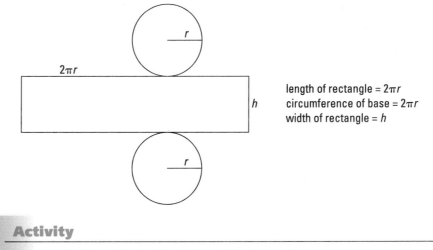

length of rectangle = $2\pi r$
circumference of base = $2\pi r$
width of rectangle = h

Activity

Make a circular cylinder, including at least one base.

Surface Area of Prisms

Instead of a circular base, you can choose a polygon for a base. Then you can form a figure known as a prism. In a net for a prism, again the length of the rectangle must equal the perimeter of the base. Then the net can be folded into the prism.

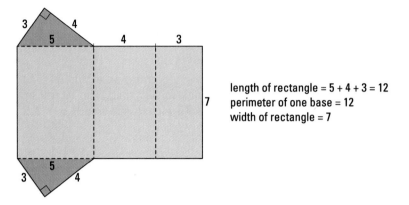

length of rectangle = 5 + 4 + 3 = 12
perimeter of one base = 12
width of rectangle = 7

Example 2

Find the total surface area of the triangular prism whose net is shown above.

Solution

The total surface area is the sum of the areas of the bases of the prism (the triangles) and the lateral area (the rectangle).

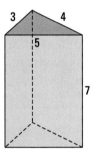

Here is how the prism of Example 2 looks when folded.

Each 3-4-5 right triangle base has area $\frac{1}{2} \cdot 3 \cdot 4$, or 6. The rectangles have area $7 \cdot 3 + 7 \cdot 4 + 7 \cdot 5 = 7(3 + 4 + 5) = 7 \cdot 12 = 84$. So the total surface area is 6 + 84 + 6, or 96 square units.

QUESTIONS

Covering the Reading

9.5 cm

5.4 cm

1. Small juice cans often have a height of 9.5 cm and a base with diameter 5.4 cm.
   a. What is the lateral area of such a can?
   b. What is the area of one of its bases?
   c. What is its total surface area?

2. Consider the circular cylinder you made in the Activity of this lesson.
   a. What is the height of the cylinder you have made?
   b. What is the radius of its base?
   c. What is its lateral area?

3. Draw a net for a circular cylinder with two bases that is twice as high as the cylinder you made in the Activity of this lesson.

4. Indicate why the net below will not form a circular cylinder with an open top.

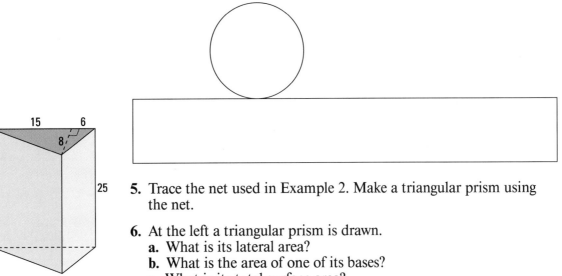

5. Trace the net used in Example 2. Make a triangular prism using the net.

6. At the left a triangular prism is drawn.
   a. What is its lateral area?
   b. What is the area of one of its bases?
   c. What is its total surface area?

7. Draw a net for the prism of Question 6.

Applying the Mathematics

8. Convert the dimensions of the cylinder of Question 1 to the nearest $\frac{1}{8}$ inch. Use these dimensions to find the surface area in square inches.

9. Many wood pencils are shaped like hexagonal prisms (prisms whose bases are hexagons). New pencils often have sides of length 4 mm and height (without eraser) of 170 mm. How much surface of the pencil is there to paint?

10. Draw a net for a pentagonal prism with a height of 4 cm. (Your pentagon may have any dimensions you wish.)

11. *Multiple choice.* Suppose each dimension of the can of Example 1 were doubled. What would happen to the total surface area?
   (a) It would be doubled.
   (b) It would be multiplied by 4.
   (c) It would be multiplied by 8.
   (d) It would be multiplied by 16.

Get a grip. *The hexagonal shape makes pencils easier to grip and less likely to roll on a flat surface.*

12. U.S. silver dollars minted from 1840 to 1978 have a diameter of 38.1 mm. U.S. quarters minted since 1831 have a diameter of 24.3 mm.
   a. Explain why a silver dollar covers more than twice the area of a quarter.
   b. Explain why a silver dollar does not have twice the circumference of a quarter. *(Lessons 12-4, 12-5)*

13. Suppose the top of a card table is a square 30″ on a side.
   a. What is the area of the tabletop?
   b. What is the perimeter of the tabletop?
   c. What is the length of a diagonal of the tabletop?
   (Lessons 3-8, 5-10, 10-8, 12-3)

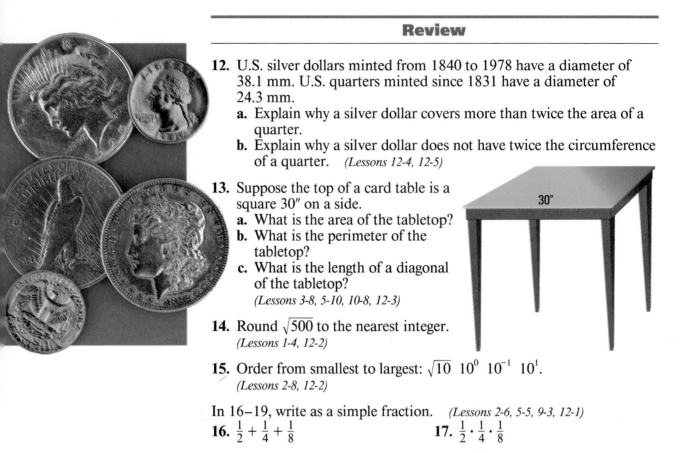

30″

14. Round $\sqrt{500}$ to the nearest integer.
   (Lessons 1-4, 12-2)

15. Order from smallest to largest: $\sqrt{10}$ 10^0 10^{-1} 10^1.
   (Lessons 2-8, 12-2)

In 16–19, write as a simple fraction. *(Lessons 2-6, 5-5, 9-3, 12-1)*

16. $\frac{1}{2} + \frac{1}{4} + \frac{1}{8}$

17. $\frac{1}{2} \cdot \frac{1}{4} \cdot \frac{1}{8}$

18. 2.48

19. $2.4\overline{8}$

20. Suppose the area of a triangle is 100 square inches, and its height is 10 inches. What is the length of the base for that height?
   (Lessons 10-2, 10-9)

21. a. Locate a cylinder or prism with dimensions different from any in this lesson. Describe what you find.
   b. Calculate its total surface area, including bases only if they are part of the object.

Volumes of Cylinders and Prisms

Silo trio. *Tower silos are often the preferred structure for storing grain crops. With a diameter of 30 ft and a height of 65 ft, one large silo can hold nearly 46,000 cubic feet, or 37,000 bushels, of grain.*

The surface area of a cylinder or prism-shaped container tells how much material is needed to cover the container, but it does not tell how much the container holds. For this the container's *volume* is needed.

In Lesson 9-2, two formulas were given for the volume of a box.

$V = abc,$ where a, b, and c are the dimensions of the box.

$V = Bh,$ where B is the area of a base and h the height to that base.

We included the second formula because a box is a type of prism, and the second formula applies to all prisms. Furthermore, all prisms and cylinders belong to the family of *cylindric surfaces,* and the formula $V = Bh$ applies to all members of that family.

Cylindric Solids

Any **cylindric solid** can be formed in the following way. Begin with a two-dimensional region F. Now translate that figure out of its plane into 3-dimensional space. Call the image F'. Connect all points of F to the corresponding points on F'. The result is the cylindric solid. The surface of a cylindric solid is called a **cylindric surface.** F and F' are the **bases.**

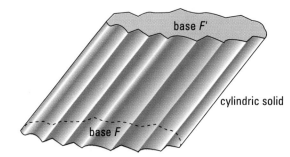

There are many kinds of cylindric solids, but in this lesson we are interested only in the simplest ones. For our solids, the translation will be perpendicular to the plane of the base F. Our base F or F' will either be a polygon or a circle. When F is a polygon, a **prism** is formed. Some bricks and some unsharpened pencils are prisms. When F is a circle, the solid is called a **circular cylinder,** or simply a **cylinder.** Aluminum juice cans are quite close to being cylinders.

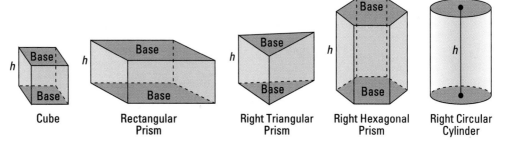

Cube   Rectangular Prism   Right Triangular Prism   Right Hexagonal Prism   Right Circular Cylinder

The distance between the bases is the **height** of the cylinder or prism. To see why the volume formula $V = Bh$ works, think of the prism or cylinder as made up of h layers of height 1 unit. If the base has area B, then each layer has volume B cubic units.

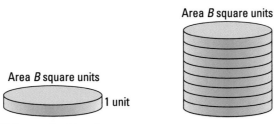

Area B square units

Area B square units

1 unit

height h units

Volume = B cubic units

Volume Formula for a Cylindric Solid
The volume of a cylindric solid with height h and base with area B is given by $V = Bh$.

Example 1

A 12-ounce aluminum juice can is about 12 centimeters high. It has a radius of about 3 centimeters. What is its volume?

Solution

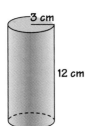

First it is always useful to draw a picture, as at the left. The base is a circle, so its area B is given by
$B = \pi r^2 = \pi \cdot 3^2 = 9\pi$ square centimeters
The volume V is given by
$V = Bh = 9\pi \cdot 12 = 108\pi$ cubic centimeters.
Now is the time to substitute for π.
$V = 108\pi \approx 339$ cubic centimeters.

Check

Information on 12-oz cans state that they contain 355 mL. Since 1 mL = 1 cm^3, the answer we found is quite close.

Example 2

Pictured here is the outline of a skyscraper that is a prism 356 ft high with a reflection-symmetric pentagon as a base. What is its volume?

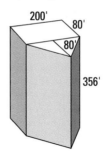

Solution

The pentagonal base, redrawn here, can be split into a triangle and a rectangle.

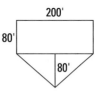

The area of the rectangle is 200 ft · 80 ft = 16,000 ft².

The area of the triangle is $\frac{1}{2}$ · 200 ft · 80 ft = 8,000 ft².

So the area of the base is 24,000 ft². This is B in the formula V = Bh.

$$V = Bh = 356 \text{ ft} \cdot 24,000 \text{ ft}^2 = 8,544,000 \text{ ft}^3$$

The volume of the skyscraper is about 8,544,000 cubic feet.

QUESTIONS

Covering the Reading

In 1 and 2, *multiple choice.*   Choose from these.
(a) volume   (b) surface area   (c) perimeter

1. how much a container holds

2. how much material it takes to cover a container

3. What is the difference between a cylindric surface and a cylindric solid?

4. A cylindric surface is a cylinder when its base is a __?__.

5. A cylindric surface is a prism when its base is a __?__.

6. Draw a prism whose base is a pentagon.

7. Draw a cylinder whose height equals the radius of its base.

8. How is the height of a cylinder or prism measured?

In 9 and 10, find the volume of the figure.

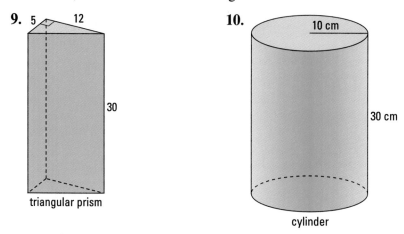

9. triangular prism

5, 12, 30

10. cylinder

10 cm, 30 cm

11. Find the volume of a cylindrical potato-chip can whose height is 10″ and with a base with diameter 3″.

12. Find the volume of a prism whose base is a hexagon with area 40 mm² and whose height is 30 mm.

Applying the Mathematics

13. Give an example of a figure with a large surface area for its volume.

14. Almost every adult human can easily fit in a space of 6 cubic feet. Could full-size replicas of the entire population of Jacksonville, Florida, about 673,000 people in 1990, fit inside the skyscraper of Example 2?

15. a. A book is a prism. Its base is a ___?___.
   b. What is the volume of a book whose base has dimensions 6″ and 9″ and whose height is $1\frac{1}{4}$″?

16. a. What is the capacity of a plastic straw that is 6 mm in diameter and 200 mm long?
   b. What is the surface area of this straw?
   c. If the plastic is 0.5 mm thick, how much plastic is needed to make the straw?
   d. Would you say the straw has a lot of surface area for its volume, or not much surface area for its volume? Explain your choice.

17. Here is a net for a cylinder with one base. Find its volume.

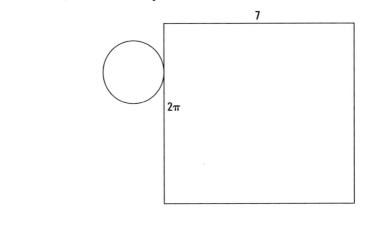

Review

18. Give the length of the hypotenuse of the base of Question 9. *(Lesson 12-3)*

19. Triangle *ABC* below is equilateral. Find each indicated measure. *(Lessons 8-6, 12-3)*

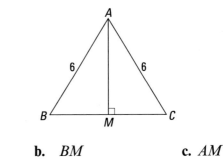

   a. *BC*   **b.** *BM*   **c.** *AM*

In 20–23, solve. *(Lessons 5-8, 7-3, 10-2, 10-5, 11-7)*

20. $-31 + x = 42$ **21.** $-31y = 42$

22. $\frac{-31}{z} = 42$ **23.** $-31w - 20 = 42$

24. If $a + a + a + a + a + a = ba$, what is the value of b?   *(Lesson 10-1)*

Exploration

25. Give dimensions for a cylindric solid whose volume is exactly 100 cubic meters and whose base is not a rectangle.

Spheres

A sporting chance. *Can you name the sport for each of these balls?*

Formulas for Spheres

A **sphere** is the set of points *in space* at a given distance (its **radius**) from a given point (its **center**). Drawn below is a sphere with radius *r*.

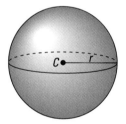

Planets and moons, baseballs, marbles, and many other objects are nearly the shape of spheres. A sphere is like a soap bubble. It doesn't include any points inside. Even the center *C* of a sphere is not a point on the sphere. A sphere together with all the points inside it is called a **ball.**

The formulas for the surface area and volume of a sphere were first discovered by the Greek mathematician Archimedes. You will learn how these formulas were found when you study geometry in more detail in a later course. Like formulas for the area and circumference of a circle, they involve the number π.

Surface Area and Volume Formulas for a Sphere
In a sphere with radius *r*, surface area *S*, and volume *V*,
$$S = 4\pi r^2$$
and $\quad V = \frac{4}{3}\pi r^3.$

The exponents in these formulas signal the units in which surface area and volume are measured. The surface area of a sphere is measured in square units, just as any other area is, even though the surface of a sphere cannot be flattened. Volume is measured in cubic units.

Example 1

Find the volume of a sphere with radius 6.

Solution 1

Find the exact value. Let V be the volume.

$$V = \frac{4}{3}\pi r^3 = \frac{4}{3}\pi \cdot 6^3 = \frac{4}{3}\pi \cdot 216 = 288\pi \text{ cubic units.}$$

288π is the exact value. To get an approximate value, substitute an estimate for π, say 3.14. Then $V \approx 288 \cdot 3.14 \approx 904$ cubic units.

Solution 2

Use a calculator. Here is a key sequence that works on many calculators.

$$4 \boxed{\div} 3 \boxed{\times} \boxed{\pi} \boxed{\times} 6 \boxed{y^x} 3 \boxed{=}$$

To the nearest hundredth, a calculator will give 904.78. This differs from the answer of Solution 1 because a different estimate is used for π.

The Earth is not exactly a sphere because its rotation has flattened it slightly at the poles. The length of the equator is about 24,902 miles, while its circumference through the poles is about 24,860 miles. Dividing these circumferences by π, we find diameters of about 7927 and 7913 miles. These lengths are double the radii, so the Earth is nearly a sphere with radius 3960 miles.

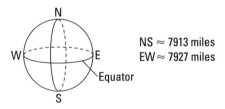

NS ≈ 7913 miles
EW ≈ 7927 miles

Shown is Jupiter, the largest planet in the solar system. It has a diameter of 142,984 km (about 88,846 mi) at its equator. A sphere the size of Jupiter could hold about 1,000 spheres the size of Earth.

Example 2

Estimate the surface area of the Earth.

Solution

The formula for surface area is $S = 4\pi r^2$. Substitute 3960 for r. Here is a key sequence that will work on many calculators.

$$4 \boxed{\times} \boxed{\pi} \boxed{\times} 3960 \boxed{x^2} \boxed{=}$$

(By using the squaring key, 3960 does not have to be entered twice.) The calculator should display a result close to 197,000,000 square miles.

The surface area of the Earth is about 197 million square miles.

QUESTIONS

Covering the Reading

1. What is a *sphere?*

2. Why is the Earth not exactly a sphere?

3. Give a formula for the surface area of a sphere.

4. Calculate the surface area of a sphere with radius 12 cm.

5. Give a formula for the volume of a sphere.

6. Calculate the volume of a sphere with radius 12 cm.

7. Who discovered the formulas for the surface area and volume of a sphere?

8. **a.** What calculator sequence will yield the volume of a sphere with radius 7?
   b. Find the volume of the sphere.

9. The moon is approximately a sphere with radius 1080 miles.
   a. Estimate the surface area of the moon to the nearest million square miles.
   b. The surface area of the Earth is how many times the surface area of the moon?

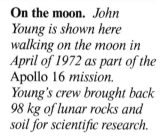

On the moon. *John Young is shown here walking on the moon in April of 1972 as part of the Apollo 16 mission. Young's crew brought back 98 kg of lunar rocks and soil for scientific research.*

Applying the Mathematics

In 10–13, tell whether the idea is more like surface area or volume.

10. how much land there is in the United States

11. how much material is in a bowling ball

12. how much material it takes to make a basketball

13. how much material is in a marble

In 14 and 15, use this information. A bowling ball is approximately 8.59 inches in diameter.

14. How much surface does a bowling ball have? (Ignore the finger holes.)

15. How much material does it take to make a bowling ball?

16. If a 12-cm diameter ball fits snugly into a box, what percent of the box is filled by the ball?

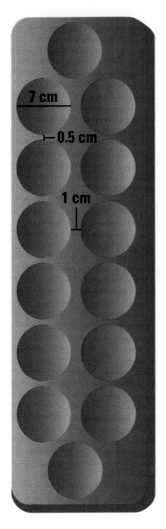

17. At the left is a drawing of a board for the game *mancala,* played throughout the world. The version popular in Ghana begins by having players place stones in the "circles" (actually halves of spheres) on their side of the board. The end circles are used primarily to store the stones you "capture" from your opponent. Traditionally, the game board is carved out of a solid piece of wood.
  a. Determine the minimum dimensions of the piece of wood you would need for the game board shown at the left.
  b. If you had two pieces of wood, one 4 cm × 16 cm × 80 cm, and the other 4 cm × 32 cm × 60 cm, which would you use to make the game board? Explain your choice.

Review

In 18 and 19, consider a tube for carrying a fishing rod that is a cylinder 8′ long and 6″ in diameter.

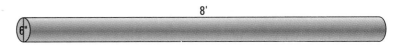

18. What is the volume of this tube?   *(Lesson 12-7)*

19. What is the total surface area of this container?   *(Lesson 12-6)*

20. From the top of a tall building, a person can see 25 kilometers away in any direction. How many square kilometers are then visible? *(Lesson 12-5)*

21. The repeating decimal $1.\overline{41}$ is within .0002 of $\sqrt{2}$. What simple fraction equals $1.\overline{41}$?   *(Lesson 12-1)*

22. The surface area of the Earth is given in this lesson. Only about 29.4% of the surface area is land.
  a. To the nearest million square miles, what is the total land area of the Earth?   *(Lesson 2-5)*
  b. The area of the United States is approximately 3,540,000 square miles. What percent of the total land area of the Earth is in the United States?   *(Lesson 11-5)*

23. According to Michael H. Hart in the book *The 100,* the 100 most influential persons in history include 37 scientists and inventors, 30 political and military leaders, 14 secular philosophers, 11 religious leaders, 6 artists and literary figures, and 2 explorers. Make a circle graph with this information.   *(Lesson 3-6)*

24. The mass of the Earth is about 6 sextillion, 588 quintillion short tons. Write this number as a decimal.   *(Lesson 2-2)*

Exploration

25. Refer to Question 22b. What three countries of the world have more land area than the United States? (Look in an almanac for this information, or look at a globe.)

A project presents an opportunity for you to extend your knowledge of a topic related to the material of this chapter. You should allow more time for a project than you do for typical homework questions.

1 Complex Numbers

In the 16th century, Italian mathematicians began to realize that it was possible to work with square roots of negative numbers. These numbers, which are not real numbers—they cannot be represented by a single decimal—were later called *imaginary numbers* and are part of the set of numbers known as *complex numbers*. Write an essay about these numbers and where they are used.

2 A 3-dimensional Pythagorean Pattern

The longest diagonal of a box stretches inside the box from one corner to the far opposite corner. If the dimensions of the box are *a, b,* and *c* and the diagonal has length *d,* then $d^2 = a^2 + b^2 + c^2$. For instance, if the box has dimensions 3, 4, and 12, then the length *d* of the longest diagonal satisfies $d^2 = 3^2 + 4^2 + 12^2 = 9 + 16 + 144 = 169$, so $d = 13$. Find boxes of other shapes in which all the dimensions and the longest diagonal have *integer* lengths.

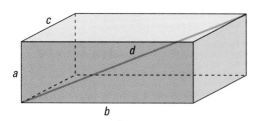

3 Cylindrical Packages

Some goods are packaged in cylindrical containers. Some of these cylinders are tall but not very wide. Others are wide but not very tall. Go to a food store. What is the largest ratio of height to diameter that you can find in a cylindrical package? What is the smallest ratio? Write down what you find in your exploration.

4 Approximations to π

The number π (pi) has a long history. People from many cultures, including the Chinese, Hindus, Egyptians, and Greeks, used approximations over 500 years ago. Find out what these approximations to π were. If possible, find out how at least one of the approximations was determined.

5 The Number of Digits in a Repetend

The number of digits in the repetend of a repeating decimal for a fraction is determined by the denominator of the fraction. If the denominator is not divisible by 2 or 5, then the length is determined by the first of the numbers 9, 99, 999, 9999, . . . into which it divides evenly. Use this information to find all the denominators of this type that have repetends with 1 digit, all with 2 digits, all with 3 digits, and all with 4 digits. Go further if you can.

6 Weights of Spheres

We come into contact with spheres of many different sizes. Tennis balls, basketballs, and ball bearings are just a few of the things with this shape. Find at least five different spheres and weigh them. Which has the greatest weight per unit volume? Which has the least weight per unit volume?

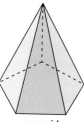

7 Cones and Pyramids

There are formulas for surface area and volume of cones and pyramids. Find out what these formulas are and how they are related to the formulas for cylinders and prisms. Make up some sample problems using these formulas.

cone   pyramid

SUMMARY

Numbers that can be represented as decimals are called real numbers. The real numbers are of two types: rational and irrational. The rational numbers are those that can be written as simple fractions. All finite and all infinite repeating decimals are rational numbers. Using algebra, it is possible to find a fraction for any infinite repeating decimal.

Irrational numbers have infinite decimals that do not repeat. The number π is irrational. If a square root of a positive integer is not an integer, the square root is irrational. Lengths, areas, and volumes may be rational or irrational.

Formulas for the areas and volumes of many common figures show the importance of irrational numbers. The following formulas are in this chapter. You should look back at the lessons for the meanings of the variables in them.

Pythagorean Theorem: $\quad a^2 + b^2 = c^2$

Circumference of a circle: $\quad C = \pi d$

Area of a circle: $\quad A = \pi r^2$

Lateral area of a cylinder: $\quad L.A. = hC$

Total surface area
of a cylinder: $\quad S.A. = 2\pi r^2 + 2\pi rh$

Area of a sector: $\quad A = \frac{m}{360} \pi r^2$

Volume of a cylindric solid: $\quad V = Bh$

Surface area of a sphere: $\quad S = 4\pi r^2$

Volume of a sphere: $\quad V = \frac{4}{3} \pi r^3$

VOCABULARY

You should be able to give a general description and a specific example for each of the following ideas.

Lesson 12-1
infinite repeating decimal

Lesson 12-2
square root
radical sign, $\sqrt{}$, $\pm$

Lesson 12-3
Pythagorean Theorem
theorem
Pythagoras

Lesson 12-4
radii
circumference
π
Formula for Circumference
of a Circle
rational number
irrational number
real number

Lesson 12-5
Area Formula for a Circle
Area formula for a Sector

Lesson 12-6
lateral area (*L.A.*)
total surface area (*S.A.*)

Lesson 12-7
cylindric surface
cylindric solid
base, height
prism
circular cylinder, cylinder
Volume Formula for a
Cylindric Solid

Lesson 12-8
sphere, ball
center, radius, diameter of
a sphere
Surface Area Formula for
a Sphere
Volume Formula for a Sphere

PROGRESS SELF-TEST

Take this test as you would take a test in class. Then check your work with the solutions in the Selected Answers section in the back of the book.

1. A circle has radius $\frac{3''}{8}$. Find its circumference and area to the nearest tenth.

2. What is the area of a square if one side has length $\sqrt{3}$?

3. If a square has area 10, what is the exact length of a side?

4. Estimate $\sqrt{33} + \sqrt{51}$ to the nearest integer.

5. Find x.

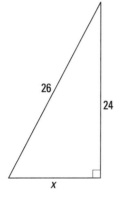

6. Find the length of $\overline{AC}$.

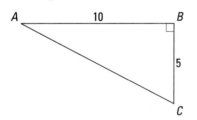

7. Which number is rational? $\sqrt{24}$ $\sqrt{25}$ $\sqrt{27}$.

In 8 and 9, Find a simple fraction in lowest terms equal to the given number.

8. -2.7   **9.** $0.\overline{15}$

10. An empty 12-oz aluminum can has a diameter of approximately 57 mm and a height of approximately 123 mm. To the nearest 10 square millimeters, about how much aluminum was used to make this can?

In 11 and 12, use the triangular prism shown below.

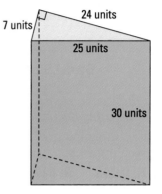

11. What is the lateral area of this prism?

12. What is the volume of this prism?

13. Name the two square roots of 16.

14. To the nearest integer, what is the circumference of a 120-mm diameter compact disc?

15. A circle has radius 16. What is its exact area?

16. When does a decimal represent an irrational number?

17. All basketballs are hollow. If one has a radius of about 12 cm, about how much material is needed to make it?

18. a. What is the exact volume of a sphere with radius 9?

   b. Round your answer to the nearest cubic unit.

19. The diameter of Mars is approximately 7000 kilometers. Estimate the surface area of Mars.

20. A circle has a diameter of 12 cm. What is the exact area of a sector of this circle if the central angle of the sector is 60°?

CHAPTER REVIEW

Questions on SPUR Objectives

SPUR stands for **S**kills, **P**roperties, **U**ses, and **R**epresentations. The Chapter Review questions are grouped according to the SPUR Objectives for this chapter.

SKILLS DEAL WITH THE PROCEDURES USED TO GET ANSWERS.

Objective A: *Find a simple fraction equal to any terminating or repeating decimal.* *(Lesson 12-1)*

In 1–4, find the simple fraction in lowest terms equal to the given number.

1. $-1.\overline{23}$ **2.** $0.0\overline{46}$ **3.** 81.55 **4.** 11.02

Objective B: *Estimate square roots of a number without a calculator.* *(Lesson 12-2)*

5. Between what two integers is $\sqrt{80}$?

6. Between what two integers is $-\sqrt{3}$?

7. Simplify $\sqrt{144 + 256}$.

8. Name the two square roots of 36.

Objective C: *Use the Pythagorean Theorem to find unknown lengths of third sides in right triangles.* *(Lesson 12-3)*

9. Find *BI* in the figure at the left below.

10. Find *HA* in the figure at the right below.

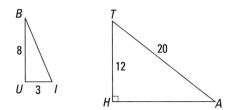

11. The legs of a right triangle have lengths 20 and 48. What is the length of the hypotenuse of this triangle?

12. Find *y* in the figure below.

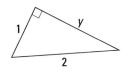

Objective D: *Find the circumference and area of a circle or sector, given its radius or diameter.* *(Lessons 12-4, 12-5)*

13. Find the circumference and area of a circle with radius 10.

14. To the nearest whole number, find the circumference and area of a circle with diameter 2 meters.

15. Give the area of a 90° sector of a circle with radius 5 feet.

16. To the nearest square unit, give the area of a 100° sector of a circle with radius 6.

Objective E: *Find the surface area and volume of cylinders and prisms.* *(Lessons 12-6, 12-7)*

In 17 and 18, a cylinder has a base of radius 5 units and height of 6 units.

17. Find its surface area.

18. Find its volume.

In 19 and 20, a prism has a base that is a right triangle with legs 20 and 21, and a height of 12.

19. Find its volume.

20. Find its total surface area.

Objective F: *Find the surface area and volume of a sphere, given its radius or diameter.* *(Lesson 12-8)*

21. A sphere has diameter 10. Give its exact surface area.

22. To the nearest cubic inch, give the volume of a sphere with radius 4″.

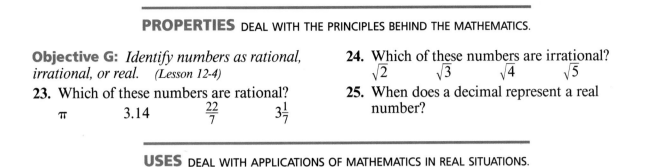

PROPERTIES DEAL WITH THE PRINCIPLES BEHIND THE MATHEMATICS.

Objective G: *Identify numbers as rational, irrational, or real.* *(Lesson 12-4)*

23. Which of these numbers are rational?
π 3.14 $\frac{22}{7}$ $3\frac{1}{7}$

24. Which of these numbers are irrational?
$\sqrt{2}$ $\sqrt{3}$ $\sqrt{4}$ $\sqrt{5}$

25. When does a decimal represent a real number?

USES DEAL WITH APPLICATIONS OF MATHEMATICS IN REAL SITUATIONS.

Objective H: *Apply formulas for the surface area and volume of cylinders and prisms in real situations.* *(Lessons 12-6, 12-7)*

26. How much wood is needed to make a cylindrical pencil 17 cm long and 1 cm in diameter? (Ignore the lead.)

27. The side of a cylindrical oil storage tank is to be painted. If the tank is 60′ high and has a circumference of 400′, how much surface area needs to be painted?

In 28 and 29, use the triangular prism ruler pictured here. The base of the prism is an equilateral triangle.

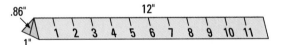

28. Give its lateral surface area.
29. Give its volume.

Objective I: *Apply formulas for the circumference of a circle and area of a circle or sector and surface area and volume of a sphere in real situations.* *(Lessons 12-4, 12-5, 12-8)*

30. A child is lost in the forest. Police decide to search every place within 2 miles of the place the child was last seen. To the nearest square mile, how much area must be searched?

31. What is the surface area of a ball 20 cm in diameter?

32. Pearl is making a color wheel for school. She wants to put 12 equal-sized sectors on the wheel. What is the area of one of those sectors if the wheel is to be 18″ in diameter?

33. The Earth goes around the sun in an orbit that is almost a circle with radius 150,000,000 km. How far does the Earth travel in one year in its orbit?

34. How much clay is needed to make a ball with a 4″ diameter?

35. To the nearest cubic centimeter, what is the volume of a soap bubble with a radius of 1.5 centimeters?

REPRESENTATIONS DEAL WITH PICTURES, GRAPHS, OR OBJECTS THAT ILLUSTRATE CONCEPTS.

Objective J: *Know how square roots and geometric squares are related.* *(Lesson 12-2)*

36. A square has area 50. What is the exact length of a side?

37. A side of a square has length $\sqrt{4.9}$. What is the area of the square?

38. A farm of area 1 square mile is square.
   a. What is the length of each side?
   b. Relate your answer in part **a** to square roots.

39. If the largest square to the right has area 4, what is the length of a side of the tilted square?

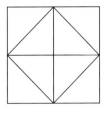

CHAPTER
13

COORDINATE GRAPHS AND EQUATIONS

It is possible to graph the formulas $C = 2\pi r$ and $A = \pi r^2$ for the circumference and area of a circle. Make r the first coordinate and C (or A) the second coordinate. The graphs are shown here.

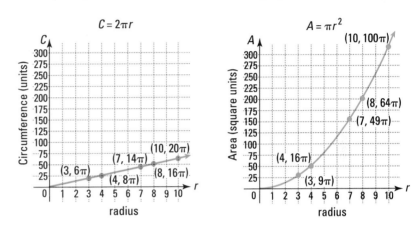

As you can see, the graphs are quite different. The graph on the left is a ray. The ray increases at a constant rate. This means the circumference of a pizza will increase the same amount when its radius is increased from 3″ to 4″ as when it is increased from 7″ to 8″. The graph on the right is part of a curve known as a *parabola*. The right graph goes up faster and faster as r increases more and more. This pictures the idea that the area of a circle increases more and more quickly as the radius gets larger and larger. So, for instance, changing the radius of a pizza from 7″ to 8″ will increase the area of the pizza quite a bit more than will changing the radius from 3″ to 4″.

In this chapter, you will study these and other graphs related to a variety of ideas you have seen in previous chapters.

13-1

Graphing
$y = ax + b$

Carried out to the letter. *More than half of all letter mail in the United States is sorted and coded by machines that are programmed with algebraic functions.*

As of early 1994, costs of mailing a first-class letter were 29¢ for the first ounce and 23¢ for each additional ounce (up to ten ounces). The costs are in the table at the left below. At the right, the ordered pairs (weight, cost) are graphed.

weight in ounces	cost in cents
1	29
2	52
3	75
4	98
5	121
6	144
7	167
8	190
9	213
10	236

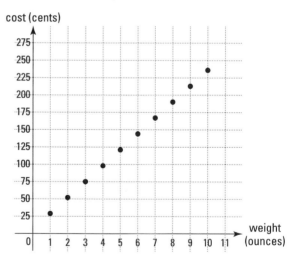

Call the weight w and the cost c. To find a formula relating w and c, you can rewrite the values of c without doing the arithmetic, a strategy used in Lesson 6-5.

w	c	
1	29	
2	29 + 23	
3	29 + 23 · 2	The number that changes in the
4	29 + 23 · 3	right column is always one less
5	29 + 23 · 4	than the value of *w*.
⋮	⋮	
w	29 + 23 · (w − 1)	

So $c = 29 + 23(w - 1)$. Using the Distributive Property,

$$c = 29 + 23w - 23 = 23w + 6.$$

This formula shows that there is a simple pattern connecting w and c. If you multiply w by 23 and add 6, you get c.

$$c = 23w + 6$$

You might say that it costs 6¢ to mail anything, plus 23¢ for each ounce. For instance, suppose you want to mail some papers weighing 6.3 ounces. The post office always rounds weights up to the next ounce. So $w = 7$. Substitute 7 for w in the formula to find the cost.

$$c = 23 \cdot 7 + 6$$
$$c = 167$$

The cost is 167 cents. Dividing by 100 converts the cents to dollars. The cost to mail these papers is $1.67. The table at the beginning of the lesson gives the same cost.

You have just seen three ways to display postal rates: in a table, with a formula, and with a graph. Each way has some advantages. The table is easiest to understand. The formula is shortest and allows for values not in the table. Also, the formula can be used by a computer. The graph pictures the rates. It shows that they go up evenly.

Benjamin Franklin, pictured on the 13-cent stamp, was responsible for providing a more extensive, frequent, and speedy mail service. As deputy postmaster general for the American colonies in 1753, he built a solid foundation for a U. S. postal service. He became the first U.S. postmaster general in 1775.

Postal Rates in Three Different Years

The graph below shows the first-class postal rates for 1965, 1975, and 1990. The graph displays the changes over time in a way that is easy to understand.

w = weight in ounces
c = cost in cents
1965: $c = 5w$
1975: $c = 11w + 2$
1990: $c = 20w + 5$

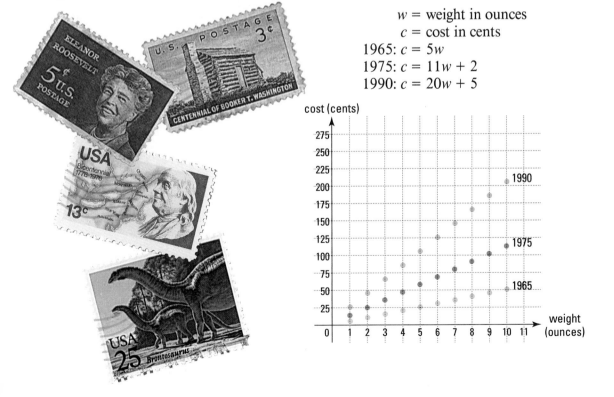

Each formula yields points on a line. The line $c = 5w$ is the lowest line. It shows that the 1965 costs were lowest. The line $c = 11w + 2$ is in the middle. The line $c = 20w + 5$ is the highest line, because costs were higher in 1990 than in 1975 or 1965.

In each of the three equations $c = 5w$, $c = 11w + 2$, and $c = 20w + 5$, w is multiplied by a number. This number indicates how fast the line goes up as you move one unit to the right. For instance, in $c = 11w + 2$, w is multiplied by 11. This causes the line to move up 11 units for every one unit increase in w. This is because, in 1975, each additional ounce of weight increased the cost of mailing a letter by 11¢.

Each of the postal rate equations is of the form $y = ax + b$, where a and b are fixed real numbers and the points (x, y) are graphed. For instance, in $c = 11w + 2$, $a = 11$, $b = 2$, and w and c take the place of x and y. Any equation of this form is a **linear equation.** The name *linear* is used because its graph is a line. When x is not restricted to be a whole number, then the graph of the equation consists of all points on the line.

Example

Graph the line with equation $y = -3x + 4$, and check that the graph is correct.

Solution

The graphs below show three steps. You should put all your work on one graph.

Step 1: Substitute numbers for x and find the corresponding values of y. For instance we choose $x = -1$. Then $y = -3 \cdot -1 + 4 = 3 + 4 = 7$. This means that $(-1, 7)$ is on the graph.

Now we choose $x = 2$. Then $y = -3 \cdot 2 + 4 = -6 + 4 = -2$. So $(2, -2)$ is on the graph.

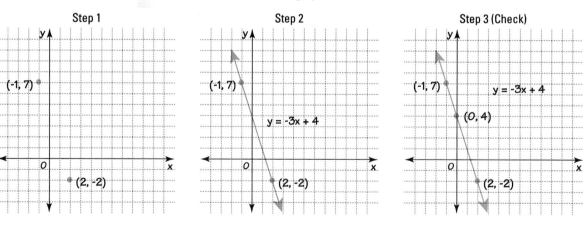

Step 2: Draw a line through the points. Label the line with the equation.

Step 3: Pick a third value for x. We pick $x = 0$. Then, according to the equation, $y = -3 \cdot 0 + 4 = 4$. This means that $(0, 4)$ should be on the graph. Is it? Yes, so the graph is correct.

694

Caution: When graphing lines, it is easy to make errors in calculation. Then a point will be incorrect. You must check with a third point. If the three points do not lie on the same line, try a fourth point. Keep trying points until you see the pattern of the graph.

QUESTIONS

Covering the Reading

In 1–4, use the first-class postal rates for 1994.

1. What formula relates the weight of a letter and the cost to mail it?

2. What is the cost of mailing a letter that weighs 6 ounces?

3. What is the cost of mailing a letter that weighs 2.4 ounces?

4. For what weights does a letter cost 98¢?

5. Convert 213¢ to dollars.

6. What is the general rule for converting cents to dollars?

7. What is an advantage of displaying postal rates in a table?

8. What is an advantage of having a formula for postal rates?

9. What is an advantage of graphing postal rates?

In 10–14, use the first-class postal rates for 1965, 1975, and 1990.

10. What was the cost of mailing a 2-oz letter in 1965?

11. In 1965, what was the cost of mailing a letter weighing 9.2 oz?

12. You could mail a 3-oz letter in 1975 for what it cost to mail a __?__-oz letter in 1965.

13. What was the lowest cost for mailing a letter in the indicated year?
    a. 1965   **b.** 1975   **c.** 1990

14. In 1975, by how much did the cost go up for each extra ounce of weight?

15. The line $y = -3x + 4$ is graphed in this lesson. Give the coordinates of two points on this line other than the points identified on the graph.

16. Graph the line with equation $y = 2x - 6$ and check your graph.

17. *Multiple choice.*   Which line could be the graph of $y = 5x - 2$?

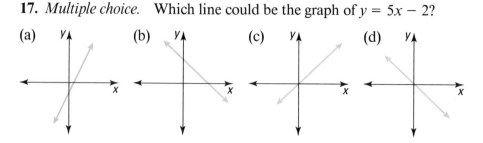

(a)   (b)   (c)   (d)

18. Today there are 400 packages of duplicating paper at a school. Each week about 12 packages are used.
   a. Make a table with two columns, "weeks from now" and "number of packages left." Fill in at least 10 rows of your table.
   b. Graph six pairs of numbers in the table. Let w, the number of weeks, be graphed on the x-axis. Let L, the number of packages left, be graphed on the y-axis.
   c. What equation relates w and L?
   d. Find the value of w when $L = 0$. What does this value mean?

19. a. Coordinates of points on what line are found by this program?

```
10 PRINT "X", "Y"
20 FOR X = 1 TO 10
30 Y = 11*X + 2
40 PRINT X, Y
50 NEXT X
60 END
```

   b. Modify the program so that it will find 100 points on the line of Question 16.

20. Consider the equation $2x - 3y = 12$.
   a. Find the coordinates of three points that satisfy this equation.
   b. If you have answered part **a** correctly, the three points should lie on the same line. Graph this line.

21. Dan de Lion is 9 years old. His mother says he is growing "like a weed," 3 inches a year. He is now 56 inches tall.
   a. Suppose Dan continues to grow at this rate. Make a table with four pairs of numbers for his age x (in years) and his height y (in inches).
   b. Graph the four pairs.
   c. An equation that relates x and y is $y = 3x + b$. Find the value of b.
   d. Explain why $y = 3x + b$ will not relate Dan's age and height when he is an adult.

22. Curling, a sport that probably originated in Scotland or the Netherlands around 400 years ago, involves sliding stones weighing over 40 pounds across ice at a target called the *house*. A point is scored by the team that gets closest to the middle of the target. Use the information below to find the area of the *tee*, or center. Then find the area of the red, white, and blue rings.   *(Lesson 12-5)*

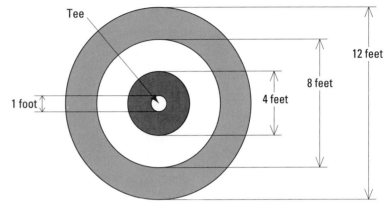

House

Smoothing the way.
Curling players sweep vigorously to smooth and clean the ice as a teammate's stone approaches the house.

In 23 and 24, solve and check.   *(Lessons 10-4, 11-6)*

23. $3(5n + 22) - 156 = 0$

24. $\frac{x - 9}{2} = \frac{4}{5}$

25. Alvin ate about $\frac{1}{4}$ of the tossed salad. Betty ate half of what was left. How much salad now remains?   *(Lesson 9-3)*

26. Diana gave P plants $\frac{1}{2}$ cup water each and Q plants $\frac{3}{4}$ cup water each. How much water did she use altogether?   *(Lessons 5-7, 9-6)*

27. A salesperson keeps a record of miles traveled for business. Last year, the salesperson drove 18,000 miles; 65% of this was for business. The company reimburses the salesperson 28¢ per mile traveled for business. How much money should the salesperson get back from the company?   *(Lessons 10-1, 10-3)*

28. Simplify.   *(Lessons 5-3, 7-2, 9-1, 11-3)*
   a. $-5 + -3$ **b.** $-5 - -3$ **c.** $-5 \cdot -3$ **d.** $\frac{-5}{-3}$

29. Use $\triangle BCD$ at the left to find m$\angle BCD$.   *(Lesson 7-9)*

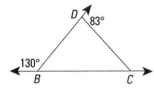

30. Consider the equation $5x + 4y = 13$.
   a. Find three points on this line in which both coordinates are integers.
   b. Describe how you could find many more points on the line in which both coordinates are integers.

Using an Automatic Grapher

IN-CLASS
ACTIVITY

The first computers were built in the 1940s. By the 1960s, software had been developed that enabled computers to print out graphs of equations. By 1985 this technology was available on relatively inexpensive hand-held calculators. Today most people graph equations with automatic graphers, either graphics calculators or computers equipped with software that can graph equations. For this activity you need to use an automatic grapher.

1 Turn on the machine you use and access the technology that enables equations to be graphed. What you need to do depends on the machine. Your teacher will help you.

2 Enter the equation that is to be graphed. You will begin by graphing the postal-rate equations from Lesson 13-1, so type in $y = 5x$. (Sometimes "$y = $" is already typed in.)

3 Determine the window for the graph. The window is the part of the coordinate plane that appears on the screen. For the postal-rate graphs, x (the weight) ranges from 0 to 10. A window for x that shows these values and leaves some room at the side is $-1 \le x \le 11$. On some programs, you would type in XMIN = -1 and XMAX = 11. (These stand for the minimum and maximum values of x.) The costs range from 5¢ to 205¢, so type in values of y that include this range. One possibility is YMIN = -50 and YMAX = 250.

4 Display the graph of the equation. With some technology, you need only press a button or move a mouse to GRAPH. You should see a line that contains the blue points in the graph on page 693.

5 Follow steps 2 and 4 to graph $y = 11x + 2$ and $y = 20x + 5$ on the same axes.

6 Look at Example 1 of Lesson 13-2. Follow steps 2 through 4 above to graph the two equations in the example.

Situations Leading to ax + b = cx + d

The algebra of renting a car. *Algebraic equations can be used to compare rates charged by different companies, such as car rental agencies, that provide the same services.*

In the last lesson, first-class postal rates for four different years were graphed and compared. In some situations there are choices of rates for the same time period. When this is the case, it is natural to want to know when a particular rate is cheaper. Example 1 shows how to compare the rates using graphs.

Example 1

A driver wants to rent a car for a day and is given a choice of two plans.

> Plan I: $30 plus 25¢ a mile
> Plan II: $20 plus 32¢ a mile

a. When is Plan I cheaper than Plan II?
b. When is Plan II cheaper than Plan I?
c. When do the two plans cost the same?

Solution

Begin by thinking about the situations. Plan I starts out with a higher cost but each mile driven costs less. This means that Plan I will be more expensive at first, but if enough miles are driven, it will cost less.

Find an equation for each plan. Let C be the cost (in dollars) for driving m miles. Here are equations relating m and C for each plan. (Do you see how these equations were found?)

> Plan I: $C = 30 + .25m$
> Plan II: $C = 20 + .32m$

▶

The graph of each equation is a line. We graph the lines on the same axes so that they can be compared easily.

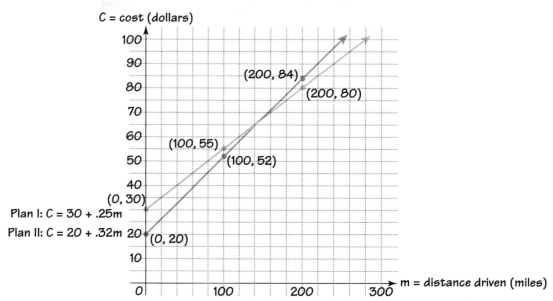

C = cost (dollars)

Plan I: C = 30 + .25m
Plan II: C = 20 + .32m

m = distance driven (miles)

When the line for one of the plans is higher, then that plan is more expensive.

a. The lines intersect when m is about 150. This means: Plan I is more expensive when m is less than 150. If the car is to be driven more than 150 miles, Plan I should be chosen.

b. Plan II is cheaper if the car is driven less than 150 miles.

c. The two plans cost the same when the car is driven about 150 miles.

There are advantages and disadvantages to using a graph. A hand-drawn graph pictures the situation very nicely, but usually it is not very accurate. So it does not give an exact answer to any of the questions asked. However, some calculators and computer software automatically graph equations. With these, you can get enough accuracy to get an exact or nearly exact solution.

The key to an exact answer to part **c** of Example 1 is the location of the point of intersection of the lines. The first coordinate of that point is found by making the two values for C equal.

$$30 + .25m = 20 + .32m$$

This is called *equating* the values of C. This equation is of the form

$$ax + b = cx + d$$

because the unknown on each side of the equation is to the first power. In the next lesson, you will learn how to solve this kind of equation without a graph.

Example 2

Peggy is spending money while her sister Vanna is saving it. At present, Peggy has $65 but she spends about $2 more than her allowance each week. Vanna has $40 but she is saving about $3 a week.

a. Use graphs to estimate when the two sisters will have the same amount of money.
b. Solving what equation will give the exact time when the sisters have the same amount?

Solution

a. First identify the variables you will use. Here we Let A be the amount of money that each sister will have after w weeks. Then

for Peggy: $A = 65 - 2w$
for Vanna: $A = 40 + 3w$.

Now graph these two equations.

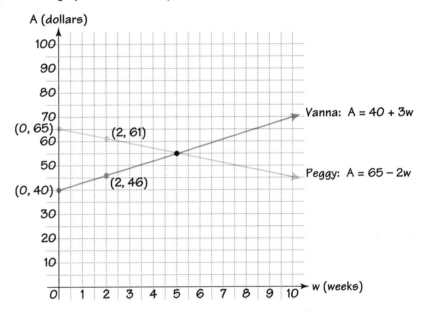

The graphs seem to intersect at (5, 55). This means that after 5 weeks, both girls will have $55. (Check that this is correct.) After that time, Vanna will have more money than Peggy will.

b. Equate the two values of A. Solving the equation

$$65 - 2w = 40 + 3w$$

will give the exact number of weeks.

QUESTIONS

Covering the Reading

In 1–7, refer to Example 1.

1. **a.** What does it cost to drive 10 miles using Plan I?
   b. Name the point on the line for Plan I which determines the answer to part **a.**

2. **a.** What does it cost to drive 30 miles using Plan II?
   b. Name the point on the line for Plan II which determines the answer to part **a.**

3. How can you tell from the graph which plan is more expensive when you drive 200 miles?

4. How can you tell from the equations which plan is more expensive when you drive 200 miles?

5. Solving what single equation tells when the plans cost the same?

6. $30 + .25m = 20 + .32m$ is an equation of the form $ax + b = cx + d$. Give the values of a, b, c, and d.

7. With an automatic grapher, graph the lines of Example 1. Use a window that will show the answer to part **c** to the nearest mile.

In 8–12, refer to Example 2.

8. After w weeks, how much money will Peggy have?

9. After w weeks, how much money will Vanna have?

10. Solving what equation tells when Peggy and Vanna have the same amount of money?

11. **a.** At what point do the lines in the graph intersect?
    b. What do the coordinates of this point mean?

12. $65 - 2w = 40 + 3w$ is an equation of the form $ax + b = cx + d$. Give the values of a, b, c, and d.

Applying the Mathematics

Valuable Bonds.
Barry Bonds of the San Francisco Giants was the unanimous choice for National League Most Valuable Player in 1993. In 1990 and 1992, he was NL MVP when he played for the Pittsburgh Pirates.

13. At the beginning of the 1994 baseball season, Barry Bonds had 222 home runs in his career and had averaged .190 home runs per game. Juan Gonzales had 120 home runs in his career and had averaged .260 home runs per game. Let P be the number of home runs in a career after g more games at these averages.
    a. For Barry, what equation relates P and g?
    b. Graph the equation you found in part **a.**
    c. For Juan, what equation relates P and g?

d. Graph the equation from part **c** on the same axes you used for part **b.**

e. From your graph, estimate the number of games it will take Juan to catch up with Barry in the number of home runs in his career.

f. Solving what equation will tell when the two players will have the same number of home runs?

14. Long-distance company I charges 20¢ for the first minute plus 15¢ for each additional minute of a long-distance call. Company II charges 26¢ for the first minute plus 13¢ for each additional minute. Let C be the cost (in cents) of an m-minute call.
a. Give an equation relating m and C for company I.
b. Give an equation relating m and C for company II.
c. Solving what equation will tell when the plans cost the same?

Review

15. Graph $y = 2x + 1$ for values of x from -10 to 10. *(Lesson 13-1)*

In 16–18, solve and check. *(Lessons 10-1, 10-2, 10-4)*

16. $99 - 3x = 150$ **17.** $\frac{2}{5}y = \frac{3}{4}$

18. $(a + 1) + (2a + 2) + (3a + 3) + (4a + 4) = 5$

19. Suppose you have an 80% chance of getting an A on your next math test, and a 70% chance of getting an A on your next science test. If your scores on these tests are independent, what is the probability that you will get an A on both tests? *(Lesson 9-4)*

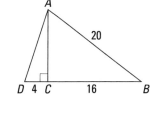

20. Use the figure at the left. *(Lessons 7-9, 12-3)*
a. What is the measure of $\angle ACB$?
b. $AC = \underline{\ \ ?\ \ }$
c. $AD = \underline{\ \ ?\ \ }$

21. Along I-64 in Virginia, the distance from Staunton to Charlottesville is 34 miles. The distance from Waynesboro to Richmond is 91 miles. The distance from Staunton to Richmond is 103 miles. What is the distance from Waynesboro to Charlottesville? *(Lesson 7-5)*

22. Write 6×10^3 as a decimal. *(Lesson 2-3)*

Exploration

23. Design a spreadsheet with the following columns.

m	$30 + .25m$	$20 + .32m$

By trying values of m between 100 and 200, make the values of $30 + .25m$ and $20 + .32m$ as close to each other as you can. Use this idea to determine m to two decimal places. This helps answer part **c** of Example 1.

13-3

Solving $ax + b = cx + d$

Saving versus spending. *Example 1 provides two solutions to the equation of Example 2 from Lesson 13-2. It tells when Vanna will have as much money as her sister.*

In Lesson 13-2, you saw how an equation of the form $ax + b = cx + d$ can be solved with graphs. Using algebra, exact solutions can be found. Notice that

$$ax + b = cx + d$$

has the unknown x on both sides of the equal sign. By adding a carefully chosen expression to both sides of this equation, it can be converted to an equation with the unknown on only one side. The first example solves the equation from Example 2 of Lesson 13-2. Notice that you have a choice for the first step.

Example 1

Solve $65 - 2w = 40 + 3w$.

Solution 1

One way to start is to add $-3w$ to both sides. This results in an equation with w on only one side.

original equation	$65 - 2w = 40 + 3w$
Add $-3w$ to both sides.	$65 - 2w + \text{-}3w = 40 + 3w + \text{-}3w$
Add like terms.	$65 - 5w = 40$

Now proceed as you have previously with equations of this type.

Add -65 to both sides.	$\text{-}65 + 65 - 5w = \text{-}65 + 40$
Simplify.	$\text{-}5w = \text{-}25$
Multiply both sides by $-\frac{1}{5}$.	$\text{-}\frac{1}{5} \cdot \text{-}5w = \text{-}\frac{1}{5} \cdot \text{-}25$
Simplify.	$w = 5$

Solution 2

Another way is to add 2*w* to both sides. This avoids negative numbers.

original equation	$65 - 2w = 40 + 3w$
Add 2*w* to both sides.	$65 - 2w + 2w = 40 + 3w + 2w$
Add like terms.	$65 = 40 + 5w$

Solve this equation as you have done in earlier lessons.

Add -40 to both sides.	$-40 + 65 = -40 + 40 + 5w$
Simplify.	$25 = 5w$
Multiply both sides by $\frac{1}{5}$.	$5 = w$

Both solutions to Example 1 confirm that after 5 weeks Peggy and Vanna will have the same amount of money.

Example 2

Solve $3x + 5 = 10x + 26$.

Solution

We add -10*x* to both sides. This gets rid of the variable on the right side.

original equation	$3x + 5 = 10x + 26$
Add -10*x* to both sides.	$-10x + 3x + 5 = -10x + 10x + 26$
Simplify.	$-7x + 5 = 26$
Now add -5 to both sides.	$-7x + 5 + -5 = 26 + -5$
Simplify.	$-7x = 21$
Multiply both sides by $-\frac{1}{7}$.	$x = -3$

Check

Substitute -3 for *x* every place it occurs in the original equation. Does $3 \cdot -3 + 5 = 10 \cdot -3 + 26$? Yes, both sides equal -4.

The equation in Example 3 may look as if it is solved for *L*. But there is an *L* on the right side. The method of Examples 1 and 2 can be used to solve equations of this type.

Example 3

Solve $L = 15 - 4L$.

Solution

	$L = 15 - 4L$
Add 4*L* to both sides.	$L + 4L = 15 - 4L + 4L$
	$5L = 15$
	$L = 3$

Check

Does $3 = 15 - 4 \cdot 3$? Yes.

The next example combines a number of ideas from this and the preceding chapters.

Example 4

Solve $\frac{x-2}{2} = \frac{x+3}{4}$.

Solution

First use the Means-Extremes Property.

$$4(x - 2) = 2(x + 3)$$

Now use the Distributive Property.

$$4x - 4 \cdot 2 = 2x + 2 \cdot 3$$
$$4x - 8 = 2x + 6$$

In this form, the equation is like those of Examples 1, 2, and 3. Add $-2x$ to both sides.

$$-2x + 4x - 8 = -2x + 2x + 6$$
$$2x - 8 = 6$$

Add 8 to both sides. $\qquad\qquad 2x = 14$

Solve in your head. $\qquad\qquad x = 7$

Check

Substitute. Does $\frac{7-2}{2} = \frac{7+3}{4}$?

Does $\qquad \frac{5}{2} = \frac{10}{4}$? Yes.

QUESTIONS

Covering the Reading

1. In solving $65 - 2w = 40 + 3w$, name two things you could do to both sides to obtain an equation with the unknown on one side.

2. Solve $3x + 5 = 10x + 26$ by first adding $-3x$ to both sides.

3. **a.** To solve $s = 18 - 35s$, first add ___?___ to both sides.
   b. Solve this equation.

In 4–9, solve and check.

4. $11A + 5 = 7A + 35$

5. $12 - 3q = 2q - 2$

6. $4 - y = 6y - 10$

7. $2(n - 4) = 3n$

8. $\frac{t-3}{4} = \frac{t+6}{12}$

9. $0.6m + 5.4 = -1.3 + 2.6m$

10. **a.** Solve $30 + .25m = 20 + .32m$.
    b. Use the answer to part **a** to answer the question of part **c** of Example 1 in Lesson 13-2.

11. Under rate plan 1 a new car costs $1000 down plus $200 per month. Under rate plan 2 the car costs $750 down and $250 per month.
    a. Write an expression for the amount paid after n months under plan 1.
    b. Write an expression for the amount paid after n months under plan 2.
    c. After how many months will the amount paid be the same for both plans?

12. Twice a number is 500 more than six times the number. What is the number?

In 13 and 14, solve and check.

13. $11p + 5(p - 1) = 9p - 12$

14. $-n + 4 - 5n + 6 = 21 + 3n$

15. In $\triangle PIN$, the measure of angle N is $4x + 36$. The measure of angle P is $10x$. If the measure of $\angle N$ equals the measure of $\angle P$, find the measures of all three angles in the triangle.

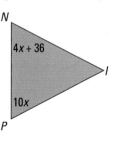

16. Hasty Harry wrote the following solution to $5x - 1 = 2x + 8$. When he checked the answer, it didn't work.
    a. In which step did Harry make a mistake?
    b. What is the correct solution?

    $$5x - 1 = 2x + 8$$
    Step 1: $-2x + 5x - 1 = -2x + 2x + 8$
    Step 2: $\quad 3x - 1 = 0 + 8$
    Step 3: $\quad\quad 3x = 8 - 1$
    Step 4: $\quad\quad 3x = 7$
    Step 5: $\quad\quad\quad x = \frac{7}{3}$

17. Let A be the amount paid after n months of the rate plans of Question 11. Graph the lines corresponding to each of these rate plans. Explain why these graphs check the answer to Question 11c. *(Lesson 13-2)*

In 18 and 19, rewrite as a simple fraction in lowest terms. *(Lesson 12-1)*

18. 0.92

19. $6.\overline{36}$

20. What is the total area of these three rectangles? *(Lesson 10-5)*

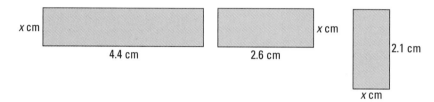

21. Trace the figure below. What word results when the figure is reflected over the given line? *(Lesson 8-6)*

22. Graph all pairs of solutions to $x - y = 4$. *(Lesson 8-4)*

23. Draw a cube. *(Lesson 3-9)*

24. In 1991 there were approximately two million, one hundred five thousand farms in the United States. *(Lessons 1-1, 2-3)*
a. Write this number as a decimal.
b. Write this number in scientific notation.

25. Use the information in Question 24. In 1930 there were about 330% as many farms as in 1991. About how many farms were there in the U.S. in 1930? *(Lessons 2-6, 10-1)*

Future Farmers.
Shown here are two members of the Future Farmers of America. FFA helps prepare students for careers not only in farming, but in such related areas as agricultural marketing, agribusiness, forestry, communications, and horticulture.

Exploration

26. Paula wanted to solve $4x + 7 = 2x - 3$, but did not know the method of this lesson. She knew she wanted the value of the left side to equal the value of the right side. So she substituted a 2 for x to see what happened.

Left side	Right side
$4 \cdot 2 + 7 = 15$	$2 \cdot 2 - 3 = 1$

The left side was bigger than the right. Paula tried -10.

$$4 \cdot \text{-}10 + 7 = \text{-}33 \qquad 2 \cdot \text{-}10 - 3 = \text{-}23$$

Now the right side was bigger than the left. She figured that the solution must be some number between -10 and 2.
a. Find the value that makes the two sides of the equation equal.
b. Use Paula's method to solve $5x - 7 = 3x + 9$.

13-4

Fractions and Relative Frequencies Revisited

Let's Make a Deal. *Game show host Bob Hilton is asking a contestant to consider trading a prize for a better prize concealed behind one of three doors. What is the probability of choosing a door that does not have the best prize?*

Any ordered pair of real numbers can be graphed. Sometimes the graphs of sets of ordered pairs present a nice geometric picture. In this lesson, we look at fractions in a way that may surprise you.

Every rational number can be expressed using many different simple fractions. Here are some fractions that equal $\frac{2}{3}$.

$$\frac{2}{3} \ = \ \frac{4}{6} \ = \ \frac{-2}{-3} \ = \ \frac{6}{9} \ = \ \frac{12}{18}$$

Now think of these fractions as ordered pairs, with the denominator as the first coordinate and the numerator as the second coordinate.

$$(3, 2) \quad (6, 4) \quad (-3, -2) \quad (9, 6) \quad (18, 12)$$

When these pairs are graphed, they all lie on the same line.

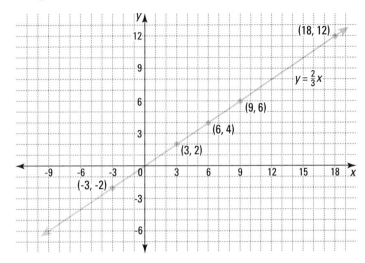

Here is the reason that all these points lie on the same line. Suppose y and x are numbers with

$$\frac{y}{x} = \frac{2}{3}.$$

Then after multiplying both sides of the equation by x,

$$y = \frac{2}{3}x.$$

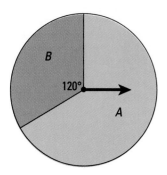

The equation $y = \frac{2}{3}x$ also appears in the following situation. Suppose that the probability that an event occurs is $\frac{2}{3}$. For instance, if the spinner pictured at the left is fair, then the probability that the spinner lands in region A is $\frac{2}{3}$. Thus, if you spin the spinner 60 times, you should expect that it will land in region A about $\frac{2}{3} \cdot 60$, or 40 times. More generally, if you spin the spinner x times, you should expect that it will land in region A about $\frac{2}{3}x$ times.

Below is a record of 150 spins of a spinner like the one pictured. After each 15 spins, we recorded the number of times that the spinner landed in region A. The data are given at the left and graphed at the right below.

Number of Spins	Spins in Region A
15	10
30	21
45	29
60	38
75	46
90	58
105	67
120	78
135	89
150	99

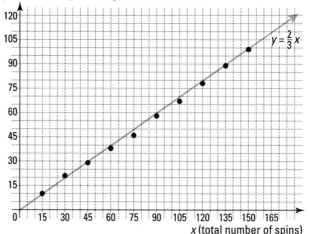

The points do not lie exactly on the line $y = \frac{2}{3}x$ because the relative frequencies are not exactly $\frac{2}{3}$, but they are close. In the long run, we would expect the relative frequencies to get closer and closer to $\frac{2}{3}$. The equation $y = \frac{2}{3}x$ is of the form $y = ax$. For any particular value of a, the graph of this equation is a line. Three such equations are graphed at the top of page 711.

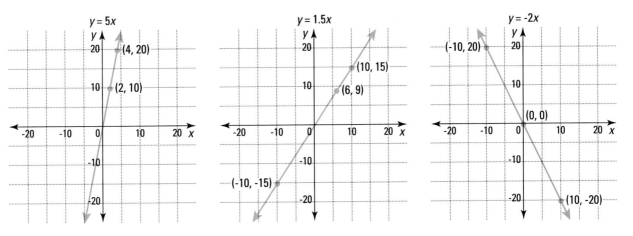

The number a in the equation $y = ax$ is called the **slope** of the line. It tells how much y changes for every increase of 1 unit by x. For instance, in $y = 5x$, $a = 5$, and as x increases from 2 to 3, y increases from 10 to 15, an increase of 5.

QUESTIONS

Covering the Reading

1. Consider the rational number $\frac{5}{4}$.
   a. Name five different fractions equal to it.
   b. For each fraction $\frac{a}{b}$ that you named in part **a**, graph the ordered pair (b, a).
   c. The points in part **b** should lie on a line. What is an equation for that line?

2. Repeat Question 1 for the number $\frac{-3}{5}$.

3. Refer to the spinner experiment in the lesson.
   a. After 90 spins, how many times did the spinner land in region A?
   b. What point on the graph was determined by the information in part **a**?
   c. Why are the points on the graph close to the line?
   d. Why are the points on the graph not all on the line?

In 4–6, an equation of a line is given.
a. Name three points on the line.
b. Graph the line.
c. As x increases from 10 to 11, by how much does y change?
d. Give the slope of the line.

4. $y = 3x$ **5.** $y = \text{-}1.5x$ **6.** $y = \frac{1}{2}x$

Applying the Mathematics

7. a. Toss a coin 100 times, recording H or T after each toss.
   b. Make a table with the total number of heads after every ten tosses.
   c. Graph the pair of numbers in each row of the table.
   d. To what line should the ten points be near? Are your points near that line?
   e. Do you think your coin was fair? Why or why not?

8. a. Graph $y = 2x$ and $y = 2x - 1$ on the same pair of axes.
   b. How are these graphs related?
   c. Generalize the idea of parts **a** and **b**.

9. a. Graph $y = 2x$ and $y = \text{-}2x$ on the same pair of axes.
   b. How are the graphs of these equations related?
   c. Generalize the idea of parts **a** and **b**.

Review

10. Refer to Question 14c of Lesson 13-2. Solve the equation. *(Lesson 13-3)*

In 11–16, solve. *(Lessons 10-5, 13-3)*

11. $12B + 5 = 10B + 17$ **12.** $3 - 4x = x + 23$

13. $18(x - 2) = 12(x + 3)$ **14.** $v - \frac{1}{2} = \frac{3}{4} - v$

15. $\frac{d}{4} = 11.22$ **16.** $\frac{30}{2x + 5} = \frac{20}{x - 1}$

17. About how long will it take to read a 312-page novel if the first 35 pages can be read in 45 minutes? *(Lesson 11-6)*

18. When their odometer read 13,486 miles, the Villareal family filled the gas tank of their car with 12.4 gallons. At 13,869 miles they again filled the tank, now with 14.6 gallons. How many miles per gallon is their car getting? *(Lesson 11-2)*

19. A small store has two full-time employees. If each employee is absent about 3% of the days the store is open, what is the probability that both employees will be absent on the same day? Assume the absences are independent events. *(Lesson 9-4)*

Serve yourself! *In 1972, a few independent gas stations began offering savings of a few cents per gallon to customers willing to pump their own gas. Today, except in states where it is illegal to pump your own gas, almost all gas stations offer self-serve pumps.*

20. Figure *ABCD* is a parallelogram. m∠*BAD* = 65°. Find the measures of as many other sides and angles as can be found from this information.   *(Lesson 7-8)*

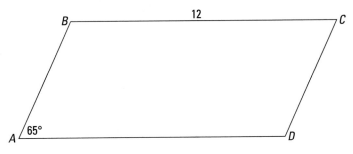

21. Name four metric units of length and tell how they are related to each other.   *(Lesson 3-4)*

22. Round 4.921685 to the nearest thousandth.   *(Lesson 1-4)*

Exploration

23. a. Perform an experiment like the spinner experiment of this lesson.
   b. Record the results of your experiment in a table.
   c. Graph the pairs of numbers in the table.
   d. Graph the line that contains the points you would get if the relative frequencies of the event equalled the probability of the event.
   e. In a few sentences describe what you found.

In 24 and 25, use an automatic grapher.

24. Graph $y = 2x$, $y = 3x$, $y = 4x$, and $y = 5x$ on the same window. Predict what the graphs of $y = 6x$, $y = 7x$, ... will look like.

25. Graph $y = \frac{1}{2}x$, $y = \frac{1}{3}x$, $y = \frac{1}{4}x$, and $y = \frac{1}{5}x$ on the same window. Predict what the graphs of $y = \frac{1}{6}x$, $y = \frac{1}{7}x$, ... will look like.

Satellite dishes. *Large dishes, like these in Eagle River, Alaska, serve as stations for satellite communication systems. Each slice of this dish parallel to its face is a circle. As the diameter of the circle gets larger, so does its circumference.*

Some formulas, like the area formula $A = \ell w$ for a rectangle, involve three variables. To graph them requires graphs in three dimensions. But other formulas, like the formula $C = \pi d$ for the circumference of a circle, involve two variables. These formulas can be graphed just as you have done with other equations in previous lessons.

Example 1

Graph $C = \pi d$, where C and d are the circumferences and diameters of circles.

Solution 1

We pick d as the first coordinate, and C as the second coordinate of each ordered pair. To plot points, use an approximation to π, such as 3.14.

When $d = 1$, $C = \pi \cdot 1 \approx 3.14 \cdot 1 = 3.14$.
When $d = 2$, $C = \pi \cdot 2 \approx 3.14 \cdot 2 = 6.28$.
When $d = 5$, $C = \pi \cdot 5 \approx 3.14 \cdot 5 = 15.7$

Plot as many points as you think you need to get a good graph. A graph is shown here.

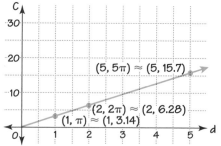

Because d and C stand for lengths, the graph contains only positive values of d and C. The graph is a ray without its endpoint.

Another formula that involves only two variables is the formula $A = s^2$ for the area of a square. The graph of this formula is *not* part of a line, as Example 2 shows.

Example 2

Let A be the area of a square with side s. Graph all possible pairs of values (s, A).

Solution

Since $A = s^2$, the values (s, A) are the same as the values (s, s^2). So the graph contains all ordered pairs in which the first coordinate is a positive number (it is a length), and the second coordinate is the square of that number. At the left is a table of some values. At the right is the graph.

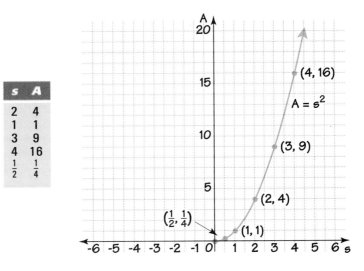

s	A
2	4
1	1
3	9
4	16
$\frac{1}{2}$	$\frac{1}{4}$

The arc in Example 2 is part of the curve known as a parabola. Another graph with this shape is found on page 691. The **parabola** is the shape of all graphs of equations of the form $y = ax^2$.

To graph all the possible pairs of numbers satisfying $y = x^2$, including possibly negative coordinates, remember that a number and its opposite have the same square. Thus $(-x)^2 = x^2$. For this reason, the graph is symmetric to the y-axis. The full graph cannot be shown because it extends forever. Its shape is the entire parabola.

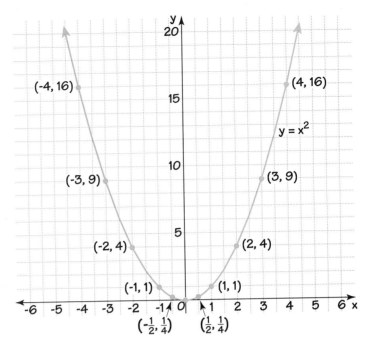

$y = x^2$

Points shown on the parabola: $(-4, 16)$, $(4, 16)$, $(-3, 9)$, $(3, 9)$, $(-2, 4)$, $(2, 4)$, $(-1, 1)$, $(1, 1)$, $\left(-\frac{1}{2}, \frac{1}{4}\right)$, $\left(\frac{1}{2}, \frac{1}{4}\right)$

Any vertical cross section of a radio telescope receiving dish, like the one shown here in Arecibo, Puerto Rico, is part of a parabola. Compare this picture to the part of the graph from $x = -1$ to $x = 1$.

The parabola has important properties. Because of these properties, it is a shape used in the manufacture of automobile headlights, satellite dishes, and telescopes.

Graphs of equations may have shapes quite different from lines or parabolas. You are asked to explore these in the questions and in the next lesson.

QUESTIONS

Covering the Reading

In 1–4, consider the graph of $C = \pi d$ in this lesson.

1. Which variable is the first coordinate of points on the graph?

2. Which variable is the second coordinate of points on the graph?

3. *Multiple choice.* Which is the best description of the graph?
    (a) line   (b) ray without its endpoint   (c) line segment

4. Name a point on the graph other than those given in Example 1.

In 5–9, consider the graph of $A = s^2$.

5. Which variable is the first coordinate of points on the graph?

6. Which variable is the second coordinate of points on the graph?

7. The graph is part of the curve known as a __?__.

8. Suppose the point $(5, t)$ is on this graph. What is t?

9. a. How does this graph differ from the graph of $y = x^2$?
    b. How is this graph the same as the graph of $y = x^2$?

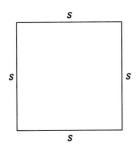

10. A formula for the perimeter p of a square is $p = 4s$, where s is the length of a side of the square.
   a. Graph 5 pairs of values (s, p) that satisfy this formula.
   b. Sketch a graph of all pairs (s, p) that satisfy $p = 4s$.
   c. Describe the graph.

11. Recall that a formula for the surface area $S.A.$ of a sphere with radius r is $S.A. = 4\pi r^2$. Let A stand for the surface area $S.A.$
   a. Graph 6 pairs of values (r, A) that satisfy this formula. (You may need to use an approximation to 4π.)
   b. Sketch a graph of all pairs (x, y) that satisfy $y = 4\pi x^2$.

12. **a.** Graph the equation $y = 3 - x^2$.
   b. Explain how this graph is related to the graph of $y = x^2$.

Review

13. Graph six points (x, y) such that $\frac{y}{x} = \frac{4}{3}$. *(Lesson 13-4)*

14. Solve $9x - 10 = 30 + x$. *(Lesson 13-3)*

15. Solve $36t = -\frac{2}{3} \cdot 12 - 28t$. *(Lesson 13-3)*

16. Triangle MIX is a size change image of $\triangle DEN$.

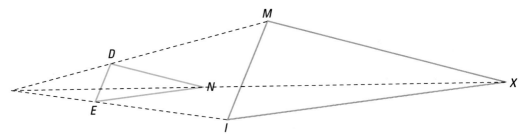

   a. *True or false.* The triangles are congruent.
   b. If $DE = 10$, $MI = 22$, and $EN = 40$, what is the value of IX?
   (Lesson 11-8)

17. In 1990, the most populous Canadian province, Ontario, had 9,747,600 people living in 344,090 square miles. The least populous province, Prince Edward Island, had 130,400 people living in 2,185 square miles. Which of these provinces was more densely populated? Explain your answer. *(Lesson 11-2)*

18. *TRAP* is a trapezoid with lengths given as shown.
   a. Is enough information given to find the area of *TRAP*?
   b. If so, find the area. If not, tell what additional information is needed.   *(Lessons 10-10, 12-3)*

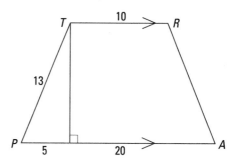

19. A box of dried fruit has 7 rows of fruit with 8 pieces in each row.
   a. How many pieces of fruit are in the box?
   b. How many pieces of fruit are *not* next to a side of the box?
   (Lessons 6-3, 9-1)

20. Graph on a number line.
   a. $2 < x < 3$
   b. $0 \le x \le 1$
   c. $-4 < x \le -3$ *(Lesson 4-10)*

Exploration

---

21. An automatic grapher is recommended for this question. The graph of $y = x^2$ contains the points $(-1, 1)$, $(0, 0)$, and $(1, 1)$.
   a. Draw an accurate picture of the part of the curve from $(-1, 1)$ to $(1, 1)$ on the window $-1 \le x \le 1$, $-1 \le y \le 1$.
   b. Graph the equation $y = x^3$ on the same window as part **a**. Describe one similarity and one difference in the graphs of $y = x^2$ and $y = x^3$.
   c. Explore these graphs with different windows. Describe what happens.

LESSON

13-6

*Graphs of
Equations
with
Symbols for
Rounding*

Stamp of approval. *The Postal Service offers a large variety of first-class stamps.*

There is a sense in which the postal-rate graphs from Lesson 13-1 are incomplete. They do not take into account weights that are not whole numbers. Below at the left is the graph of $c = 23w + 6$ from Lesson 13-1. At the right below is a more accurate graph for the situation.

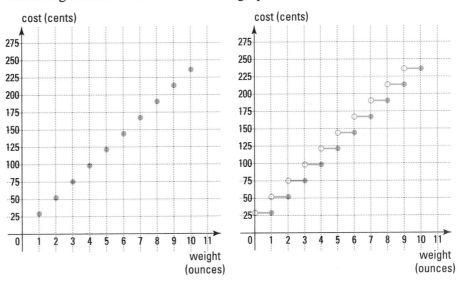

Notice how the graph at the right gets its shape. When a letter weighs less than 1 ounce, it costs 29¢ to mail. So for values of w between 0 and 1, the value of c is 29¢. This is represented by the bar that is farthest left and lowest. This bar has an open circle at its left because a weight of 0 is impossible. The closed circle at the right means that a weight of exactly 1 ounce costs 29¢.

The next bar goes from (1, 52) to (2, 52). This means that it costs 52¢ to mail a letter weighing between 1 and 2 ounces. The open circle means that (1, 52) is not included in the graph. The closed circle means that (2, 52) is included. The other bars have similar meanings.

In this lesson, we introduce two new symbols. These symbols make it possible to write an equation for the graph at the right. They are the *rounding up* and *rounding down* symbols.

$\lceil x \rceil$ is the result of rounding x up to the nearest integer.
$\lfloor x \rfloor$ is the result of rounding x down to the nearest integer.

Example 1

Evaluate each expression.
a. $\lceil 4.3 \rceil$ **b.** $\lfloor 4.3 \rfloor$ **c.** $\lceil 75 \rceil$
d. $\lfloor 75 \rfloor$ **e.** $\lceil -199.456 \rceil$ **f.** $\lfloor -199.456 \rfloor$

Solutions

a. 4.3 rounded up to the nearest integer is 5, so $\lceil 4.3 \rceil = 5$.
b. 4.3 rounded down to the nearest integer is 4, so $\lfloor 4.3 \rfloor = 4$.
c. and d. 75 is already rounded to the nearest integer. So $\lceil 75 \rceil = \lfloor 75 \rfloor = 75$.
e. -199.456 is between -199 and -200. The larger of these is -199. So, when rounded up, -199.456 goes to -199. Thus $\lceil -199.456 \rceil = -199$.
f. -199.456 rounded down to the nearest integer is -200. So $\lfloor -199.456 \rfloor = -200$.

Computers often have to deal with rounded values. So the symbols $\lfloor \ \rfloor$ and $\lceil \ \rceil$ are very commonly used in computer programming. Programmers sometimes call $\lfloor \ \rfloor$ the **floor function** symbol and call $\lceil \ \rceil$ the **ceiling function** symbol. Sometimes brackets [ ] are used for the floor function and it is called the **greatest integer function** symbol. In some programming languages and on some calculators, INT(x) has the same meaning as $\lfloor x \rfloor$.

It is possible to graph expressions using these symbols.

Example 2

Graph the pairs (x, y) that satisfy $y = \lfloor x \rfloor$ for values of x from 0 to 3.

Solution

When $x = 0$, $y = \lfloor x \rfloor = \lfloor 0 \rfloor = 0$. This yields the point $(0, 0)$ at the lower left of the graph. When x has a value between 0 and 1, then $y = \lfloor x \rfloor = 0$. This yields the lowest segment of the graph. When x is between 1 and 2, $y = \lfloor x \rfloor = 1$. This yields the middle segment. The highest segment contains the points corresponding to values of x between 2 and 3. Then $y = \lfloor x \rfloor = 2$.

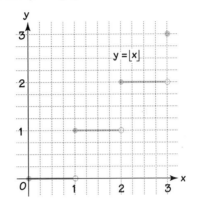

Today's graphing technology is very powerful. People now graph things that before would have been difficult, if not impossible, to graph, and newspapers and magazines now show many more graphs than they did a generation ago. As you study more mathematics, you will see graphs that a previous generation could not imagine. But the technology is not perfect. For instance, if you try to graph $y = \lfloor x \rfloor$ on today's automatic graphers, you may not get an accurate graph. Here is the output from one grapher.

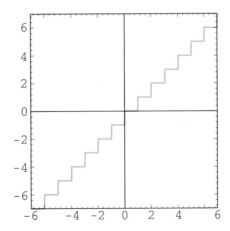

The segments of the graph are connected. This is not correct. It shows that when you use graphing technology, you should have some estimate of what an answer should be and be able to check whatever you get using a different approach. Of course, this is good advice even if you do not use technology!

Look again at the graph in the Solution of Example 2. Notice that for integer values of x, the graph of $y = \lfloor x \rfloor$ contains the same points as the graph of $y = x$. For the non-integer values, the y-values are rounded down, as they should be. The graph resembles the steps of a staircase, so sometimes the symbols $\lfloor \ \rfloor$ and $\lceil \ \rceil$ are called **step function** symbols.

Below is the graph of the full line $y = 23x + 6$ and the postage-rate graph that began this lesson. To find an equation describing the postage-rate graph, consider the steps in calculating the postage rate.

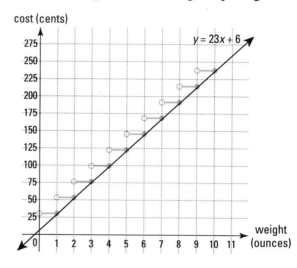

Step 1: Round the weight up to the nearest integer. If the weight was originally w, after rounding it is $\lceil w \rceil$.

Step 2: To find the postal cost, multiply the rounded weight by 23 and add 6. So the cost is $23 \lceil w \rceil + 6$.

Consequently, an equation for the postage rate graph is $c = 23 \lceil w \rceil + 6$. Notice how similar this is to the equation $y = 23w + 6$ that works for integer values of w.

Studying the rounding symbol is a fitting way to end this book. It returns us to an idea from arithmetic, rounding, that was discussed at the very beginning of the book. Graphs involving the rounding symbol have an interesting geometry. The equations with variables show how algebra is used by the programmer of the postal weighing machine to arrive at the postal rates when you weigh a letter. So, this one idea combines arithmetic, algebra, and geometry—using ideas you know and introducing you to ideas that you will encounter in your next mathematics courses.

An invention that lasts. *Jan Matzeliger was an inventor who industrialized shoe making in 1883. His invention, the shoe-lasting machine, shaped and fastened leather over a sole.*

QUESTIONS

Covering the Reading

1. Examine the right-hand graph on page 719. Describe what the top segment represents.

In 2–9, evaluate the expression.

2. $\lceil 11.3 \rceil$

3. $\lfloor 11.3 \rfloor$

4. $\lceil -8 \rceil$

5. $\lceil 0.0\overline{91} \rceil$

6. $\lfloor 600 \rfloor$

7. $\lfloor -\pi \rfloor$

8. $\lfloor \frac{3}{2} \rfloor$

9. $\lceil \frac{47}{12} \rceil$

10. Which two are equal? $\left\lceil \frac{1}{2} \right\rceil, \left\lfloor \frac{1}{2} \right\rfloor, \left\lfloor \frac{-1}{2} \right\rfloor, \left\lceil \frac{-1}{2} \right\rceil$

11. Give three names for the symbol $\lfloor \ \rfloor$.

12. Give two names for the symbol $\lceil \ \rceil$.

13. Graph the pairs (x, y) that satisfy $y = \lceil x \rceil$ for values of x from 0 to 3.

14. In 1994, the cost of mailing a letter was calculated as follows.
    Step 1: Round the weight up to the nearest ounce.
    Step 2: Multiply the rounded weight by 23, then add 6.
    Give a formula for the cost C of mailing in terms of the weight w.

Math counts! *At the annual MATHCOUNTS National Competition in Washington, D.C., mathletes from each state meet and compete. The Countdown Round, shown here, decides the individual winner from among the 10 competitors with the highest scores.*

Applying the Mathematics

15. Extend the graph of Example 2 to cover values of x from -3 to 0.

In 16 and 17, *multiple choice.* Use special cases to determine the answer.

16. A school is allowed one representative to a mathematics contest for every 250 students. (For instance, if the school has 251–500 students, it can send 2 representatives.) If the school has s students, how many representatives can it send?
    (a) $\left\lfloor \frac{s}{250} \right\rfloor$
    (b) $\left\lceil \frac{s}{250} \right\rceil$
    (c) $\frac{\lceil s \rceil}{250}$
    (d) $\frac{\lfloor s \rfloor}{250}$

17. Suppose a taxi ride costs $.75 plus $.25 for each $\frac{1}{4}$ mile. Assume the meter clicks on the quarter mile. What is the cost C of a ride of m miles?
    (a) $C = 0.75 + \lfloor m \rfloor$
    (b) $C = 0.75 + 4\lfloor 0.25m \rfloor$
    (c) $C = 0.75 + m$
    (d) $C = 0.75 + 0.25\lfloor 4m \rfloor$

18. Given that x is an integer, find x from these clues.

Clue 1: $\left\lceil \frac{x}{2} \right\rceil = 8$

Clue 2: $\left\lceil \frac{x-5}{2} \right\rceil = 5$

Review

19. If a car is traveling at s miles per hour, then once a driver steps on the brakes, it takes at least d feet to stop, where $d = \frac{s^2}{20}$. d is called the *braking distance*.
   a. Give the braking distance for four different speeds.
   b. Name four ordered pairs (s, d) that satisfy this formula.
   c. What geometric figure is the graph of the formula? *(Lesson 13-5)*

20. The graphs of $y = 3x$ and $y = 2x + 4$ intersect at what point? *(Lessons 13-1, 13-3, 13-4)*

21. Three more than half a number equals two more than a third of the same number. What is the number? *(Lesson 13-3)*

22. In a few sentences, explain why the product of two negative numbers is positive. *(Lesson 9-6)*

Exploration

23. Evaluate $\left\lfloor \frac{a}{b} \right\rfloor$ for many pairs of integers a and b, where a is quite a bit larger than b. From the results you get, tell how $\left\lfloor \frac{a}{b} \right\rfloor$ is related to the integer division of a by b.

Making the grade.
To make new cars safe and marketable, automakers put cars through a variety of tests. The car pictured here is being tested for ease of steering, maneuverability, and braking distance.

A project presents an opportunity for you to extend your knowledge of a topic related to the material of this chapter. You should allow more time for a project than you do for typical homework questions.

PROJECTS 13 CHAPTER THIRTEEN

1 Graphs of Equations with Powers

In this chapter, equations of the form $y = ax$ and $y = ax^2$ were graphed. Explore the graphs of the equations $y = ax^3$, $y = ax^4$, $y = ax^5$, $y = ax^6$, and so on, when x is any real number (positive, zero, or negative). Consider cases where a is positive and cases where a is negative.

2 Long-Distance Rates

Find out the precise long-distance rate from where you live to a place in another area you or someone in your family might call. Determine and graph at least six ordered pairs (,) based on your long-distance rate. Describe the graph with an equation using the rounding up or rounding down symbol.

3 Graphing Square Roots

a. Graph the equation $y = \sqrt{x}$. Explain how the shape of the graph compares with shapes that are mentioned in this chapter.
b. Graph $y = \sqrt{\lceil x \rceil}$ **c.** Graph $y = \lceil \sqrt{x} \rceil$.

4 An Experiment with Decimal Places

Locate a reference where a large number of digits (about 50) are given for an infinite decimal for an irrational number. (For instance, you might be able to find such a decimal for the number π or for some square roots.) Perform this experiment to see how often each of the even digits 0, 2, 4, 6, and 8 appears in the decimal. Make a table of the frequency of appearance of each digit in the first 10, 20, 30, 40, 50 places. Then graph the relative frequencies as was done in Lesson 13-4. If the digits appear at random, near what line should the points lie? Do they? Explain what has happened with the number you chose.

▶

5 How Often Is a Solution to *ax + b = cx + d* an Integer?

This project is similar to Project 2 of Chapter 10. Find four six-sided dice and call them *a, b, c,* and *d.* Toss the dice and record the numbers into the equation *ax + b = cx + d.* For instance, if the die *a* shows 3, *b* shows 1, *c* shows 4 and *d* shows 5, then the equation will be $3x + 1 = 4x + 5$. Write down 50 equations in this way. Determine the relative frequencies of the following events:

a. the solution is positive
b. the solution is an integer
c. the solution is a positive integer.
Write down what you think are the probabilities of these events, and explain how you have come to your conclusions.

6 When Will One Person Overtake Another?

In Lesson 13-2 there is a question asking when Juan Gonzales will overtake Barry Bonds in the number of home runs in his career. Find another actual example like this. You will need: two people, some count associated with each, and the count for the person who has less to be growing faster than the count for the person who has more. Determine when the person who has less will overtake the person who has more. Show this both with an equation and with a graph.

To get ideas for Project 6, think about people you know, sports and activities you like, and how people change at different rates.

SUMMARY

The solutions to any equation can be graphed. If the equation has two variables, then the graph is a set of ordered pairs that can be plotted on a coordinate graph. In this chapter, four types of equations were discussed: $y = ax$; $y = ax + b$; $y = ax^2$; and $y = \lceil x \rceil$ or $y = \lfloor x \rfloor$.

In these descriptions, x is the first coordinate of the ordered pair, y is the second coordinate, and a and b are fixed real numbers.

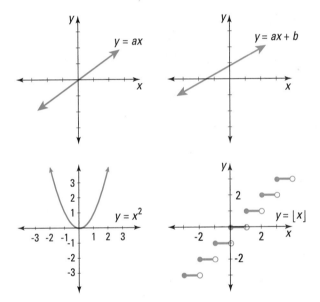

Equations of the type $y = ax$ occur in formulas for perimeter such as $p = 4s$ or $C = \pi d$. A graph of this type of an equation is a line containing the origin. This line contains all of the pairs (x, y) if $\frac{y}{x} = a$. Let an event occur y times out of x possibilities. If the event has probability a of occurring, then $\frac{y}{x}$ will be close to a, so the points (x, y) will lie close to the line $y = ax$.

Equations of the type $y = ax + b$ occur when an initial amount increases or decreases at a constant rate, such as the cost of a phone call. The graph of $y = ax + b$ is a line containing the point $(0, b)$. It is natural to compare two costs to see when one is lower. It is possible to determine which cost is lower if you know when the costs are equal. If the second equation is $y = cx + d$, then that value can be found by solving the equation $ax + b = cx + d$.

This equation, which has an unknown on both sides, can be solved by adding either $-ax$ or $-cx$ to both sides. The resulting equation has an unknown on only one side and can be solved using methods learned in earlier chapters.

Equations of the type $y = ax^2$ occur from formulas for area, like $A = s^2$ and $A = \pi r^2$. Their graphs are parabolas.

Equations involving the rounding up symbol $\lceil \ \rceil$ and the rounding down symbol $\lfloor \ \rfloor$ arise from situations in which rounding or estimating is necessary, such as when weights of letters are rounded before determining a postal rate. Their graphs resemble steps of a staircase.

VOCABULARY

You should be able to give a general description and a specific example of each of the following ideas.

Lesson 13-1
linear
linear equation

Lesson 13-2
automatic grapher, window

Lesson 13-4
slope

Lesson 13-5
parabola

Lesson 13-6
rounding up; rounding down
$\lceil x \rceil$, ceiling function
$\lfloor x \rfloor$, floor function, greatest
    integer function, INT(x)
step function

PROGRESS SELF-TEST

Take this test as you would take a test in class. You will need a calculator and graph paper. Then check your work with the solutions in the Selected Answers section in the back of the book.

In 1 and 2, solve.

1. $60 + 14y = -20 + 30y$

2. $3(2x - 8) = 4x + 9$

3. Evaluate $\lceil 987.654 \rceil$.

4. If $\lfloor x \rfloor = 8$, give two possible values of x.

5. The ordered pairs (10, 3), (20, 6), and (30, 9) all lie on the same line. What is an equation for that line?

6. Louise makes 30% of the 3-point shots she tries in basketball. If you were to graph ordered pairs (number of shots attempted, number of shots made), the points would tend to lie near what line?

In 7 and 8, use the following information. One car rental company charges $35 a day plus 48¢ a mile to rent a car. A second company charges $39 a day plus 41¢ a mile.

7. a. What will it cost to rent a car from the first company for 1 day, having driven m miles?

b. What equation can be solved to determine the number of miles m for which the car rental cost would be the same for these two companies?

8. Estimate, to the nearest mile, when the car rental cost would be the same for the two companies.

9. Graph $y = 4x$.

10. Graph $y = -3x + 5$.

In 11 and 12, use this information. A formula for the surface area of a cube is $A = 6s^2$, where s is the length of an edge of the cube.

11. a. What is the surface area of a cube whose edges have length 5?

b. What point on the graph of $A = 6s^2$ is determined by the answer to part **a?** (Let s be the first coordinate.)

12. Graph this formula.

13. Explain how you can use the graphs from Questions 9 and 10 to solve the equation $4x = -3x + 5$.

14. Solve the equation $4x = -3x + 5$ using any of the methods of this chapter.

15. *Multiple choice.* Which is the graph of $y = \lceil x \rceil$?

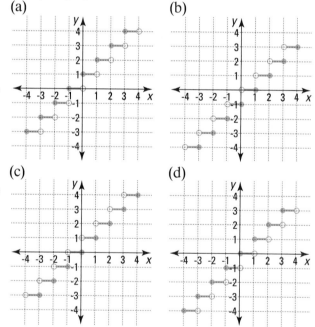

CHAPTER REVIEW

Questions on SPUR Objectives

SPUR stands for **S**kills, **P**roperties, **U**ses, and **R**epresentations. The Chapter Review questions are grouped according to the SPUR Objectives for this chapter.

SKILLS DEAL WITH THE PROCEDURES USED TO GET ANSWERS.

Objective A: *Solve equations of the form* $ax + b = cx + d.$ *(Lesson 13-3)*

In 1–8, solve.

1. $12x + 7 = 2x + 11$
2. $8m - 2 = 17m - 20$
3. $v = 3v - 1$
4. $5y = 100 - 3y$
5. $6(2 + 3r) = 9 + r$
6. $3s - (s - 15) = 4s + 5$
7. $200 - 31t = 120 + 29t$
8. $\frac{1}{2}z + \frac{1}{2} = \frac{3}{2}z - \frac{1}{2}$

Objective B: *Evaluate expressions using the symbols for rounding up or rounding down.* *(Lesson 13-6)*

In 9–12, evaluate.

9. $\lceil 2.5 \rceil - \lfloor 2.5 \rfloor$
10. $\lceil \sqrt{10} \rceil + \lceil \sqrt{11} \rceil$
11. $\lceil -40.8 \rceil$
12. $\left\lfloor -\frac{2}{3} \right\rfloor$
13. $\lceil 18.613 \rceil + 2\lfloor 4.2 \rfloor$
14. If $\lfloor x \rfloor = 10$, give two possible values of x.

PROPERTIES DEAL WITH THE PRINCIPLES BEHIND THE MATHEMATICS.

Objective C: *Find the line on which or near which the numerators and denominators of equal fractions or relative frequencies lie.* *(Lesson 13-4)*

15. **a.** Give three simple fractions equal to $\frac{7}{5}$.

    b. Think of each simple fraction $\left(\frac{y}{x}\right)$ you found in part **a** as a point (x, y) on a line. What is an equation for that line?

16. What is an equation for the line through (3, 11) and (9, 33)?

17. Tornadoes occur with about equal relative frequency on each day of the week. Suppose you were to graph the ordered pair (number of tornadoes on Sunday in 1995, total number of tornadoes in 1995). Near what line would you expect this point to lie?

18. Suppose you toss a coin you believe to be fair. Every so often you graph the points (number of heads, number of tosses). Near what line would you expect these points to be?

USES DEAL WITH APPLICATIONS OF MATHEMATICS IN REAL SITUATIONS.

Objective D: *Translate situations of constant increase or decrease that lead to sentences of the form $ax + b = cx + d$.* *(Lessons 13-2, 13-3)*

In 19–22, a situation is given.

a. Translate the given information into an equation of the form $ax + b = cx + d$ that will answer the question.

b. Solve the equation.

19. With one cellular car phone company, the monthly charge is $15, and calls cost 20¢ a minute. With another company, calls cost 34¢ a minute but the monthly charge is only $5. Determine the number of minutes for which the monthly bills from these companies would be the same.

20. Kurt and Sara are saving money to buy their parents an anniversary present. Kurt now has $35 and is saving $6 a week. Sara has $60 and is saving $4 a week. When will they have the same amount?

21. Laura is now 57 inches tall and growing at a rate of about 1 inch each 100 days, or $\frac{1}{100} \frac{\text{in.}}{\text{day}}$. Robert is 59 inches tall and growing at about 1 inch each 150 days, or $\frac{1}{150} \frac{\text{in.}}{\text{day}}$. At these rates, in how many days will Laura and Robert be the same height?

22. One football player currently has rushed for 3,000 yards in his career and gains an average of about 60 yards per game. A second player has rushed for 2,350 yards and gains an average of about 90 yards per game. If these rates continue, in how many games will the players have rushed for the same amount?

REPRESENTATIONS DEAL WITH PICTURES, GRAPHS, OR OBJECTS THAT ILLUSTRATE CONCEPTS.

Objective E: *Graph formulas for perimeter, area, and other quantities that involve two variables.* *(Lesson 13-5)*

23. In a regular hexagon, the perimeter P and the length s of a side are related by the formula $P = 6s$. Graph this formula.

24. The perimeter P and area A of a square are related by the formula $A = \frac{1}{16}P^2$. Graph this formula for values of P from 0 to 10.

25. The area A of a circle with diameter d is given by $A = \frac{\pi}{4}d^2$. Graph this formula for values of d from 0 to 10.

26. The circumference w of a person's wrist and the circumference n of that person's neck are said to be related by the formula $n = 2w$. Graph this formula.

Objective F: *Graph equations of the form* $y = ax + b$. *(Lessons 13-1, 13-4)*

In 27–30, graph.

27. $y = -\frac{2}{3}x$

28. $y = 3x$

29. $y = 2x + 4$

30. $y = -\frac{1}{2}x - 1$

Objective G: *Interpret the solution to* $ax + b = cx + d$ *graphically.* *(Lesson 13-2)*

31. Explain how Question 20 could be answered graphically. You do not have to draw the graphs.

32. Graph $y = x + 5$ and $y = 3x - 1$. Use your graph to determine the solution to the equation $x + 5 = 3x - 1$.

33. Solve $4x - 7 = 2x + 1$ using graphs.

34. *Multiple choice.* Below is a graph of $y = ax + b$ and $y = cx + d$. Which of the coordinates (r), (s), (t), or (u), is the solution to $ax + b = cx + d$?

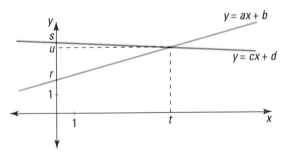

Objective H: *Interpret graphs of equations using the symbols* $\lceil\ \rceil$ *or* $\lfloor\ \rfloor$. *(Lesson 13-6)*

In 35 and 36, refer to this graph.

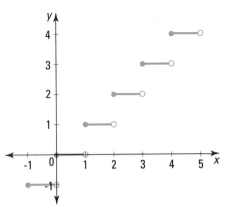

35. Is this the graph of $y = \lceil x \rceil$ or $y = \lfloor x \rfloor$? Explain how you know.

36. What does the uppermost bar on the graph describe?

In 37 and 38, refer to this graph below of postal rates in 1975.

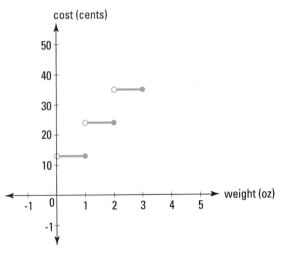

37. Is this the graph of $y = 11\lceil x \rceil + 2$ or $y = 11\lfloor x \rfloor + 2$?

38. In 1975, what did it cost to mail a letter that weighed 1.21 ounces?

LESSON 1-1 (pp. 4–9)

15. a. 50 **b.** stars **17. a.** Salinas-Seaside-Monterey, California **b.** Pensacola, Florida **19.** 600,000,000 **21.** 506 **23.** 9,999 **25. a.** 10 **b.** 100

LESSON 1-2 (pp. 10–15)

21. three hundred twenty-four and sixty-six hundredths **23.** Sample: 44.61 **25. a. and c. See below. b.** 10.39 seconds **27.** sixty-five thousandths, sixty-five, sixty-five thousand **29.** 6 **31.** 31,068 **33.** count: 76; counting unit: trombones

25. a. and c.

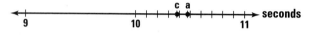

LESSON 1-3 (pp. 16–20)

11. 0.97531246 **13. a.** 6000 **b.** 5000 **c. See below. 15. a.** 1.610 **b.** 1.609 **17.** High. You want everyone to have some cake. **19.** Low. You don't want the cable to break. **21.** 5.001, 5.01, 5.1 **23.** (c) **25.** Sample: 5.99 **27. a.** Q **b.** R **c.** Y **29.** 4,030,000

13.

LESSON 1-4 (pp. 21–25)

11. 2.5 **13. a.** 3.7 **b.** 3.67 **c.** 3.667 **d.** 3.6667 **15. a.** $20.97 **b.** $21, 3¢ off **17.** $24 **19.** 921 **21. a.** 9550 **b.** 9650 **23.** $1.54 **25.** 3.7 **27.** Sample: 3.3 **29.** These numbers are equal so there is no number between them. **31.** count: 12; counting unit: eggs

LESSON 1-5 (pp. 26–30)

9. Ⓒ or (AC) or (clear) **11.** (0.1234568); 7 decimal places. Answers may vary. **13.** 0.909 **15.** 56 **17.** 99999999. Answers may vary. **19.** Answers will vary. **21. a.** .440 **b.** .441 **c.** .405 **23.** 53.5 seconds **25.** 21 **27.** three million, four hundred twelve thousand, six hundred seventy **29.** 299,997

LESSON 1-6 (pp. 31–36)

23. a. 0.1875" **b.** shorter **25. a. and b. See below. 27.** $\frac{2}{11}, \frac{2}{9}, \frac{2}{7}$ **29. a.** Sample: 59.131824 **b.** 59.132 **31.** 9.8979 **33.** 34.0079 **35.** 2916 gallons

25.

$$\vdash \quad \bullet \quad \bullet \quad \bullet \quad \bullet \quad \bullet \quad \dashv$$
$$0 \qquad \frac{1}{4} \quad \frac{1}{3} \quad \frac{1}{2}=\frac{2}{4} \quad \frac{2}{3} \quad \frac{3}{4} \qquad 1$$

LESSON 1-7 (pp. 37–40)

13. $2\frac{3}{5}, 3\frac{2}{5}, 5\frac{2}{3}$ **15.** 12.533 **17.** longer **19. See below. 21. a.** Sample: $\frac{11}{12}$ **b.** Sample: $\frac{1.4}{7}$ **23.** 100 **25.** True **27. a.** counts: 27 and 206 **b.** counting units: small bones and bones

19.

$$3.375" = 3\frac{3}{8}"$$

LESSON 1-8 (pp. 41–45)

15. a. −1 **b.** 1 **c.** −2 **d.** 2 **17.** Q **19.** none **21.** −1, 0, $\frac{1}{2}, \frac{3}{2}$ **23.** −1 **25.** −60.5 **27.** −9.3

29. 6 ⊕ 3 ⊗ 4 ⊘ 2 ⊜
⑥ ⑥. ③. ③. ④. ⑫. ②. ⑫.

31. 1.2 or $1\frac{1}{5}$ **33.** 462,000.01

LESSON 1-9 (pp. 46–49)

21. a. See below. b. 1 > −2000 **23.** > **25.** = **27.** < **29.** < **31.** 62.1 > 6.21 > 0.621 or 0.621 < 6.21 < 62.1 **33.** 99.2 < 99.8 < 100.4 or 100.4 > 99.8 > 99.2 **35.** 1, 2, 3, 4 **37.** 6.283 **39.** $26

21. a.

$$\text{dollars}$$
$$-2000 \qquad -1000 \qquad 0 \qquad 1000$$

LESSON 1-10 (pp. 50–54)

15. a. $\frac{3}{11}=\frac{21}{77}, \frac{2}{7}=\frac{22}{77}, \frac{22}{77}>\frac{21}{77}$, so $\frac{2}{7}>\frac{3}{11}$. **b.** $\frac{3}{11}=.\overline{27}$ and $\frac{2}{7}=.\overline{285714}; .\overline{285714}>.\overline{27};$ so $\frac{2}{7}>\frac{3}{11}$. **17.** $\frac{2}{8}=\frac{1}{4}$ **19.** Samples: $\frac{2}{2}, \frac{3}{3}, \frac{5}{5}$ **21.** $\frac{7}{4}$ **23. a.** 1 row of 60 chairs, 2 rows of 30 chairs, 3 rows of 20 chairs, 4 rows of 15 chairs, 5 rows of 12 chairs, 6 rows of 10 chairs, 10 rows of 6 chairs, 12 rows of 5 chairs, 15 rows of 4 chairs, 20 rows of 3 chairs, 30 rows of 2 chairs, 60 rows of 1 chair **b.** These numbers are all the factors of 60. **25.** Samples: $-\frac{1}{2}, -0.2, -0.05$ **27.** 1782.4 **29.** 4.12 **31. a.** 0.4; **b.** 0.625; **c.** 0.75; **d.** 0.$\overline{3}$; **e.** 0.8$\overline{3}$

CHAPTER 1 PROGRESS SELF-TEST (p. 58)

1. 700,000 **2.** 45.6 **3.** 0.25 **4.** $15\frac{13}{16} = 15 + \frac{13}{16} = 15 + 0.8125 = 15.8125$ **5.** 6 **6.** three-thousandths **7.** Write each number to three decimal places and compare: $.6 = .600, .66 = .660, .\overline{6} = .666 \ldots$, $.606 = .606$. This indicates that $.\overline{6}$ is the largest. **8.** Change each fraction to a decimal and compare. $\frac{1}{2} = .5, \frac{2}{5} = .4, \frac{3}{10} = .3$, and $\frac{1}{3} = .333 \ldots$. So $\frac{3}{10}$ is smallest. **9.** 98.7 **10.** 99 **11.** −80 ft > −100 ft; the negative numbers are below sea level and the > sign means "is higher than." **12.** Each interval on the number line represents $\frac{1}{4}$. Therefore M is $\frac{2}{4}$ or $\frac{1}{2}$. **13.** Each interval is $\frac{1}{4}$ or 0.25. F corresponds to −1.25. **14.** Rewrite the numbers as 16.50 and 16.60. Any decimal beginning with 16.5 . . . is between.

15. Rewrite the numbers as −2.3900 and −2.3910. Any decimal beginning with −2.390 . . . is between. **16.** < **17.** =; 24 hundredths is the same as 240 thousandths. **18.** <; $4\frac{4}{15} = 4.2666 \ldots$, and $4.2\overline{6} < 4.93$ **19.** > ; on a number line, since −4 is to the right of −5, −4 is greater than −5. **20.** $0.6 = \frac{6}{10}$. Any fraction equal to $\frac{6}{10}$ is equal to 0.6. Sample: $\frac{3}{5}$ **21.** Any number other than . . ., −3, −2, −1, 0, 1, 2, 3, ... For example, $\frac{1}{2}$, 0.43, or −5.$\overline{3}$. **22.** The store will round up. You will pay 18¢. **23.** 3 + 9 = 12 **24.** 3.456 ⊗ 2.345 ⊜ **25.** 6 ⊗ π ⊜ is a calculator sequence that yields 18.849556 . . . . Rounded to the nearest integer, the answer is 19. **26.** 47 **27.** On a number line, divide the interval between 7 and 8 into 10 equal intervals; then each interval

represents 0.1. Place a dot at 7, 7.7, and 8 and label each point. **See below for graph. 28.** Divide a vertical number line into 10 equal intervals from −5 to 5. One interval represents 1°. Place dots at −4, 0, and 5, and label each in degrees. **See right for graph.**
29. Sample: the number of people watching a parade **30.** Sample: There are 4300 people watching the parade. **31.** The numbers are 0.1, 0.000001, 0.000000001, and 0.001. Of these, 0.1 is largest.
32. 18 is divisible by 1, 2, 3, 6, 9, and 18; so these are the factors of 18. **33.** Multiply the numerator and denominator of $\frac{6}{1}$ by 5.
This gives $\frac{30}{5}$. **34.** 3 is a factor of both 12 and 21. So $\frac{12}{21} = \frac{12/3}{21/3} = \frac{4}{7}$. **35.** (c)

27.

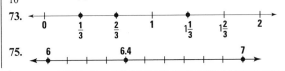

28.

The chart below keys the **Progress Self-Test** questions to the objectives in the **Chapter Review** on pages 59–61 or to the **Vocabulary** (Voc.) on page 57. This will enable you to locate those **Chapter Review** questions that correspond to questions you missed on the **Progress Self-Test.** The lesson where the material is covered is also indicated on the chart.

Question	1	2	3	4	5	6	7	8	9	10
Objective	A	A	G	F	A	A	B	B	D	D
Lesson	1-1	1-2	1-6	1-7	1-2	1-2	1-2, 1-6	1-6	1-3	1-4
Question	11	12	13	14	15	16	17	18	19	20
Objective	L	M	M	C	C	H	H	H	H	G
Lesson	1-8	1-2	1-8	1-2	1-8	1-9	1-9	1-9	1-9	1-6
Question	21	22	23	24	25	26	27	28	29	30
Objective	Voc.	K	D	E	E	I	M	M	K	Voc.
Lesson	1-8	1-4	1-4	1-5	1-5	1-6	1-2	1-8	1-3	1-1
Question	31	32	33	34	35					
Objective	B	J	J	J	N					
Lesson	1-2	1-10	1-10	1-10	1-1					

CHAPTER 1 REVIEW (pp. 59–61)

1. 4003 **3.** 120,000,000 **5.** 75.6 **7.** five hundred thousand, four hundred **9.** three and forty-one thousandths **11.** largest: 400,001; smallest: −.40000000001 **13.** 0.2, 0.19, 0, −0.2 **15.** $\frac{1}{7}, \frac{1}{9}, \frac{1}{11}$ **17.** $4\frac{1}{6}$, $3\frac{1}{3}, 2\frac{2}{3}$ **19.** Sample: 73.05 **21.** Sample: 6.995 **23.** Sample: 0 **25.** 5.84 **27.** 0.595959 **29.** 6.8 **31.** −2.5 **33.** −20 **35.** 48 **37.** 561.605 **39.** Sample: 5 (+/−) **41.** 2.2 **43.** 6.571428 **45.** .75 **47.** .2 **49.** Sample: $\frac{4}{5}$ **51.** Sample: $\frac{1}{4}$ **53.** > **55.** .667 > $\frac{2}{3}$ > .6

57. 1 **59.** $6.8\overline{9}$ **61.** Sample: $\frac{4}{14}$ and $\frac{6}{21}$ **63.** $\frac{4}{3}$ **65.** 29,000 **67.** 50¢ or \$.50 **69.** Sample: An estimate might be easier to work with. **71.** −75,000 < 10,000 **73.** See below. **75.** See below. **77. a.** .1 or $\frac{1}{10}$ **b.** 4.7 **79.** (c)

73.

75.

LESSON 2-1 (pp. 62–67)

27. Sample: You can round 1.43 up to 2 or down to 1. Multiply the 1 and 2 by 10,000. The answer should be between 10,000 and 20,000. **29. a.** 9,300,000,000 **b.** 9,320,000,000 **31.** 3.4 billion; 3,400,000,000 **33.** 5.2 billion or 5.3 billion; 5,200,000,000 or 5,300,000,000 **35.** 27.8 billion **37.** the word name "quadrillion," or next whole number 1,000,000,000,001, or \$1,000,000,000,000.01 **39. a.** .125 **b.** .1 **c.** .02 **d.** .01 **41.** 2.649

LESSON 2-2 (pp. 68–72)

19. 999 **21.** 10^4, ten thousand, 10,000; 10^5, hundred thousand, 100,000 **23. a.** 4; **b.** 4 **25.** Samples: $6^2 + 8^2 = 100$; $26^2 − 24^2 = 100$ **27.** True **29.** 2,500,000,000 **31.** 1,800,000 **33.** 13 **35.** −459.67 **37.** 2 **39.** 3 × 9 = 27

LESSON 2-3 (pp. 73–77)

15. 4.18×10^5 or 418,000 lb **17. a.** Sample: 9.9999 99 for 9.9999×10^{99} **b.** Sample: −9.9999 99 for $−9.9999 \times 10^{99}$ **19. a.** Find 3×10^5. Enter 3 (EE) 5 or 3 (×) 10 (yˣ) 5. **b.** Enter 1000 in scientific notation. Enter 1 (EE) 3. **c.** Find 2^5. Enter 2 (yˣ) 5. **21.** 6 **23. a.** Q **b.** E **c.** S **d.** C **25.** Samples: $\frac{9}{10}, \frac{27}{30}$

LESSON 2-4 (pp. 78–83)

39. $\frac{1}{100}$ or .01 **41.** 0.04387 **43.** $\frac{150}{100}$ or $1\frac{1}{2}$ or 1.50; $\frac{1795}{100}$ or $\frac{359}{20}$ or 17.95 **45.** (b) **47.** 0.001, $\frac{1}{1000}$ **49.** Volkswagen/Audi **51.** 0% **53. a.** 14,570,000; 12,472,000; 15,200,000 **b.** 1.457×10^7; 1.2472×10^7; 1.52×10^7 **55.** 8^3

LESSON 2-5 (pp. 84–88)

13. Sample: "We are with you totally." **15. a.** 50 million **b.** 75 million **c.** 100 million **d.** 125 million **17.** 36 or 37 **19.** lose; 48.8% is less than 50% which is half **21.** 1.8 million **23. a.** 990 **b.** 99 **c.** 9.9 **d.** 0.99 **25. a.** 10^{15} **b.** 10^4 **27.** 8,300,000 **29.** 4.8 **31.** $\frac{12}{25}$ **33.** 0.7853

LESSON 2-6 (pp. 89–92)

11. a. 350% **b.** $1.58 **13.** See below. **15.** $\frac{16}{5}$; 320% **17.** $\frac{27}{100}$; 27% **19.** $\frac{11}{25}$; 0.44 **21.** $\frac{3}{80}$ **23.** $\frac{1}{3}$ off **25.** 32.34 **27.** 240,000 **29.** 8.8×10^6 **31.** 5.256×10^5 **33.** one quadrillion

13.

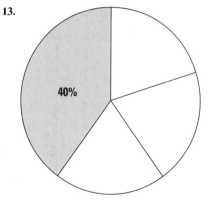

40%

LESSON 2-7 (pp. 93–98)

15. the percent of people who wake up a certain way **17.** by alarm or clock radio **19. a.** $\frac{40}{60}$ or $\frac{2}{3}$ **b.** $66\frac{2}{3}$% **c.** See below. **21. a.** 30% **b.** $\frac{25}{50}$ or $\frac{1}{2}$ **c.** See below. **23. a.** .35 **b.** 35% **25.** .125, $\frac{125}{1000}$ or $\frac{1}{8}$ **27.** If the numerator and denominator of a fraction are multiplied by the same nonzero number, the resulting fraction is equal to the original one. **29.** five and thirty-four hundredths

19. c.

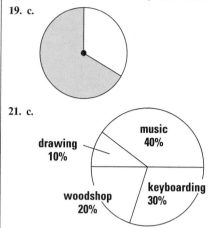

21. c.

music 40%

drawing 10%

woodshop 20%

keyboarding 30%

LESSON 2-8 (pp. 99–103)

19. (f) **21.** The zero power of any number other than 0 equals 1. **23. a.** 1 **b.** 1 **c.** When 10 to a power is multiplied by 10 to the opposite of the power, the product is 1. **25. a.** 1000 meters **b.** .01 meter or one hundredth of a meter **c.** .001 meter or one thousandth of a meter **27.** 9^2 or 3^4 **29.** 2^5 **31.** 58.0% of 13.50 million = 7.83 million or 7,830,000 **33.** West, Northeast **35. a.** $\frac{3}{16}$ **b.** 18.75% **37.** Multiply the number by .1. **39.** ninth

LESSON 2-9 (pp. 104–108)

11. a. 0.00000 001 centimeter **b.** 0.00000 000529 centimeter **13. a.** 5×10^{-2} **b.** .05 **15.** = **17. a.** 10^6 **b.** 10^{-6} **19.** millimeter, centimeter, kilometer **21.** The team lost more often. **23.** $^-19.7$ **25. a.** $0.001 or 0.1¢ **b.** $0.0001 **c.** $0.0111 **d.** $0.01 **27.** 4.5005005×10^9 **29. a.** See below. **b.** Yes **31.** Low; too much baggage would endanger the safety of the passengers.

29. a.

19    $19\frac{5}{16}$ 20

CHAPTER 2 PROGRESS SELF-TEST (pp. 112)

1. Move the decimal point 8 places to the right. 2,351,864,000 **2.** $\frac{34}{100} \times 600 = 204$ **3.** $32 \times 10^{-9} = 0.00000 0032$ **4.** Move the decimal point 5 places to the left. 0.0082459 **5.** $0.001 \times 77 = 0.077$ **6.** Move the decimal point 5 places to the right. 345,689,100 **7.** Move the decimal point 3 places to the left. 0.002816 **8.** 0.00000 01 **9.** $8\% = 8 \times .01 = 0.08$ **10.** 216 **11.** base; exponent **12. a.** 125 $\boxed{y^x}$ 6 $\boxed{=}$ **b.** $\boxed{3.8147 \ 12}$ Answers may vary. **13.** 3,814,700,000,000 **14.** radius **15.** 100% **16.** Place a decimal point between the 2 and 1. You must move the decimal point 10 places to the right to get the given number. So the scientific notation is 2.107×10^{10}. **17.** Place a decimal point to the right of the 8. Move the decimal point 8 places to the left to get the given number. So 0.00000 008 equals 8×10^{-8}. **18.** 4.5 $\boxed{EE}$ 13 or 4.5

$\boxed{\times}$ 10 $\boxed{y^x}$ 13 $\boxed{=}$ **19.** First write in scientific notation: 1.23456×10^{-7}. Then enter either 1.23456 $\boxed{EE}$ 7 $\boxed{+/-}$ or 1.23456 $\boxed{\times}$ 10 $\boxed{y^x}$ 7 $\boxed{+/-}$ $\boxed{=}$. **20.** In the order given, the numbers equal 256, 125, and 243. So from smallest to largest, they are: 5^3, 3^5, 4^4. **21.** First change to a decimal. 40% = 0.40. This number is between 0 and 1. **22. a.** $4.73 = \frac{4.73}{1} = \frac{473}{100}$ **b.** $\frac{473}{100} = 473\%$ **23.** $\frac{1}{3}$ **24. a.** cassette tapes (53.6%) **b.** 100% − 42.3% = 57.7% **c.** 3.3% of 801 million = .033 × 801 million = 26.433 million ≈ 26.4 million **25.** 30% × 150 = .3 × 150 = 45 **26.** You save 0.25 × $1699 = $424.75. Subtract from $1699 to get the sale price, $1274.25 ≈ $1274. **27.** 0.8 × 20 = 16. This is the number correct. So 20 − 16 = 4 is the number missed. **28.** 22.4 is not between 1 and 10. **29.** 6, since $10^6 = 1,000,000$ **30.** $\frac{3}{5} = 60\%$, $\frac{1}{10} = 10\%$, so $\frac{3}{5} - \frac{1}{10} = 60\% - 10\%$. **31.** 250 billion = 250,000,000,000 = 2.5×10^{11}

The chart below keys the **Progress Self-Test** questions to the objectives in the **Chapter Review** on pages 113–115 or to the **Vocabulary** (Voc.) on page 111. This will enable you to locate those **Chapter Review** questions that correspond to questions you missed on the **Progress Self-Test.** The lesson where the material is covered is also indicated on the chart.

Question	1	2	3	4	5	6	7	8	9	10
Objective	A	G, K	B	E	E	A	E	D	G	C
Lesson	2-1	2-5	2-8	2-4	2-4	2-2	2-4	2-8	2-4	2-2

Question	11	12	13	14	15	16	17	18	19	20
Objective	Voc.	N	C	J	O	F	F	N	N	C
Lesson	2-2	2-3	2-2	2-7	2-7	2-3	2-9	2-3	2-9	2-2

Question	21	22	23	24	25	26	27	28	29	30
Objective	G, H	I	H	O	M	M	M	L	D	K
Lesson	2-4	2-6	2-4	2-7	2-5	2-5	2-5	2-3	2-2	2-5

Question	31
Objective	F
Lesson	2-3

CHAPTER 2 REVIEW (pp. 113–115)

1. 320,000 **3.** 25,000 **5.** 30,000,000 **7.** 29,314,000,000
9. 0.00000 46 **11.** 64 **13.** $\approx$ 429,980,000 **15.** 100,000
17. 0.0001 **19.** one billion **21.** twelfth **23.** 0.00000 00273
25. 0.075 **27.** 8 **29.** 4.8×10^5 **31.** 1.3×10^{-4}
33. 3,000,000,000,000,000,000,000,000,000,000,000 **35.** 0.15
37. 0.09 **39.** 75 **41.** 6.2 **43.** 50% **45.** $\frac{3}{10}$ **47.** Sample: $\frac{57}{10}$

49. 86% **51.** about 42.9% **53.** $\overline{AC}, \overline{BC}$ or $\overline{DC}$ **55.** C **57.** $\overleftrightarrow{AB}$ or $\overleftrightarrow{AD}$ or $\overleftrightarrow{BD}$ **59.** If two numbers are equal, one can be substituted for the other in any computation without changing the results of the computation. **61.** $\frac{1}{2}$ and .5 **63.** 23 is not between 1 and 10
65. $15.90 **67.** $24,960 (about $25,000) **69.** Sample: 3.2 (EE) 10
71. a. Sample: 2 (y^x) 45 (=) **b.** $\approx 3.5184 \times 10^{13}$ **73.** $\approx 3.3516 \times 10^{13}$ **75.** Sample: 1.93 (EE) 15 (+/−) **77.** 6% **79.** percent of adults who sleep more than 7.5 hours up to 8.5 hours

LESSON 3-1 (pp. 116–124)
21. (d) **23.** (c) **25. See below. 27. See below. 29.** 45% **31.** 5^6

25.

6 cm $\approx 2\frac{3}{8}$ in.

27.

12.4 cm

LESSON 3-2 (pp. 125–129)
15. a. 1 rod = 16.5 ft **b.** 1 furlong = 660 ft **c.** 6600 ft
17. 120 strides **19.** 282 in. **21. a.** $3\frac{3}{8}$ in. **b.** 9 cm **23. a.** 10,000
b. $600 per student **c.** 1.5×10^8; $150,000,000 **25. a.** −2, −1.5, −1 **b.** −2 < −1.5 < −1 or −1 > −1.5 > −2 **27.** 900

LESSON 3-3 (pp. 130–134)
13. about 1,394,200,000 lb **15.** cup **17.** gallon **19.** No
21. A gallon is not a unit of length; it is a unit of volume or capacity. **23.** 2640 ft **25.** $4\frac{13}{16}$ inches **27.** 1.3837×10^{-2} **29.** 5.2
31. 1.536

LESSON 3-4 (pp. 135–139)
25. (b) **27.** Yes **29.** $\frac{1}{1000}$ or .001 second **31.** 8 times **33.** <
35. = **37.** Since $\frac{1}{16}$ inch is a smaller interval than $\frac{1}{10}$ inch, a measurement to the nearest $\frac{1}{16}$ inch is more precise. **39. a.** 25 or 26 **b.** 24 **c.** 22 **41.** 0.002 cm

LESSON 3-5 (pp. 140–144)
15. liter **17.** inch **19.** ounce **21. a.** 0.3125 **b.** 8 mm
23. $\approx$ 12.72 cups **25.** 4000 g **27.** 80 ounces **29.** 36 in.
31. a. 60 − 64 **b.** 30 − 39 **c.** 53.2 million **d.** 40% **33.** 25

LESSON 3-6 (pp. 145–151)
19. $\angle JGH$ **21 and 23. See below. 25.** 0.082 m **27. See below.**
29. $\overleftrightarrow{AC}$ **31.** 4.16×10^7

21 and 23. Sample:

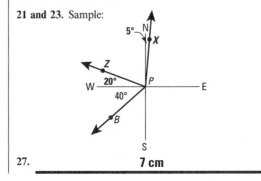

27.

7 cm

LESSON 3-7 (pp. 152–157)

17. 54°
19. See right. 21. See below.
23. a. obtuse; **b.** 134°
25. a. acute, 30°; **b.** obtuse, 120°;
c. right, 90°; **d.** acute, 15°
27. See below. 29. < **31.** < **33.** liter
35. $2\frac{1}{4}$ inches **37.** $1.20

21. a, b. Sample:

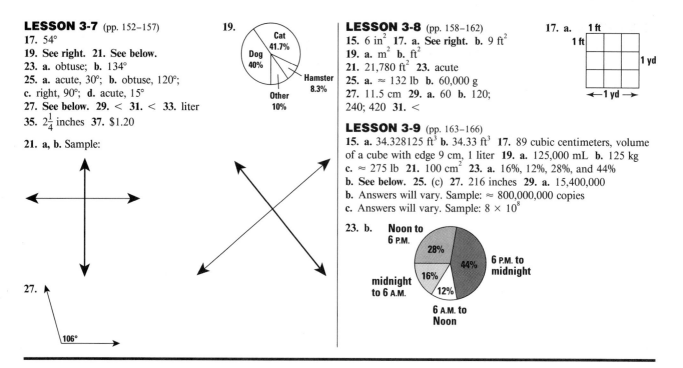

27.

LESSON 3-8 (pp. 158–162)
15. 6 in^2 **17. a. See right. b.** 9 ft^2
19. a. m^2 **b.** ft^2
21. 21,780 ft^2 **23.** acute
25. a. ≈ 132 lb **b.** 60,000 g
27. 11.5 cm **29. a.** 60 **b.** 120;
240; 420 **31.** <

17. a.

LESSON 3-9 (pp. 163–166)
15. a. 34.328125 ft^3 **b.** 34.33 ft^3 **17.** 89 cubic centimeters, volume of a cube with edge 9 cm, 1 liter **19. a.** 125,000 mL **b.** 125 kg
c. ≈ 275 lb **21.** 100 cm^2 **23. a.** 16%, 12%, 28%, and 44%
b. See below. 25. (c) **27.** 216 inches **29. a.** 15,400,000
b. Answers will vary. Sample: ≈ 800,000,000 copies
c. Answers will vary. Sample: 8×10^8

23. b.

CHAPTER 3 PROGRESS SELF-TEST (p. 170)

1. $2\frac{5}{8}''$ **2. a. See right. b.** 64 mm **3.** 1 in. = 2.54 cm
4. kilogram **5.** $\frac{3}{4}$ mile = $\frac{3}{4} \times 5280$ ft = 0.75×5280 ft = 3960 ft
6. Answers will vary: Sample: Since 1 foot = 12 inches, then 83 feet = 83 · 12 inches = 996 inches. **7.** 1000 tons **8.** 1103 mg = $1103 \times .001$ g = 1.103 g **9.** 1 quart = 2 pints, so 3 quarts = 6 pints. 1 pint = 2 cups, so 6 pints = 12 cups. Thus 3 quarts = 12 cups. **10.** 1 pound = 16 ounces, so 3.2 pounds = 3.2×16 ounces = 51.2 ounces **11.** 1 km ≈ .62 mi so 50 km ≈ $50 \times .62$ mi = 31 mi **12.** 1 kg ≈ 2.2 lb **13.** 9 liters = 9×1 liter ≈ 9×1.06 quarts = 9.54 quarts. So 10 quarts is larger. **14.** England **15.** 135° **16.** 0°; 90° **17.** 70° **18.** right **19. See right. 20.** Angles B and C are acute. There are no right angles. Angle A is obtuse. **21.** 20 violins out of 48 players is $\frac{20}{48}$ or $\frac{5}{12}$. $\frac{5}{12}$ of 360° = 150°. **22.** Area =

$(4.5 \text{ in.})^2 = 20.25 \text{ in}^2$ **23.** square centimeters **24.** 1 liter = 1000 cubic centimeters **25.** $V = (5 \text{ cm})^3 = 125 \text{ cm}^3$ **26.** Volume of the box is $(10 \text{ cm})^3 = 1000 \text{ cm}^3$. Volume of each block is $(2 \text{ cm})^3 = 8 \text{ cm}^3$. Divide the volume of the big box by the volume of one block, $\frac{1000 \text{ cm}^3}{8 \text{ cm}^3} = 125$. So 125 blocks will fit in the box.

2. a.

6.4 cm

19.

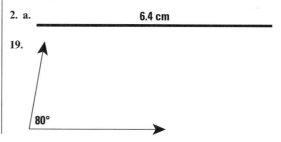

The chart below keys the **Progress Self-Test** questions to the objectives in the **Chapter Review** on pages 171–173 or to the **Vocabulary** (Voc.) on page 169. This will enable you to locate those **Chapter Review** questions that correspond to questions you missed on the **Progress Self-Test.** The lesson where the material is covered is also indicated on the chart.

Question	1	2	3	4	5	6	7	8	9	10
Objective	A	K, H	I	F	G	E	H	H	G	G
Lesson	3-1	3-1, 3-4	3-2	3-3	3-2	3-2	3-4	3-4	3-3	3-3
Question	**11**	**12**	**13**	**14**	**15**	**16**	**17**	**18**	**19**	**20**
Objective	I	I	I	N	B	Voc.	B	Voc.	L	C
Lesson	3-5	3-5	3-5	3-1	3-6	3-6	3-7	3-7	3-6	3-7
Question	**21**	**22**	**23**	**24**	**25**	**26**				
Objective	M	D	Voc.	H	D	J				
Lesson	3-7	3-8	3-8	3-4	3-9	3-9				

CHAPTER 3 REVIEW (pp. 171–173)

1. $1\frac{3}{4}$ in. **3.** 7 cm **5.** 65° **7.** 39° **9.** $\angle ABC$, $\angle BEA$, $\angle BCD$, $\angle DEC$
11. obtuse **13.** 4 cm^2 **15.** square meters **17.** 53 in^3 **19.** 1.1 lb
21. (b) **23. a.** cm **b.** in. **25. a.** liters **b.** gallons **27.** 16 ounces
29. 90 inches **31.** 10,000 lb **33.** 1 liter = 1000 cm^3 **35.** 2 m
37. 0.265 L **39.** 1 kg ≈ 2.2 lb **41.** mile **43.** 60.96 cm
45. 7.208 ≈ 7.21 qt **47.** ≈$4.44 **49.** 1728 cubes **51.** See right.
53. See right. **55.** See right. **57.** See right. **59.** More than $\frac{1}{4}$ of the
circle is shaded for lunch. **61.** France, 1790s **63.** Babylonians

51.

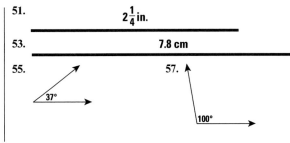

$2\frac{1}{4}$ in.

53.

7.8 cm

55. 37° **57.** 100°

LESSON 4-1 (pp. 174–181)

29. $6 \times 0.3 = 1.8$ **31.** It is not clear whether 4 is added to 2 first
or divided by 8 first. **33.** 117 **35.** 67 **37.** (c) **39.** 125 in^3
41. See below. **43. a.** The first year you will spend an additional
$7.47 since the smoke detectors cost $44.97 and the insurance
discount is only $37.50. The second through the fifth years, you
will save $37.50 per year, or $150.00 over four years. **b.** The total
savings for the five years will be $142.53.

41.

14.3 cm

LESSON 4-2 (pp. 182–186)

13. Samples: $6 \cdot 4 + 13 \cdot 4 = 19 \cdot 4$; $6 \cdot 9.7 + 13 \cdot 9.7 = 19 \cdot 9.7$;
$6 \cdot 100 + 13 \cdot 100 = 19 \cdot 100$; $6 \cdot \frac{2}{3} + 13 \cdot \frac{2}{3} = 19 \cdot \frac{2}{3}$
15. $a \cdot 0 = 0$ **17.** In n years, we expect $n \cdot 100$ more students and
$n \cdot 5$ more teachers. **19.** Yes. a and b can be the same number.
21. a. $b + .5$ is not an integer **b.** $3.5 + .5$ is an integer **c.** The
pattern is not always true because if $b = .5$, 1.5, 2.5, and so on, the
sum will be an integer. **23.** 55 **25.** 74 **27.** 36 **29.** 50/.0001 =
500,000 **31.** 3 m

LESSON 4-3 (pp. 187–192)

19. It is not clear whether $14 - 5$ or $5 + 3$ is done first.
21. $.06 \cdot C$ or $6\% \cdot C$ **23.** $x + \frac{1}{4} \cdot x$ **25.** It could mean $\frac{2}{4}$ or $\frac{4}{2}$.
27. $\frac{a}{1} = a$ **29.** $5^2 + 9^2 = 25 + 81 = 106$ **31.** $23 + .09 \times 11 =$
$23 + 0.99 = 23.99$ **33.** See below. **35.** $2\frac{4}{8}$ in. or $2\frac{1}{2}$ in.
37. about 200,000,000 White; about 30,000,000 Black; about
2,000,000 American Indian, Eskimo, or Aleut.; about 7,000,000
Asian or Pacific Islander; about 10,000,000 Other; about
22,000,000 Hispanic

33. Samples:
a. **b.** **c.**

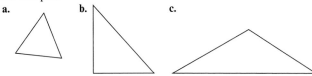

LESSON 4-4 (pp. 193–196)

15. a. 47 lb **b.** 27 lb **c.** 5 lb **17. a.** 3, 5, 7, 9, and 11 **b.** Sample:
Begin with 3; add 2 to each number to get the next number.
19. a. $4n + 9$ **b.** 49 **21.** (a) **23.** $a + 0 = a$ **25.** milli-
27. (b) and (d)

LESSON 4-5 (pp. 197–202)

25. $(a + 4)b + 3$ **27.** No **29. a.** (iii) **b.** 73.4° **31.** 436
33. $16 - (8 - 4 - 2) = 14$ **35. a.** T = number of touchdowns;
E = number of extra points; F = number of field goals;
S = number of safeties **b.** the total score **37.** 36 mi^2

LESSON 4-6 (pp. 203–207)

17. $35.\overline{6} \approx 36$ books **19. a.** $58.\overline{3}$ **b.** $91.\overline{6}$ **21.** $\frac{x + 3y}{z} + \frac{4y + z}{3x} =$
$\frac{3 + 3 \cdot 2}{1} + \frac{4 \cdot 2 + 1}{3 \cdot 3} = \frac{3 + 6}{1} + \frac{8 + 1}{9} = \frac{9}{1} + \frac{9}{9} = 9 + 1 = 10$
23. $(3 \cdot 8 - 6)/2 + 3 = 12$ **25. a.** $30 + 20$ **b.** $n + 3$ **c.** $n - .1$
27. $20.85 (It will require 3 bags; some fertilizer will be left over.)
29. a. 268,729,000,000 **b.** 269 billion **c.** 2.68729×10^{11}

LESSON 4-7 (pp. 208–212)

11. 24 m^2 **13.** 0.912 **15.** 0.500 **17.** 73° **19.** Sample: $d = 2r$ or
$r = \frac{1}{2}d$ **21.** 94 **23.** 10 **25.** (c) inside parentheses, (b) powers,
(a) multiplication or division from left to right, (d) addition or
subtraction from left to right **27.** < **29.** < **31.** 6000 g

LESSON 4-8 (pp. 213–218)

11. Relative frequency is a number $\frac{F}{N}$ from 0 to 1 that indicates
how many times (F) an event occurred out of N repetitions of an
experiment that has taken place. Probability is a number from 0 to
1 that indicates what we think is the likelihood of an event. **13.** $\frac{3}{50}$
15. a. $\frac{1}{8}$; **b.** B: $\frac{3}{8}$; C: $\frac{1}{4}$; D: $\frac{1}{4}$ **17.** 20% **19.** the 8.6 by 6.8 rectangle
21. a. 6 **b.** 2.8 **c.** 6.2 **23.** about 10.2 miles

LESSON 4-9 (pp. 219–223)

17. Sample: Los Angeles **19.** eight **21.** Sample: $^-9$ **23.** 6
25. 11 **27.** 4 **29.** 1 **31.** 3 **33.** $\frac{33}{100}$ or .33 **35.** 26,127
37. 5 km > 500 m > 50,000 mm or 50,000 mm < 500 m < 5 km

LESSON 4-10 (pp. 224–228)

25. (d) **27.** See below. **29.** See below. **31. a.** $0 \le d < 25,000$
b. Samples: 4.32, 5132, 24,999.99 **c.** See below. **33. a.** $0 \le P \le 1$
b. See below. **35.** $1.92 **37.** $1 - x$ **39.** 189 **41.** $\frac{a}{b} \cdot \frac{b}{a} = 1$

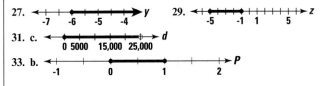

27. y: -7 -6 -5 -4 **29.** z: -5 -1 1 5
31. c. d: 0 5000 15,000 25,000
33. b. P: -1 0 1 2

1. $6 + 56 + 9 = 71$ **2.** $35 + 50 = 85$ **3.** $25 - 3 - 1 =$ $22 - 1 = 21$ **4.** $5 + 3 \cdot 16 = 5 + 48 = 53$ **5.** $\frac{100 + 10}{10 + 5} = \frac{110}{15} =$ $7.\overline{3} \approx 7$ **6.** $10 + 3 \cdot 100 = 10 + 300 = 310$ **7.** $(3 + 4)(4 - 3) =$ $7 \cdot 1 = 7$ **8.** $100 + 5[100 + 4(100 + 3)] = 100 + 5[100 + 4 \cdot 103]$ $= 100 + 5[100 + 412] = 100 + 5 \cdot 512 = 100 + 2560 = 2660$ **9.** $(4x)^2 = 64$; (a) $(4 \cdot 2)^2 = 8^2 = 64$; so 2 is a solution. (b) $(4 \cdot 4)^2$ $= 16^2 = 256 \neq 64$; so 4 is not a solution. (c) $(4 \cdot 8)^2 = 32^2 = 1024$ $\neq 64$, so 16 is not a solution. (d) $(4 \cdot 16)^2 = 64^2 \neq 64$, so 16 is not a solution. The correct answer is (a). **10.** $10 \cdot a = 6 \cdot a + 4 \cdot a$ **11.** $a + b = b + a$ **12.** In y years we expect the town to grow by $y \cdot 200$ people. **13.** $c = 23 \cdot 5 + 6 = 115 + 6 = 121$¢, or $1.21 **14.** $c = 23 \cdot 9 + 6 = 207 + 6 = 213$¢, or $2.13 **15.** $p = \$45 -$ $\$22.37 = \$22.63$ **16.** 12×16 **17.** $40 < 47$ **18.** $n \geq 0$ or $0 \leq n$ **19.** $\frac{9}{a}$ **20.** $\frac{W}{W + L} = W/(W + L)$; The correct answer is (b).

21. (c) **22.** 8 is the number of times the event occurred; 50 is the number of total outcomes. So the relative frequency is $\frac{8}{50}$, or .16. **23.** The total number of students is $5 + 7 = 12$. Since 5 students are boys, the probability a boy is chosen is $\frac{5}{12}$ or $.41\overline{6}$. **24.** (b) **25.** $6 \cdot 7 = 42$, so $x = 7$. **26.** Any number between -5 and -4 is a solution. Some sample solutions are -4.2, -4.9, $-4\frac{1}{2}$. **27.** Units in formulas must be consistent for the answer to make sense. For example, to find the area of a rectangle given the dimensions of the length and the width, if $w = 3$ inches and $\ell = 4$ feet, the area A cannot be 12 of either unit, even though $A = \ell w$. **28.** $x = 7 \cdot 3 = 21$ **29.** See below. **30.** This is written as $4.5 \leq k < 6$ **31.** Samples: If your age is 10 years, your sister's age is $10 - 5$ or 5 years. If your age is 14, your sister's age is $14 - 5$ or 9 years. **32.** Samples: $30 + 2 - 30 = 2$; $7.3 + 6.2 - 7.3 = 6.2$.

29.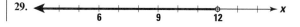

The chart below keys the **Progress Self-Test** questions to the objectives in the **Chapter Review** on pages 233–235 or to the **Vocabulary** (Voc.) on page 231. This will enable you to locate those **Chapter Review** questions that correspond to questions you missed on the **Progress Self-Test.** The lesson where the material is covered is also indicated on the chart.

Question	1	2	3	4	5	6	7	8	9	10
Objective	A	A	A	A	A	B	B	B	C	E
Lesson	4-1	4-5	4-1	4-1	4-6	4-4	4-5	4-6	4-9	4-2
Question	11	12	13	14	15	16	17	18	19	20
Objective	E	G	I	I	I	H	H	H	H	D
Lesson	4-2	4-2	4-7	4-7	4-3	4-3	4-3	4-3	4-7	4-6
Question	21	22	23	24	25	26	27	28	29	30
Objective	L	J	J	Voc.	C	C	I	I	K	K
Lesson	4-1	4-8	4-8	4-10	4-9	4-10	4-7	4-7	4-10	4-10
Question	31	32								
Objective	F	F								
Lesson	4-2	4-2								

CHAPTER 4 REVIEW (pp. 233–235)

1. 215 **3.** 83 **5.** 31 **7.** 1966 **9.** 158 **11.** 5 **13.** $\frac{14}{65} = 0.215\dots$ **15.** 24 **17.** 18 **19.** 24 **21.** 184 **23.** $\frac{8}{12}$ or $\frac{2}{3}$ or $.\overline{6}$ **25.** (b) **27.** 4 **29.** 2 **31.** powering **33.** (d) **35.** $5 \cdot x + 9 \cdot x = 14 \cdot x$ **37.** $\frac{a}{9} + \frac{b}{9}$ $= \frac{a + b}{9}$ **39.** Sample: $2 + 6 = 1 + 6 + 1$; $2 + .5 = 1 + .5 + 1$; $2 + \frac{1}{20} = 1 + \frac{1}{20} + 1$ **41.** If the weight is w ounces, the postage is 5¢ $+ w \cdot 20$¢. **43.** $18 + 27$ **45.** $4 \cdot 20 - 1$ **47.** $2x + 7$ **49.** $x < 5$ **51.** ≈ 127 **53.** 72 in² or .5 ft² **55.** 1 **57.** $\frac{2}{10}$ or $\frac{1}{5}$ or .2 **59.** $\frac{14}{30}$ or $\frac{7}{15}$ or $.4\overline{6}$ **61.** $x \geq 2$ or $2 \leq x$ **63.** See right. **65.** See right. **67.** See right. **69.** Sample: $*$, $\wedge$

63.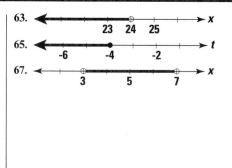

65.

67.

LESSON 5-1 (pp. 236–243)

19. Not enough information to tell. There may be overlap, so we cannot add the amounts to give an answer. The total is at least $20 and at most $35. **21.** $b + s + 1$ **23.** No, rain and sun could appear on the same day, as could snow and sun. **25.** a point 220 ft north of the starting point **27.** never positive **29.** sometimes positive **31. a.** 50% **b.** $\approx 33.9\%$ **33.** 1 **35. a.** -6 **b.** 0

LESSON 5-2 (pp. 244–247)

19. -7 **21.** b **23.** -5 **25.** $-x$ **27.** 5 **29.** 1 **31.** 9 **33. a.** The temperature is $-9°$. **b.** $-11 + 2$ **35.** $t + -35 = -60$ **37. a.** 10,000 lb **b.** 3750 lb **c.** No

LESSON 5-3 (pp. 248–252)

23. $-\frac{2}{3}$ **25.** -4 **27.** -2 **29.** 0 **31.** 0 **33. a.** True **b.** True **c.** True **35.** (b) **37.** -65.399 billion dollars or $-\$65,399,000,000$ **39. a.** 0 **b.** 15 **c.** 13 **41. a.** $-150 + 50$ **b.** -100; the person is $100 in debt. **43.** 1 in.

LESSON 5-4 (pp. 253–259)

19. $-29°$ **21.** $72°$ **23. a.** $240°$ (or $-120°$) **b.** $120°$ (or $-240°$) **25.** $100°$ **27.** $2160°$ **29.** -11 **31.** 14 **33.** -9 **35.** $45 + m = 120$ (minutes), or $\frac{3}{4} + m = 2$ (hours) **37.** $Y - 2$ **39.** 6000 m

LESSON 5-5 (pp. 260–265)

17. a. 1 tablespoon **b.** 3 teaspoons **c.** Yes. Since 1 tsp $= \frac{1}{3}$ tbsp, 3 tsp $= 3 \cdot \frac{1}{3}$ tbsp $= 1$ tbsp. **19.** $\frac{1}{16} + \frac{1}{8} = \frac{1}{16} + \frac{2}{16} = \frac{3}{16}$ **21.** $\frac{35}{8} + \frac{8}{3} = \frac{105}{24} + \frac{64}{24} = \frac{169}{24}$ **23.** $9\frac{3}{4} + \frac{1}{4} + 3 = 10 + 3 = 13$ or $\frac{13}{1}$ **25.** a counterclockwise turn of $68°$ **27.** -4.7 **29.** -8 **31. a.** Sample: 0 **b.** See below. **33. a.** 3×10^4 **b.** about 66,000 lb

31. b.

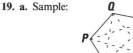

LESSON 5-6 (pp. 266–270)

15. $\frac{20}{23}$ **17.** not mutually exclusive **19.** $\frac{1}{3}$ **21.** $\frac{43}{30}$ or $1\frac{13}{30}$ **23.** $\frac{21}{4}$ lb or $5\frac{1}{4}$ lb **25.** 40 **27.** 4.351

LESSON 5-7 (pp. 271–275)

11. Your account is overdrawn. **13.** 12 **15.** 0 **17.** They are $10.25 over budget. **19.** $1°$ **21.** Associative Property of Addition **23. a.** $60°$ **b.** $120°$ **c.** $\angle AOC, \angle COD, \angle DOB$ **d.** There are none. **e.** $\angle AOD, \angle COB$ **f.** $\frac{2}{3}$ **25.** H **27.** Sample: $5 + -5 = 0$ **29.** $(t - 3)$ degrees **31.** 17.1 **33.** 32,736

LESSON 5-8 (pp. 276–281)

15. $A + 38 = 120$, $A = 82$, $82 + 43 + -5 = 120$ **17.** $1025 **19.** $4.03 **21.** $b = c + -a$ **23. a.** $\frac{11}{12}$ **b.** choosing a blue chip **c.** $b = \frac{1}{12}$ or $.08\overline{3}$ **25. a.** Sample: -4 **b.** Sample: 2 **27.** $3n$ **29.** $B/(2C)$ or $\frac{B}{2C}$ **31.** The measure of angle LNP is thirty-five degrees. **33.** 30

LESSON 5-9 (pp. 282–287)

17. nonagon **19. a.** See below for drawing. $\overline{PR}, \overline{PS}, \overline{QT}, \overline{QS}, \overline{TR}$ **b.** 5 **21.** Sample: All sides are approximately the same length. **23.** $\overline{XA}$ joins two consecutive vertices and is a side of the polygon. A diagonal is not a side of the polygon, so $\overline{XA}$ is not a diagonal. **25.** (d) **27.** $w = 83$; $300 = 172 + (45 + 83)$ **29.** $d = 8.1$; $8.1 + -1.3 = 6.8$ **31. a.** 8 ft = 96 in., so an equation is $s + 3 = 96$ **b.** $s = 93$ inches

19. a. Sample:

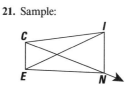

LESSON 5-10 (pp. 288–292)

15. a. $9 + 9 + 5 + 5 + x = 30$ **b.** 2 **17.** 40 and 20; 30 and 30; 20 and 40 **19. a.** Yes **b.** 20 m **21.** See below. **23. a.** $\frac{1}{50} + x = 1$ **b.** $x = \frac{49}{50}$ **25.** 16.45 **27.** See below. **29. a.** $E + -4$ **b.** $A + -4$ **c.** If $E = A$, then $E + -4 = A + -4$. **d.** Addition Property of Equality **31.** all **33.** $94°$

21. Sample:

C, I, E, N (diagram with rectangle and diagonals)

27.

0 1 ——— 5 —→ x

(number line)

CHAPTER 5 PROGRESS SELF-TEST (p. 296)

1. -7 **2.** -710 **3.** $-9.8 + -(-1) = -9.8 + 1 = -8.8$ **4.** $x + y + -x + 4 = x + -x + y + 4 = 0 + y + 4 = y + 4$ **5.** 8 **6.** $|-2 + 1| + |0| = |-1| + 0 = |-1| = 1$ **7.** $-6 + 42 + -11 + 16 + -12 = -6 + -11 + -12 + 42 = -29 + 58 = 29$ **8.** $|-(-3) + 8| = |3 + 8| = |11| = 11$ **9.** Add -43 to both sides; $x = -12$ **10.** Add 2.5 to both sides; $y = 1.3$ **11.** $8 = z + -7$; add 7 to both sides; $z = 15$ **12.** $\frac{53}{12} + \frac{11}{12} = \frac{(53 + 11)}{12} = \frac{64}{12} = \frac{16}{3}$ **13.** $\frac{5}{x} + \frac{10}{x} = \frac{(5 + 10)}{x} = \frac{15}{x}$ **14.** 9 is the common denominator. So $\frac{-8}{3} = \frac{-24}{9}$ and $\frac{17}{9} + \frac{-24}{9} = \frac{(17 + -24)}{9} = \frac{-7}{9}$. **15.** Notice that $\frac{2}{16} = \frac{1}{8}$. So 8 is the common denominator and $\frac{2}{8} + \frac{3}{8} + \frac{1}{8} = \frac{6}{8} = \frac{3}{4}$. **16.** $-20 + c = 150$ **17.** Add 20 to both sides; $c = 170$ **18.** Associative Property of Addition **19.** One instance is $19 + 0 = 19$. There are many others. **20.** hexagon **21.** Sample: A toss of a die shows a 6 and a toss of the same die shows a 2. Since these two events cannot occur at the same time, they are mutually exclusive. **22.** 3 cm + 3 cm + 4 cm + 4 cm + 4 cm = 18 cm **23.** (a) is the answer. (b) is a segment, (c) is a ray,

(d) is a line. **24.** $m + n = 50$ **25.** The probabilities must add to 100%. 50% + 45% + d = 100%; 95% + d = 100%; d = 5% **26.** 5 m + .03 m = 5.03 m **27.** positive, since the absolute value of the second addend is larger than the absolute value of the first addend **28.** $MP = MA + AP = 16 + 8 = 24$ **29.** $MA + PA = MP$, so $2.3 + PA = 3$. Add -2.3 to both sides to get $PA = 0.7$. **30.** See below. The sum is -1. **31.** The probability of getting a 1 is $\frac{1}{6}$ and the probability of getting a 2 is $\frac{1}{6}$. Since these are mutually exclusive events, you can add the probabilities. $\frac{1}{6} + \frac{1}{6} = \frac{2}{6} = \frac{1}{3}$. **32.** $-50° + 250° = 200°$; a 200° counterclockwise turn **33.** $\frac{360}{5} = 72°$ **34.** either a clockwise turn of $-72 + -72 = -144$ or a counterclockwise turn of $72 + 72 + 72 = 216°$

30.

-3
2
(number line) -5 -4 -3 -2 -1 0 1 2 3

The chart below keys the **Progress Self-Test** questions to the objectives in the **Chapter Review** on pages 297–299 or to the **Vocabulary** (Voc.) on page 295. This will enable you to locate those **Chapter Review** questions that correspond to questions you missed on the **Progress Self-Test**. The lesson where the material is covered is also indicated on the chart.

Question	1	2	3	4	5	6	7	8	9	10
Objective	A	A	C	C	B	B	A	B	D	D
Lesson	5-3	5-3	5-2	5-2	5-3	5-3	5-3	5-3	5-8	5-8

Question	11	12	13	14	15	16	17	18	19	20
Objective	D	A	A	A	A	J	D	F	F	H
Lesson	5-8	5-5	5-5	5-5	5-5	5-1	5-8	5-7	5-2	5-9

Question	21	22	23	24	25	26	27	28	29	30
Objective	G	E	Voc.	I	K	I	A, C	I	I	M
Lesson	5-6	5-10	5-10	5-1	5-6	5-10	5-2, 5-3	5-10	5-10	5-1

Question	31	32	33	34						
Objective	K	L	L	L						
Lesson	5-6	5-4	5-4	5-4						

CHAPTER 5 REVIEW (pp. 297–299)

1. -12 **3.** 9.6 **5.** 3 **7.** $\frac{46}{9}$ **9.** $\frac{7}{12}$ **11.** 12 **13.** 8 **15.** -1 **17.** -17 **19.** -40 **21.** $\frac{2}{7}$ **23.** 48 **25.** $x = 20$ **27.** $y = -3$ **29.** $c = 38$ **31.** $v = 5922$ **33.** $e = \frac{7}{6}$ **35.** 12 **37.** 16 cm **39.** $12 + 18 + 20 + 7 + x = 82$ **41.** Addition Property of Equality **43.** Commutative Property of Addition **45.** not mutually exclusive **47.** mutually exclusive **49.** $LEAK$ **51.** 4, 4 **53.** $1\frac{1}{2} + \frac{3}{5} = M$ **55.** $T = D + M + C$ **57.** $AC = x + 3$ **59. a.** $\frac{3}{8} + -\frac{1}{4}$ **b.** $\frac{1}{8}$ point **61. a.** $-60 + 25$ **b.** 35 feet below sea level **63.** 24% **65.** $\frac{144}{360} = \frac{2}{5}$ **67.** They are mutually exclusive and together include all outcomes. **69.** 135° or $-225°$ **71. a.** B **b.** 90° or $-270°$

73.

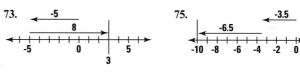

75.

LESSON 6-1 (pp. 300–306)

15. What is 34.2×5.67? **17.** 25 **19.** 13 trains

21.

quarters	dimes	nickels
2	0	1
1	3	0
1	2	2
1	1	4
1	0	6
0	4	3
0	3	5

23. 13 **25.** 125 cm^3 **27.** 26,400 feet **29.** about 54 million

LESSON 6-2 (pp. 307–310)

13. (c) **15.** 39 **17.** 35 **19.** (A regular hexagon is a 6-sided polygon whose angles all have same measure, and sides all have same length.) **See below for drawing.** **21.** Sample: 29 and 31. (Twin primes are two consecutive odd numbers that are both primes.) **23.** last year **25.** Not enough information.
If $b + e + h > c + f + g$, then last year;
if $b + e + h < c + f + g$, then this year.
27. 9 **29.** 0 **31.** 10

19.

LESSON 6-3 (pp. 311–314)

5. 56 games **7.** 20 posts **9.** 228 miles **11.** Samples: glossary, dictionary **13.** 5, 6, 7, 8 **15.** Sarah and Dana have the same number of dollars. **17.** -3 **19.** 486 **21.** 800 mm **23.** 56.83

LESSON 6-4 (pp. 315–319)

13. 147, 118, 35 **15.** 13 and 1
17. Sample:

```
10   FOR X = 1 TO 20
20   IF X ^ 3 + X ^ 2 + X = 2954 THEN PRINT X
30   NEXT X
40   END
```

19. Ali **21.** 12, 14, 15, 16, 18, 20, 21, 22 **23.** $\frac{4}{45}$ or $0.0\overline{8}$ **25.** 4.2 feet **27. a.** $1.4\overline{142857}$ **b.** 1.414 **c.** 1 and 2

LESSON 6-5 (pp. 320–324)

11. a. 98¢ **b.** $2.36 **c.** $.29 + (w - 1) \cdot .23$ dollars **13.** 28 miles **15. a.** It tests the numbers between and including 10 and 30 at .5 intervals as possible solutions to the equation $x^4 + x^2 = 360900.3125$. **b.** 41 **c.** output: $x = 24.5$ **17. a.** $5 + c = -3$ **b.** $c = -8$ pounds

LESSON 6-6 (pp. 325–331)

17. B3, D3, E3 **19.** =.05*F15
21. a.

	A	B
1	Minutes	Cost
2	1	1.19
3		
4		

Cell A3 is =A2+1 and so on; cell B3 is =B2+.79 and so on.

b. $21.73 **23. a.** $=A2^\wedge 2$ **b.** $=A3^\wedge 3$ **c.** $=A4^\wedge 4$ **d.** $=A2^\wedge 5$
25. a. $5.45 **b.** $5.80 **c.** $4.75 + (t - 1) \cdot .35$ dollars **27.** 91 and 37, or 82 and 28, or 73 and 19. **29.** $\frac{53}{12}$ or $4\frac{5}{12}$ **31.** $7\frac{1}{4}$ or $\frac{29}{4}$

LESSON 6-7 (pp. 332–337)
17. a. Sample: if $m = 2$, $-(-2 + 9) = 2 + {}^-9$, true **b.** Sample: if $m = {}^-4$, $-(-(-4) + 9) = {}^-4 + {}^-9$, true **c.** possibly true **19.** (c)
21. $R - Z$ **23.** $G - 49.95$ dollars **25.** (b)
27. $=(B3+B4+B5+B6)/4$ **29. a.** $4.50 **b.** $1.00 + (h - 1) \cdot .50$ dollars **31.** 12 stakes **33.** 70 kg

CHAPTER 6 PROGRESS SELF-TEST (p. 341)

1. any two of dictionary, glossary in a math book, encyclopedia
2. An octagon has 20 diagonals. **See right for drawing. 3.** Name the teams *A, B, C, D, E,* and *F.* Draw all the segments connecting these points. The figure created is a hexagon. The number of games is equal to the number of diagonals plus the number of sides in the hexagon. That is $9 + 6 = 15$ games. **See right for drawing. 4.** Use trial and error. The sums of each pair of factors of 24 are: $1 + 24 = 25$; $2 + 12 = 14$; $3 + 8 = 11$ and $4 + 6 = 10$. So 4 and 6 give the smallest sums. **5.** False. If you know an algorithm for an operation, you know how to answer the question. So multiplying fractions is only an exercise, not a problem. **6.** 53, 59 **7.** If one value is positive and one is negative, the pattern will be false. For instance, $|{}^-5| + |3| = 5 + 3 = 8$, but $|{}^-5 + 3| = |{}^-2| = 2$.
8. Sample: The calculations are done automatically by a computer.
9. 2 pounds at $3 per pound cost $6, so multiplication gives the answer. $1.49 per pound $\cdot$ 1.62 pounds = $2.4138, which rounds up to $2.42. **10.** True **11.** By trial and error, $(8 - 3) + (8 - 4) = 5 + 4 = 9$, so 8 is the solution. **12.** Read carefully and draw a picture. There must be 6 more dots on the right side between 3 and 10, for a total of 14 dots. **See right for drawing. 13.** Use trial and error. A 3-gon has 0 diagonals, a 4-gon has 2 diagonals, a 5-gon has 5 diagonals, a 6-gon has 9 diagonals. So $n = 6$. **14.** Use a special case. $\frac{.05}{.01} = 5$, which means that the decimal point is moved two places to the right.

15.

weeks	amount left
1	$1000 - 25$
2	$1000 - 2 \cdot 25$
3	$1000 - 3 \cdot 25$
4	$1000 - 4 \cdot 25$

16. After 31 weeks, Brianna will have $1000 - 31 \cdot 25$ dollars left. That computes to $225. **17.** After you find an answer, you should check your work. **18.** Samples: organizing and storing 16 test scores for a class; keeping track of monthly sales for a store.

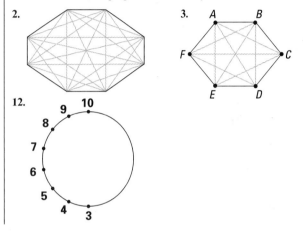

2.

3.

12.

The chart below keys the **Progress Self-Test** questions to the objectives in the **Chapter Review** on pages 342–343 or to the **Vocabulary** (Voc.) on page 340. This will enable you to locate those **Chapter Review** questions that correspond to questions you missed on the **Progress Self-Test.** The lesson where the material is covered is also indicated on the chart.

Question	1	2	3	4	5	6	7	8	9	10
Objective	D	K	I	E	A, Voc.	C	G	J	H	G
Lesson	6-2	6-3	6-3	6-5	6-1	6-2	6-7	6-6	6-7	6-7

Question	11	12	13	14	15	16	17	18
Objective	B	I	E	F	E	E	A	J
Lesson	6-4	6-3	6-5	6-7	6-5	6-5	6-1	6-6

CHAPTER 6 REVIEW (pp. 342–343)

1. (b) **3.** Sample: Divide answer by 1487 or 309. **5.** (b) **7.** 1036
9. composite; 7 is a factor **11.** 21, 22, 24, 25, 26, 27, 28
13. a number that equals the sum of its divisors except itself
15. 2   1,2
     4   1, 2, 4
     8   1, 2, 4, 8
    16   1, 2, 4, 8, 16
    32   1, 2, 4, 8, 16, 32
   256 has 9 factors.
17. 6, left **19.** Sample: If $x = 2$ and $y = 3$, then $5x + 5y \neq 10xy$.
21. $8.69 **23.** 20 games **25.** Pueblo

27.

	A	B
1	Minutes	Cost
2	0	18.75
3	15	29.75
4	30	40.75
5	45	51.75
6	60	62.75

Cell	Equation
A3	$= A2 + 15$
B3	$= B2 + 11$
.	.
.	.
.	.
A26	$= A25 + 15$
B26	$= B25 + 11$

29. 35 **31.** 7 triangles

LESSON 7-1 (pp. 344–351)
13. $10 - x$ meters **15. a.** A and C **b.** BC **17. a.** \$470.75
b. $510.75 - W$ dollars **19. a.** 17,558,165 **b.** 469,557 **c.** Answer
depends on where you are. **21.** $b^2 - a^2$ **23.** 24.64 square units
25. Sample: Let $a = 5$, $b = -2$. Then $-(a + b) = -(5 + -2) = -3$
and $-b + -a = -(-2) + -5 = -3$. Let $a = -9$ and $b = -4$. Then
$-(a + b) = -(-13) = 13$ and $-b + -a = 4 + 9 = 13$. It seems true.
27. 1 **29. a.** -5 **b.** 3.4 **c.** 0 **31.** 637

LESSON 7-2 (pp. 352–357)
25. 2 **27.** (c) **29. a.** $a - b = b - a$ **b.** No **31.** -7 **33.** $43 - m$
35. 7 **37.** 6°

LESSON 7-3 (pp. 358–361)
15. \$327.61 **17.** $2\frac{1}{8}$ **19.** -17 **21.** 632,118 **23.** \$88 **25.** 70 cm
27. 10° **29.** 1.5 **31.** 5,000,000,000,000 **33.** .75

LESSON 7-4 (pp. 362–366)
19. $c = s - p$ **21. a.** $14 - d = -3$ **b.** $d = 17$ **c.** The temperature
has decreased by 17°. **23.** $x = -120$ **25.** $K = 13$ **27.** (c)
29. Algebraic Definition of Subtraction; Addition Property of
Equality; Associative Property of Addition; Property of Opposites;
Additive Identity Property of Zero; Addition Property of Equality;
Associative Property of Addition; Property of Opposites; Additive
Identity Property of Zero. **31.** $y = 48$ **33.** 58 or 59 **35.** $d = 90 - c$

37. a. **b.** right angles

LESSON 7-5 (pp. 367–373)
11. a. Gabriella, Lakara, Jason, Kamal, Martha **b.** Lakara, Kamal,
Martha, Jason **c.** Jason **d.** Contest 2 and Contest 3 **13.** 22
15. a. $\frac{1}{2}$ or 50% **b.** $\frac{33}{100}$ or 33% **c.** 67% **17. a.** $x - 39.725 = 0.12$
b. 39.845 **19. a.** Queen Victoria, 64 years **b.** 2016 **21.** (b)
23. ≈ 810 **25.** 2-liter watering can

LESSON 7-6 (pp. 374–380)
23. If the ramp is perpendicular to expressway, the driver has to
accelerate into fast-moving traffic while making a 90° turn; if the
ramp is nearly parallel, acceleration is much easier. **25.** $\angle NOM$
and $\angle POM$ **27. a.** False **b.** Vertical angles have the same measure.
If one is acute, so is the other.
29. $x = y$ **31.** 140°
33. a. See right. **b.** 24 **35.** $y = 59$
37. $c = 300$

33. a.
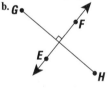
blue jeans   sneakers

LESSON 7-7 (pp. 381–386)
23. m∠1 = 90°, m∠2 = m∠4 = 40°, m∠3 = 140° **25.** 360°
27. See below. 29. See below. 31. $m = -\frac{1}{3}$ **33. a.** 16 square units
b. Samples: area surrounding a house; area of a path around a
garden **35. a.** $C - 69.55$ dollars **b.** $C - W + D$ dollars

27. Sample:

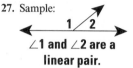

∠1 and ∠2 are a
linear pair.

29. Sample:

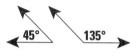

LESSON 7-8 (pp. 387–392)
17. rhombuses and squares **19.** parallelograms, rhombuses,
rectangles, and squares **21.** 11 **23.** not enough information
25. a. 147° **b.** 57° **c.** 123° **d.** 33° **27.** ABG **29. a.** $\angle DBA$, $\angle CBD$,
$\angle FBA$, $\angle GBC$, $\angle GBA$, $\angle GBD$, $\angle GBF$ **b.** m∠DBA = 30°;
m∠CBD = m∠FBA = 150°; m∠GBC = 60°; m∠ABG = 120°;
m∠GBD = m∠GBF = 90° **31.** ⊥ **33. a.** from 1950 to 1960,
103,932; from 1960 to 1970, $-23,952$; from 1970 to 1980, $-81,160$;
from 1980 to 1990, $-8,124$ **b.** $-9,304$ **c.** From 1950 to 1990, the
population of Milwaukee decreased by 9,304 people.

LESSON 7-9 (pp. 392–398)
9. 74° **11.** m∠1 = 80°, m∠2 = 100°, m∠3 = 80°, m∠4 = 20°,
m∠5 = m∠6 = m∠7 = 90°, m∠8 = 100° **13.** (a) **15.** If a
triangle had two obtuse angles, the sum of their measures would be
more than 180°. This contradicts the Triangle-Sum Property.
17. (d) **19. a.** True **b.** True **21.** $\overline{AB}$ and $\overline{AD}$ are perpendicular.
23. The angles have the same measure. **25. See below. 27.** $x = 0$
29. 910

25. a. Line EF is perpendicular to line segment GH.

b.

CHAPTER 7 PROGRESS SELF-TEST (p. 402)

1. $5 - (-5) = 5 + 5 = 10$ **2. See p. 743 for drawing.** $-6 - 22 = -28$
3. $\frac{9}{12} + -\frac{10}{12} = -\frac{1}{12}$ **4.** $5 - 13 + 2 - -11 = 5 + -13 + 2 + 11 = 5$
5. $x + -y + 5$ **6.** Use the comparison model. $V - N = L$, or
$N + L = V$, or $V - L = N$. **7.** Use the comparison model or try
simpler numbers. $Z - 67$ inches **8.** Use the Take-Away Model:
Area of outer square = 8^2 square meters; area of inner square =
4^2 square meters; $8^2 - 4^2 = 64 - 16 = 48$ m². **9. a.** Use the slide
model. $t° - 7° = -3°$ **b.** Add 7 to both sides. $t = 4°$

10. $y + -14 = -24$; add 14 to both sides; $y = -10$ **11.** Add x to
both sides; $x + -50 = 37$; add 50 to both sides; $x = 87$
12. $g + -3.2 = -2$; add 3.2 to both sides; $g = 1.2$ **13.** $c + -a = b$;
add a to both sides; $c = b + a$; add $-b$ to both sides; $c + -b = a$
14. (c); all others are equivalent to $x + y = 180$ **15.** $50 - 15 = 35$
people add something to their coffee. $25 + 20 - z = 35$ where z is
the overlap, people drinking coffee with cream and sugar. $z = 10$,
10 people drink coffee with cream and sugar. **16.** m∠CBD =
90° − m∠ABD = 90° − 25° = 65° **17.** Angles ABE and ABD are a
linear pair, so their measures add to 180. So m∠ABE + 25° = 180°,

from which m∠ABE = 155°. **18.** ∠5 and ∠6 are corresponding angles, so have the same measure, 74°. ∠2 forms a linear pair with ∠6, so 74° + m∠2 = 180°, from which m∠2 = 106°. **19.** ∠5, a corresponding angle; ∠3, a vertical angle; and ∠4, the vertical angle to ∠5 **20.** Angles 1 and 8 each form a linear pair with ∠5, so are supplementary. Angles 2 and 7, having the same measures as angles 1 and 8, are therefore also supplementary. **21.** 55° + 4° + x = 180°; 59° + x = 180°; x = 121° **22.** The sum of the angle

measures would be 270°. By the Triangle Sum Property, the sum must be 180°. **23.** ∠CBE, an alternate interior angle **24.** 130° **25.** rhombuses

2.

$$-6 - 22 = -28$$

The chart below keys the **Progress Self-Test** questions to the objectives in the **Chapter Review** on pages 403–405 or to the **Vocabulary** (Voc.) on page 401. This will enable you to locate those **Chapter Review** questions that correspond to questions students missed on the **Progress Self-Test.** The lesson where the material is covered is also indicated on the chart.

Question	1	2	3	4	5	6	7	8	9	10
Objective	A	O	A	A	G	M	M	K	L	B
Lesson	7-2	7-2	7-2	7-2	7-2	7-1	7-1	7-1	7-3	7-3
Question	11	12	13	14	15	16	17	18	19	20
Objective	B	B	B	G	N, P	C	C	D	H	H
Lesson	7-4	7-3	7-4	7-4	7-5	7-6	7-6	7-7	7-6, 7-7	7-6, 7-7
Question	21	22	23	24	25	26	27	28	29	30
Objective	E	J	D	F, I	I					
Lesson	7-9	7-9	7-7	7-8	7-8					

CHAPTER 7 REVIEW (pages 403–405)

1. −28 **3.** −0.7 **5.** $\frac{22}{15}$ or $1\frac{7}{15}$ **7.** 13.7 **9.** x = 72 **11.** V = −7.2 **13.** b = 197 **15.** x = 8 **17.** c = e + 45 **19.** 90 **21.** 180° − x° **23.** angles 7, 1, and 3 **25.** 180° − y° **27.** 50° **29.** 38° **31.** Yes, 10 cm **33.** 4 **35.** −3 **37.** a + c = b **39.** 140° **41.** 180° − x° **43.** angles 1, 8, 4, and 5 **45.** angles 1, 7, 5, and 3 **47.** rectangles and squares **49.** all **51.** Sample: Yes, angles could have measures 40°, 60°, and 80°, which add to 180°. **53.** 5500 square feet **55.** 21 **57.** (b) **59.** 2.5 million **61.** F − R = L **63.** 70%

65.

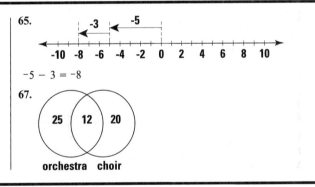

$$-5 - 3 = -8$$

67.

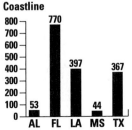

orchestra choir

LESSON 8-1 (pp. 406–414)

15. a.

Total hours of Democratic National Convention TV coverage by ABC, CBS, and NBC

year	hours
1960	85.6
1964	about 70
1968	about 85
1972	about 90
1976	about 70
1980	about 55
1984	about 35
1988	about 35
1992	24.0

b. Sample: The graph is easier to read and shows the decreasing trend more dramatically. **17.** about 6% **19.** 64° **21.** Samples: 10 + −7 + −1 = −(7 + −10 + 1); (−3) + −(4) + −(5) = −(−4 + −(3) + (−5)); probably true. **23.** G < 24 or G > 28 **25. a.** The segment should be approximately 1.7 cm long. **b.** The segment should be approximately 14 cm long. **c.** The segment should be approximately 3.1 cm long.

LESSON 8-2 (pp. 415–420)

11. a. Yes **b.** 2 **13. a.** No **b.** The first interval has length 1. The other intervals have length 0.05. **15.** 3.4 **17.** Samples: 1% or 0.5% **19. a.** Sample: 100 miles **b. See below for graph.** **21.** Samples: Graphs can be easier to read than tables or prose, and can take less space. **23. a.** x = 30 **b.** Sample: A triangle has two angles with measures 105° and 45°. What is the measure of the third angle? **25.** 3.99% **27.** $1.22 in 1990; $1.20 in 1991 **29.** 1,900,000; 1,400,000; 800,000; 600,000

19. b.

Miles of Coastline

[Bar graph showing: AL 53, FL 770, LA 397, MS 44, TX 367]

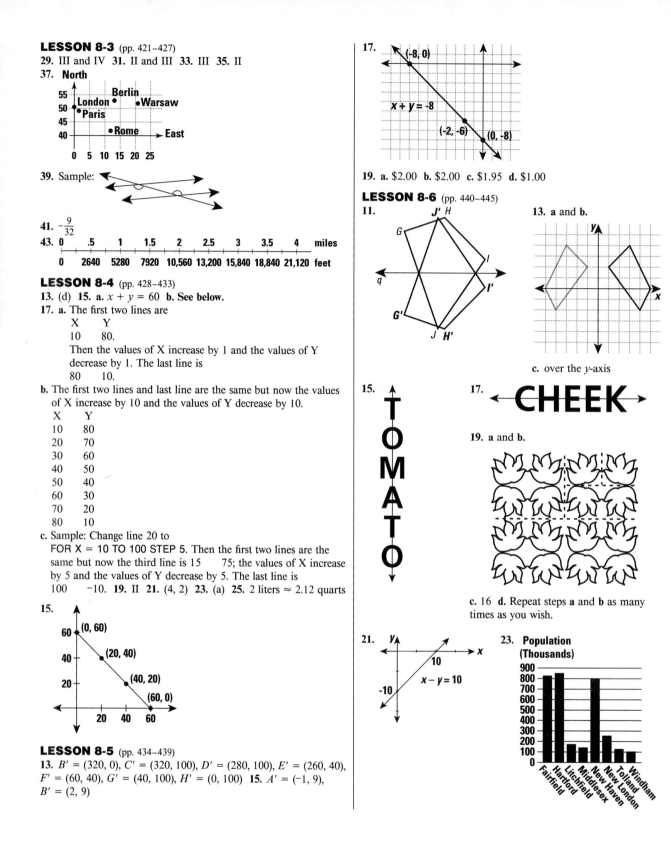

29. III and IV **31.** II and III **33.** III **35.** II

37. North

39. Sample:

41. $-\frac{9}{32}$

43.

| miles | 0 | .5 | 1 | 1.5 | 2 | 2.5 | 3 | 3.5 | 4 |

| feet | 0 | 2640 | 5280 | 7920 | 10,560 | 13,200 | 15,840 | 18,840 | 21,120 |

13. (d) **15. a.** $x + y = 60$ **b. See below.**

17. a. The first two lines are

X   Y
10   80.

Then the values of X increase by 1 and the values of Y decrease by 1. The last line is

80   10.

b. The first two lines and last line are the same but now the values of X increase by 10 and the values of Y decrease by 10.

X	Y
10	80
20	70
30	60
40	50
50	40
60	30
70	20
80	10

c. Sample: Change line 20 to

FOR X = 10 TO 100 STEP 5. Then the first two lines are the same but now the third line is 15   75; the values of X increase by 5 and the values of Y decrease by 5. The last line is

100   −10. **19.** II **21.** (4, 2) **23.** (a) **25.** 2 liters ≈ 2.12 quarts

15.

13. $B' = (320, 0)$, $C' = (320, 100)$, $D' = (280, 100)$, $E' = (260, 40)$, $F' = (60, 40)$, $G' = (40, 100)$, $H' = (0, 100)$ **15.** $A' = (−1, 9)$, $B' = (2, 9)$

17.

$x + y = -8$

(-8, 0)   (-2, -6)   (0, -8)

19. a. $2.00 **b.** $2.00 **c.** $1.95 **d.** $1.00

11.

13. a and b.

c. over the y-axis

15.

T
O
M
A
T
O

17. CHEEK

19. a and b.

c. 16 **d.** Repeat steps **a** and **b** as many times as you wish.

21.

$x - y = 10$

23. Population (Thousands)

SELECTED ANSWERS

LESSON 8-7 (pp. 446–451)

9.
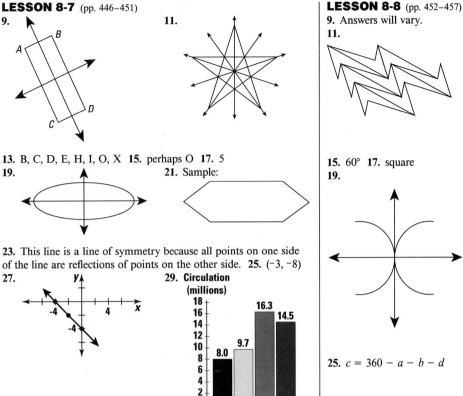

11.

13. B, C, D, E, H, I, O, X **15.** perhaps O **17.** 5
19.

21. Sample:

23. This line is a line of symmetry because all points on one side of the line are reflections of points on the other side. **25.** (−3, −8)
27.

29. Circulation (millions)

LESSON 8-8 (pp. 452–457)

9. Answers will vary.
11.

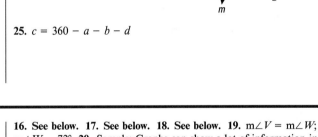

13.

15. 60° **17.** square **21.** (c)
19.

23.

25. $c = 360 - a - b - d$

CHAPTER 8 PROGRESS SELF-TEST (pp. 461–462)

1. the week of November 9 **2.** about 3 million barrels
3. 0.4 million barrels **4.** The vertical axis does not begin at 0, so the bars look more different than they actually are. **5. See below.**
6. Add 10 to the second coordinate to get (2, 15). **7. See below.**
8. a. E **b.** A **9.** E **10.** B, since −4 + 2 = −2. **11.** M = (3, 6)
12. The hole is directly above (9, 3) so its first coordinate is 9. Its second coordinate is halfway between 9 and 12, so it is 10.5.
Hole = (9, 10.5) **13. See below.** **14. See below.** **15. See below.**

16. See below. 17. See below. 18. See below. 19. m∠V = m∠W;
m∠W = 72° **20.** Sample: Graphs can show a lot of information in a small place. Graphs can picture relationships. **21.** when they have the same size and shape
22. 4 | 9 8 6
5 | 7 7 7 7 8 7 4 1 0 2 6 4
6 | 1 1 8 4 5
23. a. 57 **b.** 68 − 46 = 22 **c.** 57

5.
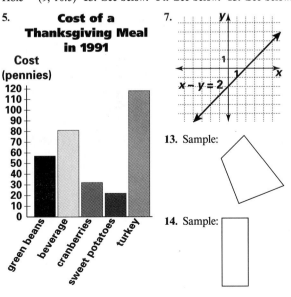

7.

13. Sample:

14. Sample:

15.
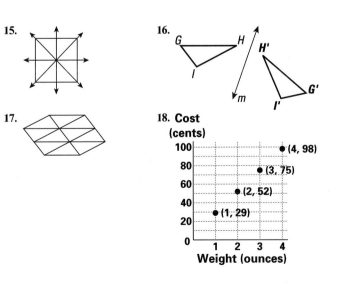

16.

17.

18. Cost (cents)

745

The chart below keys the **Progress Self-Test** questions to the objectives in the **Chapter Review** on pages 463–465 or to the **Vocabulary** (Voc.) on page 460. This will enable you to locate those **Chapter Review** questions that correspond to questions students missed on the **Progress Self-Test**. The lesson where the material is covered is also indicated on the chart.

Question	1	2	3	4	5	6	7	8	9	10
Objective	F	F	F	F	F	J	J	J	J	K
Lesson	8-2	8-2	8-2	8-2	8-2	8-3	8-3, 8-4	8-4	8-5	8-4

Question	11	12	13	14	15	16	17	18	19	20
Objective	G	G	C	C	C	B	D	G	L	H
Lesson	8-4	8-4	8-7	8-7	8-7	8-6	8-8	8-3	8-6	8-1

Question	21	22	23
Objective	E	I	A
Lesson	8-6	8-1	8-1

CHAPTER 8 REVIEW (pages 463–465)

1. median = 355, range = 222, mode = 350 **3.** median = 6, range = 4, mode = 6 **5.** and **7. See below. 9. See below. 11. See below. 13. See below. 15.** 90° **17.** Yes, a translation image is always the same size and shape as its preimage. **19.** 2″ **21.** 4′11″ **23.** The axis does not begin at 0, so the differences between the average heights of boys and girls are magnified. **25. a.** Sample: 1 million **b. See right. 27.** 1983 **29.** 1988–1990 **31. See right.** **33.** Sample: The graph can show trends and sway the reader.

35.
```
0 | 8 7
1 | 1 0 4 9 9
2 | 4 4
3 | 1 6
4 | 4
5 | 0
```

37. $B = (-2, 4); C = (-2, 0)$ **39., 41. See below right.**
43. $x - y = 3$ **45.** $(-2, -3)$ **47.** $(9, 3); (7, 7); (10, 5)$

5.

7.

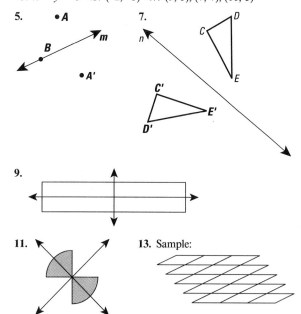

9.

11.

13. Sample:

25. b. Sample:

U.S. Population

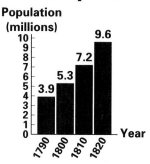

31. **U.S. Population**

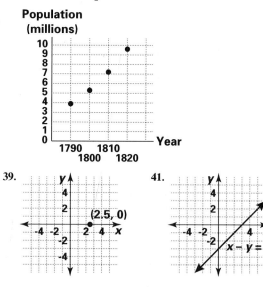

39.

41.

$x - y = 3$ (2.5, 0)

LESSON 9-1 (pp. 466–475)
13. 71 ft^2 **15.** Samples: 1 in. and 15 in., 2.5 in. and 6 in. **17.** (b)
19. 146.5 ft^2 **21.** 714 square units **23. a.** 24 miles
b. 36 square miles **25.** -25 **27.** 2.2 lb **29.** 1,440,000

LESSON 9-2 (pp. 476–481)
13. 864 ft^3 **15. a. See drawing below. b.** cube **c. See below.**
17. (e) **19.** 262,080 dots **21. a.** fourth (or IV) **b.** Sample:
$y = x - 5$ **23.** $x = -27$ **25.** 595 **27.** < **29.** >

15. a.

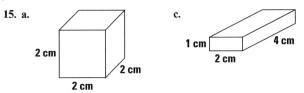

LESSON 9-3 (pp. 482–487)
23. $0.625 \cdot 1.6 = 1$ **25.** Sample: The reciprocal would be $\frac{1}{0}$, but
you cannot divide by 0. **27. a.** $\frac{1}{24}$ **b.** Sample: $\frac{1}{2}$ of $\frac{1}{3}$ is less than $\frac{1}{3}$;
$\frac{1}{4}$ of that is even smaller. **29. a.** $9\frac{5}{8}$ **b.** Sample: Convert fractions to
decimals and multiply. **31.** 210 square units **33.** m$\angle SRT = 23°$;
m$\angle S = 135°$; m$\angle T = 22°$ **35.** 57%

LESSON 9-4 (pp. 488–492)
9. $\frac{8}{15}$ or about 53% **11.** No. Since $P(A) = \frac{9}{10}$ and $P(B) = \frac{17}{24}$, $P(A) \cdot$
$P(B) = \frac{51}{80}$. However, $\frac{3}{4}$ of the right-handed people are right-eyed, so
$P(A \text{ and } B) = \frac{3}{4} \cdot \frac{9}{10} = \frac{27}{40}$. This means $P(A \text{ and } B) \neq P(A) \cdot P(B)$.
13. $\approx 23.3\%$ **15.** 3 **17.** $\frac{3}{5}$ **19.** 1 **21.** area is multiplied by 5^2 or 25
examples: $2 \cdot 3 = 6$ $5(2) \cdot 5(3) = 25(6) = 150$; $4 \cdot 1 = 4$
$5(4) \cdot 5(1) = 25(4) = 100$ Sample pattern: If dimensions of a
rectangle are multiplied by n, the area is multiplied by n^2.

LESSON 9-5 (pp. 493–498)
15. 100,800 beats **17.** 1299.6 pesetas **19.** $\frac{8}{35}$ **21.** $\frac{1}{12}$ **23.** $1\frac{1}{4}$ mi^2
25. 21 games **27.** lb **29.** mL

LESSON 9-6 (pp. 499–504)
25. a. -10 **b.** -5 **c.** 0 **d.** 5 **e.** 10 **27.** Sample: Multiplication by
zero annihilates the number, making it 0. **29.** -64 **31.** 1
33. km/sec **35.** 600 minutes, or 10 hours **37.** $(90\%)^{10} \approx 35\%$
39. It is multiplied by 27. **41.** Sample: $-2 + 3 = 1$

LESSON 9-7 (pp. 505–510)
13. a. 15 mm **b.** 150 **15.** 3 **17.** 96 **19.** -10 **21.** -10 **23.** .1
25. 101 **27.** 2% **29.** 45.3 **31.** Any number between -1 and 0 is
correct.

LESSON 9-8 (pp. 511–515)
11. Yes **13.** (b) **15. a.** 2 **b.** If figure A is the image of figure B
under a size change of magnitude k, then figure B is the image of
figure A under a size change of magnitude $\frac{1}{k}$. **17.** 62.5 cm
19. a. $\approx 10.8\%$ **b.** Being poor and being white are not independent
events. A white person has a slightly smaller chance (about 17%
less) of being poor. **21.** The graph is translated five units down.
23. Sample: Graphs can contain a lot of information in a small
space, can show trends visually, and can be used to sway a reader.
25. $\angle 2$ and $\angle 6$, $\angle 3$ and $\angle 7$ **27. a.** $17 **b.** $5 + 2 \cdot w$ dollars
29. $3\frac{23}{24}$ miles

LESSON 9-9 (pp. 516–520)
11. True **13.** 2 **15.** $b - c + d$ **19.**
17. $\approx .29 \cdot \frac{1}{7} \approx .04 \approx 4\%$
19. See right. 21. 16, 18, 20,
21, 22, 24 **23.** hexagon
25. pentagon **27.** -35
29. a. 945 **b.** equal to **31.** -0.6

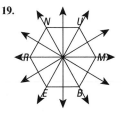

CHAPTER 9 PROGRESS SELF-TEST (p. 524)
1. $\frac{4}{3} \cdot \frac{5}{4} = \frac{5}{3}$, or $1\frac{2}{3}$ **2.** $\frac{7}{16x}$ **3.** -45 (positive $\cdot$ negative = negative)
4. $-9 + 4 = -5$ **5.** $a + a + b + 0 + -21c + c = 2a + b + -20c =$
$2a + b - 20c$ **6.** Associative Property of Multiplication
7. Property of Reciprocals **8.** $r \cdot s$ (or rs) is the total number of
seats. If 5 are broken, then the number of other seats is $rs - 5$.
9. Area equals $\frac{1}{2}$ base times height, so $\frac{1}{2} \cdot 17.3 \cdot 6.8 = 58.82$ m^2.
10. 4 hours $\cdot -3 \frac{\text{centimeters}}{\text{hours}} = -12$ centimeters **11.** $V = a \cdot b \cdot c =$
3 ft $\cdot 4$ ft $\cdot 5$ ft $= 60$ ft^3 **12.** It would probably fit since the volume
of the drawer is 21,600 cm^3, much larger than the volume of the
jewelry box. **13.** Two positive numbers whose product is 16.
Samples: 1 and 16, 1.6 and 10, 3.2 and 5. **14.** $\frac{1}{38} \approx 0.0263158 \approx$
0.026 **15.** $(4 \cdot 8, 4 \cdot 2) = (32, 8)$ **16.** $5.80 \cdot 1\frac{1}{2} = 5.80 \cdot 1.5 =$
$8.70 per hour of overtime. **17. a.** $\frac{2}{3} \cdot \frac{1}{4}$ **b.** $\frac{2}{12}$ or $\frac{1}{6}$
18. a. $8.50 \frac{\text{dollars}}{\text{hour}} \cdot 37.5 \frac{\text{hours}}{\text{week}} = 318.75 \frac{\text{dollars}}{\text{week}}$ **b.** The person earns
$318.75 per week.

19.

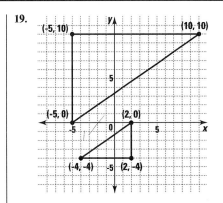

20. Sample: The corresponding angles in the preimage and image
are equal. The corresponding sides in the preimage and image are
not equal. **21.** $\frac{3}{25}$ of the 850 students are in the band. $\frac{3}{25} \cdot 850 =$
102 students. **22.** $\frac{2}{3}$ of the $8400 is the value of the scholarship.
$\frac{2}{3} \cdot 8400 = 5600$. **23.** $\frac{1}{10}$ chance for 3 independent digits =
$\frac{1}{10} \cdot \frac{1}{10} \cdot \frac{1}{10} = \frac{1}{1000} = 1$ in 1000. **24.** These are independent events.
$\frac{2}{3} \cdot \frac{2}{5} = \frac{4}{15}$, 4 in 15 of getting an A on both tests. **25.** (c) $\frac{1}{1.25} = \frac{4}{5} =$
$\frac{8}{10} = 0.8$ **26.** 1 mile = 5280 feet; $\frac{1 \text{ mi}}{5280 \text{ ft}} \cdot 19,340$ ft $=$
$3.6629 \ldots \approx 3.66$ mi

The chart below keys the **Progress Self-Test** questions to the objectives in the **Chapter Review** on pages 525–527 or to the **Vocabulary** (Voc.) on page 523. This will enable you to locate those **Chapter Review** questions that correspond to questions students missed on the **Progress Self-Test.** The lesson where the material is covered is also indicated on the chart.

Question	1	2	3	4	5	6	7	8	9	10
Objective	C	C	D	D	F	E	E	G	A	J
Lesson	9-3	9-3	9-6	9-6	9-6	9-2	9-3	9-1	9-1	9-5
Question	**11**	**12**	**13**	**14**	**15**	**16**	**17**	**18**	**19**	**20**
Objective	B	H	A, M	F	N	L	M	J	N	N
Lesson	9-2	9-2	9-1	9-3	9-7	9-7	9-3	9-5	9-9	9-9
Question	**21**	**22**	**23**	**24**	**25**	**26**				
Objective	L	L	I	I	E	K				
Lesson	9-8	9-8	9-4	9-4	9-3	9-5				

CHAPTER 9 REVIEW (pp. 525–527)

1. 24.5 cm^2 **3.** 27 square units **5.** 1875 cubic units **7.** $\frac{1}{2}$ **9.** $\frac{75}{2}$, or $37\frac{1}{2}$ **11.** $\frac{405}{4}$, or $101\frac{1}{4}$ **13.** $\frac{2}{3}$ **15.** $\frac{8}{5}$, or $1\frac{3}{5}$ **17.** -87 **19.** -11
21. -320 **23.** For any x, $x \cdot 0 = 0$. **25.** Commutative Property of Multiplication **27.** False **29.** 0 **31.** 1 **33.** $\frac{x}{2}$ **35. a.** $\frac{1}{40}$
b. $0.125 \times 0.2 = 0.025$ **37.** 48 **39.** 42 in^2 **41.** 85,680 cm^3
43. $5\frac{1}{16} \approx 5$ cubic inches **45.** $\approx 98.34\%$ **47.** $\frac{1}{7}$ **49. a.** 125 miles
b. Sample: How far can you drive in 5 hours if the average speed is 25 miles per hour? **51.** 360 people **53.** $\frac{1\ \text{ft}}{30.48\ \text{cm}}$; $\frac{30.48\ \text{cm}}{1\ \text{ft}}$
55. $6\frac{1}{4}$ days **57.** $18.75 **59.** $7.13 **61, 63.** See below. **65.** See right. **67.** contraction **69.** $(-8, 16)$ **71.** See right.

61.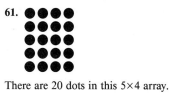

There are 20 dots in this 5×4 array.

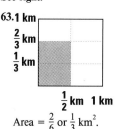

63.1 km

$\frac{2}{3}$ km

$\frac{1}{3}$ km

$\frac{1}{2}$ km   1 km

Area $= \frac{2}{6}$ or $\frac{1}{3}$ km^2.

65.

71.

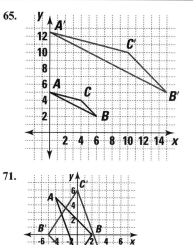

LESSON 10-1 (pp. 528–534)
11. $28s$ **13.** $3e$ **15. a.** $-\frac{1}{2} + -\frac{1}{2} + -\frac{1}{2} + -\frac{1}{2} + -\frac{1}{2} + -\frac{1}{2} + -\frac{1}{2} + -\frac{1}{2}$
b. -4 **17.** $u = -75$ **19.** (a), (b), (c), (d), (g) **21.** Associative Property of Multiplication **23.** Multiplication Property of Equality **25. a.** Yes **b.** $\ell w - ab$ **27. a.** $-1°$F **b.** $w - z°$F

LESSON 10-2 (pp. 535–539)
15. $y = 2.3$; $16.56 = 7.2 \cdot 2.3$ **17.** $x = 1.912$; $1.912 + 1.912 + 1.912 = 5.736$ **19.** 675 **21.** $y + 3z$ **23.** $53.1\overline{6}$
25. a. 2 **b.** 6 **c.** 18 **d.** $2n - 2$ **27.** 152°

LESSON 10-3 (pp. 540–543)   **17.**
7. ≈ 13.27 hours **9.** 342
11. a. $T = 4w$ **b.** $T = 3600$
c. $T = 6w$ **d.** $T = 5400$
13. $a = 7.5$ **15.** -8
17. See right. **19.** 960 **21.** .02 m

LESSON 10-4 (pp. 544–549)
21. a. $1500 + 90n$ dollars **b.** no longer than 11 days
23. a. Addition Property of Equality **b.** Associative Property of Addition **c.** Property of Opposites **d.** Additive Identity Property of 0 **e.** Multiplication Property of Equality **f.** Associative Property of Multiplication **g.** Property of Reciprocals
h. Multiplication Property of 1 **25. a.** $-n$ **b.** $x = \frac{p + -n}{m}$
27. $x = 121$ **29.** contraction **31.** $\frac{3}{130}$ **33.** Sample: $2500 - D + B - S = 2300$

LESSON 10-5 (pp. 550–554)
19. $n = -2.7$; $6 - 30(-2.7) - 18 = 6 + 81 - 18 = 87 - 18 = 69$
21. $b = 4$; $2(4) + 2 + 3(4) + 3 - 4(4) - 4 = 8 + 2 + 12 + 3 - 16 - 4 = 25 - 20 = 5$ **23.** $22\frac{1}{2}$ days **25.** y is positive
27. $u = 29$ **29.** They will have the same shape, but the length of the sides of the image will be 4 times the lengths of the corresponding sides of the preimage, and the image will be rotated 180° from the position of the preimage. **31.** $13,468
33. m$\angle HEG = 98°$; m$\angle H = 50°$; m$\angle G = 32°$ **35.** $\frac{1}{18}$

LESSON 10-6 (pp. 555–560)
23. $3x + 2x$, or $5x$ cents **25.** $a^2 + a^3$
27. $732(1,000,000,000,000,000 - 1) = 732 \cdot 10^{15} - 732 \cdot 1 = 732,000,000,000,000,000 - 732 = 731,999,999,999,999,268$
29. -6 **31.** $x = 3$ **33.** $y = -\frac{7}{2}$ or -3.5 **35. a.** $\frac{7}{6}$ **b.** $x = 98$

LESSON 10-7 (pp. 561–565)
9. a. ℓhw **b.** $2\ell w + 2\ell h + 2hw$ **11. a.** 16 in. by 14 in. **b.** 60 in^2
13. Move one of the top rectangles to the bottom. **15.** 1360 cm^2
17. $11x$ **19.** $-12m + n$ **21.** $x = \frac{1}{2}$ **23.** $-\frac{1}{9}$ **25.** Commutative Property of Addition **27.** 125 cubic units

LESSON 10-8 (pp. 566–570)
15. $4\ell + 4w + 4h$ **17.** surface area **19.** volume **21.** (c) **23.** (c)
25. $c = 28.5$ **27., 29., 31. See below.**
33. a. $33.\overline{3}\%$ **b.** $66.\overline{6}\%$ **c.** 80%

27.
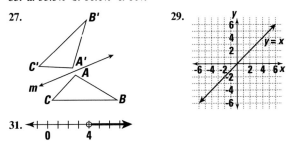

29.
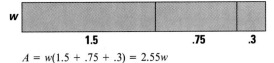

31.

LESSON 10-9 (pp. 571–576)
9. 15 cm **11.** 132 square units **13.** $a^2 + 2a$ **15.** 113.42
17. 12 numbers **19.** A, B, C, D **21.** A, D **23.** $\frac{59}{60}$
25. three hundred forty-five and twenty-nine hundredths

LESSON 10-10 (pp. 577–581)
13. 1,100,000 square kilometers **15.** 7 meters
17. a. One example of an altitude is the dashed segment between the parallel sides. **See below for sample figure. b.** $AB \approx 21$mm, $DC \approx 39$mm, $AE \approx 25$mm; Area ≈ 750 mm^2. **19. a.** $x = 10$
b. $y = -10$ **c.** $z = -13$ **21.** 96 square ft **23.** 10

17.
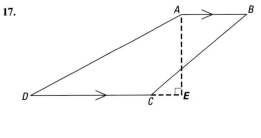

CHAPTER 10 PROGRESS SELF-TEST (pp. 585–586)

1. $(-y + y + y + y + y) + 3 = 3y + 3$ **2.** $-2m$ **3.** $(7k - 4k) + (-2j + j) = 3k - j$ **4.** $10x$ **5.** $b + b + 40 + b + b + 40 = (b + b + b + b) + 40 + 40 = 4b + 80$. **6.** Think of $49 \cdot 7$ as $(50 - 1) \cdot 7$. Then using the Distributive Property, we can mentally calculate: $50 \cdot 7 - 1 \cdot 7 = 350 - 7 = 343$. **7.** Between lines 5 and 6. $5x + -2x = (5 + -2)x$. **8.** km^2 (area) **9.** (a)
10. $A = \frac{1}{2}hb$; $400 = \frac{1}{2} \cdot 25 \cdot b$; $\frac{2}{25} \cdot 400 = \frac{2}{25} \cdot \frac{25}{2} \cdot b$. The base of the triangle is 32 inches.
11. Sample: The rectangles can be rearranged as shown.

| w | 1.5 | .75 | .3 |

$A = w(1.5 + .75 + .3) = 2.55w$

12. Front or back $= 6 \cdot 11$, sides $= 3 \cdot 11$, top or bottom $6 \cdot 3$
Surface area $= 6 \cdot 11 + 6 \cdot 11 + 3 \cdot 11 + 3 \cdot 11 + 6 \cdot 3 + 6 \cdot 3 = 66 + 66 + 33 + 33 + 18 + 18 = 234$ cm^2 **13.** $-\frac{1}{9} \cdot 35.1 = -\frac{1}{9} \cdot -9t$; $-3.9 = t$ **14.** $\frac{5}{2} \cdot \frac{2}{5}m = \frac{5}{2} \cdot -\frac{3}{4}$; $m = \frac{-15}{8}$, or $-1\frac{7}{8}$
15. $2 - 2 + 3a = 17 - 2$; $\frac{1}{3} \cdot 3a = \frac{1}{3} \cdot 15$; $a = 5$ **16.** $12 - 4h = 10$; $-12 + 12 - 4h = 10 + -12$; $-\frac{1}{4} \cdot -4h = -\frac{1}{4} \cdot -2$; $h = \frac{1}{2}$
17. $2000 = .08I$; $25000 = I$. Last year's income was $25,000.
18. $A = \frac{1}{2}hb$; $A = \frac{1}{2} \cdot 24 \cdot 7$; $A = 84$ cm^2 **19.** $A = \frac{1}{2}hb$; $A = \frac{1}{2} \cdot 30 \cdot 36$; $A = 540$ square units **20.** $A = \frac{1}{2}h(b_1 + b_2)$; $A = \frac{1}{2} \cdot 8(9 + 30)$; $A = \frac{1}{2} \cdot 8 \cdot 39$; $A = 156$ square units **21. a.** $4 - 5 \cdot .02 = 4 - .1 = 3.9$ meters **b.** $4 - .02x = 2.5$; $-.02x = 2.5 - 4$; $x = \frac{-1.5}{-.02} = 75$; after 75 seconds **22.** $3w + 1 = 10$

The chart below keys the **Progress Self-Test** questions to the objectives in the **Chapter Review** on pages 587–589 or to the **Vocabulary** (Voc.) on page 584. This will enable you to locate those **Chapter Review** questions that correspond to questions students missed on the **Progress Self-Test**. The lesson where the material is covered is also indicated on the chart.

Question	1	2	3	4	5	6	7	8	9
Objective	C	C	C	C	C	F	F	I	Voc.
Lesson	10-1	10-1	10-1	10-1	10-1	10-6	10-6	10-8	10-10
Question	10	11	12	13	14	15	16	17	18
Objective	A, D	K	H	A	A	B	B	G	D
Lesson	10-2, 10-9	10-6	10-7	10-5	10-2	10-4	10-5	10-3	10-9
Question	19	20	21	22					
Objective	D	E	G	J					
Lesson	10-9	10-10	10-4	10-4					

1. $t = 75$; $40(175) = 3000$ **3.** $v = 40$; $.02(40) = .8$ **5.** $y = 7$; $49 = -7(-7)$ **7.** $n = 15$; $\frac{4}{5}(15) = 12$ **9.** $m = 2$; $8 \cdot 2 + 2 = 18$
11. $u = 3$; $11 - 6 \cdot 3 = -7$ **13.** $x = 2.4$; $2 \cdot 2.4 + 3 \cdot 2.4 + 5 = 17$ **15.** $x = 0.25$; $1.3 = 0.8 + 2 \cdot 0.25$ **17.** $3x + 2y + 2z$ **19.** $2x$
21. $-7 - 2m$ **23.** $6a - 6b + 12c$ **25.** 150 square units
27. a. 12 square units **b.** 48 square units **29.** 24,600 square units
31. 28,800 square units **33.** Multiplication Property of Equality
35. from line 2 to line 3 **37.** 87.5 ft deep **39.** 12 cm high

41. a. 37.5 liters **b.** $4\frac{2}{3}$ hr or 4 hours, 40 minutes **43.** The area needed to cover the mattress is $7 \cdot 6 + 2 \cdot 6 \cdot \frac{3}{4} + 2 \cdot 7 \cdot \frac{3}{4} = 61.5$ square feet, which is larger than 48 square feet. **45.** square feet **47.** Sample: in^3, km^3
49. a. See right. **b.** $w = 3$
51. $x(20 + 5) = 20x + 5$

49. a. ■ = w
● = kilogram weights

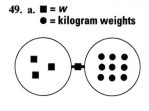

LESSON 11-1 (pp. 590–596)
9. a. $365 = 52 \cdot 7 + 1$ **b.** It shows that there are 52 full weeks in a 365-day year, with one day left over. **11.** Each student gets 43 sheets and 11 sheets are left over. **13.** Yes, $x = 14$.
15. See below. 17. a. Cesarean section **b.** 391.5 **19. a.** 0.1875
b. 0.01875 **c.** 0.001875 **d.** 0.0001875

15.

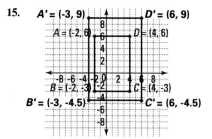

$A' = (-3, 9)$ $D' = (6, 9)$
$A = (-2, 6)$ $D = (4, 6)$
$B = (-2, -3)$ $C = (4, -3)$
$B' = (-3, -4.5)$ $C' = (6, -4.5)$

LESSON 11-2 (pp. 597–600)
13. \$6.53 per hour **15.** California **17.** Sample: 84 sheets in 10 days **19. a.** 3 **b.** 9 **c.** 27 full square yards; 7 square feet are left over **21.** angles 3 and 7; angles 2 and 6 **23.** 75° **25.** 4

LESSON 11-3 (pp. 601–605)
11. a. $\frac{35}{117}$ **b.** $5\frac{4}{7}$ is greater than $1\frac{2}{3}$. **c.** $1.\overline{6} \div 5.\overline{571428} \approx .2991$ and $\frac{35}{117} \approx .2991$ **d.** The decimals are repeating, so it is easier to work with the fractions. **13.** $\frac{2y}{x}$ **15.** 50 **17.** 2116.8 cm^2
19. $G = \frac{3}{4}$ or .75 **21.** $G = 1$ **23. a.** $\frac{17}{20}$ **b.** .85 **c.** 85% **25.** $\frac{3}{4}$

LESSON 11-4 (pp. 606–610)
17. a. 95,000 people per year
b. 20, less **c.** $\frac{-1,900,000}{-20}$; 95,000
19. $-24, 0, 144, 1$ **21.** $\frac{1}{6}, -\frac{5}{6}, -\frac{1}{6}, -\frac{2}{3}$
23. -58 **25.** ≈ 142.9 miles per day **27. a.** 1250 **b.** 200

29. Sample:

LESSON 11-5 (pp. 611–614)
11. 10.8 **13.** 22.5% **15. a.** 5 **b.** -1 **c.** $\frac{1}{2}$ **d.** $\frac{1}{10}$ **e.** $-\frac{8}{9}$ **f.** $\frac{9}{8}$
17. \$9.80 **19.** 6.25 miles **21.** $12C - 200$ dollars

LESSON 11-6 (pp. 615–619)
13. \$56.25 **15.** $2\frac{7}{9}$ (almost 3) teaspoons **17. a.** Army: 37.5%; Air Force: 32.8%; Navy: 25.5%; Marine Corps: 4.2%
b.

Female Armed Forces Personnel

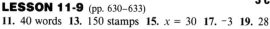

Air Force 32.8%
Army 37.5%
Marine Corps 4.2%
Navy 25.5%

19. Neither. Both fold $\frac{4}{5}$ napkin per minute. **21.** 5400 cm^2

LESSON 11-7 (pp. 620–624)
11. True **13.** False **15.** 29 cans **17.** The equation is not a proportion. **19. a.** 49.58% **b.** about 12,400 households **c.** $\frac{1}{2}$ or .5 **21.** $\frac{49}{20}$ **23.** $\frac{5}{24}$ feet/year, or 2.5 inches/year **25. a.** Answers will vary. **b.** The coordinates of the vertices of the image should be the result of multiplying the coordinates of the vertices of the first quadrilateral by -3.

LESSON 11-8 (pp. 625–629)
9. a. See right. **b.** 12 cm, 15 cm
c. The triangle should show corresponding sides of 9 cm, 12 cm, and 15 cm. **11.** 88 feet
13. 12.5 **15.** $6\frac{2}{3}$ **17.** (c), (d), (e), (f) **19.** \$15 **21.** 3 **23.** $-\frac{8}{7}$
25. $3.8 \cdot 10^9$ **27.** 5.125, $5\frac{1}{8}$

9.

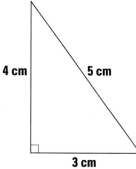

4 cm   5 cm   3 cm

LESSON 11-9 (pp. 630–633)
11. 40 words **13.** 150 stamps **15.** $x = 30$ **17.** -3 **19.** 28

CHAPTER 11 PROGRESS SELF-TEST (p. 637)

1. 60 words per minute (300 ÷ 5 words per minute)
2. $\frac{k}{h}$ kilometers per hour **3.** $\frac{4}{9} \cdot \frac{3}{1} = \frac{4}{3}$ **4.** $\frac{-42}{-24} = \frac{42}{24} = \frac{7}{4}$
5. $\frac{3}{5} \div \frac{6}{5} = \frac{3}{5} \cdot \frac{5}{6} = \frac{15}{30} = \frac{1}{2}$ **6.** $\frac{-2(-5)}{8 + -9} = \frac{10}{-1} = -10$ **7.** $\frac{a}{b}, \frac{-a}{-b}$
8. Sample: Gunther lost 8 pounds in two months. What was the rate of change of his weight? **9.** $40 \cdot 5 = 8x$; $200 = 8x$; $x = 25$
10. $5 \cdot 3 = 12 \cdot p$; $15 = 12p$; $p = \frac{15}{12} = \frac{5}{4} = 1\frac{1}{4}$ **11.** $\frac{14}{150} \approx 9.33 \approx$
9% **12.** $\frac{.30}{4.00} = 7.5\%$ **13.** $.60x = 30$; $x = \frac{30}{.60}$; $x = \frac{300}{6}$; $x = 50$
14. $\frac{45 \text{ miles}}{1 \text{ hour}} = \frac{189 \text{ miles}}{x \text{ hours}}$; $45x = 189$; $x = 4.2$ hours

15. $\frac{640 \text{ acres}}{1 \text{ sq. mile}} = \frac{x}{10,000 \text{ sq. miles}}$; $640 \cdot 10,000 = x$;
$x = 6,400,000$ acres **16.** In any true proportion $\frac{a}{b} = \frac{x}{y}$, $ay = bx$.
17. b and x are the means. **18.** $\frac{144}{64} = \frac{x}{40}$; $144 \cdot 40 = 64x$;
$5760 \div 64 = x$; $x = 90$ **19.** $\frac{144}{64} = \frac{180}{y}$; $144y = 64 \cdot 180$;
$y = 11520 \div 144 = 80$ **20.** $\frac{4}{\frac{1}{4}} = \frac{10}{x}$; $4x = 10 \cdot \frac{1}{4}$; $x =$
$\frac{5}{8}$ teaspoon of pepper **21.** $245 \div 15 = 16$ with a remainder of 5;
15 pages for 16 days, 5 pages for the 17th day. **22.** $n = d \cdot q + r$ is translated as $245 = 15 \cdot 16 + 5$.

The chart below keys the **Progress Self-Test** questions to the objectives in the **Chapter Review** on pages 638–639 or to the **Vocabulary** (Voc.) on page 636. This will enable you to locate those **Chapter Review** questions that correspond to questions students missed on the **Progress Self-Test**. The lesson where the material is covered is also indicated on the chart.

Question	1	2	3	4	5	6	7	8	9	10
Objective	G	G	A	B	A	B	E	G	C	C
Lesson	11-2	11-2	11-3	11-4	11-3	11-4	11-4	11-4	11-9	11-6
Question	11	12	13	14	15	16	17	18	19	20
Objective	H	H	H	I	I	D	D	J	J	I
Lesson	11-5	11-5	11-5	11-9	11-9	11-7	11-7	11-8	11-8	11-6
Question	21	22								
Objective	F	F								
Lesson	11-1	11-1								

CHAPTER 11 CHAPTER REVIEW (pp. 638–639)

1. 18 **3.** $\frac{1}{2}$ **5.** $\frac{bx}{ay}$ **7.** $\frac{7}{4}$ **9.** -5 **11.** $\frac{4}{3}$ **13.** No ($t = 8$) **15.** $y = \frac{18}{7}$
17. $G = 21$ **19.** $CD = \frac{3}{5}$ **21.** $6x$ **23.** (d) **25. a.** $6\frac{1}{4}$ gallons
b. $25 = 4 \cdot 6 + 1$ **27. a.** 5; 10 **b.** $70 = 12 \cdot 5 + 10$ **29.** 50 $\frac{\text{km}}{\text{hr}}$ or

50 kilometers per hour **31.** $\frac{10 \text{ pounds}}{30 \text{ days}} = \frac{1 \text{ pound}}{3 \text{ day}}$ or $\frac{1}{3}$ pound each day **33.** \$180 **35.** Sample: How fast have they been repaving the road? $\frac{-6 \text{ km}}{-3 \text{ day}} = 2\frac{\text{km}}{\text{day}}$ **37.** 20% missed **39.** 1.2 times as many Continentals **41.** $\frac{20}{9}$ cups **43.** 15.6 and 16.9 **45.** $IK = 31.5$

LESSON 12-1 (pp. 640–645)

15. Sample: $353 \div 6$ **17.** Let $x = 0.\overline{9}$; $10x = 9.\overline{9}$; $10x - x =$
$9.\overline{9} - 0.\overline{9}$; $9x = 9$; $x = 1$; So $x = 0.\overline{9} = 1$. **19.** $131\frac{3}{5}°$ **21.** none
23. a. s^2 **b.** 2 square units

LESSON 12-2 (pp. 646–650)

25. 10 square units **27. a.** 8 square units **b.** $\sqrt{8}$ units **29.** 16.9
31. 16.0 **33.** $\frac{422}{99}$
35. a.
196	9
197	4
198	27792978198799919
199	02021
b. 1988.5 **c.** 1989 **d.** 1992 or 1993
(It was published in 1992.)

LESSON 12-3 (pp. 651–656)

11. a. 11 km **b.** $\sqrt{85}$ km or about 9.2 km **c.** about 1.8 km
13. $\sqrt{5}$ **15.** (b) **17.** $\sqrt{13} \approx 3.6056$ **19. a.** 0 **b.** 0
21. 10/35 hours or about 17 minutes **23.** (a)

LESSON 12-4 (pp. 657–663)

21. 1.2π inches $\approx 3.77''$ **23.** $25\pi \approx 78.5$ units **25.** 16''
27. a. $3\frac{16}{113}$ **b.** $\frac{355}{113}$ **29.** $\sqrt{10}$ **31.** 9 **33.** 20% **35.** $\frac{m}{20}$ hours
37. 1704

LESSON 12-5 (pp. 664–669)

9. Sample: $\boxed{\pi}\ \boxed{\times}\ 50\ \boxed{x^2}\ \boxed{=}$ **11. a.** 0.3125π or about 0.98 in^2
b. the ring **13.** 141 calories **15.** one 14'' pizza **17.** when it is infinite and does not repeat **19.** $\sqrt{5}$, $\sqrt{6}$, $2\sqrt{3}$, $3\sqrt{12}$ **21.** $36.\overline{36}$
23. $5x - 5y + 60$

LESSON 12-6 (pp. 670–674)

9. 4080 mm^2 **11.** (b) **13. a.** 900 in^2 **b.** 120 in. **c.** about 42.4 in.
15. 10^{-1}, 10^0, $\sqrt{10}$, 10^1 **17.** $\frac{1}{64}$ **19.** $\frac{112}{45}$

LESSON 12-7 (pp. 675–679)

13. Sample: a compact disc **15. a.** rectangle **b.** 67.5 in^3
17. $7\pi \approx 22$ cubic units **19. a.** 6 **b.** 3 **c.** $\sqrt{27}$ or ≈ 5.2
21. $y = -\frac{42}{31}$ **23.** $w = -2$

LESSON 12-8 (pp. 680–683)

11. volume **13.** volume **15.** about 332 cubic inches
17. a. 58.5 cm by 15 cm **b.** Answers may vary. Sample: Use the 4 cm × 30 cm × 60 cm piece to make two game boards with no waste. The other piece of wood will make one game board with unusable waste left over. **19.** about 1866 in^2 or about 13 ft^2
21. $\frac{140}{99}$

23. Sample: **100 Most Influential Persons in History**

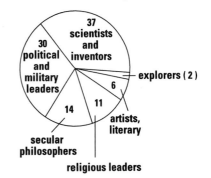

CHAPTER 12 PROGRESS SELF-TEST (p. 687)

1. $C = \pi d = \pi\left(\frac{3}{4}\right) \approx 2.4$ inches $A = \pi r^2 = \pi\left(\frac{3}{8}\right)^2 \approx$ 0.44 square inches **2.** $A = s^2$; $A = (\sqrt{3})^2$; $A = 3$ square units **3.** $A = s^2$; $10 = s^2$, $\sqrt{10}$ units **4.** $\sqrt{33} \approx 6$ and $\sqrt{51} \approx 7$; $\sqrt{33} + \sqrt{51} \approx 13$ **5.** $x^2 + 24^2 = 26^2$ (Pythagorean Theorem); $x^2 = 100$; $x = 10$ **6.** $5^2 + 10^2 = (AC)^2$; $125 = (AC)^2$; $\sqrt{125} = AC$ **7.** $\sqrt{25} = 5$; the other numbers are non-terminating, non-repeating decimals. **8.** $-2\frac{7}{10} = -\frac{27}{10}$ **9.** $n = 0.\overline{15}$; $100n = 15.\overline{15}$; $100n - n = 15.\overline{15} - .\overline{15}$; $99n = 15$; $n = \frac{15}{99} = \frac{5}{33}$ **10.** Determine surface area by adding the lateral area ($C \cdot h$) and the area of the 2 bases. $S.A. = 57\pi \cdot 123 + 2 \cdot (28.5)^2 \cdot \pi \approx 27{,}130$ mm^2

11. $L.A.$ = perimeter $\cdot$ height; $(7 + 24 + 25) \cdot 30 = 56 \cdot 30 = 1680$ square units **12.** V = area of base $\cdot$ h; $84 \cdot 30 = 2520$ cubic units **13.** 4 and $^{-}4$ **14.** $C = \pi d \approx 120\pi \approx 377$ mm **15.** $A = \pi r^2 = \pi (16)^2 = 256\pi$ units^2 **16.** when the decimal does not terminate or infinitely repeat **17.** $S = 4\pi r^2 \approx 4\pi (12)^2 \approx 1809.6$ cm$^2 \approx 1810$ cm^2 or $4\pi (.12)^2 \approx .18$ m^2 **18. a.** $V = \frac{4}{3}\pi r^3 = \frac{4}{3}\pi (9)^3 = 972\pi$ cubic units **b.** $972\pi \approx 3054$ cubic units **19.** $S.A. = 4\pi r^2 = 4\pi (3500)^2 = 49{,}000{,}000\pi \approx 1.5394 \cdot 10^8 \approx 153{,}950{,}000$ km$^2 \approx 154{,}000{,}000$ km^2 **20.** $A = \frac{60}{360}\pi \cdot 6^2 = 6\pi$ cm^2

The chart below keys the **Progress Self-Test** questions to the objectives in the **Chapter Review** on pages 688–689 or to the **Vocabulary** (Voc.) on page 686. This will enable you to locate those **Chapter Review** questions that correspond to questions students missed on the **Progress Self-Test**. The lesson where the material is covered is also indicated on the chart.

Question	1	2	3	4	5	6	7	8	9	10
Objective	D	J	J	B	C	C	G	A	A	H
Lesson	12-4, 12-5	12-2	12-2	12-2	12-3	12-3	12-2	12-1	12-1	12-6
Question	**11**	**12**	**13**	**14**	**15**	**16**	**17**	**18**	**19**	**20**
Objective	E	E	B	I	D	G	I	F	I	D
Lesson	12-6	12-7	12-2	12-4	12-5	12-2	12-8	12-8	12-8	12-5

CHAPTER 12 CHAPTER REVIEW (pages 688–689)

1. $-\frac{122}{99}$ **3.** $\frac{1631}{20}$ **5.** 8 and 9 **7.** 20 **9.** $BI = \sqrt{73}$ **11.** $\sqrt{2704} = 52$ **13.** $C = 20\pi$ units; $A = 100\pi$ square units **15.** $A = \frac{25}{4}\pi$ square feet **17.** 110π square units **19.** 2520 units^3

21. 100π units^2 **23.** 3.14, $\frac{22}{7}$, $3\frac{1}{7}$ **25.** always **27.** 24,000 ft^2 **29.** 5.16 in^3 **31.** 400π or ≈ 1256.6 cm^2 **33.** $\approx 942{,}000{,}000$ km **35.** $4.5\pi \approx 14.1$ cm^3 **37.** 4.9 **39.** $\sqrt{2}$

LESSON 13-1 (pp. 690–697)

17. (a) **19. a.** $y = 11x + 2$
b. 20 FOR X = 1 TO 100
   30 Y = 2 * X − 6
21. a. Sample:

age	height
9 yr	56″
10 yr	59″
11 yr	62″
12 yr	65″

b. See below. **c.** $b = 29$ **d.** From the equation, at age 25, for example, height would be 104″ or 8′8″, an unacceptable human height. **23.** $n = 6$; $3(5(6) + 22) - 156 = 0$ **25.** $\frac{3}{8}$ of the salad **27.** \$3276 **29.** 33°

21. b. *inches*

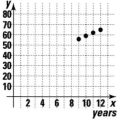

LESSON 13-2 (pp. 698–703)

13. a. $P = 222 + .190g$ **b.** See below. **c.** $P = 120 + .260g$ **d.** See below. **e.** Sample: about 1450 games **f.** $222 + .19g = 120 + .26g$ **15.** See below. **17.** $y = \frac{15}{8}$, $\frac{2}{5}\left(\frac{15}{8}\right) = \frac{3}{4}$ **19.** $0.56 = 56\%$ **21.** 22 miles

13. b., d.

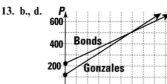

15.

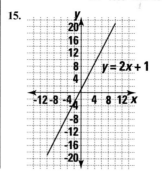

$y = 2x + 1$

LESSON 13-3 (pp. 704–708)

11. a. $P = 1000 + 200n$ **b.** $P = 750 + 250n$ **c.** 5 **13.** $p = -1$; $11(-1) + 5(-1 - 1) = -21$ and $9(-1) - 12 = -21$ **15.** $m\angle P = 60$; $m\angle N = 60$; $m\angle I = 60$ **17.** The lines intersect when $n = 5$, indicating when the amount paid is equal. **See below for graph.** **19.** $\frac{70}{11}$ **21.** BOOK **23. See below. 25.** about 6,946,500

17.

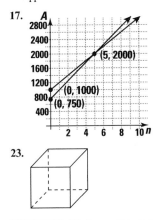

23.

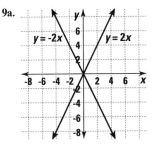

LESSON 13-4 (pp. 709–713)

7. a., b., c. Answers will vary. **d.** $y = \frac{1}{2}x$; answers may vary.
e. Answers will vary. **9. a. See below. b.** They intersect at (0, 0).
c. $y = ax$ and $y = -ax$ are reflection images of each other over the x-axis (or over the y-axis). **11.** $B = 6$ **13.** $x = 12$
15. $d = 44.88$ **17.** ≈ 401 minutes ≈ 6 hours and 41 minutes
19. .0009 or .09% **21.** Sample: 1 km = 1000 m = 100,000 cm = 1,000,000 mm

9a.

LESSON 13-5 (pp. 714–718)

11. a., b. See below. 13. Sample: (3, 4), (6, 8), (9, 12), (−3, −4), (−6, −8) **See below for graph. 15.** $t = -\frac{1}{8}$ **17.** Prince Edward Island, with about 60 people per square mile, was more densely populated than Ontario, with about 28 people per square mile.
19. a. 56 pieces **b.** 30 pieces

11.

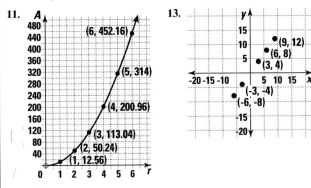

13.

LESSON 13-6 (pp. 719–724)

15. See below. 17. (d) **19. a.** Samples: At 10 mph, braking distance is 5 feet; at 20 mph, 20 feet; at 50 mph, 125 feet; at 60 mph, 180 feet. **b.** Sample: (10, 5), (20, 20), (50, 125), (60, 180) **c.** part of a parabola **21.** −6

15.

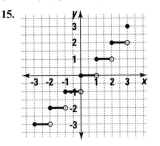

CHAPTER 13 PROGRESS SELF-TEST (page 728)

1. $60 + 14y - 14y = -20 + 30y - 14y$; $60 = -20 + 16y$; $60 + 20 = -20 + 20 + 16y$; $80 = 16y$; $5 = y$ **2.** $6x - 24 = 4x + 9$; $6x - 4x - 24 = 4x - 4x + 9$; $2x - 24 = 9 + 24 + 24$; $2x = 33$; $x = \frac{33}{2}$ **3.** 987.654 rounded up to the nearest integer is 988. **4.** Samples: 8, 8.3; any number between 8 and 9, including 8

5.

x	y
10	3
20	6
30	9
x	$\frac{3}{10}x$

$y = \frac{3}{10}x$ **6.** $y = \frac{3}{10}x$ **7. a.** $c = 35 + .48m$ **b.** $35 + .48m = 39 + .41m$ **8.** ≈ 57 miles **9. See right. 10. See right.**
11. a. $A = 6(5)^2$; $A = 150$ square units **b.** (5, 150) **12. See below.**
13. Sample: The point where the lines meet is the solution to $y = 4x$ and $y = -3x + 5$, hence $4x = -3x + 5$ will determine the x-coordinate of the point where the lines meet. **14.** Sample: $4x = -3x + 5$; $4x + 3x = -3x + 3x + 5$; $7x = 5$; $x = \frac{5}{7}$ **15.** (c)

9.

10.

12.

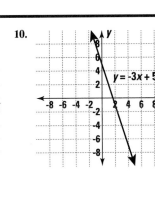

The chart below keys the **Progress Self-Test** questions to the objectives in the **Chapter Review** on pages 729–731 or to the **Vocabulary** (Voc.) on page 727. This will enable you to locate those **Chapter Review** questions that correspond to questions students missed on the **Progress Self-Test**. The lesson where the material is covered is also indicated on the chart.

Question	1	2	3	4	5	6	7	8	9	10
Objective	A	A	B	B	C	C	D	A	F	F
Lesson	13-3	13-3	13-6	13-6	13-4	13-4	13-2	13-3	13-1	13-1

Question	11	12	13	14	15
Objective	E	E	G	A	H
Lesson	13-5	13-5	13-2	13-3	13-6

CHAPTER 13 CHAPTER REVIEW (pp. 729–731)

1. $x = \frac{2}{5}$ **3.** $v = \frac{1}{2}$ **5.** $r = -\frac{3}{17}$ **7.** $t = \frac{4}{3}$ **9.** 1 **11.** -40 **13.** 27 **15. a.** Samples: $\frac{-7}{-5}, \frac{14}{10}, \frac{49}{35}$ **b.** $y = \frac{7}{5}x$ **17.** $y = 7x$ **19. a.** $15 + .20m = 5 + .34m$ **b.** ≈ 71 minutes **21. a.** $57 + \frac{1}{100}d = 59 + \frac{1}{150}d$ **b.** after 600 days **23. See below. 25. See below.**

27. See below. 29. See below. 31. Graph $y = 35 + 6w$ for Kurt and $y = 60 + 4w$ for Sara. The w-coordinate of the point where the lines intersect is the solution. **33. See graph below.** Conclusion: $x = 4$ **35.** $y = \lfloor x \rfloor$; in each bar the values of x are rounded down to the nearest integer. **37.** $y = 11\lceil x \rceil + 2$

23.

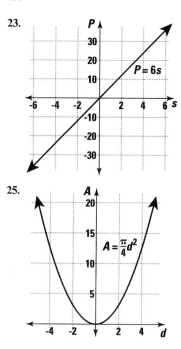

25.

27.

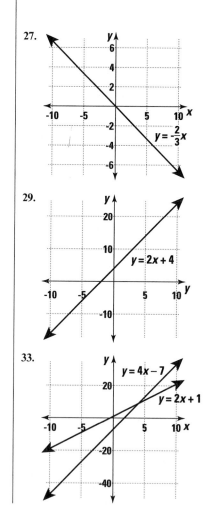

29.

33.

absolute value The distance that a number is from 0 on the number line. (248)

acre (a) A unit of area equal to 43,560 square feet. (162)

acute angle An angle with a measure less than 90°. (154)

addend A number to be added. In $a + b$, a and b are addends. (239)

Adding Fractions Property For all numbers a, b, and c, with $c \neq 0$, $\frac{a}{c} + \frac{b}{c} = \frac{a + b}{c}$. (260)

Addition Property of Equality If $a = b$, then $a + c = b + c$. (277)

additive identity The number zero. (244)

Additive Identity Property of Zero For any number n, $n + 0 = n$. (244)

additive inverse A number whose sum with a given number is 0. Also called *opposite*. (245)

adjacent angles Angles that share a common side and have no common interior points. (256)

algebra See *elementary algebra*.

Algebraic Definition of Division For all numbers a and nonzero numbers b, $a \cdot \frac{1}{b} = \frac{a}{b} = a \div b$. (483)

Algebraic Definition of Subtraction For any numbers x and y, $x - y = x + -y$. Also called the *Adding Opposites Property* or the *Add-Opp Property*. (353)

algebraic expression An expression that contains either a variable alone or with number and operation symbols. (187)

algorithm A finite sequence of steps that leads to a desired result. (302)

alternate exterior angles Angles formed by two lines and a transversal that are on opposite sides of the transversal but not between the two given lines. (382)

alternate interior angles Angles formed by two lines and a transversal that are between the two given lines and on opposite sides of the transversal. (382)

altitude The perpendicular segment from a vertex of a triangle to the opposite side of the triangle. Also see *height*. (572)

ambiguous Having more than one possible meaning. (189)

angle The union of two rays with the same endpoint. (145)

Angle Addition Property If $\overrightarrow{OB}$ is in the interior of $\angle AOC$, then $m\angle AOB + m\angle BOC = m\angle AOC$. (256)

Arabic numerals See *Hindu-Arabic numerals*.

arc of a circle A part of a circle connecting two points (called the **endpoints** of the arc) on the circle. (94)

area A measure of the space inside a two-dimensional figure. (158)

Area Formula for a Circle Let A be the area of a circle with radius r. Then $A = \pi r^2$. (666)

Area Formula for a Sector Let A be the area of a sector of a circle with radius r. Let m be the measure of the central angle of the sector in degrees. Then $A = \frac{m}{360} \cdot \pi r^2$. (667)

Area Formula for a Trapezoid Let A be the area of a trapezoid with bases b_1 and b_2 and height h. Then $A = \frac{1}{2} \cdot h(b_1 + b_2)$. (579)

Area Formula for Any Triangle Let b be the length of a side of a triangle with area A. Let h be the length of the altitude drawn to that side. Then $A = \frac{1}{2}bh$. (572)

Area Formula for a Right Triangle Let A be the area of a right triangle with legs of lengths a and b. Then $A = \frac{1}{2}ab$. (470)

Area Model for Multiplication The area of a rectangle with length ℓ units and width w units is $\ell \cdot w$ (or ℓw) square units. (469)

array See *rectangular array*.

Associative Property of Addition For any numbers a, b, and c, $(a + b) + c = a + (b + c)$. (272)

Associative Property of Multiplication For any numbers a, b, and c, $(ab)c = a(bc)$. (479)

automatic drawer Technology that can be used to draw geometric figures. (400)

automatic grapher A computer program or calculator that is used to make coordinate graphs. (458)

average See *mean*.

ball A sphere together with all the points inside it. (680)

bar graph A display in which numbers are represented by bars whose lengths correspond to the magnitude of the numbers being represented. (415)

base **1** In a power x^y, x. (68) **2** The side of a triangle perpendicular to an altitude. (572) **3** The bottom of a box, or rectangular solid. (476) **4** In a trapezoid, one of the parallel sides. (578) **5** In a cylindric solid, one of the plane regions translated to form the solid. (675)

base line of a protractor A segment on a protractor connecting a 0° mark with a 180° mark opposite which is placed over one side of the angle being measured. (147)

billion In the United States, a word name for the number 1,000,000,000 or 10^9. (65)

billionth A word name for the number 0.00000 0001 or 10^{-9}. (101)

box See *rectangular solid*.

brackets [ ] **1** Grouping symbols that indicate that all arithmetic operations inside are to be done first. As grouping symbols, brackets mean exactly the same thing as parentheses. (203) **2** A notation sometimes used for the greatest integer function. (720)

capacity See *volume*.

ceiling function symbol See *rounding up symbol*.

cell In a spreadsheet, the intersection of a column and a row. (325)

center of a circle See *circle*.

center of a protractor The midpoint of the base line on a protractor. The center is placed over the vertex of the angle being measured. (147)

center of a sphere See *sphere.*

center of gravity The center of gravity of a set of points is the point whose first coordinate is the mean of the first coordinates of the given points, and whose second coordinate is the mean of the second coordinates of the given points. (609)

centi- A prefix meaning $\frac{1}{100}$. (137)

central angle of a circle An angle whose vertex is the center of the circle. (152)

certain event An event with probability one. (214)

circle The set of all points in a plane a given distance (its **radius**) from a fixed point in the plane (its **center**). (94)

circle graph A display in which parts of a whole are represented by sectors whose areas are proportional to the fraction of the whole taken up by each part. Also called *pie chart.* (93)

circular cylinder A cylindric solid whose bases are circles. Also called *cylinder.* (670)

circumference A circle's perimeter. (659)

clockwise The same direction as that in which the hands of a clock normally move. (253)

coincide To occupy the same position as. (442)

column One of the vertical arrangements of items in a table, rectangular array, or spreadsheet. (325)

common denominator A number that is a multiple of the denominator of every fraction in a collection of fractions. (262)

Commutative Property of Addition For any numbers *a* and *b*, $a + b = b + a$. (271)

Commutative Property of Multiplication For any numbers *a* and *b*, $ab = ba$. (470)

Comparison Model for Subtraction $x - y$ is how much more *x* is than *y*. (348)

complement of an event The event consisting of all outcomes not in the given event. (268)

complementary angles Two angles whose measures add to 90°. (376)

composite number Any positive integer exactly divisible by one or more positive integers other than itself or 1; opposed to *prime number.* (307)

congruent figures Figures having the same size and shape. Figures that are the image of each other under a reflection, rotation, or translation, or any combination of these. (436)

contraction A size change with a magnitude whose absolute value is between 0 and 1. (511)

conversion factor A rate factor that equals the number 1, used for converting quantities given in one unit to a quantity given in another unit. (495)

convex polygon A polygon in which every diagonal lies inside the polygon. (451)

coordinate graph A display in which an ordered pair of numbers is displayed as a dot located relative to a fixed point called the **origin.** (420)

corresponding angles Any pair of angles in similar locations in relation to a transversal intersecting two lines. (382)

corresponding sides Any pair of sides in the same relative positions in two figures. (626)

count A whole number tally of particular things. (7)

counterclockwise The direction opposite to that in which the hands of a clock normally move. (253)

counting unit The name of the particular things being tallied in a count. (7)

cube A three-dimensional figure with six square faces. (163)

cup (c) A unit of capacity in the U.S. system of measurement equal to 8 fluid ounces. (130)

customary system See *U.S. system of measurement.*

cylinder See *circular cylinder.*

cylindric solid The set of points between a plane region (its **base**) and its translation image in space, including the region and its image. (675)

cylindric surface The surface of a cylindric solid. (675)

data disk A disk used to store computer information. (328)

decagon A ten-sided polygon. (284)

deci- A prefix meaning $\frac{1}{10}$. (137)

decimal notation A notation in which numbers are written using ten digits and each place stands for a power of 10. (6)

decimal system A system in which numbers are written in decimal notation. (6)

degree A unit of angle measure equal to $\frac{1}{360}$ of a full revolution. (146)

deka- A prefix meaning 10. (137)

denominator In a fraction, the divisor. The number *b* in the fraction $\frac{a}{b}$. (31)

diagonal of a polygon A segment that connects two vertices of the polygon but is not a side of the polygon. (286)

diameter **1** Twice the radius of a circle or sphere. **2** A segment connecting two points on a circle that contains the center of the circle. (94) **3** A segment connecting two points on a sphere that contains the center of the sphere. (680)

diamond See *rhombus.*

difference The result of a subtraction. (347)

digit One of the ten symbols 0, 1, 2, 3, 4, 5, 6, 7, 8, 9 used in the decimal system to stand for the first ten whole numbers. (3)

dimension **1** The length or width of a rectangle. (469) **2** The number of rows or the number of columns in an array. (469) **3** The length, width, or height of a box. (476)

display **1** Numbers or other symbols that appear on the screen of a calculator or computer. (26) **2** A diagram, graph, table, or other visual arrangement used to organize, present, or represent information. (407)

Distributive Property of Multiplication over Addition For any numbers *a*, *b*, and *x*, $ax + bx = (a + b)x$ and $x(a + b) = xa + xb$. (556)

Distributive Property of Multiplication over Subtraction For any numbers *a*, *b*, and *x*, $ax - bx = (a - b)x$ and $x(a - b) = xa - xb$. (557)

dividend The number in a quotient that is being divided. In the division $a \div b$, *a* is the dividend. (31)

divisible An integer *a* is divisible by an integer *b* when $\frac{a}{b}$ is an integer. Also called *evenly divisible*. (34)

divisor **1** The number by which you divide in a quotient. In the division $a \div b$, *b* is the divisor. (31) **2** A number that divides into another number with zero remainder. Also called *factor*. (308)

dodecagon A 12-sided polygon (319)

double bar graph A display in which two bar graphs are combined into a single display. (416)

edges The sides of the faces of a rectangular solid. (476)

elementary algebra The study of variables and the operations of arithmetic with variables. (184)

ending decimal See *terminating decimal.*

endpoint of a ray See *ray.*

entering Pressing a key on a calculator to enter numbers, operations, or other symbols. (27)

Equal Fractions Property If the numerator and denominator of a fraction are both multiplied (or divided) by the same nonzero number, then the resulting fraction is equal to the original one. (51)

equally likely outcomes Outcomes in a situation where the likelihood of each outcome is assumed to be equal. (215)

equation A sentence with an equal sign. (219)

equilateral triangle A triangle whose sides all have the same length and whose angles all have the same measure. (452)

equivalent equations Equations that have the same solutions. (363)

equivalent formulas Formulas that have the same solutions. (364)

estimate **1** A number which is near another number. Also called *approximation.* **2** To obtain an estimate. (16)

evaluating an algebraic expression Replacing each variable with a particular number (the **value of the variable**) and then evaluating the resulting numerical expression. (193)

evaluating a numerical expression Carrying out the operations in a numerical expression to find the value of the expression. (176)

evenly divisible See *divisible.*

event Any collection of outcomes from the same experiment. (216)

exercise A question that you know how to answer. (302)

expansion A size change with a magnitude whose absolute value is greater than 1. (505)

exponent In a power x^y, *y*. (68)

exponential form A number written as a power. The form x^y. (99)

exterior angles Angles formed by two lines and a transversal that have no points between the two given lines. (382)

extremes In the proportion $\frac{a}{b} = \frac{c}{d}$, the numbers *a* and *d*. (620)

faces See *rectangular solid.*

factor **1** A number that divides evenly into another number. Also called *divisor.* (308) **2** To find the factors of a particular number. (51)

fair An object or experiment in which all outcomes are equally likely. Also called *unbiased.* (215)

finite decimal See *terminating decimal.*

first coordinate See *ordered pair.*

floor function symbol See *rounding down symbol.*

fluid ounce (fl oz) A unit of capacity in the U.S. system of measurement equal to $\frac{1}{128}$ of a gallon. Often called simply *ounce.* (130)

foot (ft) A unit of length in the U.S. system of measurement equal to twelve inches. (125)

formula A sentence in which one variable is written in terms of other variables. (208)

Formula for Circumference of a Circle In a circle with diameter *d* and circumference *C*, $C = \pi d$. (659)

fraction A symbol of the form $\frac{a}{b}$ which represents the quotient when *a* is divided by *b*. (31)

fraction bar The horizontal line in a fraction that indicates the operation of division. (31)

frequency of an outcome The number of times an outcome occurs. (213)

full turn A turn of 360°. Also called *revolution.* (253)

Fundamental Property of Turns If a turn of magnitude *x* is followed by a turn of magnitude *y*, the result is a turn of magnitude $x + y$. (255)

fundamental region A figure whose congruent copies make up a tessellation. Also called *fundamental shape.* (452)

fundamental shape See *fundamental region.*

furlong (fur) A unit of length in the U.S. system of measurement equal to 40 rods. (128)

gallon (gal) A unit of capacity in the U.S. system of measurement equal to 4 quarts. One gallon is equal to 231 cubic inches. (130)

generalization A statement that is true about many instances. (321)

giga- A prefix meaning one billion. (137)

gram (g) A unit of mass in the metric system equal to $\frac{1}{1000}$ of a kilogram. (136)

greatest common factor (GCF) The largest number that is a factor of every number in a given collection of numbers. (52)

greatest integer function symbol See *rounding down symbol.*

gross ton A unit of weight in the U.S. system of measurement equal to 2240 pounds. Also called *long ton.* (132)

grouping symbols Symbols such as parentheses, brackets, or fraction bars that indicate that certain arithmetic operations are to be done before others in an expression. (203)

half turn A turn of magnitude 180°. (253)

hectare (ha) A metric unit of area equal to 10,000 m². (159)

hecto- A prefix meaning 100. (137)

height **1** The perpendicular distance from any vertex of a triangle to the side opposite that vertex. (572) **2** The distance between the bases of the trapezoid. (578) **3** The distance between the bases of a cylinder or prism. (676) Also called *altitude.*

heptagon A seven-sided polygon. (284)

hexagon A six-sided polygon. (284)

Hindu-Arabic numerals The numerals 0, 1, 2, 3, 4, 5, 6, 7, 8, 9 used throughout the world today to represent numbers in the decimal system. (3)

hour (h) A unit of time equal to 60 minutes. (130)

hundred A word name for the number 100 or 10². (65)

hundred-thousandth A word name for 0.00001 or 10^{-5}. (12)

hundredth A word name for 0.01 or 10^{-2}. (11)

hypotenuse The longest side of a right triangle. (470)

image The result of applying a transformation to a figure. (435)

impossible event An event with probability zero. (214)

inch (in.) The basic unit of length in the U.S. system of measurement. One inch equals 2.54 centimeters exactly. (119)

independent events Two events *A* and *B* are independent events when and only when *Prob(A and B)* = *Prob(A)* · *Prob(B).* (489)

inequality A sentence with one or more of the symbols $\neq, <, \leq, >, \geq$. (224)

inequality symbols Any one of the symbols $\neq$ (not equal to), $<$ (is less than), $\leq$ (is less than or equal to), $>$ (is greater than), or $\geq$ (is greater than or equal to). (224)

infinite repeating decimal A decimal in which a digit or group of digits to the right of the decimal point repeat forever. (33)

instance A particular example of a pattern. (182)

integer A number which is a whole number or the opposite of a whole number. Any of the numbers 0, 1, -1, 2, -2, 3, -3, ... (43)

integer division Dividing an integer *a* by a positive integer *b* to yield an integer quotient and an integer remainder less than *b.* (592)

interior angles Angles formed by two lines and a transversal that have some points between the two given lines. (382)

international system of measurement See *metric system.*

irrational number A real number that cannot be written as a simple fraction. (660)

key in To press a key on a calculator to enter numbers or perform operations. (27)

key sequence A description of the calculator keys and the order in which they must be pressed to perform a particular computation. (27)

kilo- A prefix meaning 1000. (137)

kilogram (kg) The base unit of mass in the metric system. (136)

lateral area The area of the surface of a cylindric solid excluding the bases. (670)

least common denominator (LCD) The least common multiple of the denominators of every fraction in a given collection of fractions. (262)

least common multiple (LCM) The smallest number that is a multiple of every number in a given collection of integers. (262)

legs of a right triangle The sides of a right triangle that are on the sides of the right angle, or the lengths of these sides. (470)

light-year The distance that light travels in a year. (74)

like terms Terms that involve the same variables to the same powers. (532)

linear equation An equation of the form $y = ax + b$. An equation equivalent to $ax + b = cx + d$. (694)

linear pair Two angles that share a common side and whose non-common sides are opposite rays. (374)

line of symmetry A line over which the reflection image of a figure coincides with its preimage. Also called *symmetry line.* (446)

line segment A pair of given points *A* and *B* (the **endpoints** of the line segment) along with all the points on $\overleftrightarrow{AB}$ between *A* and *B*. Also called *segment.* (283)

liter (L) A unit of volume in the metric system equal to 1000 cm³. (136)

long ton See *gross ton.*

lowest terms A fraction is in lowest terms when the numerator and denominator have no factors in common except 1. (51)

magic square A square array of numbers in which the sum along every row, column, or main diagonal is the same. (281)

magnitude **1** An amount measuring the size of a turn. (253) **2** The distance between a point and its translation image. (435) **3** A size change factor. (507)

make a table A problem-solving strategy in which information is organized in a table to make it easier to find patterns and generalizations. (320)

mean The sum of the numbers in a collection divided by the number of numbers in the collection. Also called *average.* (206)

means In the proportion $\frac{a}{b} = \frac{c}{d}$, the numbers *b* and *c.* (620)

Means-Extremes Property In any true proportion, the product of the means equals the product of the extremes. If $\frac{a}{b} = \frac{c}{d}$, then $b \cdot c = a \cdot d$. (621)

measure A real number indicating an amount of some quantity. (10)

median In a collection consisting of an odd number of numbers arranged in numerical order, the middle number. In a collection of an even number of numbers arranged in numerical order, the average of the middle two numbers. (411)

mega- A prefix meaning one million. (137)

meter (m) The basic unit of length in the metric system, defined as the length that light in vacuum travels in $\frac{1}{299,792,458}$ of a second. (119)

metric system A decimal system of measurement with the meter as its base unit of length and the kilogram as its base unit of mass. Also called *international system of measurement*. (118)

micro- A prefix meaning one millionth. (137)

mile (mi) A unit of length in the U.S. system of measurement equal to 5280 feet. (125)

milli- A prefix meaning $\frac{1}{1000}$. (137)

million A word name for the number 1,000,000 or 10^6. (65)

millionth A word name for 0.000001 or 10^{-6}. (12)

minuend In a subtraction problem, the number from which another number is subtracted. In $a - b$, the minuend is a. (347)

minute A unit of time equal to 60 seconds. (130)

mirror See *reflection image*.

mixed number The sum of a whole number and a fraction. (37)

mode The mode of a collection of objects is the object that appears most often. (411)

Multiplication of Fractions Property For all numbers a and c, and nonzero numbers b and d, $\frac{a}{b} \cdot \frac{c}{d} = \frac{ac}{bd}$. (484)

Multiplication of Unit Fractions Property For all nonzero numbers a and b, $\frac{1}{a} \cdot \frac{1}{b}$. (483)

Multiplication Property of Equality When equal quantities are multiplied by the same number, the resulting quantities are equal. (125)

Multiplication Property of -1 For any number x, $-1 \cdot x = -x$. (501)

Multiplication Property of Zero any number x, $0 \cdot x = 0$. (502)

Multiplicative Identity Property of One For any number x, $1 \cdot x = x$. (507)

multiplicative inverse A number whose product with a given number is 1. Also called *reciprocal*. (483)

mutually exclusive events Events that cannot occur at the same time. (267)

nano- A prefix meaning one billionth. (137)

natural number Any of the numbers 1, 2, 3, ... Also called *positive integer*. (43) (Some people like to include 0 as a natural number.)

negative integer Any of the numbers -1, -2, -3, ... (43)

negative numbers The opposites of the positive numbers. Numbers that are graphed to the left of zero on a horizontal number line. (41)

nested parentheses Parentheses inside parentheses. (199)

net A plane pattern for a three-dimensional figure. (561)

n-gon A polygon with n sides. (284)

nonagon A nine-sided polygon. (284)

numerator In a fraction, the dividend. The number a in the fraction $\frac{a}{b}$. (31)

numerical expression A combination of symbols for numbers and operations that stands for a number. (176)

obtuse angle An angle whose measure is between 90° and 180°. (154)

octagon An eight-sided polygon. (284)

odds against an event The ratio of the number of ways an event cannot occur to the number of ways that event can occur. (218)

odds for an event The ratio of the number of ways an event can occur to the number of ways that event cannot occur. (218)

pattern A genera... there are many e...

pentagon A five...

percent A num... percent sign %... indicates that th... should be mult...

perfect num... for which the... except n, is...

perimeter of the side...

perpendi... or lines t...

pi (π) to the...

pico- trillion...

pie c...

pint U.S. to...

opp... two a... (388)

opposite the same... is a line. (37...

opposite side... sides that do n...

ordered pair A ... objects (x, y) in wh... x is designated the ... and the second numb... the **second coordinate**...

Order of Operations T... carrying out arithmetic op... 1. Work first inside parenth... other grouping symbols. 2. ... grouping symbols or if there ... grouping symbols: **a.** First take... powers; **b.** Second, do all multiplications or divisions in orde... from left to right; **c.** Finally, do all additions or subtractions in order, from left to right. (177)

origin The point (0, 0) in a coordinate graph. (424)

ounce (oz) A unit of weight in the U.S. system of measurement, equal to $\frac{1}{16}$ of a pound. (130)

outcome A possible result of an experiment. (213)

palindrome An integer which reads the same forward or backward. (433)

parabola The shape of all graphs of equations of the form $y = ax^2$. (691)

parallel Two lines in a plane are parallel if they have no points in common or are identical. (381)

parallelogram A quadrilateral with two pairs of parallel sides. (388)

parentheses ( ) Grouping symbols that indicate that all arithmetic operations inside are to be done first. (197)

idea for which
amples. (182)

-sided polygon. (284)

ber written using the
The percent sign
e number preceding it
plied by 0.01 or $\frac{1}{100}$. (79)

er   A positive integer n
sum of all its factors,
equal to n. (310)

The sum of the lengths
s of a polygon. (289)

cular   Two rays, segments,
at form right angles. (153)

he ratio of the circumference
iameter of any circle. (12)

A prefix meaning one
th. (137)

art   See *circle graph*.

(pt)   A unit of capacity in the
system of measurement equal
cups. (130)

ace value   The number that each
git in a decimal stands for. For
stance, in the decimal 123.45 the
place value of the digit 2 is twenty. (7)

polygon   A union of segments (its
sides) in which each segment
intersects two others, one at each of
its endpoints. (283)

positive integer   Any of the numbers
1, 2, 3, ... Also called *natural
number*. (43)

positive numbers   Numbers that are
graphed to the right of zero on a
horizontal number line. (43)

pound (lb)   A unit of weight in the
U.S. system of measurement equal
to 16 ounces. One pound is
approximately 455 grams. (130)

power   The result of x^y. (68)

preimage   A figure to which a
transformation has been applied. (435)

prime number   A positive integer
other than 1 that is divisible only by
itself and 1. (307)

prism   A cylindric solid whose
bases are polygons. (671)

Probabilities of Complements   The
sum of the probabilities of any event
E and its complement *not E* is 1. (268)

probability   A number from 0 to 1
that tells how likely an event is to
happen. (214)

**Probability Formula for Equally Likely
Outcomes**   Suppose a situation has
N equally likely possible outcomes
and an event includes E of these. Let
P be the probability that the event
will occur. Then $P = \frac{E}{N}$. (216)

Probability of *A or B*   If A and B are
mutually exclusive events, the
probability of A plus the probability
of B. (267)

problem   A question for which you
do not have an algorithm to arrive at
an answer. (302)

problem-solving strategies
Strategies such as *read carefully,
make a table, trial and error, draw
a picture, try special cases,* and
check work that good problem
solvers use when solving
problems. (303)

product   The result of multiplication.
(175)

Properties of Parallelograms   In a
parallelogram, opposite sides have
the same length; opposite angles
have the same measure. (388)

Property of Opposites   For any
number n, $n + -n = 0$. Sometimes
called the *additive inverse property*.
(245)

Property of Reciprocals   For any
nonzero number a, a and $\frac{1}{a}$ are
reciprocals. That is, $a \cdot \frac{1}{a} = 1$. (484)

proportion   A statement that two
fractions are equal. (615)

proportional thinking   The ability to
get or estimate an answer to a
proportion without solving an
equation. (630)

protractor   An instrument for
measuring angles. (147)

**Putting-Together Model for
Addition**   Suppose a count or
measure x is put together with a
count or measure y with the same
units. If there is no overlap, then
the result has count or measure
$x + y$. (238)

**Putting-Together with Overlap
Model**   If a quantity x is put
together with a quantity y, and there
is overlap z, the result is the quantity
$x + y - z$. (368)

Pythagorean Theorem   Let the legs
of a right triangle have lengths a and
b. Let the hypotenuse have length c.
Then $a^2 + b^2 = c^2$. (652)

quadrant   One of the four regions
of the coordinate plane determined
by the x- and y- axes. (424)

quadrilateral   A four-sided polygon.
(284)

quadrillion   A word name for the
number 1,000,000,000,000,000 or
10^{15}. (65)

quadrillionth   A word name for the
number 0.00000 00000 00001 or
10^{-15}. (101)

quart (qt)   A unit of capacity in the
U.S. system of measurement equal
to 2 pints. (130)

quarter turn   A turn of 90°. (253)

quintillion   A word name for the
number 1,000,000,000,000,000,000
or 10^{18}. (65)

quintillionth   A word name for the
number 0.00000 00000 00000 001 or
10^{-18}. (101)

quotient   The result of dividing one
number by another. (31)

Quotient-Remainder Formula   Let n
be an integer and d be a positive
integer. Let q be the integer quotient
and let r be the remainder after
dividing n by d. Then $n = d \cdot q + r$.
(593)

quotient-remainder form   The result
of an integer division expressed as
an integer quotient and an integer
remainder. (593)

radical sign   The symbol $\sqrt{}$ used
to denote a positive square root.
(647)

radius   **1** The distance that all
points in a circle are from the center
of the circle. **2** A segment connecting
the center of a circle to any point on
the circle. (94) **3** The distance that all
points on a sphere are from the
center of the sphere. **4** A segment
connecting the center of a sphere to
any point on the sphere. (680)

random experiment   An experiment
in which all outcomes are equally
likely. (215)

range   In a collection of numbers,
the difference between the largest
number and the smallest number.
(411)

rate   The quotient of two quantities
with different units. A quantity whose
unit contains the word ''per'' or ''for
each'' or some synonym. (493)

rate factor A rate used in a multiplication. (494)

Rate Factor Model for Multiplication The product of (a $unit_1$) and $\left(b\ \frac{unit_2}{unit_1}\right)$ is (ab $unit_2$), signifying the total amount of $unit_2$ in the situation. (494)

Rate Model for Division If a and b are quantities with different units, then $\frac{a}{b}$ is the amount of quantity a per quantity b. (598)

rate unit The unit in a rate. (493)

ratio The quotient of two quantities with the same units. (611)

Ratio Comparison Model for Division If a and b are quantities with the same units, then $\frac{a}{b}$ compares a to b. (611)

rational number A number that can be written as a simple fraction. (660)

ray A part of a line starting at one point (the **endpoint** of the ray) and continuing forever in a particular direction. (145)

real-number division Dividing one real number a by a nonzero real number b to yield the single real number $\frac{a}{b}$. (592)

real number A number that can be written as a decimal. (641)

reciprocal A number whose product with a given number is 1. Also called *multiplicative inverse.* (483)

rectangle A parallelogram with a right angle. (388)

rectangular array An arrangement of objects into rows and columns. Also called *array.* (469)

rectangular solid A three-dimensional figure with six **faces** that are all rectangles. Also called *box.* (476)

reflection A transformation in which the image of each point lies on a line perpendicular to a fixed line called the **reflection line,** or **mirror,** and is the same distance from the reflection line as the preimage point. (440)

reflection line See *reflection.*

reflection symmetry The property that a figure coincides with its image under a reflection over a line called a **line of symmetry.** Also called *symmetry with respect to a line.* (446)

regular polygon A convex polygon in which all sides have the same length and all angles have the same measure. (338)

relative frequency The frequency of a particular outcome divided by the total number of times an experiment is performed. (213)

Repeated Addition Property of Multiplication If n is a positive integer, then $nx = x + x + \ldots + x$. (531) (n addends)

repeating decimal A decimal in which a digit or group of digits to the right of the decimal point repeat forever. (33)

repetend The digits in a repeating decimal that repeat forever. (33)

revolution A turn of 360°. Also called *full turn.* (253)

rhombus A quadrilateral with all sides the same length. Also called *diamond.* (388)

right angle An angle with a measure of 90°. (153)

right triangle A triangle with a right angle. (154)

rod (rd) A unit of length in the U.S. system of measurement equal to $5\frac{1}{2}$ yards. (128)

rotation symmetry The property that a figure coincides with its own image under a rotation of less than 360°. (451)

rounding down Making an estimate that is smaller than the actual value. (16)

rounding down symbol The symbol ⌊ ⌋ used to indicate that the number inside is to be rounded down to the preceding integer. Also called the *floor function symbol* or the *greatest integer function symbol.* (720)

rounding to the nearest Making an estimate to a particular decimal place by either rounding up or rounding down, depending on which estimate is closer to the actual value. (21)

rounding up Making an estimate that is larger than the actual value. (16)

rounding up symbol The symbol ⌈ ⌉ used to indicate that the number inside is to be rounded up to the next integer. Also called the *ceiling function symbol.* (720)

row One of the horizontal arrangements of items in a table, rectangular array, or spreadsheet. (325)

scale of a map The conversion factor used to convert real-world distances to distances on the map. (515)

scientific calculator A calculator which displays very large or very small numbers in scientific notation and which has powering, factorial, square root, negative, reciprocal, and other keys. (26)

scientific notation A notation in which a number is expressed as a number greater than or equal to 1 and less than 10 multiplied by an integer power of 10. A notation in which a number is written in the form $decimal \times 10^{exponent}$ with $1 \le decimal < 10$. (74)

second coordinate See *ordered pair.*

second The fundamental unit of time in the metric and customary systems of measurement. (134)

sector A region bounded by an arc of a circle and the two radii to the endpoints of the arc. (94)

segment See *line segment.*

short ton A unit of weight in the U.S. system of measurement equal to 2000 pounds. Sometimes simply called *ton.* (130)

side **1** In an angle, one of the two rays that make up an angle. (145) **2** In a polygon, one of the line segments that make up the polygon, or the length of a side. (145) **3** In a rectangular solid, a face of the solid. (476)

similar figures Two figures that have the same shape, but not necessarily the same size. (507)

simple fraction A fraction with a whole number as its numerator and a nonzero whole number as its denominator. (31)

size change factor The number k by which the coordinates of the preimage are multiplied in a size change. (507)

Size Change Model for Multiplication (two-dimensional version) Under a size change of magnitude $k,$ the image of (x, y) is (kx, ky). (507)

Size Change Model for Multiplication Let k be a nonzero number without a unit. Then ka is the result of applying a **size change of magnitude k** to the quantity a. (506)

slash The symbol / used to indicate division. (31)

slide image See *translation image*.

Slide Model for Addition If a slide x is followed by a slide y, the result is a slide $x + y$. (240)

Slide Model for Subtraction If a quantity x is decreased by an amount y, the resulting quantity is $x - y$. (352)

slope The amount of change in the height of a line as you move one unit to the right. For a line with equation $y = ax + b$, the slope is a. (711)

solution to an open sentence A value of the variable or variables in an open sentence that makes the sentence true. (219)

solve a proportion To find the values of all variables in a proportion that make the proportion true. (616)

solving a sentence Finding the values of the unknown or unknowns that make the sentence true. (220)

special case An instance of a pattern used for some definite purpose. (332)

sphere The set of points in space at a given distance (its **radius**) from a given point (its **center**). (680)

spreadsheet **1** A table. **2** A computer program in which data is presented in a table, and calculations upon entries in the table can be made. (325)

square A four-sided figure with four right angles and four sides of equal length. (158)

square root If $A = s^2$, then s is called a square root of A. (646)

statistic A number used to describe a set of numbers. (411)

stem-and-leaf display A display of a collection of numbers in which the digits in certain place values are designated as the **stems** and digits in lower place values are designated as the **leaves** and placed side-by-side next to the stems. (409)

step function symbols The symbols $\lfloor \; \rfloor$ and $\lceil \; \rceil$. (722)

Substitution Principle If two numbers are equal, then one can be substituted for the other in any computation without changing the results of the computation. (84)

subtrahend The number being subtracted in a subtraction. In $a - b$, the subtrahend is b. (347)

sum The result of an addition. (175)

supplementary angles Two angles whose measures add to 180°. (376)

Surface Area and Volume Formulas for a Sphere In a sphere with radius r, surface area S, and volume V, $S = 4\pi r^2$ and $V = \frac{4}{3}\pi r^3$. (680)

surface area The sum of the areas of the faces of a solid. (562)

symmetric figure A figure that coincides with its image under a reflection or rotation. (446)

symmetry line A reflection line over which the image of a reflection symmetric figure coincides with the preimage. Also called *line of symmetry*. (446)

symmetry with respect to a line See *reflection symmetry*.

table An arrangement of numbers or symbols into rows and columns. (320)

tablespoon (T) A unit of capacity in the U.S. system of measurement equal to $\frac{1}{2}$ of a fluid ounce. (264)

Take-Away Model for Subtraction If a quantity y is taken away from an original quantity x with the same units, the quantity left is $x - y$. (346)

teaspoon (tsp) A unit of capacity in the U.S. system of measurement equal to $\frac{1}{3}$ of a tablespoon. (264)

ten A word name for the number 10. (65)

ten-millionth A word name for 0.0000001 or 10^{-7}. (12)

ten-thousandth A word name for 0.0001 or 10^{-4}. (12)

tenth A word name for 0.1 or 10^{-1}. (11)

tenths place In a decimal, the first position to the right of the decimal point. (11)

tera- A prefix meaning one trillion. (137)

terminating decimal A decimal with only a finite number of nonzero decimal places. Also called *ending decimal* or *finite decimal*. (89)

tessellation A filling up of a two-dimensional space by congruent copies of a figure that do not overlap. (452)

test a special case A problem-solving strategy in which one or more special cases are examined to formulate a generalization or determine whether a property or generalization is true. (332)

theorem A statement that follows logically from other statements. (652)

thousand A word name for the number 1,000 or 10^3. (65)

thousandth A word name for 0.001 or 10^{-3}. (12)

ton (t) A unit of weight in the U.S. system equal to 2000 pounds. Also called *short ton*. (130)

transformation A one-to-one correspondence between a first set (the **preimage**) and a second set (the **image**). (440)

translation image The result of adding a number h to each first coordinate and a number k to each second coordinate of all the points of a figure. Also called *slide image*. (435)

transversal A line intersecting both of a pair of lines. (382)

trapezoid A quadrilateral with at least one pair of parallel sides. (577)

trial and error A problem-solving strategy in which potential solutions are tried and discarded repeatedly until a correct solution is found. (315)

triangle A three-sided polygon. (154)

Triangle-Sum Property In any triangle, the sum of the measures of the angles is 180°. (394)

trillion A word name for the number 1,000,000,000,000 or 10^{12}. (65)

trillionth A word name for the number 0.00000 00000 01 or 10^{-12}. (101)

truncate To discard all digits to the right of a particular place. To round down to a particular decimal place. (17)

try simpler numbers A problem-solving strategy in which numbers in a problem are replaced with numbers that are easier to work with so that a general method of solution can be found. (334)

twin primes Two consecutive odd numbers that are both prime numbers. (310)

unbiased See *fair*.

undecagon An eleven-sided polygon. (324)

uniform scale A scale in which the numbers are equally spaced so that each interval represents the same value. (416)

unit cube A cube whose edges are 1 unit long. (477)

unit fraction A fraction with a 1 in its denominator and a nonzero integer in its denominator. (483)

unit of measure A standardized amount with which measures can be compared. (10)

unit square A square whose sides are 1 unit long. (468)

U.S. system of measurement A system of measurement evolved from the British Imperial system in which length is measured in inches, feet, yards, miles and other units, weight is measured in ounces, pounds, tons, and other units, and capacity is measured in ounces, cups, pints, quarts, gallons, and other units. Also called *customary system*. (119)

unknown A variable in an open sentence for which the sentence is to be solved. (220)

unlike terms Terms that involve different variables or the same variable with different exponents. (532)

value of a numerical expression The result of carrying out the operations in a numerical expression. (176)

variable A symbol that can stand for any one of a set of numbers or other objects. (182)

Venn diagram A diagram using circles to represent sets and the relationships between them. (367)

vertex **1** The common endpoint of the two rays that make up the angle. (145) **2** A point common to two sides of a polygon. (283) **3** One of the points at which two or more edges meet. (476)

vertical angles Angles formed by two intersecting lines, but that are not a linear pair. (375)

Volume Formula for a Cylindric Solid The volume of a cylindric solid with height h and base with area B is given by $V = Bh$. (676)

volume The measure of the space inside a three-dimensional figure. Also called *capacity*. (163)

whole number Any of the numbers 0, 1, 2, 3, ... (6)

window The part of the coordinate plane that appears on the screen of an automatic grapher. (698)

x-axis The first (usually horizontal) number line in a coordinate graph. (424)

x-coordinate The first coordinate of an ordered pair. (424)

y-axis The second (usually vertical) number line in a coordinate graph. (424)

y-coordinate The second coordinate of an ordered pair. (424)

yard (yd) A unit of length in the U.S. system of measurement equal to 3 feet. (125)

INDEX

$>$	is greater than	$\overleftrightarrow{AB}$	line through A and B
$<$	is less than	AB	length of segment from A to B
$=$	is equal to	$\overrightarrow{AB}$	ray starting at A and containing B
$\neq$	is not equal to	$\overline{AB}$	segment with endpoints A and B
$\leq$	is less than or equal to	$\angle ABC$	angle ABC
$\geq$	is greater than or equal to	$m\angle ABC$	measure of angle ABC
$\approx$	is approximately equal to	$\triangle ABC$	triangle ABC
$+$	plus sign	$\sqrt{n}$	positive square root of n
$-$	subtraction sign	$\lfloor x \rfloor$	floor of x; greatest integer less than or equal to x
$\times, \cdot$	multiplication signs		
$\div, \overline{)}\,, /$	division signs	$\lceil x \rceil$	ceiling of x; least integer greater than or equal to x
$\%$	percent		
ft	abbreviation for foot	$\pm$	positive or negative
yd	abbreviation for yard	π	Greek letter pi; $= 3.141592...$ or $\approx \frac{22}{7}$.
mi	abbreviation for mile		
in.	abbreviation for inch	?	computer input or PRINT command
oz	abbreviation for ounce	INT( )	computer command rounding down to the nearest integer
lb	abbreviation for pound		
qt	abbreviation for quart	2*3	computer command for $2\cdot3$
gal	abbreviation for gallon	4/3	computer command for $4 \div 3$
$n°$	n degrees	3^5	computer command for 3^5
( )	parentheses	SQR(N)	computer command for $\sqrt{n}$
[ ]	brackets	$\boxed{y^x}$ or $\boxed{x^y}$	calculator powering key
$|x|$	absolute value of x	$\boxed{x!}$	calculator factorial key
$-x$	opposite of x	$\boxed{\sqrt{n}}$	calculator square root key
$\perp$	is perpendicular to	$\boxed{\pm}$ or $\boxed{+/-}$	calculator negative key
$//$	is parallel to	$\boxed{\pi}$	calculator pi key
$\ulcorner$	right angle symbol	$\boxed{1/x}$	calculator reciprocal key
(x, y)	ordered pair x, y	$\boxed{INV}$, $\boxed{2nd}$, or $\boxed{F}$	calculator second function key
A'	image of point A		
$\overparen{AB}$	arc AB	$\boxed{EE}$ or $\boxed{EXP}$	calculator scientific notation key
$\overparen{ABC}$	arc ABC	$\boxed{123456789.}$	calculator display

Acknowledgments

Unless otherwise acknowledged, all photographs are the property of Scott, Foresman and Company. Page abbreviations are as follows: (T) top, (C) center, (B) bottom, (L) left, (R) right.

COVER & TITLE PAGE Melvin L. Prueitt / Los Alamos National Laboratory **vi(L)** Ralph Mercer **vi(R)** Joe Vanos / The Image Bank **vii(L)** Roger Tully / Tony Stone Images **vii(R)** Scott Morgan / West Light **viii** Pictor / Uniphoto **ix** Charly Franklin / FPG **x** Michael Simpson / FPG **3T** Charlie Westerman / Gamma-Liaison **3B** Courtesy The Robinson Foundation **4T** Jon Davison / The Image Bank **4CL** Steven E. Sutton / Duomo Photography Inc. **4CC** Ralph Mercer **4CR** Index Stock International **4B** Paul Berger / Tony Stone Images **6** Brent Jones **8** P. Vandermark / Stock Boston **10** Focus On Sports **11** Photo: Mary Jane Koznick / Courtesy Special Olympics **15** Tadanori Saito / Photo Researchers **16** AP / Wide World **19** Bill Freeman / Photo Edit **21** Roger Tully / Tony Stone Images **24** Bob Daemmrich **29** Courtesy Texas Instruments **30** Dennis MacDonald / Photo Edit **35** Tony Freeman / Photo Edit **40** Carol Zacny **42** Carl Purcell / Photo Researchers **44** Focus On Sports **46** Michel Tcherevkoff / The Image Bank **48** Courtesy The Robinson Foundation **55C** Courtesy Evelyn Boyd Granville **55R** Courtesy The University of Michigan **61TL** Murray Alcosser / The Image Bank **61BL** Courtesy of the Trustees of the British Museum **61C** Robert Frerck / Odyssey Productions, Chicago **62T** John Turner / Tony Stone Images **62C** Royal Observatory, Edinburgh / AATB / Photo Researchers (SPL) **62BL** Tony Brain / SPL / Photo Researchers **62BR** Joe Vanos / The Image Bank **63L** Patrice Loiez,Cern / SPL / Photo Researchers **63R** ©1984 / Tribune Media Services **64** Focus On Sports **68** Melvin Prueitt / Los Alamos National Laboratory **72** Peter Arnold, Inc. **73** Tony Stone Images **75** Vic Cox / Peter Arnold, Inc. **78** AP / Wide World **80** Greg McRill **84** Bob Daemmrich **86** Milt & Joan Mann / Cameramann International, Ltd. **88** Takehide Kazami / Peter Arnold, Inc. **89** SCRABBLE® is a registered trademark of The Milton Bradley Company. © 1993 MBC, a division of Hasbro, Inc. Used with permission. **91** Everett Collection **92** Robert Frerck / Stock Boston **98** Bettmann **99** Steve Vidler / Leo de Wys **103** Beryl Goldberg **105** Courtesy, Sandia Laboratories **107** CNRI / SPL / Photo Researchers **108** Courtesy Penreal Estate Agency, Cornwall **110T** Focus On Sports **110TC** Everett Collection **110** Francolon / Gamma-Liaison **110BL** Everett Collection **110BR** Indy Car **114** Airphoto Jim Wark / Peter Arnold, Inc. **116T** Klaus Mitteldorf / The Image Bank **116C** Steve Satushey / The Image Bank **116BL** Roger Tully / Tony Stone Images **116BR** Chuck O'Rear / West Light **117T** Gerard Del Vecchio / Tony Stone Images **117BR** Alan Becker / The Image Bank **118** National Museum of American Art / Art Resource, NY **121** Everett Collection **125** Carol Zacny **128** Robert Frerck / Odyssey Productions, Chicago **130** Ed Pritchard / Tony Stone Images **134** U.S. Dept. of Commerce, National Institute of Standards and Technology **136** U.S. Dept. of Agriculture **138** David R. Frazier Photolibrary **140** Milt & Joan Mann / Cameramann International, Ltd. **141** Joseph Viesti **142** Bob Daemmrich **145** Jonathan Wright / Bruce Coleman Inc. **153** Wingstock / Comstock Inc. **155** Laura Dwight **156** Vince Streano / Tony Stone Images **158** Pugliano / Gamma-Liaison **162T** Mark Linfield / Green Umbrella Ltd. **163** NASA **165** U.S. Navy Photo / U.S. Department of Defense **167** N. Parfitt / Tony Stone Images **168T** Steve Vidler / Leo de Wys **168B** Orion SVC / TRDNG 1992 / FPG **171** National Portrait Gallery, London **174T** Richard Gorbun / Leo de Wys **174TC** Comstock Inc. **174BC** Hans R. Uthoff / Tony Stone Images **174B** Comstock Inc. **175B** Scott Morgan / West Light **182L** Smithsonian Institution / Courtesy of: National Museum of the American Indian **182R** Jake Rajs / The Image Bank **184** Library of Congress **185** Rooraid / Photo Edit **187** Will & Deni McIntyre / Photo Researchers **188** Bob Daemmrich / Stock Boston **189** Robert Fried / Stock Boston **191** The British Library **193** Courtesy The University of Chicago **196** Tony Stone Images **197** Joseph Viesti **201** Focus On Sports **207** Vandermark / Stock Boston **208** Ken Whitmore / Tony Stone Images **211** Ian Halperin **214** Bob Daemmrich / Stock Boston **218** Marka / Leo de Wys **219** Brent Jones **222** Mark Antman / The Image Works **224** Bill Gillette / Stock Boston **228** Everett Collection **229TL** K. Scholz / H. Armstrong Roberts **229TC** C. Ursillo / H. Armstrong Roberts **229TR** NASA **229BL** Addison Gallery of American Art Phillips Academy, Andover, MA **229BR** FPG **230TL** The Image Bank **230CL** Annie Griffiths Belt / West Light **230CR** Roy Gumpel / Mon-Tresor / Panoramic Stock Images **230B** Mark Segal / Panoramic Stock Images **230TR** Harald Sund / The Image Bank **235BL** Stokes Collection / New York Public Library, Astor, Lenox and Tilden Foundations **235TL** Bettmann Archive **235TC** Brown Brothers **235TR** NASA **235BR** Courtesy American Antiquarian Society **236TL** Comstock Inc. **236TR** Spencer Jones / FPG **236C** Hans N. Eleman / The Image Bank **236BL** H. Armstrong Roberts **236BR** Pictor / Uniphoto **238** Breck Kent / Earth Scenes **240** Les Stone / Sygma **242** Tony Freeman / Photo Edit **243** Michael Newman / Photo Edit **244** Carol Zacny **245** Joseph Viesti **247** D. MacDonald / Photo Edit **248** Vito Palmisano **252** Warren Jacobs / Tony Stone Images **254** G. Vandystadt / ALLSPORT USA **257** David Young-Wolff / Photo Edit **259** Chicago Historical Society **260** Bob Daemmrich **264** Bettmann **265** The Kobal Collection **266** Brent Jones **269** Tony Freeman / Photo Edit **271** Everett Collection **274** Milt & Joan Mann / Cameramann International, Ltd. **275** Milt & Joan Mann / Cameramann International, Ltd. **281** ©1959 / Cambridge University Press from SCIENCE AND CIVILISATION IN CHINA Vol. 3 by Joseph Needham **283** John Running **285** U.S. Government **288** Mike Powell / ALLSPORT USA **291** The Russian Museum, St. Petersburg "The Mowers" 1887 by Grigory Miasoyedov **293T** Mark Stephenson / West Light **293B** Picturesque **294TL** Pierre-Yves Goavec / The Image Bank **294C** Dave Burgering / Duomo Photography Inc. **294B** Sports Photo Masters, Inc. **300T** Pictor / Uniphoto **300C** Index Stock International **300BL** Comstock Inc. **300BR** Comstock Inc. **301CL** Gary Buss / FPG **302** Courtesy Lebanon Valley College Photo: Dwayne Arehart **304** Stanford University **307** William S. Favata **308** Tom McCarthy / Photo Edit **311** Courtesy Ford Motor Company **313** Bob Daemmrich / The Image Works **315** David Young-Wolff / Photo Edit **322** Hutchings / Photo Edit **323** Beryl Goldberg **325** Van Elton / Photo Edit **327** Tony Freeman / Photo Edit **334** A. Gyori / Sygma **336** Courtesy Ford Motor Company **338T** Peter Till / The Image Bank **338BL & BR(background)** Gary Commer / Index Stock International **338BC** C. Moore / West Light **338BR** W. Cody / West

Light **339T** Illustration from OLD POSSUM'S BOOK OF PRACTICAL CATS by T. S. Eliot, illustration copyright © 1982 by Edward Gorey, reproduced by permission of Harcourt Brace & Company. **344TL** Natural Selection **344TR** Digital Art / West Light **344C** Index Stock International **344B** Rick Rusing / Leo de Wys **345B** Ralph Mercer **346** Courtesy Diamond B. Lumber Company / Photo: Milt & Joan Mann / Cameramann International, Ltd. **349** Courtesy United Airlines **350** Bettmann Archive **352** Barnwell / Stock Boston **355** Chip Peterson **357** Courtesy Chicago Youth Symphony Orchestra **358** Cornwell / Pacific Stock **359** Roessler / ANIMALS ANIMALS **360** Jerome Academia / Photo Edit **361** Nancy Rabener / Barker **362** Paul Conklin **366** Library of Congress **367** MacDonald / Photo Edit **369** Hutchings / Photo Edit **372L** Kennard / Stock Boston **372R** Hulton Picture Library **375** Holland / The Image Bank **380** ©1984 / UNIVERSAL PRESS SYNDICATE. Reprinted with permission. All rights reserved. **381** Joseph Viesti **385** Milt & Joan Mann / Cameramann International, Ltd. **387** J. L. Atlan / Sygma **393** Tribune Media Services **400T** Jim Richardson / West Light **406T** SuperStock, Inc. **406C** Ed Honowitz / Tony Stone Images **407T** Craig Aurness / West Light **407C** Kathleen Campbell / Gamma-Liaison **408** Yvonne Hemsey / Gamma-Liaison **410 411** Drawings by John Tenniel **415** Bob Daemmrich **421** Chip Peterson **427** Georgia Dept. of Tourism **428** Scott Berner / Visuals Unlimited **435** R. Saunders / Leo de Wys **439** David R. Frazier Photolibrary **440** David Boyle / Earth Scenes **442** David Wells / The Image Works **446** John Running / Stock Boston **449L** Doug Wechsler / Earth Scenes **452** D. Specker / ANIMALS ANIMALS **454** ©1941 / Escher Foundation-Haags Gemeentemuseum-The Hague **455** ©1994 / Escher Foundation-Haags Gemeentemuseum-The Hague **457** Adam Woolfit / Woodfin Camp & Associates **458TL** SuperStock, Inc. **458TC** Kathleen Campbell / Gamma-Liaison **458R** David Sutherland / Tony Stone Images **459T** Telegraph Colour Library / FPG **466TL** Andrew Sacks / Tony Stone Images **466TR** Comstock Inc. **466CL** Uniphoto **466B** Craig Aurness / West Light **467C** Paul Horsted / Direct Stock **471** Milt & Joan Mann / Cameramann International, Ltd. **473** Freeman / Photo Edit **476** D. Burnett / Contact / The Stock Market **482** David Spangler **484** Larry Lefever / Grant Heilman Photography **486** S. Grotta / The Stock Market **487** C. Mishler / AlaskaStock Images **488** T. Freeman / Photo Edit **491** MacDonald / Photo Edit **492** Focus On Sports **493** Courtesy British Airways **495T** Pollack / The Stock Market **495B** AP / Wide World **497** McCarthy / Photo Edit **498** Alex MacLean / Landslides **499** Laura Dwight **503** T. Freeman / Photo Edit **505** J. Maher / The Stock Market **506** M. Newman / Photo Edit **508** V. Beller / The Stock Market **510** The Kobal Collection **511T** Janice Travia / Tony Stone Images **511C** Janice Travia / Tony Stone Images **511B** Janice Travia / Tony Stone Images **513** Charles Osgood / Copyrighted, Chicago Tribune Company, all rights reserved **515** P Beck / The Stock Market **519** NASA **521CL** S. Krasemann / Photo Researchers **521CL** S. Krasemann / Photo Researchers **521BL** S. Krasemann / Photo Researchers **521BR** S Krasemann / Photo Researchers **528T** Charly Franklin / FPG **528B** Freeman Patterson / Masterfile **529T** J Amos / H. Armstrong Roberts **529B** Steven E Sutton / Duomo Photography Inc. **530T** McCarthy / Photo Edit **535T** Jeff Speilman / The Image Bank **539** Markowitz / Sygma **541** M McVay / The Stock Market **542** D Young-Wolff / Photo Edit **544T** Walter Chandoha **546** Brenner / Photo Edit **548** The Image Bank **550** D Young-Wolff / The Image Bank **555** Stichting Beeldrecht Haags Gemeentemuseum, by Piet Mondrian **559** Simon Jauncey / Tony Stone Images **561** Courtesy California Raisin Advisory Board **562** Courtesy California Raisin Advisory Borad **564** Werner Bokelberg / The Image Bank **566** Thomas Tampy / The Image Bank **570** J R Eyerman 1973 Time, Inc. / TIME Magazine **571** The Walt Disney Company / Walt Disney Productions **575** Klaus Mitteldorf / The Image Bank **577** Carol Zacny **582T** Ric Ergenbright Photography (582 & 583) **582B** Brian Tolbert / brt Photo **583BL** Syguest **583BR** Bettmann Archive **590 TL** Robert Marien / RO-MA Stock **590TR** H. Abernathy / H. Armstrong Roberts **590BL** Steve Gottleib **590BL** David M. Philips / SS / Photo Researchers **591** David Madison **592** Savino / The Image Works **593** Bob Daemmrich **594** Frederick Grassle / Woods Hole Oceanographic Institution **596** Laura Dwight **597** Brent Jones **598** Bob Daemmrich / The Image Works **599** The Kobal Collection **600** Gillette / Stock Boston **602** Myrleen Ferguson / Photo Edit **605** Grant Heilman Photography **606** David Young-Wolff / Photo Edit **607** Anthony Bannister / ANIMALS ANIMALS **609** J. R. Williams / Earth Scenes **610** A. Tilley / Tony Stone Images **611** D. Jacobs / Tony Stone Images **612** Bill Brooks / Masterfile **614** Bob Daemmrich / The Image Works **615** Mary Kate Denny / Photo Edit **618** Bob Daemmrich / The Image Works **620** Kunsthistorisches Museum, / Art Resource, NY Gemaeldegalerie, Vienna / Photo: Erich Lessing, detail **621** Kunsthistorisches Museum, / Art Resource, NY Gemaeldegalerie, Vienna / Photo: Erich Lessing, detail **623** Courtesy of the Trustees of the British Museum **624** Bettmann **628** Brian Seed / Tony Stone Images **630** Grant Heilman Photography **633** Aresa Pryor-Adams / Earth Scenes **634T** C. Shotwell / Mon-Tresor **634 635** David Madison **635T** David Madison / Duomo Photography Inc. **640T** Jaime Villaseca / The Image Bank **640C** Karageorge / H. Armstrong Roberts **640B** Jim Barber Studios **641T** Pete Saloutos / The Stock Market **641B** J. McDermott / Tony Stone Images **642** David Madison **646** Walt Disney Productions **647** Rita Boserup **652** Milt & Joan Mann / Cameramann International, Ltd. **656** Walt Disney Productions **658** Bob Daemmrich / Stock Boston **662** by Samuel B. Waugh / The Museum of the City of New York **663** by Andre Thevet, Keruert et / The Burndy Library Chaudiere, Paris, 1584 **664** Grant Heilman Photography **670** Courtesy R. R. Donnelley **675** W. Johnson / Stock Boston **681** NASA **682** NASA **684T** Jaime Villaseca / The Image Bank **684B** D. Sutherland / Tony Stone Images **685** Neveux / H. Armstrong Roberts **690T** Eugen Gebhardt / FPG **690BL** Baron Wolman **690BR** Comstock Inc. **691T** Michael Simpson / FPG **692** Milt & Joan Mann / Cameramann International, Ltd. **695** Milt & Joan Mann / Cameramann International, Ltd. **697** David Madison / David Madison **699** Milt & Joan Mann / Cameramann International, Ltd. **702** David Madison **707** Bachmann / The Image Works **708** Courtesy Future Farmers of America **709** Everett Collection **714** Milt & Joan Mann / Cameramann International, Ltd. **716** Stephanie Maze / Woodfin Camp, Inc. **723** Courtesy Illinois Society of Professional Engineers **724** Courtesy Road & Track Magazine **725** Scott Morgan / West Light **726T** M. Angelo / West Light **726TC** The Stock Shop **726BC** Richard Price / West Light **726BL** Tony Freeman / Photo Edit **726BR** D. Degnan / H. Armstrong Roberts